A-Level

Biology

Exam Board: AQA

Everyone knows that life is complicated. As a Biology student, you'll know that life is *really* complicated... and you have to understand how it all works for the exams.

Never mind — at least this CGP book will make your revision a whole lot simpler. It has everything you'll need for a top grade in the AQA A-Level course, including crystal-clear study notes, realistic exam questions (with answers) and more!

There's also a free Online Edition you can read on your computer or tablet, so you don't even have to strain your myofibrils by carrying the book around.

How to access your free Online Edition

This book includes a free Online Edition to read on your PC, Mac or tablet. You'll just need to go to **cgpbooks.co.uk/extras** and enter this code:

1704 9830 3954 2599

By the way, this code only works for one person. If somebody else has used this book before you, they might have already claimed the Online Edition.

A-Level revision? It has to be CGP!

Published by CGP

From original material by Richard Parsons.

Editors:
Daniel Fielding, Andy Park, Rachael Rogers, Camilla Simson.

Contributors:
Sophie Anderson, Gloria Barnett, Jessica Egan, Mark Ellingham, Paddy Gannon, Derek Harvey, Emily Lucas, David Martindill, Liz Masters, Adrian Schmit.

ISBN: 978 1 78908 026 1

With thanks to Phil Armstrong, Charlotte Burrows and Hayley Thompson for the proofreading.
With thanks to Jan Greenway for the copyright research.

Clipart from Corel®
Printed by Elanders Ltd, Newcastle upon Tyne.

Contents

If you're revising for the **AS exams**, you'll need to revise Topics 1-4, and the Practical Skills section at the back. If you're revising for the **A-level exams**, you'll need to revise the **whole book**.

Topic 3B — More Exchange and Transport Systems

Topic 4A — DNA, RNA and Protein Synthesis

Topic 4B — Diversity, Classification and Variation

Topic 5A — Photosynthesis and Respiration

Topic 5B — Energy Transfer and Nutrient Cycles

Topic 6A — Stimuli and Responses

Topic 6B — Nervous Coordination

Topic 6C — Homeostasis

Topic 7A — Genetics

Topic 7B — Populations and Evolution

Topic 7C — Populations in Ecosystems

Topic 8A — Mutations and Gene Expression

Topic 8B — Genome Projects and Gene Technologies

Practical Skills (AS)

Do Well In Your Exams

Specification Map

This specification map tells you where each bit of the AQA specification is covered in this book.

Specification Map

Carbohydrates

Even though there is, and has been, a huge variety of different organisms on Earth, they all share some biochemistry — for example, they all contain a few carbon-based compounds that interact in similar ways.

Most Carbohydrates are Polymers

1) Most **carbohydrates** (as well as proteins and nucleic acids) are **polymers**.
2) Polymers are **large**, **complex molecules** composed of **long chains** of **monomers** joined together.
3) **Monomers** are **small**, **basic molecular units**.
4) Examples of monomers include **monosaccharides**, **amino acids** and **nucleotides**.

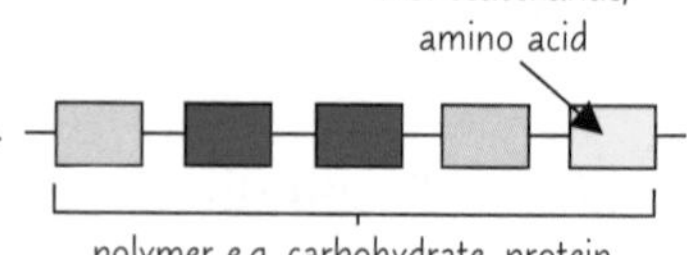

Carbohydrates are Made from Monosaccharides

1) All carbohydrates contain the elements **C**, **H** and **O**.
2) The **monomers** that they're made from are **monosaccharides**, e.g. **glucose**, **fructose** and **galactose**.

1) Glucose is a **hexose** sugar — a monosaccharide with **six carbon atoms** in each molecule.
2) There are **two types** of **glucose**, **alpha** (α) and **beta** (β) — they're **isomers** (molecules with the same molecular formula as each other, but with the atoms connected in a different way).
3) You need to know the structures of **both types** of glucose for your exam — it's pretty easy because there's only one difference between the two:

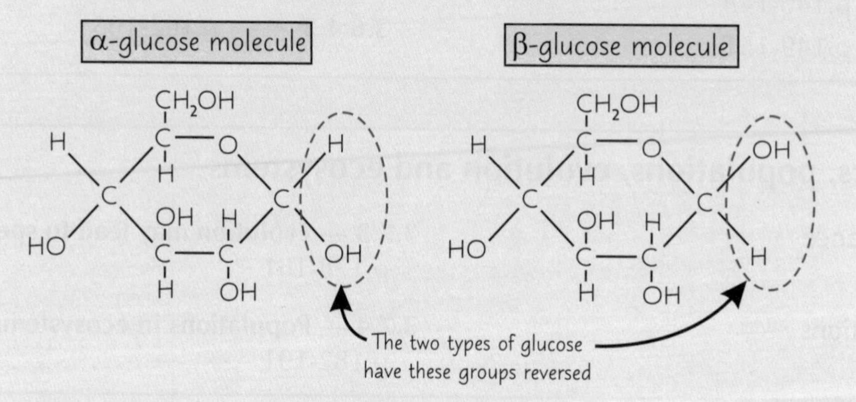

Condensation Reactions Join Monosaccharides Together

1) A **condensation reaction** is when two molecules join together with the formation of a new **chemical bond**, and a **water** molecule is released when the bond is formed.
2) Monosaccharides are **joined together** by **condensation reactions**.
3) A **glycosidic bond** forms between the two monosaccharides as a molecule of water is released.
4) A **disaccharide** is formed when **two monosaccharides** join together.

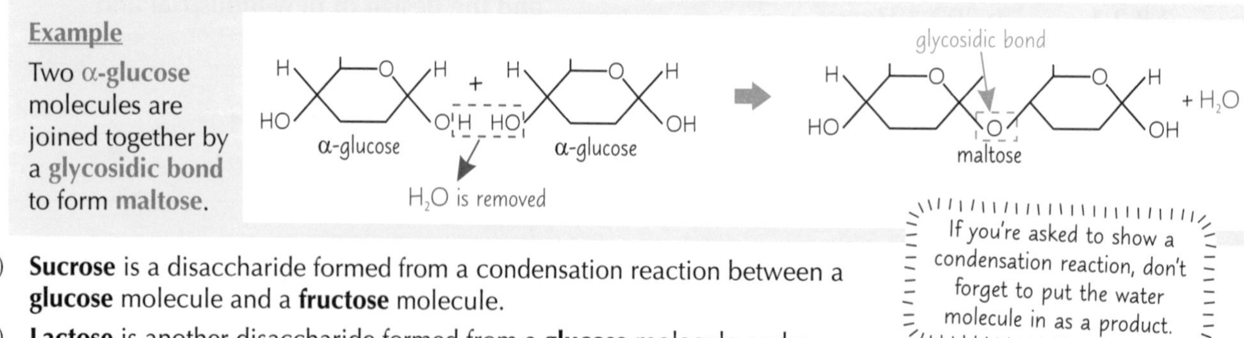

5) **Sucrose** is a disaccharide formed from a condensation reaction between a **glucose** molecule and a **fructose** molecule.
6) **Lactose** is another disaccharide formed from a **glucose** molecule and a **galactose** molecule.

If you're asked to show a condensation reaction, don't forget to put the water molecule in as a product.

Carbohydrates

Hydrolysis Reactions Break Polymers Apart

1) **Polymers** can be broken down into **monomers** by **hydrolysis reactions**.
2) A **hydrolysis reaction breaks** the **chemical bond** between monomers using a **water molecule**. It's basically the **opposite** of a condensation reaction.
3) For example, **carbohydrates** can be broken down into their constituent **monosaccharides** by **hydrolysis** reactions.

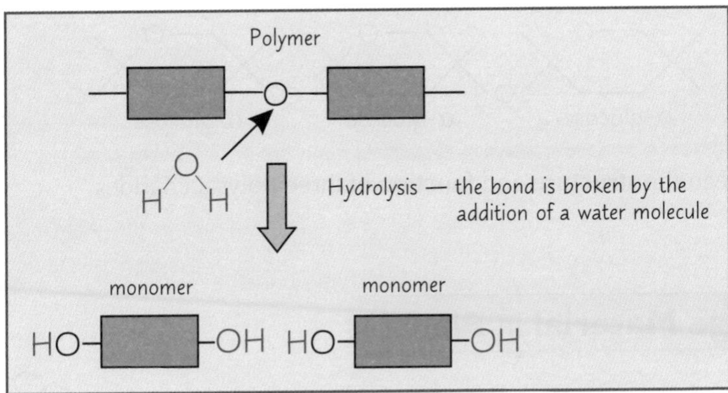

Even hydrolysis couldn't break this bond.

Use the Benedict's Test for Sugars

Sugar is a general term for **monosaccharides** and **disaccharides**. All sugars can be classified as **reducing** or **non-reducing**. The **Benedict's test** tests for sugars — it **differs** depending on the **type** of sugar you are testing for.

REDUCING SUGARS

1) Reducing sugars include **all monosaccharides** (e.g. glucose) and **some disaccharides** (e.g. maltose and lactose).
2) You add **Benedict's reagent** (which is **blue**) to a sample and **heat it** in a water bath that's been brought to the **boil**.
3) If the test's **positive** it will form a **coloured precipitate** (solid particles suspended in the solution).

   The colour of the precipitate changes from:

   blue ⇒ green ⇒ yellow ⇒ orange ⇒ brick red

4) The higher the concentration of reducing sugar, the further the colour change goes — you can use this to **compare** the amount of reducing sugar in different solutions. A more accurate way of doing this is to **filter** the solution and **weigh the precipitate**.

Always use an excess of Benedict's solution — this makes sure that all the sugar reacts.

NON-REDUCING SUGARS

1) If the result of the reducing sugars test is **negative**, there could still be a non-reducing sugar present. To test for **non-reducing sugars**, like sucrose, first you have to break them down into monosaccharides.
2) You do this by getting a new sample of the test solution, adding **dilute hydrochloric acid** and carefully heating it in a water bath that's been brought to the **boil**. You then **neutralise** it with **sodium hydrogencarbonate**. Then just carry out the **Benedict's test** as you would for a reducing sugar.
3) If the test's **positive** it will form a **coloured precipitate** (as for the reducing sugars test). If the test's **negative** the solution will **stay blue**, which means it **doesn't contain any sugar** (either reducing or non-reducing).

Carbohydrates

So, you've already looked at monosaccharides and disaccharides... now it's time to give polysaccharides some love.

Polysaccharides are **Loads of Sugars** Joined Together

A **polysaccharide** is formed when **more than two monosaccharides** are joined together by **condensation reactions**.

Example: Lots of **α–glucose** molecules are joined together by **glycosidic bonds** to form **amylose**.

glycosidic bonds

α-glucose α-glucose α-glucose α-glucose α-glucose

You need to know about the relationship between the **structure** and **function** of three polysaccharides — starch, glycogen and cellulose.

Starch is the Main **Energy Storage Material** in **Plants**

1) Cells get **energy** from **glucose**. Plants **store** excess glucose as **starch** (when a plant **needs more glucose** for energy, it **breaks down** starch to release the glucose).
2) Starch is a mixture of **two** polysaccharides of **alpha-glucose** — **amylose** and **amylopectin**:
   - **Amylose** — a long, **unbranched chain** of α–glucose. The angles of the glycosidic bonds give it a **coiled structure**, almost like a cylinder. This makes it **compact**, so it's really **good for storage** because you can **fit more in** to a small space.
   - **Amylopectin** — a long, **branched chain** of α–glucose. Its **side branches** allow the **enzymes** that break down the molecule to get at the **glycosidic bonds easily**. This means that the glucose can be **released quickly**.
3) Starch is **insoluble** in water and doesn't affect **water potential** (see page 42), so it **doesn't** cause water to enter cells by **osmosis**, which would make them swell. This makes it good for **storage**.

Amylose

one alpha-glucose molecule

Amylopectin

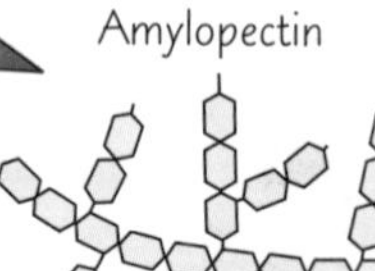

Use the **Iodine Test** for **Starch**

If you do any **experiment** on the **digestion** of **starch** and want to find out if any is **left**, you'll need the **iodine test**.

Just add **iodine dissolved in potassium iodide solution** to the test sample. If there is **starch present**, the sample changes from **browny-orange** to a dark, **blue-black** colour.

Make sure you always talk about iodine in potassium iodide solution, not just iodine.

Glycogen is the Main **Energy Storage Material** in **Animals**

1) Animal cells get **energy** from **glucose** too. But animals **store** excess glucose as **glycogen** — another polysaccharide of **alpha-glucose**.
2) Its structure is very similar to amylopectin, except that it has **loads** more **side branches** coming off it. Loads of branches means that stored glucose can be **released quickly**, which is **important for energy release** in animals.
3) It's also a very **compact** molecule, so it's good for storage.

After throwing and fetching the ball no less than 312 times, Chappy and Stuart were finally out of glycogen.

Carbohydrates

Cellulose is the Major Component of Cell Walls in Plants

1) Cellulose is made of **long**, **unbranched** chains of **beta–glucose**.
2) When **beta–glucose** molecules **bond**, they form straight cellulose chains.
3) The cellulose chains are linked together by **hydrogen bonds** to form strong fibres called **microfibrils**. The strong fibres mean cellulose provides **structural support** for cells (e.g. in plant cell walls).

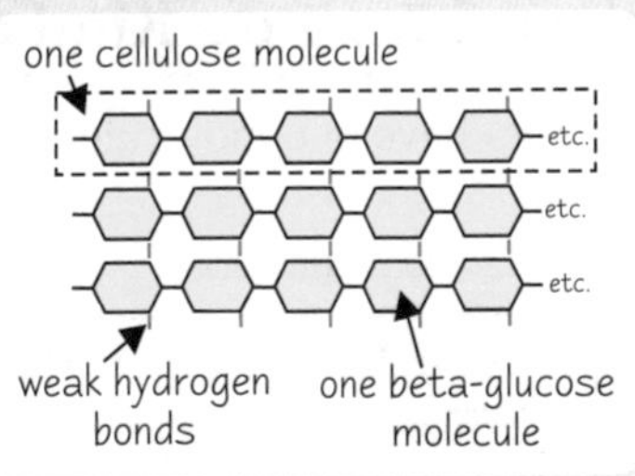

Warm-Up Questions

Q1 What is a polymer?

Q2 Draw the structure of α-glucose.

Q3 What type of bond holds monosaccharide molecules together in a polysaccharide?

Q4 Name the two polysaccharides present in starch.

Q5 Describe the iodine test for starch.

Exam Questions

Q1 Maltose is a sugar. Describe how a molecule of maltose is formed. *[3 marks]*

Q2 Sugars can be classed as reducing or non-reducing. Describe the test used to identify a non-reducing sugar. Include the different results you would expect to see if the test was positive or negative. *[5 marks]*

Q3 Read the following passage:

Chitin is a structural polysaccharide, similar to cellulose in plants, that is found in the exoskeletons of insects and crustaceans, as well as in the cell walls of fungi. It is made up of chains of the monosaccharide N-acetylglucosamine, which is derived from glucose. The polysaccharide chains are long, unbranched and linked together by weak hydrogen bonds.

Chitin can be broken down by enzymes called chitinases, which catalyse hydrolysis reactions. Some organisms are able to make their own chitinases. Amongst these are yeasts, such as *Saccharomyces cerevisiae*. In yeast reproduction, a newly formed yeast cell 'buds off' from the cell wall of its parent cell to become a new independent organism. This requires the separation of the cell wall of the new cell from the cell wall of the parent cell. *Saccharomyces cerevisiae* uses a chitinase for this purpose.

Use information from the passage and your own knowledge to answer the following questions:

a) Explain why chitin can be described as a polysaccharide (line 1). *[1 mark]*

b) Chitin is similar to cellulose in plants (line 1). Describe the ways in which cellulose and chitin are similar. *[3 marks]*

c) Chitin can be broken down by enzymes called chitinases, which catalyse hydrolysis reactions (line 5). Explain how these hydrolysis reactions break down chitin. *[2 marks]*

d) Some organisms are able to make their own chitinases (line 5 and 6). Explain how it would be beneficial for plants to make and secrete chitinases as a defence system. *[4 marks]*

Starch — I thought that was just for shirt collars...

Every cell in an organism is adapted to perform a function — you can always trace some of its features back to its function. Different cells even use the exact same molecules to do completely different things. Take glucose, for example — all plant cells use it to make cellulose, but they can also make starch from it if they need to store energy. Smashing.

Lipids

Lipids are really nice. Without them, we'd have no cell membranes. You owe it to them to make sure you can remember all of the stuff about them on these pages. It'll help you and your membranes get a good grade.

Triglycerides are a **Kind** of **Lipid**

Triglycerides have one molecule of **glycerol** with **three fatty acids** attached to it.

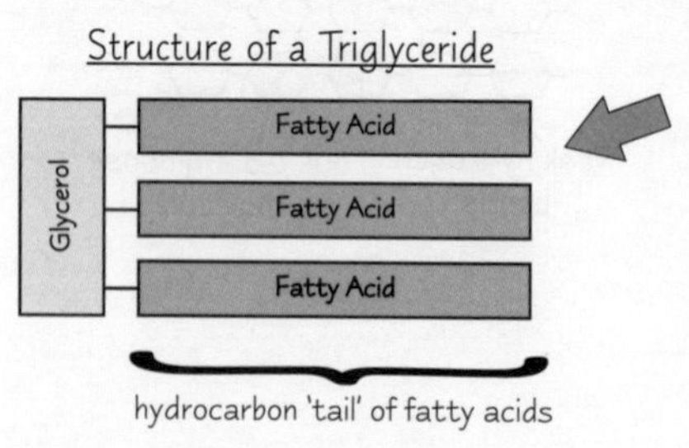

Fatty acid molecules have long 'tails' made of **hydrocarbons**. The tails are '**hydrophobic**' (they repel water molecules). These tails make lipids **insoluble** in **water**. All **fatty acids** have the same basic structure, but the **hydrocarbon tail varies**.

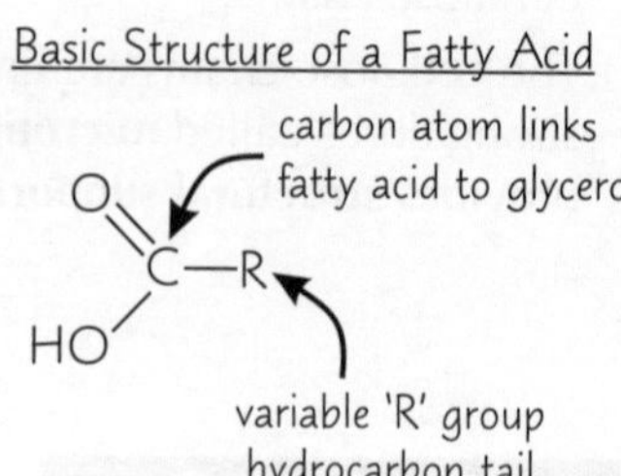

Triglycerides are **Formed** by **Condensation Reactions**

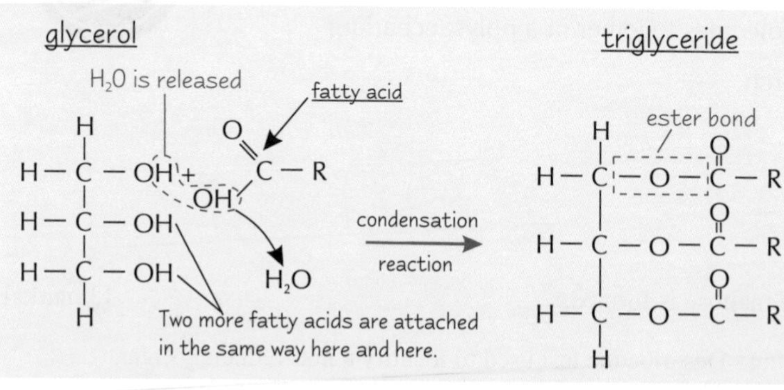

The diagram shows a **fatty acid** joining to a **glycerol molecule**. When the **ester bond** is formed a molecule of **water** is **released**. — it's a **condensation reaction**. This process happens twice more to form a **triglyceride**.

Fatty Acids can be **Saturated** or **Unsaturated**

There are **two** kinds of fatty acids — **saturated** and **unsaturated**.
The difference is in their **hydrocarbon tails (R group)**.

Saturated fatty acids **don't** have any **double bonds** between their **carbon atoms**. The fatty acid is 'saturated' with hydrogen.

saturated hydrocarbon tail

Unsaturated fatty acids have **at least one** double bond between **carbon atoms**, which cause the chain to kink.

unsaturated hydrocarbon tail

Phospholipids are **Similar** to **Triglycerides**

1) The lipids found in cell membranes **aren't** triglycerides — they're **phospholipids**.
2) Phospholipids are pretty **similar** to triglycerides except one of the fatty acid molecules is replaced by a **phosphate group**.
3) The phosphate group is **hydrophilic** (attracts water). The fatty acid tails are **hydrophobic** (repel water). This is important in the cell membrane (see next page to find out why).

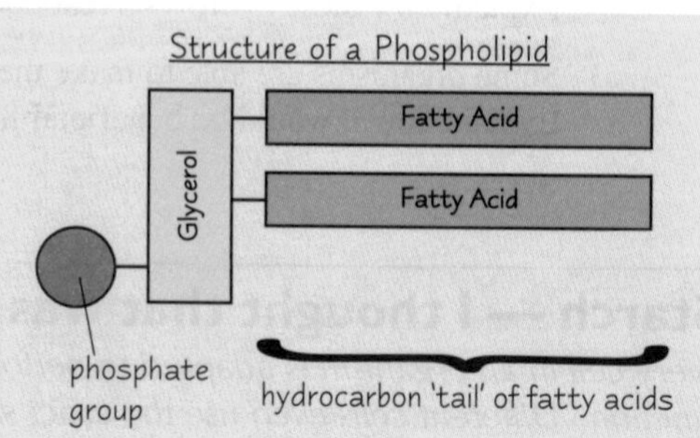

Lipids

The **Structures** of Lipids Relate to Their **Functions**

You need to know how the **structures** of **triglycerides** and **phospholipids** are related to their **functions**:

Triglycerides

Triglycerides are mainly used as **energy storage molecules**. They're good for this because:

1) The **long hydrocarbon tails** of the fatty acids contain lots of **chemical energy** — a load of energy is **released** when they're **broken down**. Because of these tails, lipids contain about **twice** as much energy per gram as carbohydrates.
2) They're **insoluble**, so they don't affect the **water potential** (see p. 42) of the cell and cause water to enter the cells by **osmosis** (which would make them swell). The triglycerides clump together as **insoluble droplets** in cells because the fatty acid tails are **hydrophobic** (water-repelling) — the tails **face inwards**, shielding themselves from water with their glycerol heads.

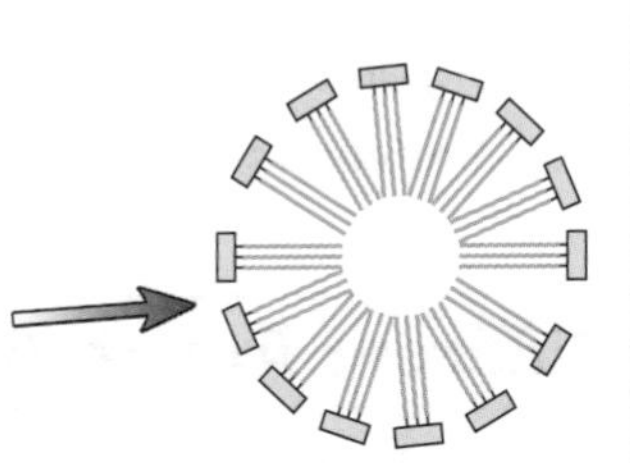

Phospholipids

Phospholipids make up the **bilayer** of **cell membranes** (see p. 38).
Cell membranes **control** what **enters and leaves a cell**.

1) Their heads are **hydrophilic** and their tails are **hydrophobic**, so they form a **double** layer with their heads facing **out** towards the water on either side.
2) The **centre** of the bilayer is **hydrophobic**, so water-soluble substances **can't** easily pass through it — the membrane acts as a **barrier** to those substances.

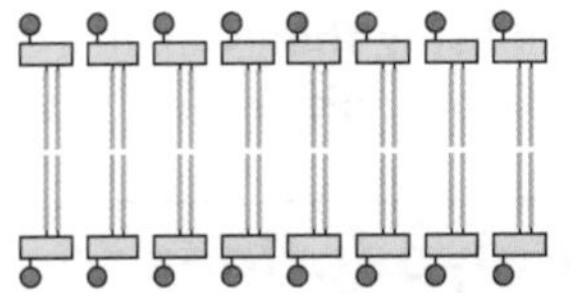

Use the **Emulsion Test** for **Lipids**

If you wanted to find out if there was any **fat** in a particular **food** you could do the **emulsion test**:

1) **Shake** the test substance with **ethanol** for about a minute so that it dissolves, then **pour** the solution into **water**.
2) Any lipid will show up as a **milky emulsion**.
3) The more lipid there is, the more noticeable the milky colour will be.

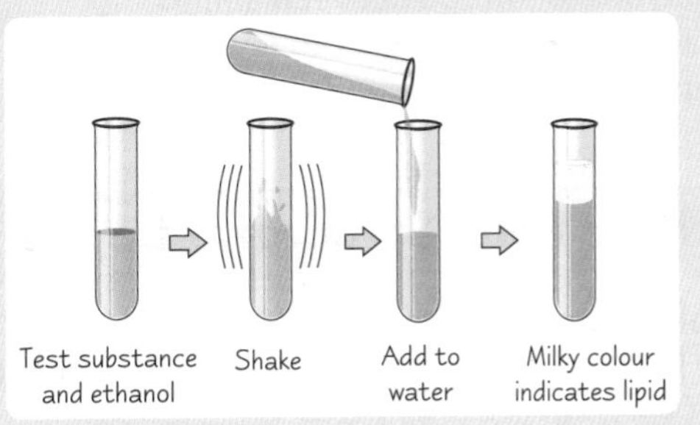

Warm-Up Questions

Q1 What type of bond is made from a condensation reaction between glycerol and a fatty acid molecule?

Q2 Describe how you would test for lipids in a solution.

Exam Questions

Q1 Triglycerides have a hydrophobic tail. Explain how this feature of a lipid is important for its function. [2 marks]

Q2 Cell membranes contain phospholipids.

a) Describe the structure of a phospholipid. [3 marks]

b) Explain the difference between a saturated fatty acid and an unsaturated fatty acid. [2 marks]

The test for lipids — stick them in a can of paint...

Not really. Otherwise you might upset your Biology teacher a bit. Instead, why not sit and contemplate all those phospholipids jumping around in your plasma membranes... their water-loving, phosphate heads poking out of the cell and into the cytoplasm, and their water-hating, hydrocarbon tails forming an impenetrable layer in between...

Proteins

There are loads of different proteins with loads of different functions. But what are proteins? What do they look like? Well, for your enjoyment, here are the answers to all those questions and many, many more...

Proteins are Made from **Long Chains** of **Amino Acids**

1) The **monomers** of proteins are **amino acids**.
2) A **dipeptide** is formed when **two** amino acids join together.
3) A **polypeptide** is formed when **more than two** amino acids join together.
4) **Proteins** are made up of **one or more polypeptides**.

Grant's cries of "die peptide, die" could be heard for miles around. He'd never forgiven it for sleeping with his wife.

Different Amino Acids Have **Different Variable Groups**

Amino acids have the same general structure — a **carboxyl group** (-COOH), an **amine** or **amino group** ($-NH_2$) and an **R group** (also known as a **variable** side group).

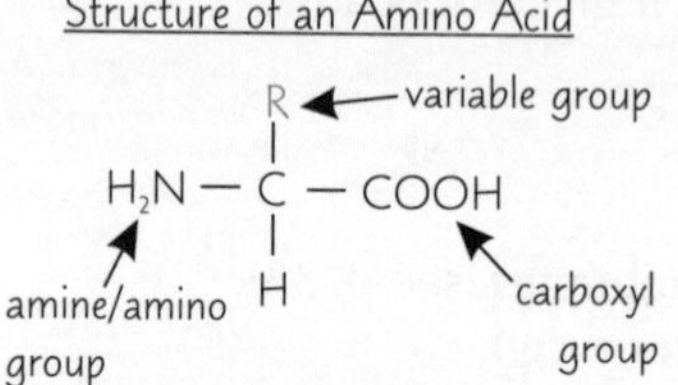

E.g. Structure of Alanine

CH_3
$H_2N - C - COOH$
H

All living things share a bank of only **20 amino acids**.
The only **difference** between them is what makes up their **R group**.

Glycine is the only amino acid that doesn't have carbon in its side group. Its R group consists of just one hydrogen atom.

Polypeptides are Formed by **Condensation Reactions**

Amino acids are linked together by **condensation** reactions to form polypeptides. A molecule of **water** is **released** during the reaction. The bonds formed between amino acids are called **peptide bonds**. The reverse reaction happens during digestion.

amino acid 1, amino acid 2 — condensation ⇌ hydrolysis — dipeptide

H_2O — a molecule of water is formed during condensation.

peptide bond

Proteins Have **Four Structural Levels**

Proteins are **big**, **complicated** molecules. They're much easier to explain if you describe their structure in four 'levels'. These levels are a protein's **primary**, **secondary**, **tertiary** and **quaternary** structures.

<u>Primary Structure</u> — this is the **sequence** of **amino acids** in the **polypeptide chain.**

<u>Secondary Structure</u> — the polypeptide chain doesn't remain flat and straight. **Hydrogen bonds** form between the amino acids in the chain. This makes it automatically **coil** into an **alpha** (α) **helix** or **fold** into a **beta** (β) **pleated sheet** — this is the secondary structure.

<u>Tertiary Structure</u> — the coiled or folded chain of amino acids is often **coiled** and **folded further**. More **bonds** form between different parts of the polypeptide chain, including **hydrogen bonds** and **ionic bonds** (**attractions** between **negative** and **positive** charges on different parts of the molecule). **Disulfide bridges** also form whenever two molecules of the amino acid **cysteine** come **close together** — the **sulfur** atom in one cysteine **bonds** to the sulfur atom in the other. For proteins made from a **single** polypeptide chain, the tertiary structure forms their **final 3D structure.**

<u>Quaternary Structure</u> — some proteins are made of **several different polypeptide chains** held together by **bonds**. The **quaternary structure** is the way these polypeptide chains are assembled together. For proteins made from more than one polypeptide chain (e.g. haemoglobin, insulin, collagen), the quaternary structure is the protein's **final 3D structure.**

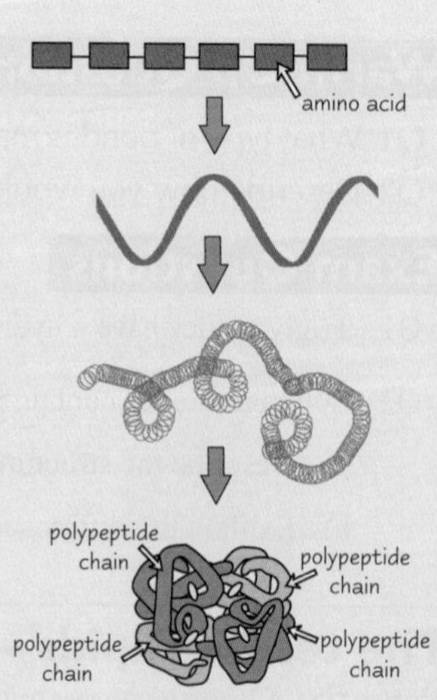

Proteins

Proteins have a **Variety** of **Functions**

There are **loads** of different **proteins** found in **living organisms**. They've all got **different structures** and **shapes**, which makes them **specialised** to carry out particular **jobs**. For example:

1) **Enzymes** — they're usually roughly **spherical** in shape due to the **tight folding** of the polypeptide chains. They're **soluble** and often have roles in **metabolism**, e.g. some enzymes break down large food molecules (**digestive enzymes**, see pages 70-71) and other enzymes help to **synthesise** (make) large molecules.

2) **Antibodies** — are involved in the **immune response**. They're made up of **two light** (short) polypeptide chains and **two heavy** (long) polypeptide chains bonded together. Antibodies have **variable regions** (see p. 46) — the **amino acid sequences** in these regions **vary** greatly.

3) **Transport proteins** — e.g. channel proteins are present in **cell membranes** (p. 40). Channel proteins contain **hydrophobic** (water hating) and **hydrophilic** (water loving) amino acids, which cause the protein to **fold up** and form a **channel**. These proteins **transport molecules** and **ions across** membranes.

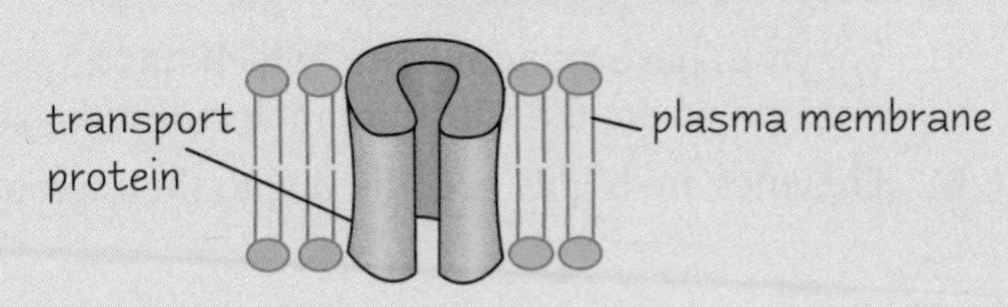

4) **Structural proteins** — are physically **strong**. They consist of **long polypeptide chains** lying **parallel** to each other with **cross-links** between them. Structural proteins include **keratin** (found in hair and nails) and **collagen** (found in connective tissue).

Use the **Biuret Test** for **Proteins**

If you needed to find out if a substance, e.g. a **food sample**, contained **protein** you'd use the **biuret test**.

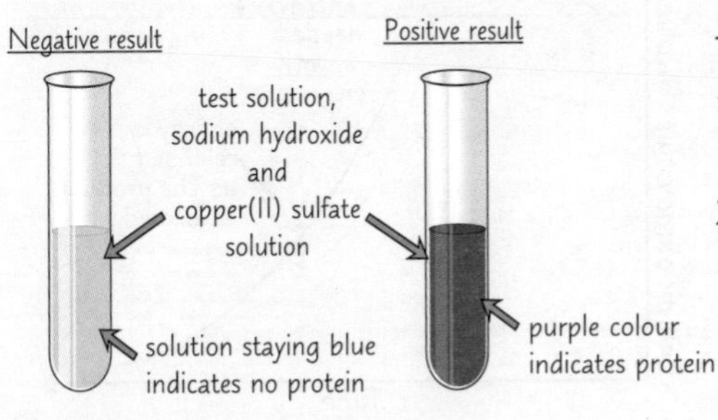

There are **two stages** to this test.

1) The test solution needs to be **alkaline**, so first you add a few drops of **sodium hydroxide solution**.
2) Then you add some **copper(II) sulfate solution**.
   - If protein **is** present the solution turns **purple**.
   - If there's **no protein**, the solution will **stay blue**. The colours are pale, so you need to look carefully.

Warm-Up Questions

Q1 What groups do all amino acid molecules have in common?

Q2 Give three functions of proteins.

Q3 Describe how you would test for the presence of protein in a sample.

Exam Questions

Q1 Leucyl-alanine is a dipeptide. Describe how a dipeptide is formed. [3 marks]

Q2 Myoglobin is a protein formed from a single polypeptide chain.
Describe the tertiary structure of a protein like myoglobin. [2 marks]

Condensation — I can see the reaction happening on my car windows...

Protein structure is hard to imagine. I think of a Slinky® — the wire's the primary structure, it coils up to form the secondary structure and if you coil the Slinky around your arm, that's the tertiary structure. When a few Slinkies get tangled up, that's like the quaternary structure. I need to get out more. I wish I had more than a Slinky for company.

Enzyme Action

Enzymes crop up loads in biology — they're really useful 'cos they make reactions work quickly. So, whether you feel the need for some speed or not, read on — because you really need to know this basic stuff about enzymes.

Enzymes are Biological Catalysts

Enzymes **speed up chemical reactions** by acting as **biological catalysts**.

A catalyst is a substance that speeds up a chemical reaction without being used up in the reaction itself.

1) Enzymes catalyse **metabolic reactions** — both at a **cellular level** (e.g. **respiration**) and for the **organism** as a **whole** (e.g. **digestion** in mammals).
2) Enzymes can affect **structures** in an organism (e.g. enzymes are involved in the production of **collagen**, an important protein in the **connective tissues** of animals) as well as **functions** (like **respiration**).
3) Enzyme action can be **intracellular** — **within** cells, or **extracellular** — **outside** cells.
4) Enzymes are **proteins** (see previous page).
5) Enzymes have an **active site**, which has a **specific shape**. The active site is the part of the enzyme where the **substrate** molecules (the substance that the enzyme interacts with) **bind to**.
6) Enzymes are **highly specific** due to their tertiary structure (see next page).

Enzymes Lower the Activation Energy of a Reaction

In a chemical reaction, a certain amount of **energy** needs to be supplied to the chemicals before the reaction will **start**. This is called the **activation energy** — it's often provided as **heat**. Enzymes **lower** the amount of activation energy that's needed, often making reactions happen at a **lower temperature** than they could without an enzyme. This **speeds up** the **rate of reaction**.

When a substrate fits into the enzyme's active site it forms an **enzyme-substrate complex** — it's this that lowers the activation energy. Here are two reasons why:

1) If two substrate molecules need to be **joined**, being attached to the enzyme holds them **close together**, **reducing** any **repulsion** between the molecules so they can bond more easily.
2) If the enzyme is catalysing a **breakdown reaction**, fitting into the active site puts a **strain** on bonds in the substrate, so the substrate molecule **breaks up** more easily.

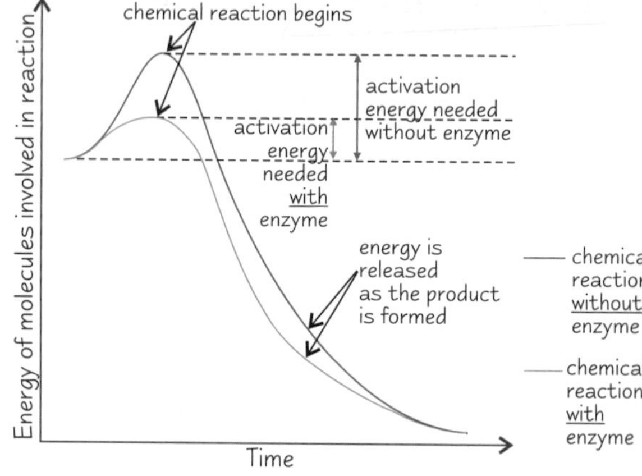

The 'Lock and Key' Model is a Good Start...

Enzymes are a bit picky — they only work with substrates that fit their active site. Early scientists studying the action of enzymes came up with the '**lock and key**' model. This is where the **substrate fits** into the **enzyme** in the same way that a **key fits** into a **lock**.

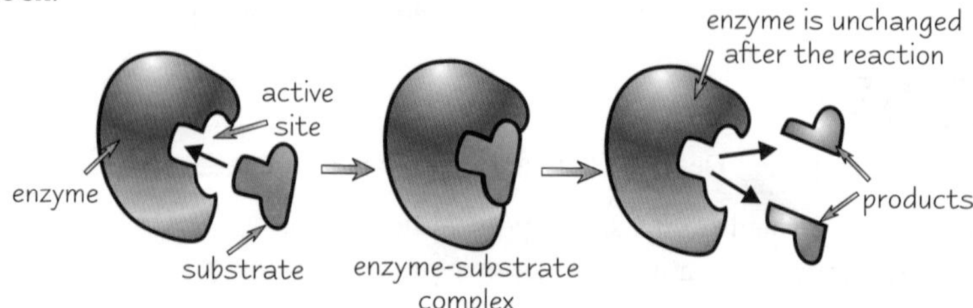

Scientists soon realised that the lock and key model didn't give the full story. The enzyme and substrate do have to fit together in the first place, but new evidence showed that the **enzyme-substrate complex changed shape** slightly to complete the fit. This **locks** the substrate even more tightly to the enzyme. Scientists modified the old lock and key model and came up with the '**induced fit**' model.

Enzyme Action

...but the 'Induced Fit' Model is a Better Theory

The '**induced fit**' model helps to explain why enzymes are so **specific** and only bond to one particular substrate. The substrate doesn't only have to be the right shape to fit the active site, it has to make the active site **change shape** in the right way as well. This is a prime example of how a widely accepted theory can **change** when **new evidence** comes along. The 'induced fit' model is still widely accepted — for now, anyway.

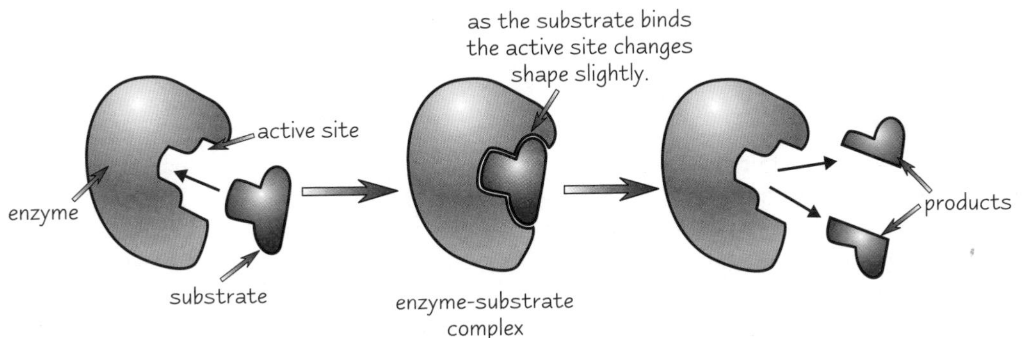

The 'Luminous Tights' model was popular in the 1980s but has since been found to be grossly inappropriate.

Enzyme Properties Relate to Their Tertiary Structure

1) Enzymes are **very specific** — they usually only catalyse **one** reaction, e.g. maltase only breaks down maltose, sucrase only breaks down sucrose.
2) This is because **only one complementary substrate will fit** into the active site.
3) The active site's **shape** is determined by the enzyme's **tertiary structure** (which is determined by the enzyme's **primary structure**).
4) Each **different enzyme** has a **different tertiary structure** and so a **different shaped active site**. If the substrate shape doesn't match the active site, an enzyme-substrate complex **won't** be formed and the reaction won't be catalysed.
5) If the tertiary structure of the enzyme is **altered** in any way, the **shape** of the active site will **change**. This means the **substrate won't fit** into the active site, an enzyme-substrate complex **won't** be formed and the enzyme will no longer be able to carry out its function.
6) The tertiary structure of an enzyme may be **altered** by changes in **pH** or **temperature** (see next page).
7) The **primary structure** (amino acid sequence) of a protein is determined by a **gene**. If a mutation occurs in that gene (see p. 97), it could change the tertiary structure of the enzyme **produced**.

Warm-Up Questions

Q1 What is an enzyme?

Q2 What is the name given to the amount of energy needed to start a reaction?

Q3 What is an enzyme-substrate complex?

Q4 Why can an enzyme only bind to one substance?

Exam Questions

Q1 Describe the 'induced fit' model of enzyme action. [4 marks]

Q2 Explain how a change in the amino acid sequence of an enzyme may prevent it from functioning properly. [2 marks]

But why is the enzyme-substrate complex?

So enzymes lower the activation energy of a reaction. I like to think of it as an assault course (bear with me). Suppose the assault course starts with a massive wall — enzymes are like the person who gives you a leg up over the wall (see?). Without it you'd need lots of energy to get over the wall yourself and complete the rest of the course. Unlikely.

Factors Affecting Enzyme Activity

Now you know what enzymes are and how they work, let's take a look at what makes them tick. Humans need things like money and the newest mobile phone, but enzymes are quite content with the right temperature and pH.

Temperature has a **Big Influence** on Enzyme Activity

Like any chemical reaction, the **rate** of an enzyme-controlled reaction **increases** when the **temperature's increased**. More heat means **more kinetic energy**, so molecules **move faster**. This makes the enzymes **more likely** to **collide** with the substrate molecules. The **energy** of these collisions also **increases**, which means each collision is more likely to **result** in a **reaction**. But, if the temperature gets too high, the **reaction stops**.

1) The rise in temperature makes the enzyme's molecules **vibrate more**.
2) If the temperature goes above a certain level, this vibration **breaks** some of the **bonds** that hold the enzyme in shape.
3) The **active site changes shape** and the enzyme and substrate **no longer fit together**.
4) At this point, the enzyme is **denatured** — it no longer functions as a catalyst.

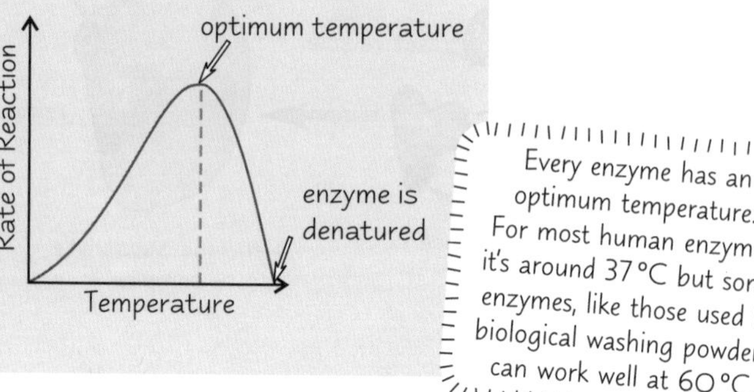

Every enzyme has an optimum temperature. For most human enzymes it's around 37 °C but some enzymes, like those used in biological washing powders, can work well at 60 °C.

pH Also Affects Enzyme **Activity**

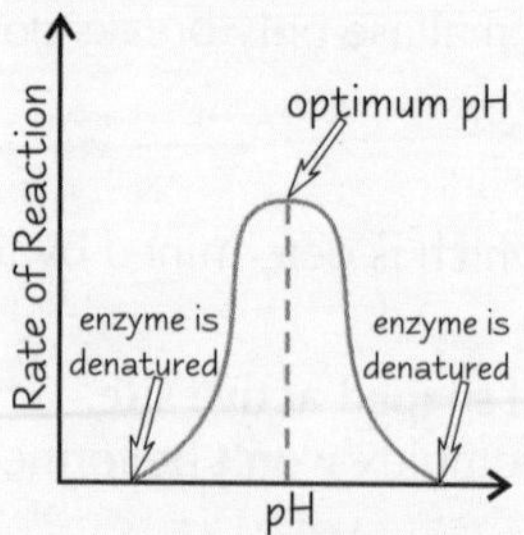

All enzymes have an **optimum pH value**. Most human enzymes work best at pH 7 (neutral), but there are exceptions. **Pepsin**, for example, works best at acidic pH 2, which is useful because it's found in the stomach. Above and below the optimum pH, the H^+ and OH^- ions found in acids and alkalis can mess up the **ionic bonds** and **hydrogen bonds** that hold the enzyme's tertiary structure in place. This makes the active site change shape, so the enzyme is **denatured**.

Enzyme Concentration Affects the Rate of Reaction

1) The **more enzyme molecules** there are in a solution, the more likely a substrate molecule is to **collide** with one and form an **enzyme-substrate complex**. So increasing the concentration of the enzyme **increases** the **rate of reaction**.
2) But, if the amount of **substrate** is **limited**, there comes a point when there's more than enough enzyme molecules to deal with all the available substrate, so adding more enzyme has **no further effect**.

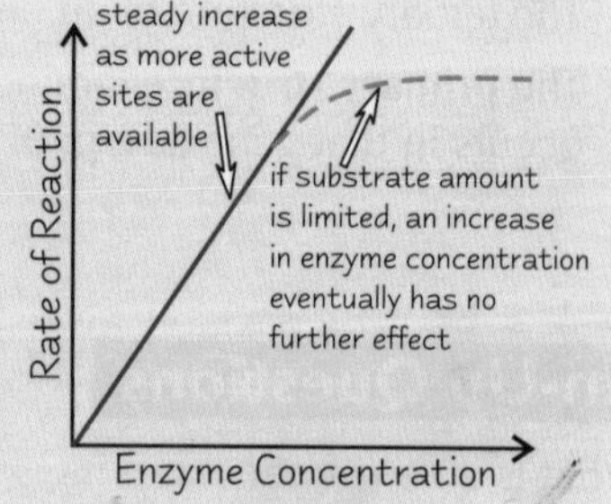

Substrate Concentration Affects the Rate of Reaction **Up to a Point**

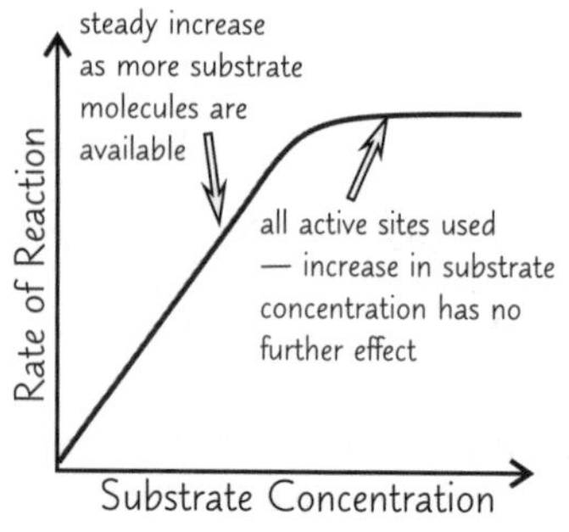

1) The **higher** the substrate concentration, the **faster** the reaction — more substrate molecules means a **collision** between substrate and enzyme is **more likely** and so more active sites will be used. This is only true up until a **'saturation'** point though. After that, there are so many substrate molecules that the enzymes have about as much as they can cope with (all the **active sites are full**), and adding more **makes no difference**.
2) Substrate concentration **decreases** with **time** during a reaction (unless more substrate is added to the reaction mixture), so if no other variables are changed, the **rate of reaction will decrease over time** too. This makes the **initial** rate of reaction (the reaction rate at the **start**) the **highest** rate of reaction.

Factors Affecting Enzyme Activity

Enzyme Activity can be Inhibited

Enzyme activity can be prevented by **enzyme inhibitors** — molecules that **bind to the enzyme** that they inhibit. Inhibition can be **competitive** or **non-competitive**.

COMPETITIVE INHIBITION

1) **Competitive inhibitor** molecules have a **similar shape** to that of the **substrate** molecules.
2) They **compete** with the substrate molecules to **bind** to the **active site**, but **no reaction** takes place.
3) Instead they **block** the active site, so **no substrate** molecules can **fit** in it.

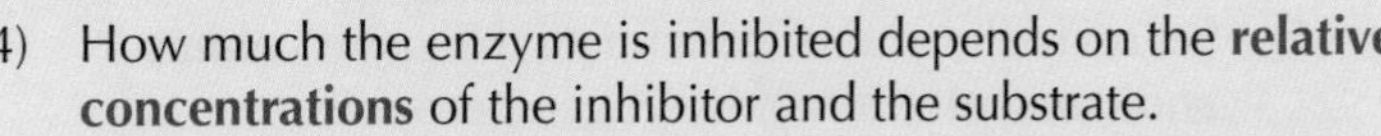

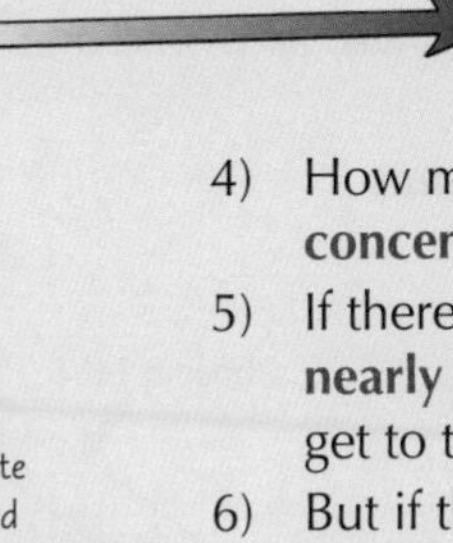

4) How much the enzyme is inhibited depends on the **relative concentrations** of the inhibitor and the substrate.
5) If there's a **high concentration** of the **inhibitor**, it'll take up **nearly all** the **active sites** and hardly any of the substrate will get to the enzyme.
6) But if there's a **higher concentration** of **substrate**, then the substrate's chances of getting to an active site before the inhibitor **increase**. So **increasing** the concentration of **substrate** will **increase** the **rate of reaction** (up to a point).

NON-COMPETITIVE INHIBITION

1) **Non-competitive inhibitor** molecules bind to the enzyme **away from its active site**.
2) This causes the active site to **change shape** so the substrate molecules can **no longer bind** to it.
3) They don't 'compete' with the substrate molecules to bind to the active site because they are a different shape.
4) **Increasing** the concentration of **substrate won't** make any difference to the reaction rate — enzyme activity will still be inhibited.

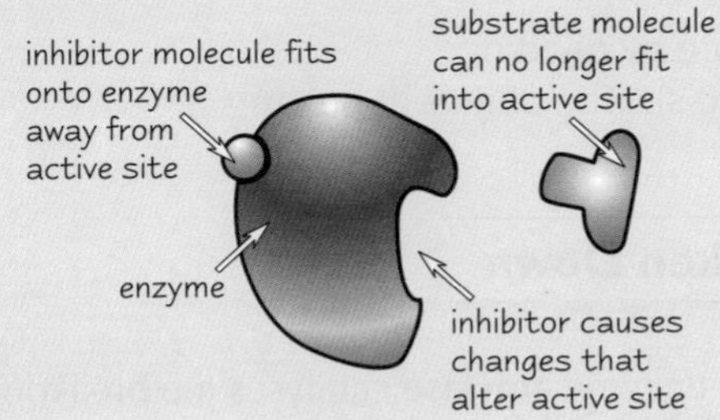

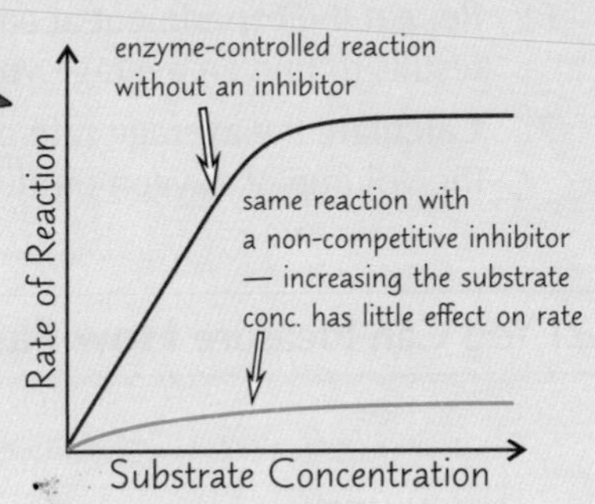

Warm-Up Questions

Q1 Draw a graph to show the effect of temperature on enzyme activity.

Q2 Draw a graph to show the effect of pH on enzyme activity.

Q3 Explain the effect of increasing substrate concentration on the rate of an enzyme-catalysed reaction.

Exam Question

Q1 Inhibitors prevent enzymes from working properly. They can be competitive or non-competitive.

a) Explain how a competitive inhibitor works. [3 marks]

b) Explain how a non-competitive inhibitor works. [2 marks]

Activity — mine is usually inhibited by pizza and a movie...

Human enzymes work well under normal body conditions — a neutral pH and body temp of 37 °C. Many poisons are enzyme inhibitors, e.g. cyanide. Even though there are thousands of enzymes in our bodies, inhibiting just one of them can cause severe problems. Some drugs are enzyme inhibitors though, e.g. penicillin, so they're not all bad.

Enzyme-Controlled Reactions

Science isn't all about words and theory, it's also about getting your pipette dirty and making bad smells (in the name of discovery of course). These pages show you how to measure the rate of an enzyme-controlled reaction.

You can **Measure** the **Rate** of an **Enzyme-Controlled** Reaction

Here are two ways of measuring the **rate** of an enzyme-controlled reaction:

1) You Can Measure **How Fast** the **Product** of the Reaction is **Made**

Catalase catalyses the **breakdown** of **hydrogen peroxide** into **water** and **oxygen**. It's easy to measure the volume of oxygen produced and to work out **how fast** it's given off. The diagram below shows the **apparatus** you'll need. The oxygen released **displaces** the water from the measuring cylinder. (A **stand** and **clamp** would also be pretty useful to hold the cylinder upside down, as would a **stopwatch** and a **water bath**.)

Here's how to carry out the experiment:

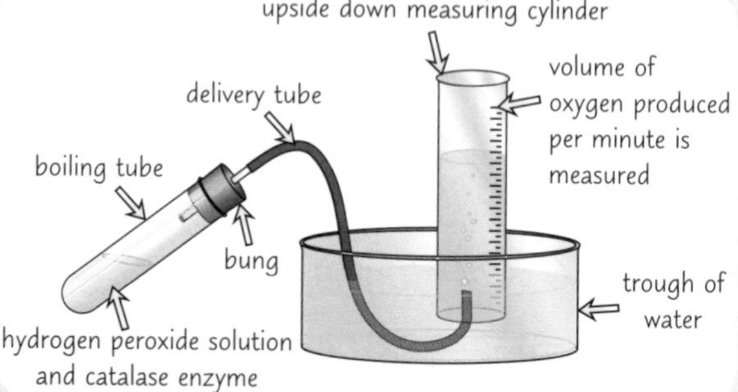

1) Set up boiling tubes containing the **same volume** and **concentration** of **hydrogen peroxide**. To keep the pH constant, add **equal volumes** of a suitable **buffer solution** to each tube. (A buffer solution is able to resist changes in pH when small amounts of acid or alkali are added.)
2) Set up the rest of the **apparatus** as shown in the diagram.
3) Put each boiling tube in a **water bath** set to a different temperature (e.g. 10 °C, 20 °C, 30 °C and 40 °C) along with another tube containing **catalase** (wait 5 minutes before moving onto the next step so the enzyme gets up to temperature).
4) Use a pipette to add the **same volume** and **concentration** of **catalase** to each boiling tube. Then **quickly attach** the **bung** and **delivery tube**.
5) **Record** how much oxygen is produced in the **first minute** (60 s) of the reaction. Use a **stopwatch** to measure the time.
6) **Repeat** the experiment at each temperature three times, and use the results to find an **average volume of oxygen produced**.
7) **Calculate** the **average rate of reaction** at each temperature by dividing the volume of oxygen produced by the time taken (i.e. 60 s). The units will be cm^3s^{-1}.

A negative control reaction, i.e. a boiling tube not containing catalase, should also be carried out at each temperature.

2) You Can Measure **How Fast** the **Substrate** is **Broken Down**

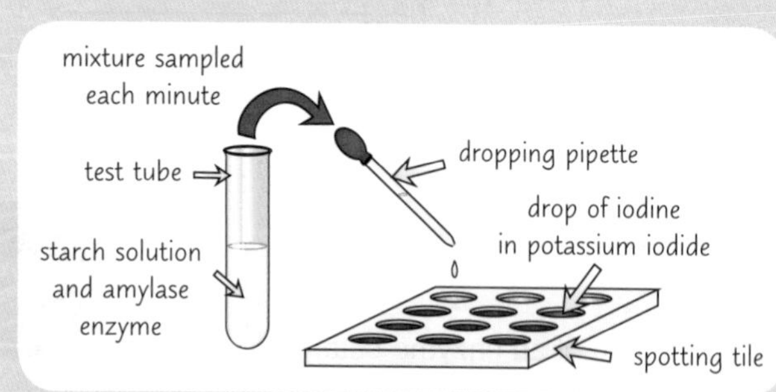

The enzyme **amylase** catalyses the breakdown of **starch** to **maltose**. The diagram shows how the experiment can be **set up**. You'll need the **apparatus** shown in the diagram as well as a **stopwatch**. A drop of **iodine in potassium iodide** is put into each well on a **spotting tile**. A known concentration of **amylase** and **starch** are then mixed together in a test tube. A **dropping pipette** is used to put a drop of this mixture into one of the wells containing the iodine solution on the spotting tile at **regular intervals** and the resulting colour is observed. The iodine solution goes **dark blue-black** when **starch** is **present** but remains its normal **browny-orange** colour when there's **no starch** around. You can see how fast **amylase** is working by **recording** how long it takes for the iodine solution to **no longer** turn blue-black when starch/amylase mixture is added. **Repeat** the experiment using **different concentrations** of **amylase**. Make sure that you also **repeat** the experiment three times at **each** amylase concentration.

The **experiments above** show you how you can investigate the effects of **temperature** and **enzyme concentration** on the rate of enzyme-controlled reactions. You can also **alter** these experiments to investigate the effect of a **different variable**, such as **pH** (by adding a **buffer solution** with a **different pH** to each test tube) or **substrate concentration** (you could use **serial dilutions** to make substrate solutions with **different** concentrations). The key to experiments like this is to remember to **only** change **one variable** — **everything else** should stay the **same**.

Enzyme-Controlled Reactions

You Need to be Able to **Interpret Graphs** of **Enzyme-Controlled** Reactions

The results of enzyme-controlled reactions are usually shown in **line graphs**. You might be asked to **interpret** the **graph** of an **enzyme-controlled** reaction in the exam. The graph below shows the **release of a product over time**:

1. First look at the start of the graph and compare the rates of reaction here. E.g. the rate of reaction is fastest at 65 °C. Use what you know about factors affecting enzyme activity to explain why (see p. 12). You might have to work out the initial rate of reaction (see below).

Volume of product released by an enzyme-controlled reaction at different temperatures

—65 °C —37 °C —25 °C

Volume of product released (cm³)

0 10 20 30 40 50

0 10 20 30 40 50 60

Time (s)

2. Now look at what else the graphs are showing you and make comparisons between the different temperatures.

At **37 °C** the graph has plateaued (flattened out) because all the substrate has been used up.

At **65 °C** the graph has plateaued earlier than at 37 °C, because the high temperature caused the enzyme to denature, so the reaction stopped sooner. Not as much product was made because not all the substrate was converted to product before the enzyme was denatured, so there is still substrate left.

At 25 °C the rate of reaction is remaining constant and the volume of product is continuing to increase because not all of the substrate has been used up.

The graph in your exam could be based on any variable — e.g. pH, temperature, enzyme concentration or substrate concentration. You'll have to use your knowledge of enzymes to explain what's going on.

You Can Use a **Tangent** to **Calculate** the **Initial Rate of Reaction**

The **initial** rate of reaction is the rate of reaction right at the **start** of the reaction, close to **time equals zero** (t = 0) on the graph. To work out the initial rate of reaction carry out the following steps:

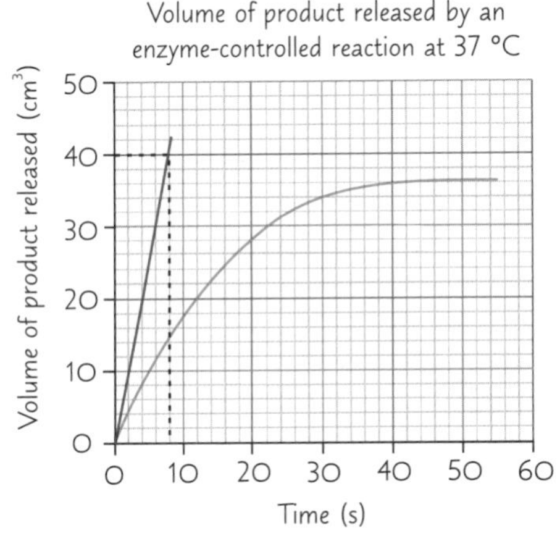

1) **Draw** a **tangent** to the curve at **t = 0**, using a ruler. Do this by positioning the ruler so it's an **equal distance** from the curve at **both sides** of where it's touching it. Here you'll have to **estimate** where the curve would **continue** if it carried on **below zero**. Then draw a **line** along the ruler. (For more on drawing tangents see p. 228.)
2) Then calculate the **gradient** of the **tangent** — this is the **initial rate of reaction**. Gradient = change in y axis ÷ change in x axis In this graph it's: 40 cm^3 ÷ 8 s = **5 cm^3 s^{-1}**

If you're comparing the initial rate of reaction for two different reactions, you can work out the ratio of the rates to give you a quick and easy comparison.

PRACTICE QUESTIONS

Warm-Up Question

Q1 You are testing the effects of pH on the action of an enzyme. What other variables must you keep constant?

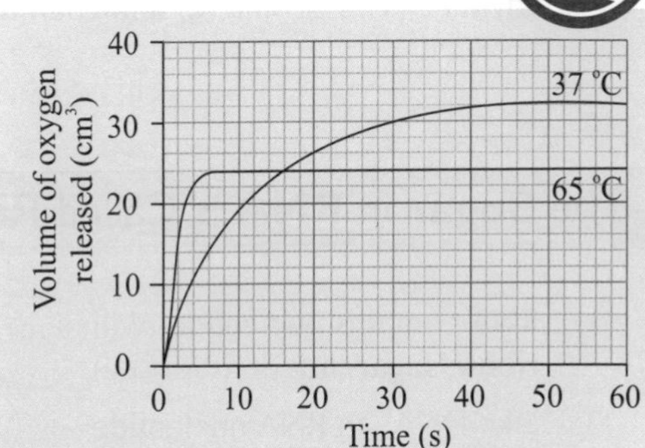

Exam Question

Q1 A student carries out an enzyme-controlled reaction at 37 °C and 65 °C. Her results are shown in the graph above. Draw a tangent to find the initial rate of reaction at 65 °C. Show your working. [1 mark]

My rate of reaction depends on what time of day it is...

In your exam, you could get asked about methods used to measure the rate of an enzyme-controlled reaction or to calculate the rate from a graph. It's worth your time to memorise the examples and learn the maths on these pages.

DNA and RNA

...ges are all about nucleic acids — DNA and RNA. These molecules are needed to build proteins, ...ich are required for the cells in living organisms to function. They're right handy little things.

DNA and RNA Carry Important Information

DNA and RNA are both types of **nucleic acid**. They're found in **all living cells** and they both carry **information**.

1) **DNA** (deoxyribonucleic acid) is used to store **genetic information** — that's **all the instructions** an organism needs to **grow and develop** from a fertilised egg to a fully grown adult.
2) **RNA** (ribonucleic acid) is similar in structure to DNA. One of its main functions is to **transfer** genetic information from the **DNA** to the **ribosomes**. Ribosomes are the body's **'protein factories'** — they read the RNA to make **polypeptides** (proteins) in a process called **translation** (see p. 91). Ribosomes themselves are made from **RNA** and **proteins**.

DNA and RNA are Polymers of Nucleotides

1) A **nucleotide** is a type of biological molecule. It's made from:

- a **pentose sugar** (that's a sugar with **5** carbon atoms),
- a **nitrogen-containing** organic **base**,
- a **phosphate** group.

'Organic' means that it contains carbon.

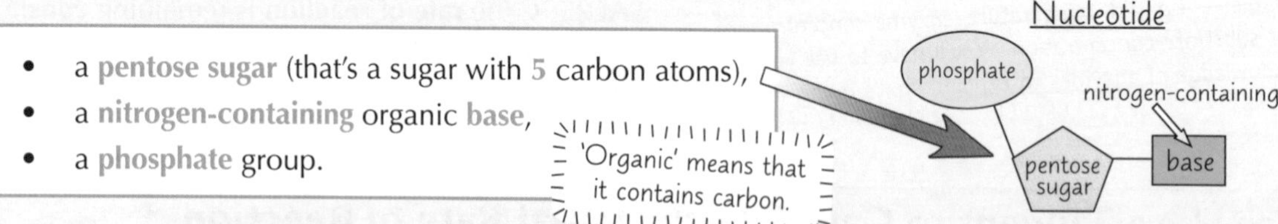

2) Nucleotides are really **important**. For a start, they're the **monomers** (see p. 2) that make up **DNA** and **RNA**.

The Sugar in DNA is Called Deoxyribose

1) The **pentose sugar** in a **DNA nucleotide** is called **deoxyribose**.
2) Each DNA nucleotide has the **same sugar** and a **phosphate group**. The **base** on each nucleotide can **vary** though.
3) There are **four** possible bases — adenine (**A**), thymine (**T**), cytosine (**C**) and guanine (**G**).

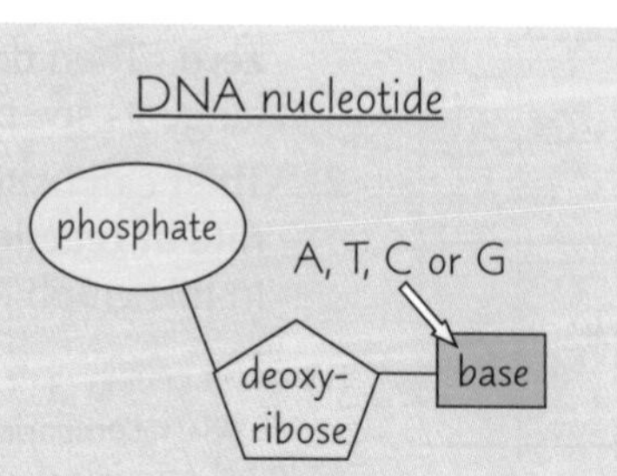

The Sugar in RNA is Called Ribose

1) **RNA** contains nucleotides with a **ribose sugar** (not deoxyribose).
2) Like DNA, an RNA nucleotide also has a **phosphate group** and one of **four** different **bases**.
3) In RNA though, **uracil** (U) replaces **thymine** as a base.

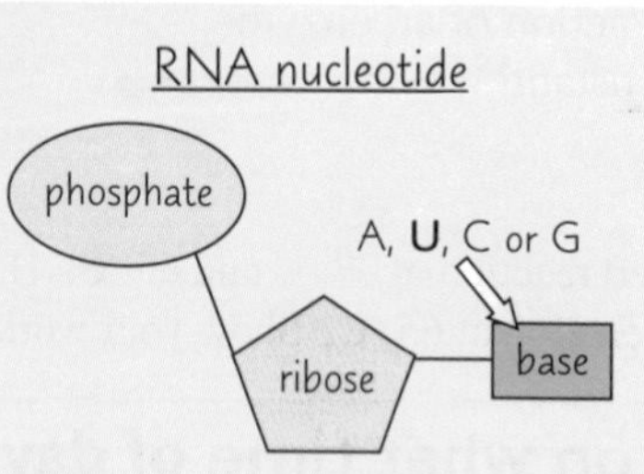

Mary didn't care if it was ribose or deoxyribose, she just wanted her cuppa.

DNA and RNA

Nucleotides Join Together to Form Polynucleotides

1) A **polynucleotide** is a **polymer** of **nucleotides**. Both DNA and RNA nucleotides form polynucleotides.
2) The nucleotides join up via a **condensation reaction** (see p. 2) between the **phosphate** group of one nucleotide and the **sugar** of another.
3) This forms a **phosphodiester bond** (consisting of the phosphate group and two ester bonds).
4) The chain of sugars and phosphates is known as the **sugar-phosphate backbone**.

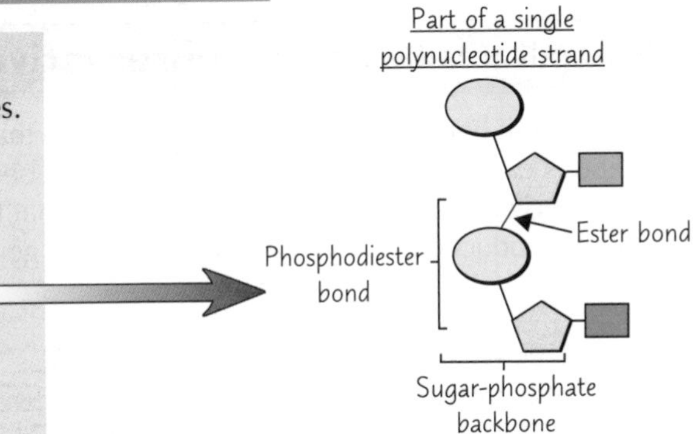

DNA is Made of Two Polynucleotide Chains in a Double-Helix Structure

1) **Two DNA** polynucleotide strands join together by **hydrogen bonding** between the bases.
2) Each base can only join with one particular partner — this is called **complementary base pairing** (or specific base pairing).
3) **Adenine** always pairs with **thymine** (**A - T**) and **cytosine** always pairs with **guanine** (**C - G**). This means that there are always **equal amounts** of adenine and thymine in a DNA molecule and **equal amounts** of cytosine and guanine.
4) **Two** hydrogen bonds form between **A and T**, and **three** hydrogen bonds form between **C and G**.
5) Two **antiparallel** (running in opposite directions) polynucleotide strands **twist** to form the **DNA double-helix**.
6) DNA was first observed in the 1800s, but lots of scientists at the time **doubted** that it could carry the **genetic code** because it has a **relatively simple chemical composition**. Some argued that genetic information must be carried by **proteins** — which are much more **chemically varied**.
7) By 1953, experiments had shown that DNA was the carrier of the genetic code. This was also the year in which the **double-helix structure**, which helps DNA to carry out its function, was determined by **Watson** and **Crick**.

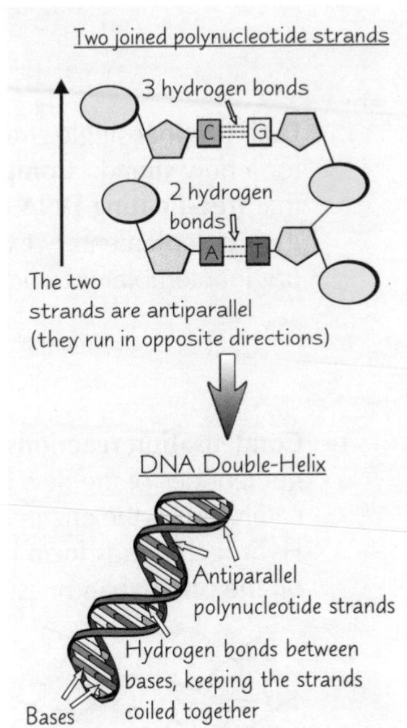

RNA is a Relatively Short Polynucleotide Chain

RNA is made from a **single** polynucleotide chain (not a double one). It's much **shorter** than most DNA polynucleotides.

Warm-Up Questions

PRACTICE QUESTIONS

Q1 Name the bases in RNA.

Exam Questions

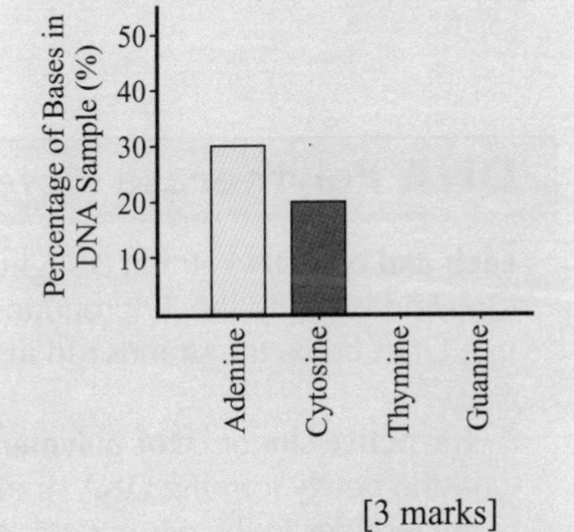

Q1 The bar chart shows the percentage of the bases in a DNA sample that are adenine and cytosine. On the chart, sketch bars to show the percentages of thymine and guanine in the sample. [2 marks]

Q2 a) Describe how nucleotides are joined together in DNA. [3 marks]

b) Describe how two single polynucleotide strands are joined to make a double helix. [3 marks]

Give me a D, give me an N, give me an A! What do you get? — confused...

You need to learn the structure of DNA — the polynucleotide strands, the hydrogen bonds, and don't forget complementary base pairing. Make sure you know the differences between RNA and DNA too — interesting stuff.

DNA Replication

DNA has the amazing ability to replicate (copy) itself. These pages cover the facts behind the replication mechanism, as well as some of the history behind its discovery. This stuff is really clever. Honest.

DNA Replicates by Semi-Conservative Replication

DNA **copies** itself **before** cell division (see p. 34) so that each **new** cell has the **full** amount of **DNA**. The method is called **semi-conservative replication** because **half** of the strands in **each new DNA molecule** are from the **original** DNA molecule. This means that there's **genetic continuity** between generations of cells (i.e. the cells produced by cell division inherit their genes from their parent cells).

1) The enzyme **DNA helicase** **breaks** the **hydrogen bonds** between bases on the two **polynucleotide** DNA strands. This makes the helix **unwind** to form two single strands.

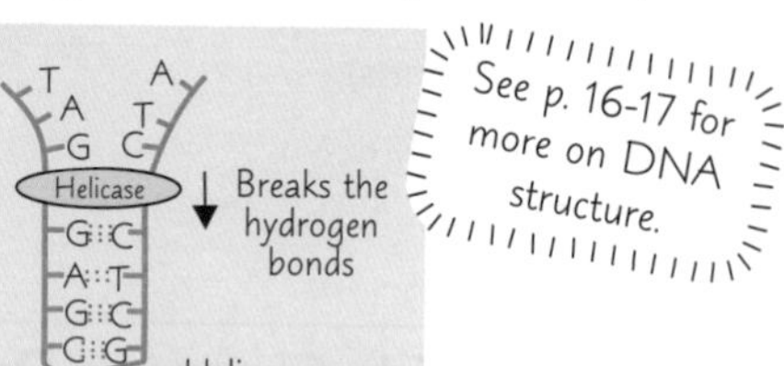

See p. 16-17 for more on DNA structure.

Gerald doesn't need helicase to unwind. He just needs a beach full of seals.

2) Each **original** single strand acts as a **template** for a new strand. **Complementary base pairing** means that **free-floating DNA nucleotides** are attracted to their complementary **exposed bases** on each original template strand — A with T and C with G.

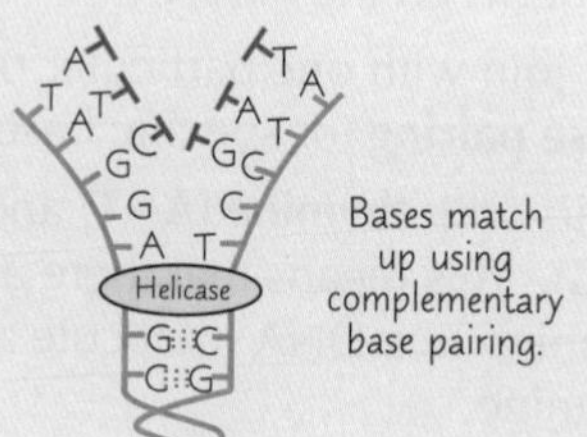

3) **Condensation reactions** join the nucleotides of the new strands together — catalysed by the enzyme **DNA polymerase**. Hydrogen bonds **form** between the bases on the original and new strands.

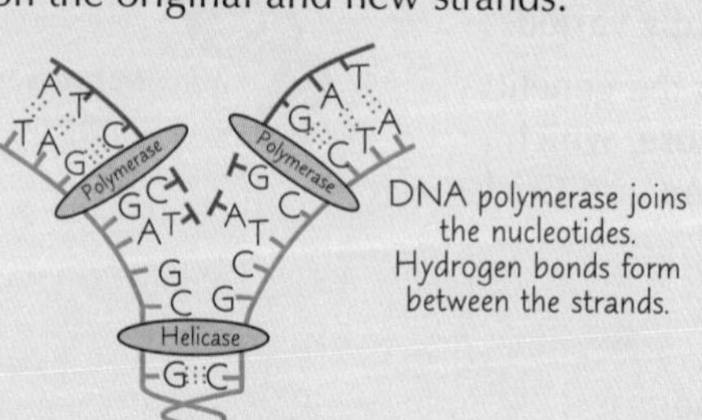

4) Each new DNA molecule contains **one strand** from the **original** DNA molecule and one **new strand**.

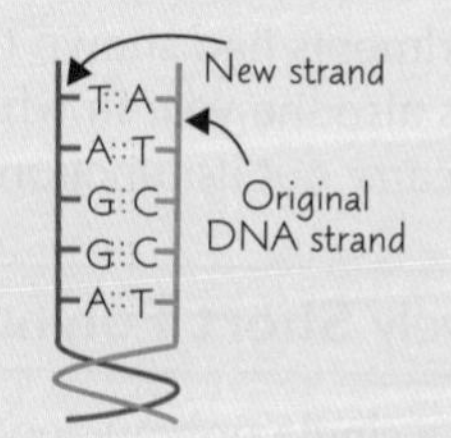

DNA Polymerase Moves in Opposite Ways Along Antiparallel DNA Strands

Each end of a DNA strand is slightly **different** in its structure.
One end is called the **3′** (pronounced 'three prime') end and one end is called the **5′** (five prime) end.
In a DNA helix, the strands run in **opposite** directions — they're **antiparallel**.

The **active site** of **DNA polymerase** is only **complementary** to the **3′ end** of the newly forming DNA strand — so the enzyme can **only add nucleotides** to the new strand at the **3′ end**.

This means that the **new strand** is made in a **5′ to 3′** direction and that DNA polymerase moves down the **template strand** in a **3′ to 5′** direction. Because the strands in the double-helix are **antiparallel**, the DNA polymerase working on **one** of the template strands **moves** in the **opposite direction** to the DNA polymerase working on the **other** template strand.

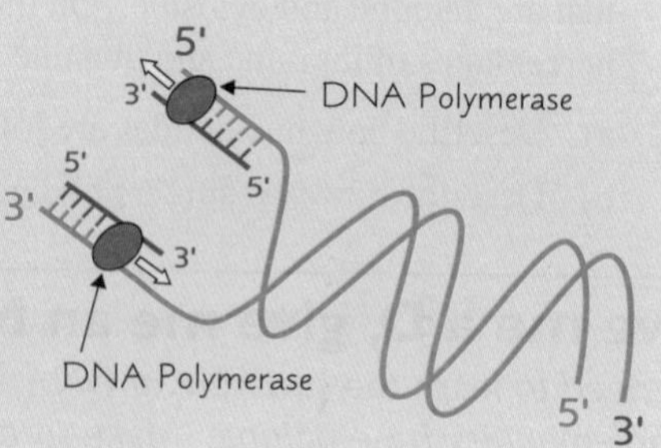

DNA Replication

Meselson and Stahl Provided Evidence for Semi-Conservative Replication

1) You might remember from page 17 that **Watson** and **Crick** determined the structure of DNA. They also came up with the theory of **semi-conservative DNA replication**.
2) It wasn't until **Meselson** and **Stahl's** experiment a few years **later** that this theory was **validated**. Before that, people were unsure whether DNA replication was **semi-conservative** or **conservative**. If the method was **conservative**, the original DNA strands would **stay together** and the new DNA molecules would contain **two new strands**.
3) Meselson and Stahl showed that DNA is replicated using the **semi-conservative method**. Their experiment used two **isotopes** of **nitrogen** (DNA contains nitrogen) — **heavy** nitrogen (^{15}N) and **light** nitrogen (^{14}N). Here's how it worked:

Isotopes are different forms of the same element.

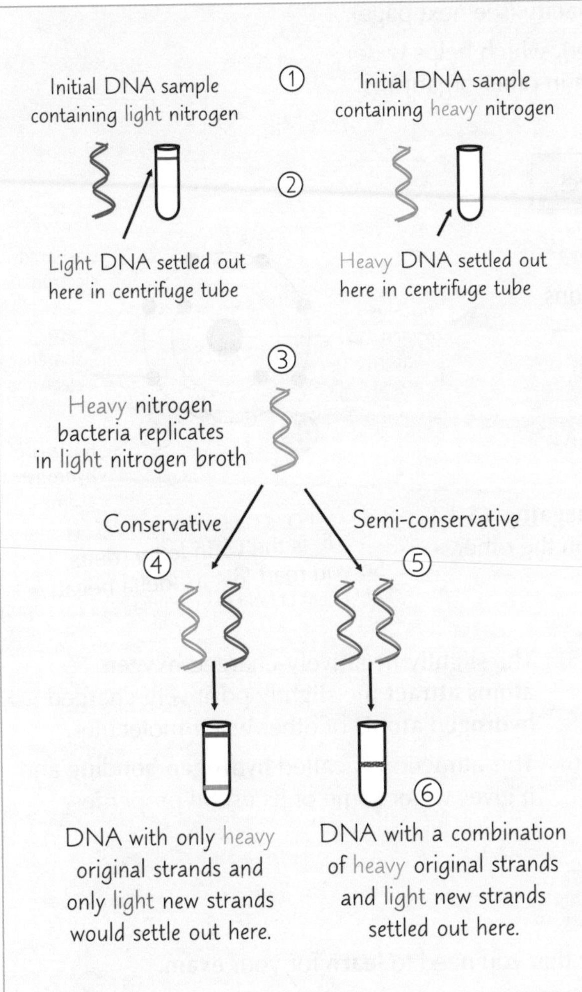

1) Two samples of bacteria were grown — one in a nutrient broth containing **light** nitrogen, and one in a broth with **heavy** nitrogen. As the **bacteria reproduced**, they **took up nitrogen** from the broth to help make nucleotides for new DNA. So the nitrogen gradually became part of the bacteria's DNA.
2) A **sample of DNA** was taken from each batch of bacteria, and spun in a **centrifuge**. The DNA from the **heavy** nitrogen bacteria settled **lower** down the **centrifuge tube** than the DNA from the **light** nitrogen bacteria — because it's **heavier**.
3) Then the bacteria grown in the **heavy** nitrogen broth were **taken out** and put in a broth containing only **light nitrogen**. The bacteria were left for **one round of DNA replication**, and then **another DNA sample** was taken out and spun in the centrifuge.
4) If replication was **conservative**, the original **heavy** DNA, which would still be together, would settle at the bottom and the new **light** DNA would settle at the top.
5) If replication was **semi-conservative**, the new bacterial DNA molecules would contain **one strand** of the **old DNA** containing **heavy** nitrogen and **one strand** of **new DNA** containing **light** nitrogen. So the DNA would settle out **between** where the **light** nitrogen DNA settled out and where the **heavy** nitrogen DNA settled out.
6) As it turned out, the DNA settled out in the **middle**, showing that the DNA molecules contained a **mixture** of **heavy** and **light** nitrogen. The bacterial DNA had **replicated semi-conservatively** in the **light** nitrogen.

Once Meselson and Stahl had confirmed that **DNA replication** in **bacteria** was **semi-conservative**, other scientists carried out experiments to show that it was the **universal method** for DNA replication in **all living things**.

Warm-Up Questions

PRACTICE QUESTIONS

Q1 What is the role of DNA helicase in DNA replication?

Q2 What's the key difference between the conservative and semi-conservative theories of DNA replication?

Exam Question

Q1 Describe the process of semi-conservative DNA replication. [5 marks]

DNA DNA Replication Replication is is Semi-Conservative Semi-Conservative

Make sure you can recall the mechanism of DNA replication — you might be asked for it in your exam. You might also be asked to evaluate the work of the scientists who validated Watson and Crick's theory of semi-conservative replication.

Water

Your body needs lots of molecules to stay alive, and these pages cover one of the most important — water.

Water is Vital to Living Organisms

Water makes up about 80% of a cell's contents. It has loads of important **functions**, inside and outside of cells:

1) Water is a metabolite in loads of important metabolic reactions, including condensation and hydrolysis reactions (see below).
2) Water is a solvent, which means some substances dissolve in it. Most metabolic reactions take place in solution (e.g. in the cytoplasm of eukaryotic and prokaryotic cells) so water's pretty essential.
3) Water helps with temperature control because it has a high latent heat of vaporisation (see below) and a high specific heat capacity (see next page).
4) Water molecules are very cohesive (they stick together), which helps water transport in plants (see next page) as well as transport in other organisms.

A metabolic reaction is a chemical reaction that happens in a living organism to keep the organism alive. A metabolite is a substance involved in a metabolic reaction.

Water Molecules Have a Simple Structure

1) A molecule of **water** (**H_2O**) is **one atom of oxygen** (**O**) joined to **two atoms of hydrogen** (**H_2**) by **shared electrons**.
2) Because the **shared negative** hydrogen electrons are **pulled towards** the oxygen atom, the other side of each hydrogen atom is left with a **slight positive charge**.
3) The **unshared** negative electrons on the oxygen atom give it a **slight negative charge**.
4) This makes water a **polar** molecule — it has a **partial negative** (δ–) charge on one side and a **partial positive** (δ+) charge on the other.

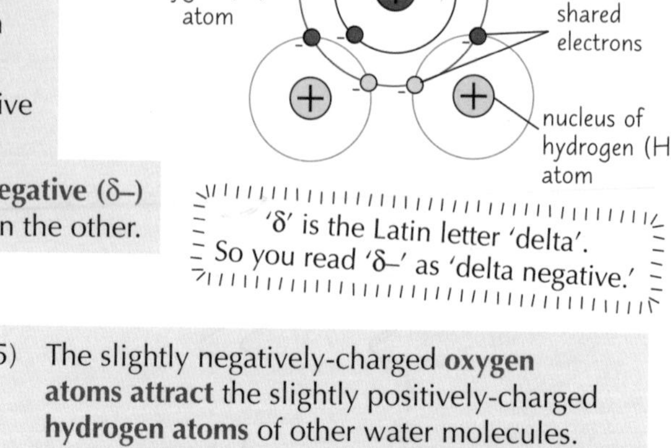

'δ' is the Latin letter 'delta'. So you read 'δ–' as 'delta negative.'

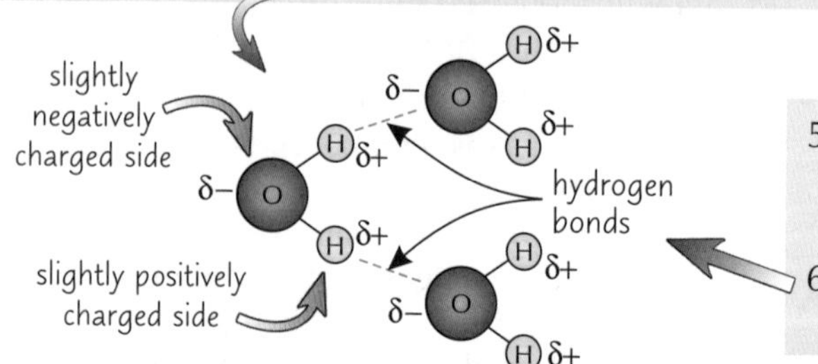

5) The slightly negatively-charged **oxygen atoms attract** the slightly positively-charged **hydrogen atoms** of other water molecules.
6) This attraction is called **hydrogen bonding** and it gives water some of its useful properties.

Water Has Some Really Useful Properties

Here's a bit more about each of the **useful properties** of **water** that you need to **learn** for your **exam**.

Water is an Important Metabolite

1) Many metabolic reactions involve a **condensation** or **hydrolysis** reaction.
2) A **hydrolysis** reaction requires a **molecule of water** to **break** a **bond**. A **condensation** reaction releases a molecule of water as a **new bond** is formed.
3) For example, **amino acids** are **joined** together to make **polypeptides** (proteins) by **condensation** reactions (see page 8). **Energy** from **ATP** is released through a **hydrolysis** reaction (see page 22).

Water has a High Latent Heat of Vaporisation

1) It takes a lot of **energy** (**heat**) to **break** the hydrogen bonds between water molecules.
2) So water has a **high latent heat of vaporisation** — a lot of energy is used up when water **evaporates** (vaporises).
3) This is useful for living organisms because it means they can use water loss through evaporation to **cool down** (e.g. humans **sweat** to cool down) without losing too much water.

Water

Water Can **Buffer** (Resist) Changes in **Temperature**

1) The **hydrogen bonds** between water molecules can **absorb** a **lot** of energy.
2) So water has a **high specific heat capacity** — it takes a lot of energy to heat it up.
3) This is useful for living organisms because it means that water **doesn't** experience **rapid temperature changes**. This makes water a **good habitat** because the temperature **under water** is likely to be **more stable** than on land. The water **inside** organisms also remains at a fairly **stable** temperature — helping them to **maintain** a **constant** internal **body temperature**.

Water is a **Good Solvent**

1) A lot of important substances in metabolic reactions are **ionic** (like **salt**, for example). This means they're made from **one positively charged** atom or molecule and **one negatively charged** atom or molecule (e.g. salt is made from a positive sodium ion and a negative chloride ion).
2) Because water is polar, the **positive end** of a water molecule will be attracted to the **negative ion**, and the **negative end** of a water molecule will be attracted to the **positive ion**.
3) This means the ions will get **totally surrounded** by water molecules — in other words, they'll **dissolve**.

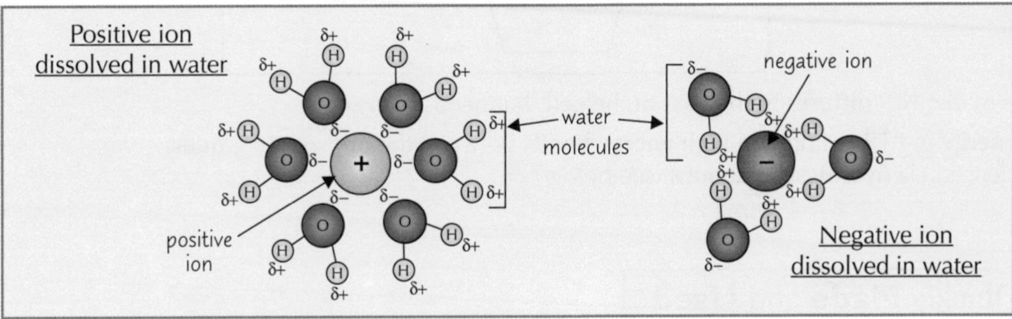

4) So water's **polarity** makes it a useful **solvent**.

There's Strong **Cohesion** Between **Water** Molecules

1) Cohesion is the **attraction** between molecules of the same type (e.g. two water molecules). Water molecules are **very cohesive** (they tend to stick together) because they're **polar**.
2) Strong cohesion helps water to **flow**, making it great for **transporting substances**. For example, it's how water travels in **columns** up the **xylem** (tube-like transport cells) in **plants** (see p. 82).
3) Strong cohesion also means that water has a **high surface tension** when it comes into contact with **air**. This is the reason why sweat forms **droplets**, which evaporate from the skin to cool an organism down. It's also the reason that **pond skaters**, and some other insects, can 'walk' on the surface of a pond.

Warm-Up Questions

Q1 Briefly describe the structure of a water molecule.
Q2 Why is water's high specific heat capacity useful for living organisms?
Q3 Describe how a positive ion dissolves in water.

PRACTICE QUESTIONS

Exam Question

Q1 In hot temperatures, elephants commonly spray themselves with water. With reference to the structure and properties of water, explain:

a) why this behaviour acts as a cooling mechanism for the elephant. [3 marks]

b) why water forms droplets when the elephant sprays it from its trunk. [2 marks]

Psssssssssssssssssssssssssssssssssssssss — need the loo yet?

Water is pretty darn useful really. It looks so, well, dull — but in fact it's scientifically amazing. It's essential for all kinds of jobs — keeping cool, transporting things, enabling reactions, etc. You need to learn all of its properties and functions.

ATP

ATP is an important molecule in all living things. Without it, we wouldn't be able to function. On the plus side, that would mean no exams — but on the other hand, we wouldn't know the great smell of freshly baked bread. Ahh...

ATP is the Immediate Source of Energy in a Cell

1) Plant and animal cells **release energy** from **glucose** — this process is called **respiration**.
2) A cell **can't** get its energy **directly** from glucose.
3) So, in respiration, the **energy released** from glucose is used to **make ATP** (adenosine triphosphate).
4) ATP is made from the nucleotide base **adenine**, combined with a **ribose sugar** and **three phosphate groups**. It's what's known as a **nucleotide derivative** because it's a **modified** form of a **nucleotide**:

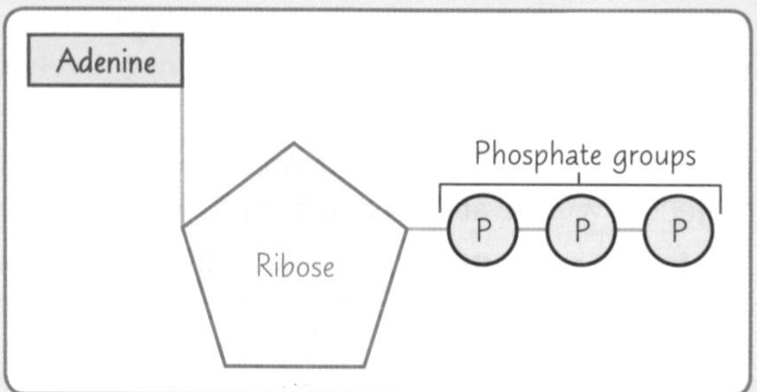

5) Once made, ATP **diffuses** to the **part** of the **cell** that needs **energy**.
6) The **energy** in ATP is stored in **high energy bonds** between the **phosphate groups**. It's released via **hydrolysis reactions** (see below).

ATP is Quickly Made and Used

1) When **energy** is needed by a cell, **ATP** is **broken down** into **ADP** (adenosine diphosphate) and $\mathbf{P_i}$ (inorganic phosphate).
2) This is a **hydrolysis reaction**. A **phosphate bond** is **broken** and **energy** is **released**. The reaction is **catalysed** by the enzyme **ATP hydrolase**.
3) **ATP hydrolysis** can be '**coupled**' to other **energy-requiring reactions** in the cell — this means the energy released can be **used directly** to make the **coupled reaction** happen, rather than being **lost** as heat.
4) The released **inorganic phosphate** can also be put to use — it can be **added** to another **compound** (this is known as **phosphorylation**), which often makes the compound **more reactive**.
5) **ATP** can be **re-synthesised** in a **condensation reaction** between **ADP** and $\mathbf{P_i}$. This happens during both **respiration** and **photosynthesis**, and is **catalysed** by the enzyme **ATP synthase**.

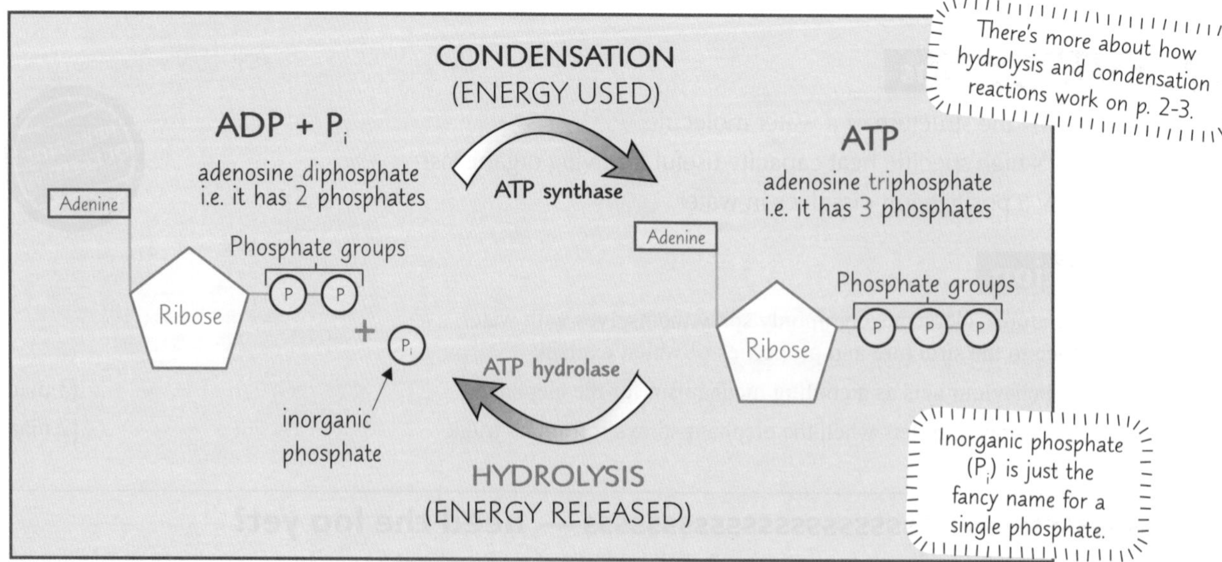

There's more about how hydrolysis and condensation reactions work on p. 2-3.

Inorganic phosphate (P_i) is just the fancy name for a single phosphate.

Inorganic Ions

Inorganic Ions Have an Electric Charge

1) An **ion** is an atom (or group of atoms) that has an **electric charge**.
2) An ion with a **positive charge** is called a **cation**.
3) An ion with a **negative charge** is called an **anion**.
4) An **inorganic** ion is one which **doesn't contain carbon** (although there are a few exceptions to this rule).
5) There are inorganic ions, in **solution**, in the **cytoplasms of cells** and in the **body fluids of organisms**. **Each** ion has a **specific role**, depending on its **properties**. An ion's role determines whether it is found in **high** or **low concentrations**.

Iron Ions Are an Important Part of Haemoglobin

See page 72 for more on haemoglobin.

- **Haemoglobin** is a large protein that carries **oxygen** around the **body**, in the **red blood cells**.
- It's made up of **four** different polypeptide chains, each with an **iron ion (Fe^{2+})** in the centre.
- It's the Fe^{2+} that actually **binds** to the **oxygen** in haemoglobin — so it's a pretty key component.
- When oxygen **is** bound, the Fe^{2+} ion temporarily becomes an Fe^{3+} ion, until oxygen is released.

Hydrogen Ions (H^+) Determine pH

pH is calculated based on the **concentration** of **hydrogen ions** (H^+) in the environment. The **more** H^+ present, the **lower** the pH (and the more **acidic** the environment). Enzyme-controlled reactions are all affected by pH.

Sodium Ions (Na^+) Help Transport Glucose and Amino Acids Across Membranes

- **Glucose** and **amino acids** need a bit of help crossing cell membranes.
- A molecule of **glucose** or an **amino acid** can be transported into a cell (across the cell-surface membrane) alongside **sodium** ions (Na^+). This is known as **co-transport** (see pages 44 and 45 for more).

Phosphate Ions Are an Essential Component of ATP and DNA

- When a **phosphate ion** (PO_4^{3-}) is attached to another molecule, it's known as a **phosphate group**.
- **DNA**, **RNA** and **ATP** all contain phosphate groups.
- It's the bonds between **phosphate groups** that store energy in **ATP** (see previous page).
- The phosphate groups in **DNA** and **RNA** allow **nucleotides** to join up to form the **polynucleotides** (see p. 17).

Warm-Up Questions

Q1 Draw a molecule of ATP.

Q2 How many phosphate groups does ADP have?

Q3 How are hydrogen ions related to the pH of an environment?

Exam Questions

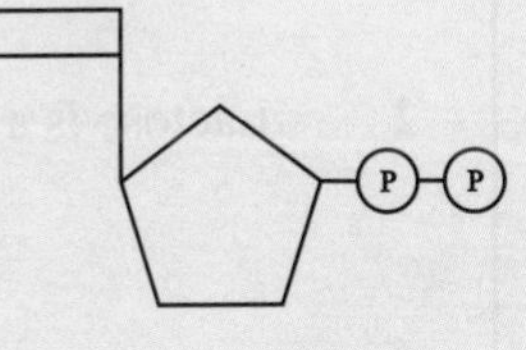

Q1 The diagram on the right shows a molecule involved in the synthesis of ATP. Describe how ATP is synthesised from this molecule. [3 marks]

Q2 Free inorganic ions can play very important roles in the body.

a) Describe how iron ions carry oxygen to where it is needed in the body. [3 marks]

b) Explain the role of phosphate ions in providing energy for cellular reactions. [2 marks]

Oh dear, I've used up all my ATP on these two pages...

You need to learn the roles of ATP hydrolase and ATP synthase. Remember — ATP hydrolase catalyses the hydrolysis of ATP, and ATP synthase catalyses the synthesis of ATP from ADP and P_i. Don't say science hasn't made it easy for you in the naming stakes. Inorganic ions should provide you with a little light relief at any rate.

Extra Exam Practice

That's Topic 1 about done. Give these questions a go to see if you can link ideas from the topic together.

- Have a look at this example of how to answer a tricky exam question.
- Then check how much you've understood from Topic 1 by having a go at the questions that follow.

When you're feeling confident with all the topics in this book, there are some synoptic questions on p.233-241 that'll test your knowledge of the whole course.

1 DNA polymerase is one enzyme involved in DNA replication.

1.1 Explain how DNA replication could be affected by a change in the amino acid sequence of DNA polymerase.

(4 marks)

Figure 1 shows a section of double-stranded DNA undergoing replication.

Figure 1

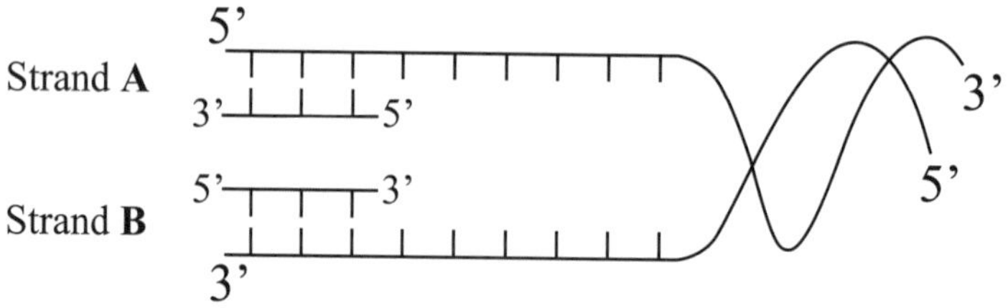

Strand **A**
Strand **B**
5' 3' 5' 3' 5' 3' 3' 5'

1.2 DNA polymerase's active site is only complementary to the 3' end of the newly forming strand. Explain why nucleotides can be continuously added to one strand but not the other.

(3 marks)

To get you started, think about the role of DNA polymerase in DNA replication when it's working normally — this'll help you to work out what will happen when it's not working normally.

1.1

Nucleotides may not be joined together as they are added to the new strand during DNA replication. This is because changes to the amino acid sequence of DNA polymerase could alter the tertiary structure of the enzyme. If the tertiary structure is altered then the active site could change shape. This would mean that the shape of DNA polymerase's active site is no longer complementary to the substrate (the nucleotides), so it could not catalyse the condensation reaction that forms phosphodiester bonds between nucleotides.

1.2

Nucleotides can only be added to the 3' end of the newly forming DNA strand by DNA polymerase, as **its active site is only complementary to the 3' end**. **In strand B**, the 3' end of the developing strand is exposed as the template strand is unwound from the double-helix, so nucleotides can continuously be added. **However, in strand A**, the 5' end of the developing strand is exposed as the template strand is unwound, so nucleotides can't be added continuously.

Make sure you compare the two strands.

Don't panic if you're given an unfamiliar diagram like Figure 1 in 1.2 — just look at it carefully, along with any information you're given in the question to work out what it is showing.

2 Lactose is a sugar found in milk. **Figure 2** shows the structure of lactose.

Figure 2

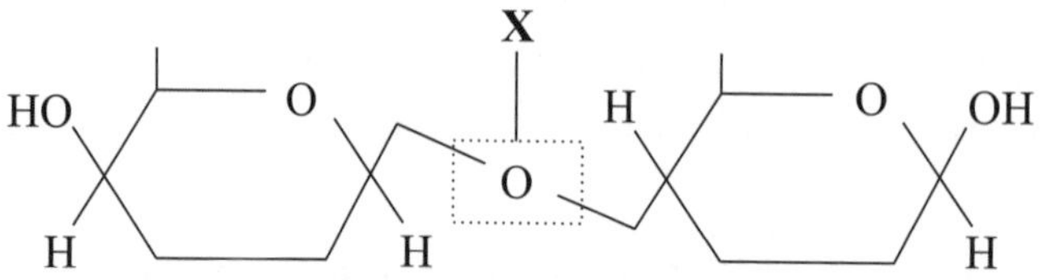

X
HO O H O OH
O
H H H

2.1 Identify the bond labelled **X**, in **Figure 2**.

(1 mark)

2.2 Using the information in **Figure 2**, draw the structure of a galactose molecule.

(1 mark)

Lactase is an enzyme that catalyses the breakdown of lactose. Some people do not produce this enzyme, so they cannot digest milk and are lactose-intolerant. Lactose-free milk is an alternative for people who are lactose-intolerant and can be produced using the process shown in **Figure 3**.

Figure 3

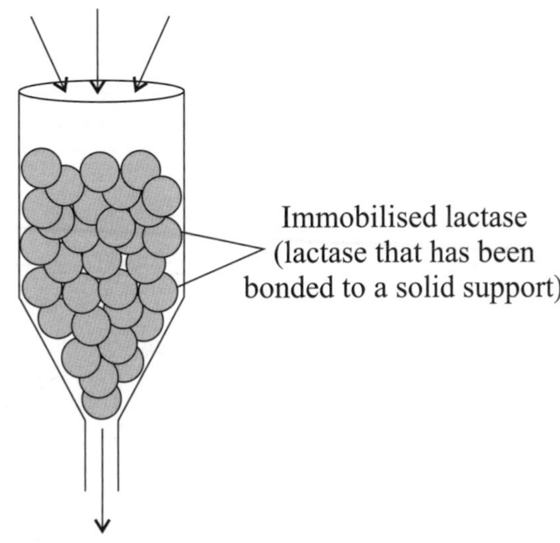

2.3 Studies have shown that galactose can be a competitive inhibitor of lactase. Using this information, explain why the rate of flow of milk through the column should be controlled.

(3 marks)

Free lactase is lactase which has not been immobilised. It can also be used to produce lactose-free milk. A group of students investigated the effect of temperature on the yield of glucose produced by both free lactase and immobilised lactase, relative to the yield produced at the optimum temperature for lactase. The results of their investigation are shown in **Table 1**.

Table 1

Temperature / °C	Relative yield of glucose from free lactase / %	Relative yield of glucose from immobilised lactase / %
40	100	100
50	65	90
60	20	60
70	0	25

2.4 State **one** factor that the students should have controlled in this investigation.

(1 mark)

One of the students suggested that immobilisation strengthens the tertiary structure of the enzyme.

2.5 Explain how the results shown in **Table 1** support this suggestion.

(3 marks)

Eukaryotic Cells and Organelles

There are two types of cell — prokaryotic and eukaryotic. The next few pages are about eukaryotic cells and their organelles (all the tiny bits and bobs that you can only see in detail with a fancy microscope)...

Organisms can be **Prokaryotes** or **Eukaryotes**

1) Prokaryotic organisms are **prokaryotic cells** (i.e. they're single-celled organisms) and eukaryotic organisms are made up of **eukaryotic cells**.
2) Both types of cells contain **organelles**. Organelles are **parts** of cells — each one has a **specific function**.

1) Eukaryotic cells are **complex** and include all **animal** and **plant** cells, as well as all cells in **algae** and **fungi**.
2) Prokaryotic cells are **smaller** and **simpler**, e.g. bacteria. See page 30 for more.

You Need to Know the Structure of **Eukaryotic Cells**

Eukaryotic cells are generally a **bit more complicated** than prokaryotic cells. You've probably been looking at **animal** and **plant cell** diagrams for years, so hopefully you'll be familiar with some of the bits and pieces...

Animal Cell

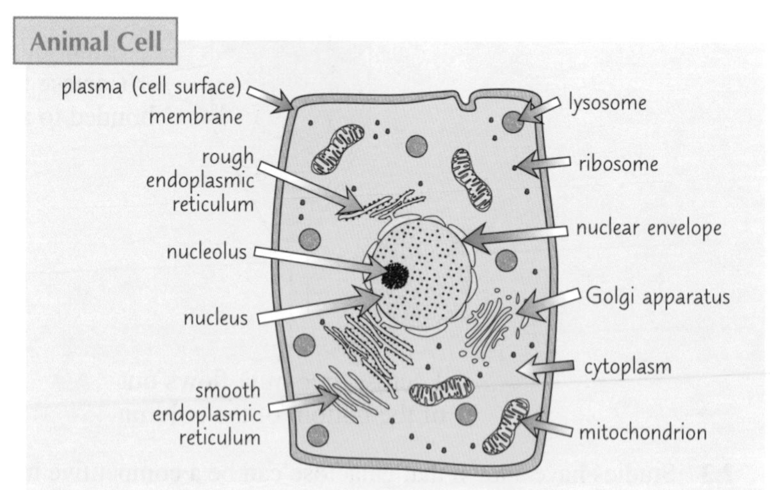

Plant Cell

Plant cells have all the **same organelles** as animal cells, but with a few **added extras**:

- a **cellulose cell wall** with **plasmodesmata** ('channels' for exchanging substances with adjacent cells),
- a **vacuole** (compartment that contains cell sap),
- and of course good old **chloroplasts**.

You might also see starch grains in plant cells, although they're not organelles. Plants use starch grains to store excess sugars.

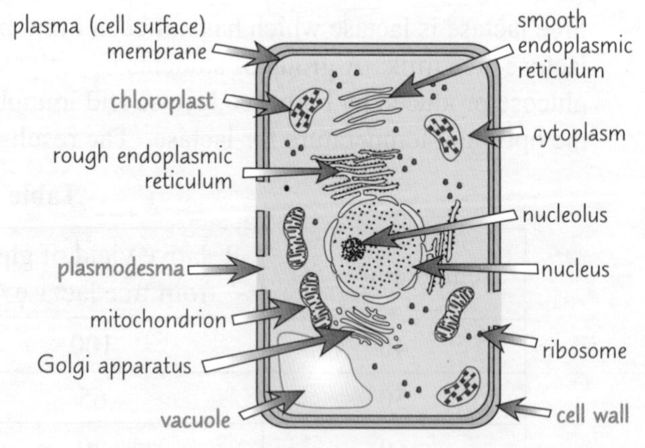

Algal and Fungal Cells

1) **Algal** cells are a lot like **plant cells** — they have all the **same** organelles, including a **cell wall** and **chloroplasts**.
2) **Fungal** cells are also a lot like plant cells, but with two key **differences**:
   - their cell walls are made of **chitin**, not cellulose.
   - they **don't have chloroplasts** (because they don't photosynthesise).

Algae carry out photosynthesis, like plants, but can be single-celled or multicellular. Fungi include mushrooms and yeast.

Eukaryotic Cells and Organelles

Different Organelles have **Different Functions**

This giant table contains a big list of organelles — you need to know the **structure** and **function** of them all. Sorry. Most organelles are surrounded by **membranes**, which sometimes causes confusion — don't make the mistake of thinking that a diagram of an organelle is a diagram of a whole cell. They're not cells — they're **parts of** cells.

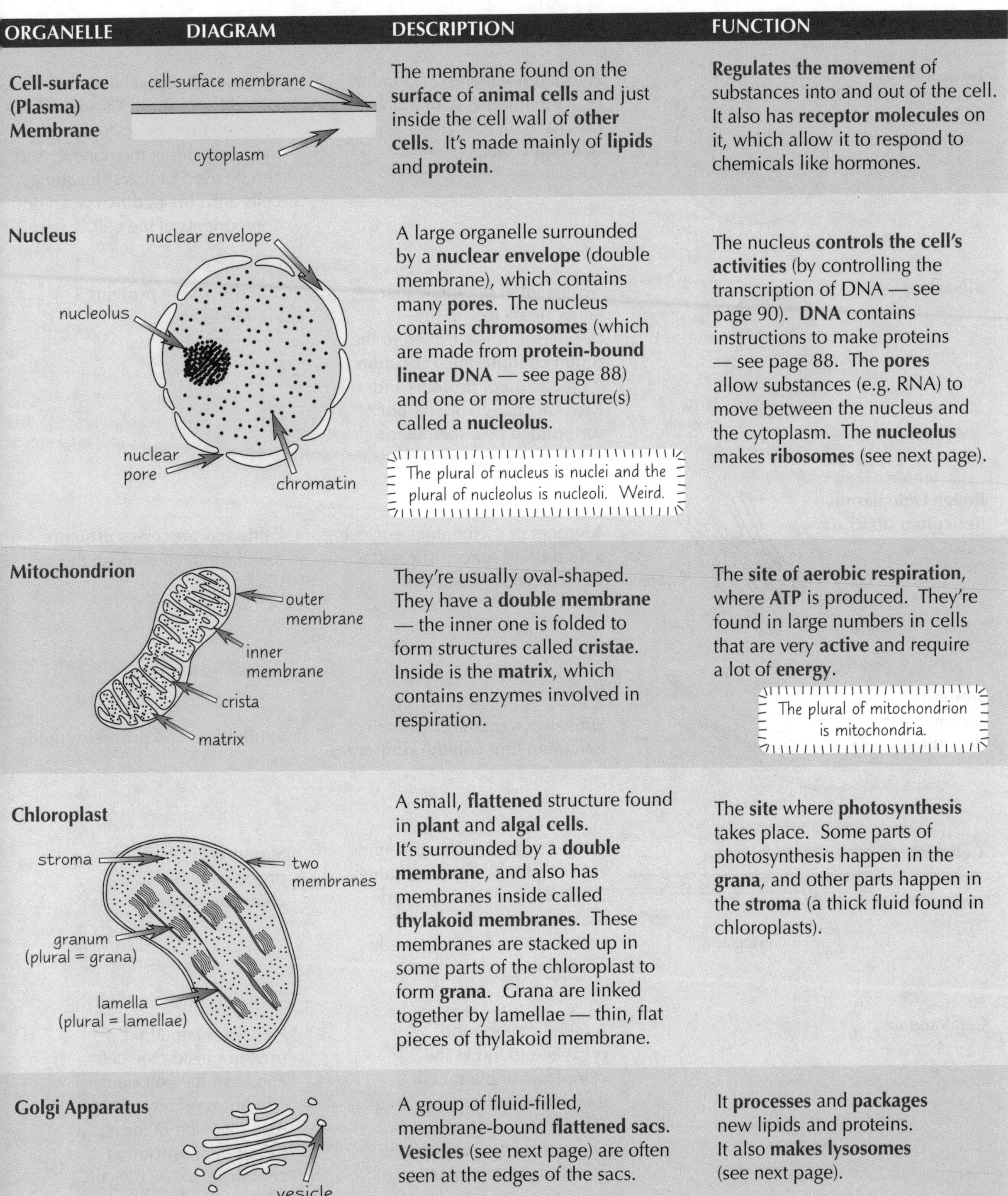

ORGANELLE	DIAGRAM	DESCRIPTION	FUNCTION
Cell-surface (Plasma) Membrane	cell-surface membrane; cytoplasm	The membrane found on the **surface** of **animal cells** and just inside the cell wall of **other cells**. It's made mainly of **lipids** and **protein**.	**Regulates the movement** of substances into and out of the cell. It also has **receptor molecules** on it, which allow it to respond to chemicals like hormones.
Nucleus	nuclear envelope; nucleolus; nuclear pore; chromatin	A large organelle surrounded by a **nuclear envelope** (double membrane), which contains many **pores**. The nucleus contains **chromosomes** (which are made from **protein-bound linear DNA** — see page 88) and one or more structure(s) called a **nucleolus**. The plural of nucleus is nuclei and the plural of nucleolus is nucleoli. Weird.	The nucleus **controls the cell's activities** (by controlling the transcription of DNA — see page 90). **DNA** contains instructions to make proteins — see page 88. The **pores** allow substances (e.g. RNA) to move between the nucleus and the cytoplasm. The **nucleolus** makes **ribosomes** (see next page).
Mitochondrion	outer membrane; inner membrane; crista; matrix	They're usually oval-shaped. They have a **double membrane** — the inner one is folded to form structures called **cristae**. Inside is the **matrix**, which contains enzymes involved in respiration.	The **site of aerobic respiration**, where **ATP** is produced. They're found in large numbers in cells that are very **active** and require a lot of **energy**. The plural of mitochondrion is mitochondria.
Chloroplast	stroma; two membranes; granum (plural = grana); lamella (plural = lamellae)	A small, **flattened** structure found in **plant** and **algal cells**. It's surrounded by a **double membrane**, and also has membranes inside called **thylakoid membranes**. These membranes are stacked up in some parts of the chloroplast to form **grana**. Grana are linked together by lamellae — thin, flat pieces of thylakoid membrane.	The **site** where **photosynthesis** takes place. Some parts of photosynthesis happen in the **grana**, and other parts happen in the **stroma** (a thick fluid found in chloroplasts).
Golgi Apparatus	vesicle	A group of fluid-filled, membrane-bound **flattened sacs**. **Vesicles** (see next page) are often seen at the edges of the sacs.	It **processes** and **packages** new lipids and proteins. It also **makes lysosomes** (see next page).

Eukaryotic Cells and Organelles

ORGANELLE	DIAGRAM	DESCRIPTION	FUNCTION
Golgi Vesicle	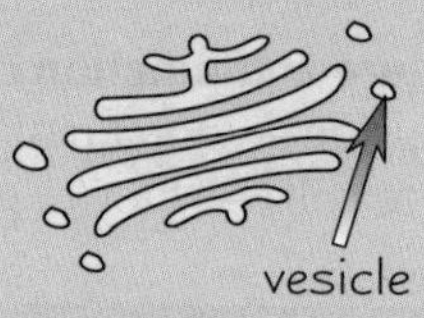 vesicle	A small **fluid-filled sac** in the cytoplasm, surrounded by a membrane and produced by the **Golgi apparatus**.	**Stores** lipids and proteins made by the Golgi apparatus and **transports** them out of the cell (via the cell-surface membrane).
Lysosome	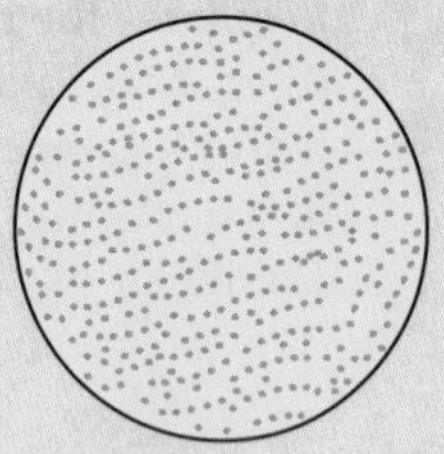	A **round organelle** surrounded by a **membrane**, with no clear internal structure. It's a type of **Golgi vesicle**.	Contains **digestive enzymes** called **lysozymes**. These are kept separate from the cytoplasm by the surrounding membrane, and can be used to **digest invading cells** or to **break down** worn out components of the cell.
Ribosome	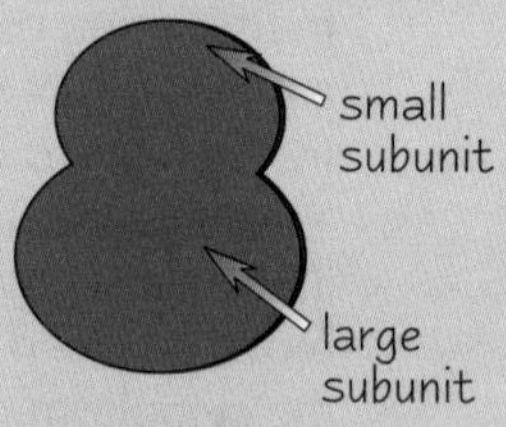 small subunit large subunit	A **very small organelle** that either **floats free** in the cytoplasm or is attached to the **rough endoplasmic reticulum**. It's made up of **proteins** and **RNA** (see page 16). It's **not** surrounded by a membrane.	The **site** where **proteins** are made.
Rough Endoplasmic Reticulum (RER)	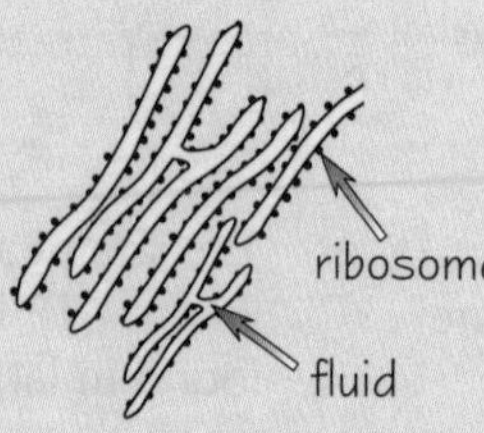 ribosome fluid	A system of membranes enclosing a fluid-filled space. The surface is **covered with ribosomes**.	**Folds** and **processes proteins** that have been made at the ribosomes.
Smooth Endoplasmic Reticulum (SER)	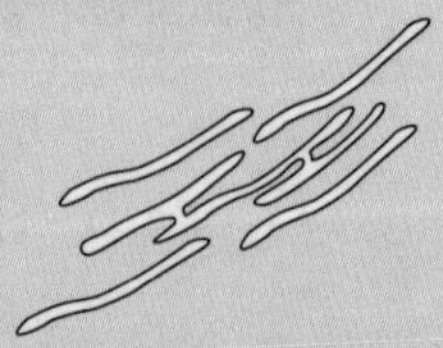	Similar to rough endoplasmic reticulum, but with no **ribosomes**.	**Synthesises** and **processes lipids**.
Cell Wall	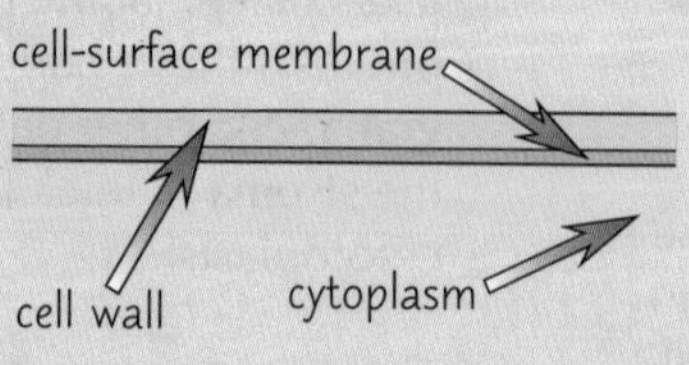 cell-surface membrane cell wall cytoplasm	A rigid structure that surrounds cells in **plants**, **algae** and **fungi**. In plants and algae it's made mainly of the carbohydrate **cellulose**. In fungi, it's made of **chitin**.	**Supports** cells and prevents them from **changing shape**.
Cell Vacuole	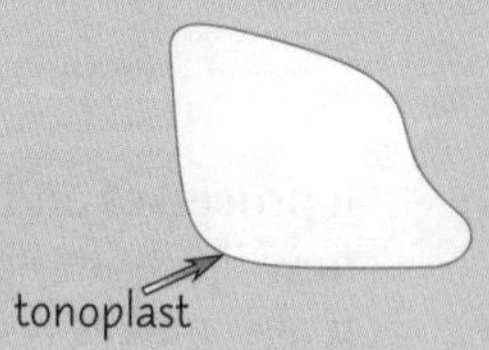 tonoplast	A membrane-bound organelle found in the **cytoplasm** of **plant cells**. It contains **cell sap** — a weak solution of sugar and salts. The surrounding membrane is called the **tonoplast**.	Helps to maintain **pressure** inside the cell and keep the cell **rigid**. This **stops** plants **wilting**. Also involved in the **isolation** of **unwanted chemicals** inside the cell.

Eukaryotic Cells and Organelles

The **Organelles** in **Specialised** Cells **Vary**

1) In **multicellular** eukaryotic organisms, cells become **specialised** to carry out **specific functions**.
2) A cell's **structure** (i.e. its shape and the organelles it contains) helps it to carry out its **function** — so depending on what job it does, a specialised cell can look very different to the cells you saw on page 26.
3) In the exam, you might get a question where you need to apply your knowledge of **organelles** to explain why a specialised cell is particularly **suited** to its **function**. You'll need to think about **what organelles** the cell **needs** to do its **job** — e.g. if the cell uses a lot of **energy**, it'll need lots of **mitochondria**. If it makes a lot of **proteins** it'll need a lot of **ribosomes**.

Example: **Epithelial cells** in the **small intestine** are specialised to **absorb food efficiently**.

1) The walls of the small intestine have lots of finger-like projections called **villi**. These **increase surface area** for absorption.
2) The **epithelial cells** on the surface of the villi have **folds** in their **cell-surface membranes**, called **microvilli**. Microvilli increase surface area even more.
3) They also have **lots of mitochondria** — to provide **energy** for the transport of digested food molecules into the cell.

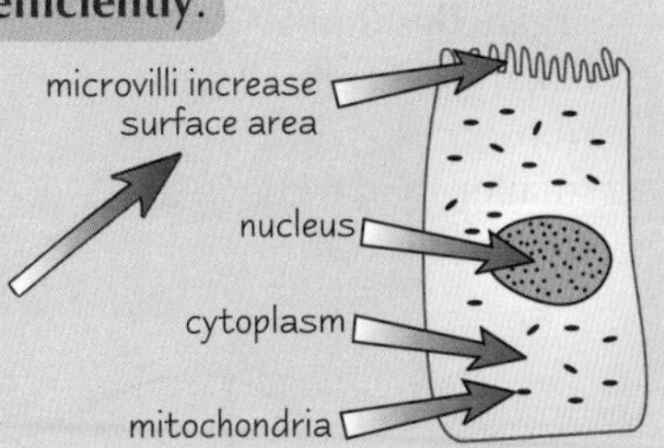
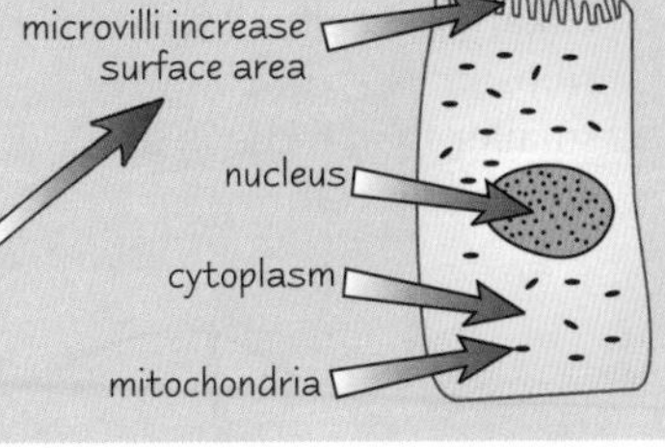

Specialised Cells are **Organised** into **Tissues, Organs** and **Organ Systems**

1) In multicellular eukaryotic organisms, **specialised cells** are grouped together to form **tissues**.
2) A tissue is a group of cells **working together** to perform a particular **function**.
3) Different **tissues** work together to form **organs**. Different organs make up an **organ system**.

For example...

Epithelial cells make up **epithelial tissue**. Epithelial tissue, **muscular tissue** and **glandular tissue** (which secretes chemicals) all work together to form the **stomach** — an **organ**. The stomach is part of the **digestive system** — this is an **organ system** made up of all the organs involved in the digestion and absorption of food (including the **small intestine**, **large intestine** and **liver**).

Warm-Up Questions

PRACTICE QUESTIONS

Q1 What is a plant cell wall made of? What about a fungal cell wall?
Q2 What is the function of a mitochondrion?
Q3 What is the function of a ribosome?
Q4 How does the structure of rough endoplasmic reticulum differ from that of smooth endoplasmic reticulum?
Q5 In multicellular organisms, what is a tissue?

Exam Questions

Q1 Plant cells have a vacuole, but animal cells do not.
   a) Give two functions of a plant cell vacuole. [2 marks]
   b) Name two other organelles found in plant cells but not in animal cells. [2 marks]

Q2 Cilia are hair-like structures found on lung epithelial cells. Their function is to beat and move mucus out of the lungs. Beating requires energy. Suggest how ciliated cells are adapted to their function in terms of the organelles they contain. Explain your answer. [2 marks]

Q3 Pancreatic cells make and secrete hormones (made of protein) into the blood. From production to secretion, list, in order, four organelles involved in making hormones. [4 marks]

Organelles — not a church girl band...

Not the most exciting pages in the world, but you need to know what all the organelles listed do. I'm afraid they'll keep popping up throughout the rest of the book — mitochondria are needed for respiration, the cell-surface membrane is essential for controlling the movement of things in and out of the cell, and all the DNA stuff happens in the nucleus.

Prokaryotic Cells and Viruses

Now we're on to prokaryotic cells and viruses. They're much smaller than eukaryotic cells — and, luckily for both of us, so is the section on them in this book. Nevertheless, you need to know everything in it for your exams...

The Structure of **Prokaryotic** Cells is Different to **Eukaryotic** Cells

Remember, prokaryotic cells are **smaller** and **simpler** than eukaryotic cells (see page 26). **Bacteria** are examples of prokaryotic cells. You need to know the **structure** of a prokaryotic cell and what all the different organelles do.

The **cytoplasm** of a prokaryotic cell has **no membrane-bound organelles** (unlike a eukaryotic cell). It has **ribosomes** — but they're **smaller** than those in a eukaryotic cell.

Just like in a eukaryotic cell, the **plasma membrane** is mainly made of lipids and proteins. It controls the movement of substances into and out of the cell.

See pages 27-28 for more on organelles.

The **cell wall supports** the cell and prevents it from changing shape. It's made of a polymer called **murein**. Murein is a **glycoprotein** (a protein with a carbohydrate attached).

The **flagellum** (plural **flagella**) is a long, hair-like structure that rotates to make the prokaryotic cell **move**. **Not all** prokaryotes have a flagellum. **Some** have **more than one**.

Some prokaryotes, e.g. bacteria, also have a **capsule** made up of secreted **slime**. It helps to **protect** bacteria from attack by cells of the immune system.

Unlike a eukaryotic cell, a prokaryotic cell **doesn't** have a nucleus. Instead, the **DNA** floats free in the cytoplasm. It's **circular DNA**, present as one long coiled-up strand. It's **not attached** to any **histone proteins** (see p. 88).

Plasmids are **small loops of DNA** that aren't part of the main circular DNA molecule. Plasmids contain genes for things like **antibiotic resistance**, and can be passed between prokaryotes. Plasmids are **not always** present in prokaryotic cells. **Some** prokaryotic cells have **several**.

Theo went the wrong way about getting practical experience in understanding cell structure.

Viruses are **Acellular** — They're **Not Cells**

Viruses are just **nucleic acids** surrounded by **protein** — they're **not even alive**.

- They're even **smaller** than bacteria — e.g. HIV is about 0.1 µm across.
- **Unlike** bacteria, viruses have **no** plasma membrane, **no** cytoplasm and **no** ribosomes.
- **All** viruses invade and reproduce **inside** the cells of **other** organisms. These cells are known as **host cells**.

Viruses contain a **core** of **genetic material** — either **DNA** or **RNA**.

DNA and RNA are nucleic acids — see page 16.

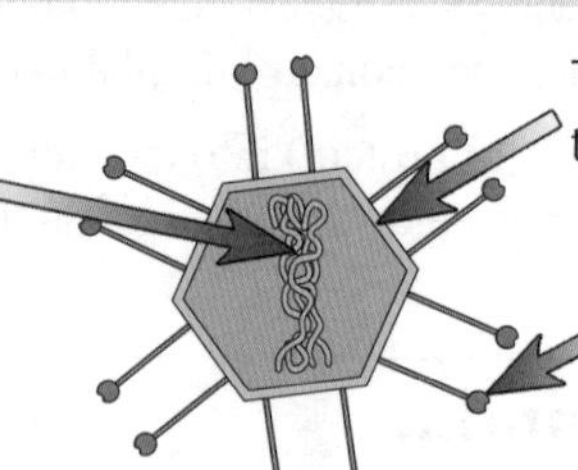

The **protein coat** around the core is called the **capsid**.

Attachment proteins stick out from the edge of the capsid. These let the virus cling on to a suitable host cell.

Prokaryotic Cells and Viruses

Prokaryotic Cells **Replicate** by **Binary Fission**

In binary fission, the cell **replicates** (makes copies of) its genetic material, before physically **splitting** into **two daughter cells**:

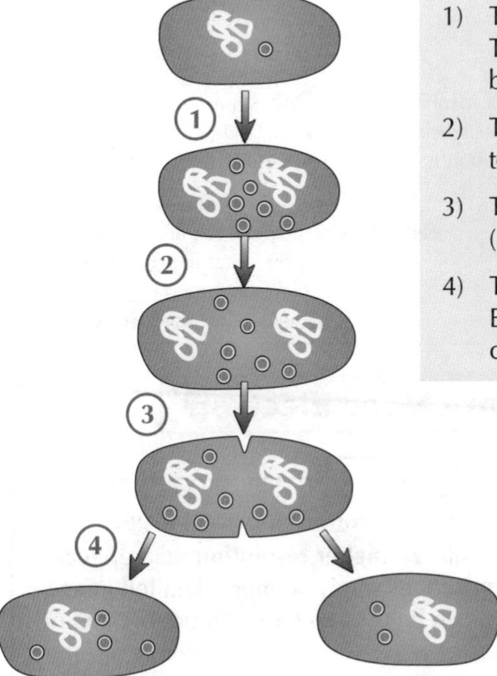

1) The circular DNA and plasmid(s) **replicate**. The main **DNA loop** is only replicated **once**, but **plasmids** can be replicated **loads of times**.
2) The cell gets bigger and the **DNA loops** move to **opposite 'poles'** (ends) of the cell.
3) The **cytoplasm** begins to **divide** (and **new cell walls** begin to form).
4) The cytoplasm **divides** and two **daughter cells** are produced. Each daughter cell has **one copy** of the **circular DNA**, but can have a **variable** number of copies of the **plasmid(s)**.

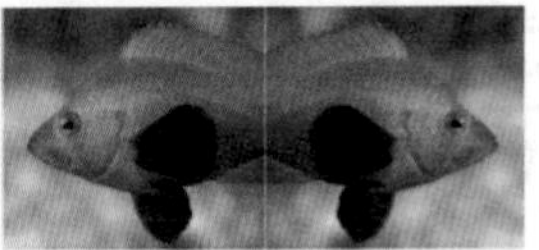

Binary fishin'

Viruses Use **Host Cells** to Replicate Themselves

1) Viruses use their **attachment proteins** to **bind** to **complementary receptor proteins** on the surface of **host cells**.
2) Different viruses have different attachment proteins and therefore require **different** receptor proteins on host cells. As a result, some viruses can only infect **one type of cell** (others can infect lots of different cells).
3) Because they're not alive, viruses **don't** undergo cell division. Instead, they **inject** their **DNA** or **RNA** into the host cell — this hijacked cell then uses its own 'machinery' (e.g. enzymes, ribosomes) to do the virus's dirty work and **replicate the viral particles**.

Remember — viruses are <u>acellular</u> — they're not cells.

Warm-Up Questions

Q1 What is a plasmid?

Q2 What is a flagellum?

Q3 What is the protein coat around the core of a virus called?

PRACTICE QUESTIONS

Exam Question

Q1 Cholera is a disease caused by the prokaryotic organism *Vibrio cholerae*.

a) Name the polymer that makes up the cell wall of *Vibrio cholerae*. [1 mark]

b) Outline the process by which *Vibrio cholerae* replicates. [3 marks]

c) There are different strains of *Vibrio cholerae*. One strain has a capsule. Another does not. Suggest how having a capsule might benefit *Vibrio cholerae*. [1 mark]

Viruses and binary fission — nothing to do with computers...

You need to know the differences between eukaryotic and prokaryotic cells. Make sure you spend plenty of time memorising them (see page 26 for more on eukaryotic cells). Remember that binary fission is only how prokaryotic cells replicate — eukaryotic cells and viruses use different techniques. Remember viruses aren't cells or prokaryotes

Analysis of Cell Components

You can use microscopes to look at all the lovely organelles you've been learning about...

Magnification is Size, Resolution is Detail

We all know that microscopes produce a **magnified image** of a sample, but **resolution** is just as important...

1) MAGNIFICATION is how much **bigger** the image is than the specimen (the sample you're looking at). It's calculated using this formula:

$$\text{magnification} = \frac{\text{size of image}}{\text{size of real object}}$$

For example:
If you have a magnified image that's 5 mm wide and your specimen is 0.05 mm wide, the magnification is: 5 ÷ 0.05 = × 100.

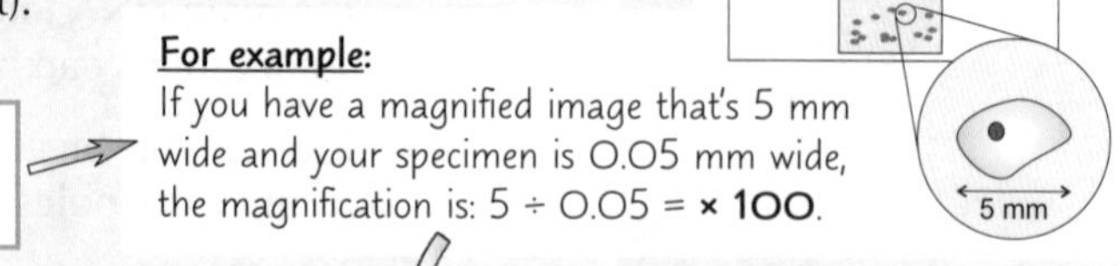

2) RESOLUTION is how **detailed** the image is. More specifically, it's how well a microscope **distinguishes** between **two points** that are **close together**. If a microscope lens can't separate two objects, then increasing the magnification won't help.

If you're given the size of the image and the size of the object in different units in the exam, make sure you convert them into the same units before using the formula.

There are Two Main Types of Microscope — Optical and Electron

Optical (light) microscopes

1) They use **light** to form an image.
2) They have a maximum resolution of about **0.2 micrometres** (μm). This means you can't use an optical microscope to view organelles smaller than 0.2 μm. That includes **ribosomes**, the **endoplasmic reticulum** and **lysosomes**. You may be able to make out **mitochondria** — but not in perfect detail. You can also see the **nucleus**.
3) The maximum useful **magnification** of an optical microscope is about **× 1500**.

Electron microscopes

1) They use **electrons** to form an image.
2) They have a **higher resolution** than optical microscopes so give a **more detailed image** (and can be used to look at more organelles).
3) They have a maximum resolution of about **0.0002 micrometres** (μm). (About 1000 times higher than optical microscopes.)
4) The maximum useful **magnification** of an electron microscope is about **× 1 500 000**.

A micrometre (μm) is three orders of magnitude smaller than a millimetre (1 μm = 0.001 mm). To convert from μm to mm, divide by 1000.

Electron Microscopes are either 'Scanning' or 'Transmission'

Transmission electron microscopes (TEMs)

1) TEMs use **electromagnets** to focus a **beam of electrons**, which is then transmitted **through** the specimen.
2) **Denser** parts of the specimen absorb **more electrons**, which makes them look **darker** on the image you end up with.
3) TEMs are good because they give **high resolution images**, so you see the **internal structure** of **organelles** like chloroplasts.
4) But they can only be used on **thin specimens**.

Scanning electron microscopes (SEMs)

1) SEMs **scan** a beam of electrons across the specimen. This **knocks off** electrons from the **specimen**, which are gathered in a **cathode ray tube** to form an **image**.
2) The images you end up with show the **surface** of the specimen and they can be **3-D**.
3) SEMs are good because they can be used on **thick specimens**.
4) But they give **lower resolution images** than TEMs.

You View Specimens Under an Optical Microscope Using Slides

Here's how to prepare a '**temporary mount**' of a specimen on a slide:

- Start by pipetting a small **drop of water** onto the **slide** (a strip of clear glass or plastic). Then use **tweezers** to place a **thin section** of your specimen on **top** of the water drop.
- Add a drop of a **stain**. Stains are used to **highlight objects** in a cell. For example, **eosin** is used to make the **cytoplasm** show up. **Iodine in potassium iodide solution** (see p. 4) is used to stain **starch grains** in plant cells.
- Finally, add the **cover slip** (a square of clear plastic that protects the specimen). To do so, stand the slip **upright** on the slide, next to the water droplet. Then carefully **tilt** and lower it so it covers the specimen. Try **not** to get any **air bubbles** under there — they'll obstruct your view of the specimen (see page 37).

Analysis of Cell Components

Cell Fractionation Separates Organelles

Suppose you wanted to look at some **organelles** under an **electron microscope**. First you'd need to **separate** them from the **rest of the cell** — you can do this by **cell fractionation**. There are **three** steps to this technique:

1) Homogenisation — Breaking Up the Cells

Homogenisation can be done in several **different ways**, e.g. by vibrating the cells or by grinding the cells up in a blender. This **breaks up** the **plasma membrane** and **releases** the **organelles** into solution. The solution must be kept **ice-cold**, to reduce the activity of enzymes that break down organelles. The solution should also be **isotonic** — this means it should have the **same concentration** of **chemicals** as the cells being broken down, to prevent damage to the organelles through **osmosis**. A **buffer solution** should be added to **maintain** the **pH**.

2) Filtration — Getting Rid of the Big Bits

Next, the homogenised cell solution is **filtered** through a **gauze** to separate any **large cell debris** or **tissue debris**, like connective tissue, from the organelles. The organelles are much **smaller** than the debris, so they pass through the gauze.

3) Ultracentrifugation — Separating the Organelles

After filtration, you're left with a solution containing a **mixture** of organelles. To separate a particular organelle from all the others you use **ultracentrifugation**.

1) The cell fragments are poured into a **tube**. The tube is put into a **centrifuge** (a machine that separates material by spinning) and is spun at a **low speed**. The **heaviest organelles**, like nuclei, get flung to the **bottom** of the tube by the centrifuge. They form a **thick sediment** at the bottom — the **pellet**. The rest of the organelles stay suspended in the fluid above the sediment — the **supernatant**.
2) The supernatant is **drained off**, poured into **another tube**, and spun in the centrifuge at a **higher speed**. Again, the **heaviest organelles**, this time the mitochondria, form a pellet at the bottom of the tube. The supernatant containing the rest of the organelles is drained off and spun in the centrifuge at an **even higher speed**.
3) This process is **repeated** at higher and higher speeds, until all the organelles are **separated out**. Each time, the pellet at the bottom of the tube is made up of lighter and lighter organelles.

As the ride got faster, everyone felt their nuclei sink to their toes...

In plant cells, the chloroplasts come out after the nuclei, but before the mitochondria.

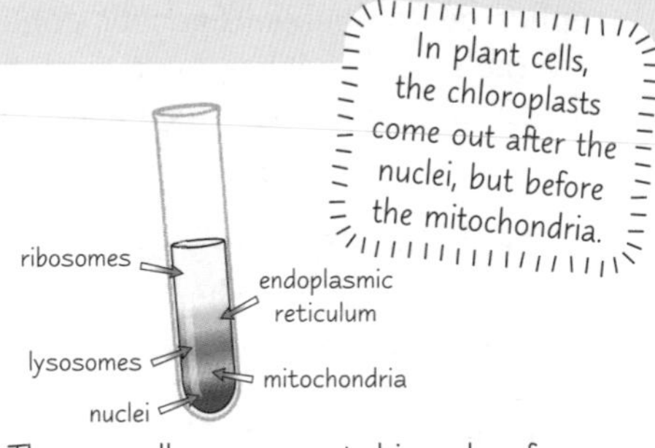

The organelles are separated in order of mass (from heaviest to lightest) — this order is usually: nuclei, then mitochondria, then lysosomes, then endoplasmic reticulum, and finally ribosomes.

Warm-Up Questions

PRACTICE QUESTIONS

Q1 What is meant by a microscope's magnification?

Q2 What is meant by a microscope's resolution?

Exam Questions

organelle	diameter / µm
lysosome	0.1
mitochondrion	2
nucleus	5
ribosome	0.02
vesicle	0.05

Q1 The table shows the dimensions of some different organelles found in animal cells. Name those organelles in the table that would be visible using a good quality light microscope. Explain your answer. [3 marks]

Q2 Explain why a homogenised cell solution should be kept ice-cold and isotonic. [2 marks]

Cell fractionation — sounds more like maths to me...

So, if you fancy getting up close and personal with mitochondria remember to homogenise, filter and ultracentrifuge first. Then decide if you want to use an SEM or TEM to view them, taking into account each of their limitations.

Cell Division — Mitosis

cell division, we'd still only be one cell big. If it wasn't for pies, my favourite jeans would still fit.

Mitosis is **Cell Division** that Produces **Genetically Identical Cells**

There are two types of cell division in **eukaryotes** — **mitosis** and **meiosis** (see pages 94-95 for more on meiosis).

1) In **mitosis** a **parent cell** divides to produce **two genetically identical daughter cells** (they contain an **exact copy** of the **DNA** of the parent cell).
2) Mitosis is needed for the **growth** of multicellular organisms (like us) and for **repairing damaged tissues**.
3) In multicellular organisms, not all cells keep their ability to divide (see next page). The ones that do, follow a **cell cycle**. Mitosis is part of the cell cycle:

The cell cycle consists of a period of **cell growth** and **DNA replication** called **interphase**. **Mitosis** happens after that. Interphase (cell growth) is subdivided into three separate growth stages. These are called $\mathbf{G_1}$, **S** and $\mathbf{G_2}$.

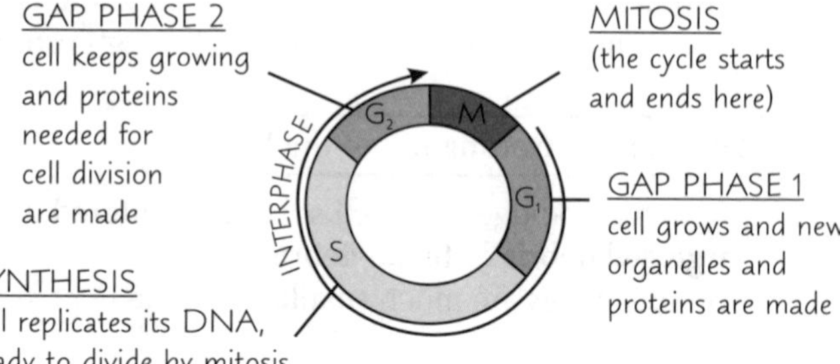

Mitosis has **Four Division Stages**

Mitosis is really one **continuous process**, but it's described as a series of **division stages** — prophase, metaphase, anaphase and telophase. **Interphase** comes **before** mitosis in the cell cycle.

Interphase — The cell carries out normal functions, but also prepares to divide. The cell's **DNA** is **unravelled** and **replicated**, to double its genetic content. The **organelles** are also **replicated** so it has spare ones, and its ATP content is increased (ATP provides the energy needed for cell division).

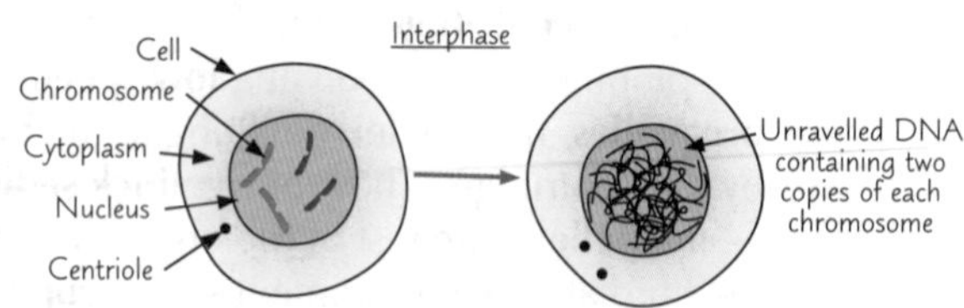

1) **Prophase** — The **chromosomes condense**, getting **shorter** and **fatter**. Tiny bundles of protein called **centrioles** start moving to opposite ends of the cell, forming a network of protein fibres across it called the **spindle**. The **nuclear envelope** (the membrane around the nucleus) **breaks down** and chromosomes lie free in the cytoplasm.

Centrioles move to opposite ends of the cell. Nuclear envelope starts to break down. Centromere.

As mitosis begins, the chromosomes are made of two strands joined in the middle by a centromere. The separate strands are called chromatids. There are two strands because each chromosome has already made an identical copy of itself during interphase. When mitosis is over, the chromatids end up as one-strand chromosomes in the daughter cells.

One chromatid. Centromere. Sister chromatids.

2) **Metaphase** — The chromosomes (each with two chromatids) **line up** along the middle of the cell and become **attached** to the **spindle** by their **centromere**.

Spindle fibres. Centromeres on spindle equator.

3) **Anaphase** — The **centromeres divide**, **separating** each pair of sister **chromatids**. The **spindles contract**, pulling chromatids to **opposite poles** (ends) of the **spindle**, centromere first. This makes the chromatids appear **v-shaped**.

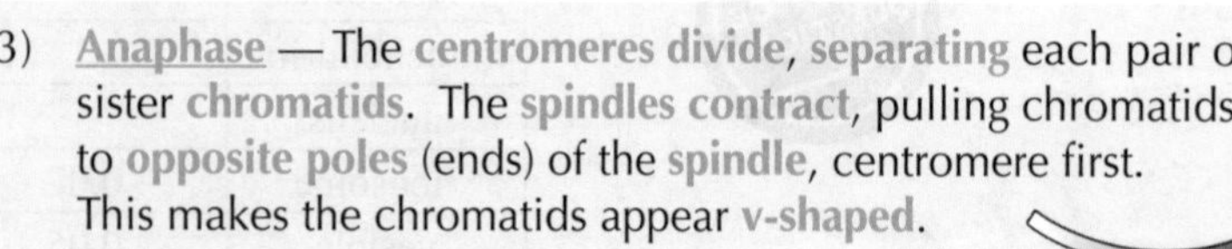

Sister chromatids moving to opposite poles of the spindle.

You need to be able to explain the appearance of cells at each stage of mitosis for your exam.

4) **Telophase** — The chromatids reach the **opposite poles** on the spindle. They **uncoil** and become **long** and **thin** again. They're now called **chromosomes** again. A **nuclear envelope** forms around each group of chromosomes, so there are now **two nuclei**. The **cytoplasm divides** (**cytokinesis**) and there are now **two daughter cells** that are **genetically identical** to the original cell and to each other. Mitosis is finished and each daughter cell starts the **interphase** part of the cell cycle to get ready for the next round of mitosis.

Centriole. Cytoplasm beginning to divide.

Cell Division — Mitosis

The **Time Taken** for Each **Stage** of Mitosis **Varies**

You can **calculate** how long each stage of mitosis lasts if you're given the right information.

Example: **A scientist observes a section of growing tissue under the microscope. He counts 100 cells undergoing mitosis. Of those, 10 cells are in metaphase. One complete cell cycle of the tissue lasts 15 hours. How long do the cells spend in metaphase? Give your answer in minutes.**

1) The scientist has observed that **10 out of 100 cells** are in **metaphase**. This suggests that the **proportion** of time the cells spend in metaphase must be **10/100th** of the **cell cycle**.
2) You're told that the cell cycle in these cells lasts **15 hours**. That's (15 × 60 =) **900 minutes**.
3) So the cells spend: $\frac{10}{100} \times 900 =$ **90 minutes** in metaphase.

Cancer is the Result of **Uncontrolled Cell Division**

Mutations are changes in the base sequence of an organism's DNA (see p. 97).

1) **Mitosis** and the **cell cycle** are **controlled by genes**.
2) Normally, when cells have divided enough times to make **enough new cells**, they stop. But if there's a **mutation** in a gene that controls cell division, the cells can **grow out of control**.
3) The cells **keep on dividing** to make more and more cells, which form a **tumour**.
4) **Cancer** is a tumour that **invades** surrounding tissue.

Some **Cancer Treatments** Target the **Cell Cycle**

Some treatments for cancer are designed to **control** the **rate of cell division** in tumour cells by **disrupting** the **cell cycle**. This **kills** the **tumour cells**. These treatments don't **distinguish** tumour cells from normal cells though — they also **kill normal body cells** that are dividing. However, tumour cells **divide much more frequently** than normal cells, so the treatments are **more likely** to kill tumour cells. Some cell cycle **targets** of cancer treatments include:

1) **G1 (cell growth and protein production)** — Some chemical drugs (chemotherapy) prevent the **synthesis of enzymes** needed for DNA replication. If these aren't produced, the cell is unable to enter the **synthesis phase** (S), disrupting the cell cycle and forcing the cell to **kill itself**.
2) **S phase (DNA replication)** — **Radiation** and some drugs **damage DNA**. At several points in the cell cycle (including just before and during S phase) the DNA in the cell is **checked** for damage. If severe damage is detected, the **cell** will **kill itself** — **preventing** further **tumour growth**.

Warm-Up Questions

Q1 Give the two main functions of mitosis.

Q2 List the four stages of mitosis.

Q3 Describe how tumours are formed.

Q4 Give one example of how a cancer treatment can target the cell cycle.

Exam Question

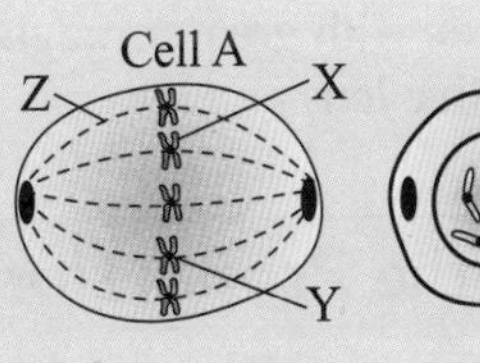

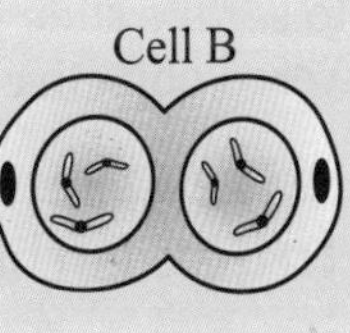

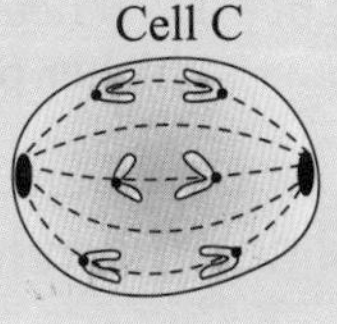

Q1 The diagrams show cells at different stages of mitosis.

a) For each of the cells A, B and C, name the stage of mitosis. [3 marks]

b) Name the structures labelled X, Y and Z in cell A. [3 marks]

Doctor, I'm getting short and fat — don't worry, it's just a phase...

Quite a lot to learn on these pages — but it's all important stuff, so no slacking. Mitosis is vital — it's how cells multiply and how organisms like us grow. Don't forget — the best way to learn is to get drawing those diagrams.

Cell Division — Investigating Mitosis

It's time to dust off your lab coat and get out your safety specs. Here are all the techniques you need to study mitosis. You'll need to know how to stain root cells on slides and how to use an optical microscope and graticules.

Root Tips Can be Stained and Squashed to Observe Mitosis

You need to know how to **prepare** and **stain** a **root tip** in order to observe the **stages of mitosis**. Make sure you're wearing **safety goggles** and a **lab coat** before you start. You should also wear **gloves** when using **stains**.

1) **Cut** 1 cm from the **tip** from a **growing root** (e.g. of an onion). It needs to be the **tip** because that's where **growth** occurs (and so that's where **mitosis** takes place).
If you're using ethano-orcein to stain the cells, the tips will also need to be fixed in ethanoic acid.
2) **Prepare** a boiling tube containing **1 M hydrochloric acid** and put it in a **water bath** at **60 °C**.
3) **Transfer** the **root tip** into the **boiling tube** and incubate for about **5 minutes**.
4) Use a pipette to **rinse** the **root tip** well with **cold water**. Leave the tip to **dry** on a **paper towel**.
5) Place the root tip on a **microscope slide** and cut **2 mm** from the **very tip** of it. Get **rid** of the **rest**.
6) Use a **mounted needle** to **break** the tip **open** and **spread** the cells out thinly.
7) **Add** a few drops of **stain** and leave it for a few minutes. The stain will make the **chromosomes easier** to **see** under a microscope. There are loads of different stains, all with crazy names (**toluidine blue O, ethano-orcein, Feulgen stain...**
If you're using the Feulgen stain, you'll need an extra rinse.
8) **Place** a **cover slip** over the cells and **push** down firmly to **squash** the tissue. This will make the tissue **thinner** and allow **light** to pass through it. **Don't smear** the cover slip sideways (or you'll damage the chromosomes).
9) Now you can look at all the stages of mitosis under an **optical microscope** (see below). You should see something that looks like the photograph on the right.

Stained Root Cells

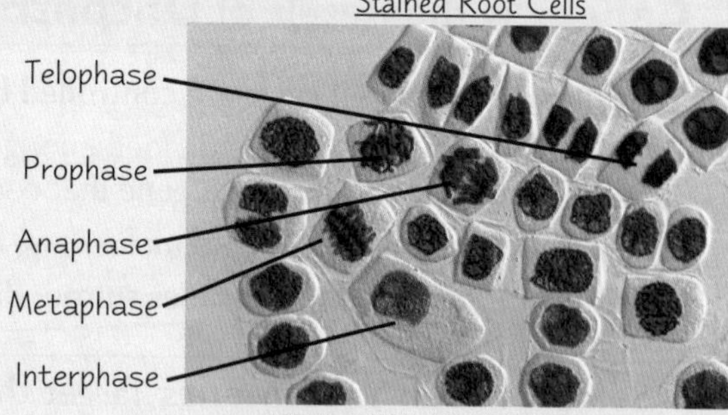

HERVE CONGE, ISM/SCIENCE PHOTO LIBRARY

You need to be able to recognise cells in the different stages of mitosis — see p. 34 for more info.

You Can Observe Cells Using an Optical Microscope

You need to know how to use an optical microscope to **observe** your prepared root tip cells:

1) Start by clipping the **slide** you've prepared onto the **stage**.
2) Select the **lowest-powered objective lens** (i.e. the one that produces the lowest magnification).
3) Use the **coarse adjustment knob** to bring the stage up to just below the objective lens.
4) Look down the **eyepiece** (which contains the ocular lens). Use the **coarse adjustment knob** to move the stage downwards, away from the objective lens until the image is roughly in focus.
5) Adjust the **focus** with the **fine adjustment knob**, until you get a **clear image** of what's on the slide.
6) If you need to see the slide with **greater magnification**, swap to a **higher-powered objective lens** and refocus.

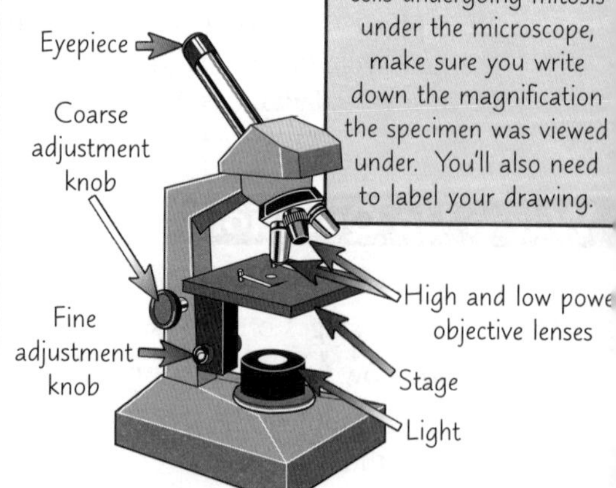

If you're asked to draw cells undergoing mitosis under the microscope, make sure you write down the magnification the specimen was viewed under. You'll also need to label your drawing.

The Mitotic Index Is the Proportion of Cells Undergoing Mitosis

You can **calculate** the **mitotic index** of your cells using this **formula**:

$$\text{mitotic index} = \frac{\text{number of cells with visible chromosomes}}{\text{total number of cells observed}}$$

This lets you work out how quickly the **tissue** is growing and if there's anything **weird** going on. A **plant root tip** is constantly **growing**, so you'd expect a **high mitotic index** (i.e. **lots** of cells in **mitosis**). In other tissue samples, a high mitotic index could mean that **tissue repair** is taking place or that there is **cancerous growth** in the tissue.

Cell Division — Investigating Mitosis

You Can Use A **Graticule** and **Micrometer** to Calculate the **Size** of **Cells**...

1) You need to be able to calculate the **size** of the cells you're looking at. That's where the **eyepiece graticule** and **stage micrometer** come in — they're a bit like **rulers**.
2) An **eyepiece graticule** is fitted onto the **eyepiece**. It's like a transparent ruler with **numbers**, but **no units**.
3) The **stage micrometer** is placed on the **stage** — it is a microscope slide with an **accurate scale** (it has units) and it's used to work out the **value** of the divisions on the **eyepiece graticule** at a **particular magnification**.
4) This means that when you take the stage micrometer away and replace it with the slide containing your tissue sample, you'll be able to **measure** the size of the cells. Here's an **example**:

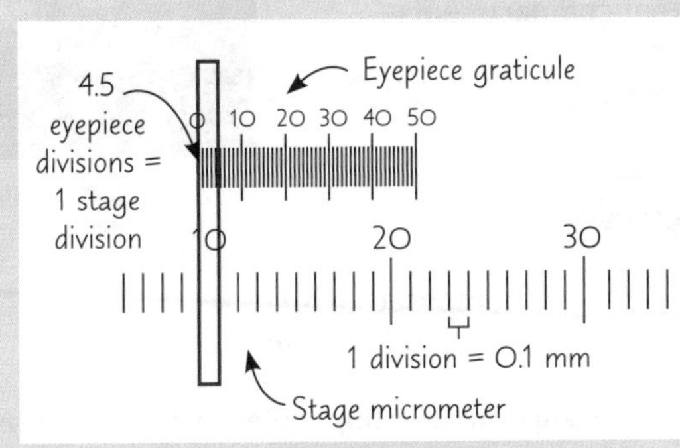

1) Line up the **eyepiece** graticule and the **stage** micrometer.
2) Each **division** on the stage micrometer is **0.1 mm** long.
3) At this magnification, **1 division** on the **stage micrometer** is the same as **4.5 divisions** on the **eyepiece graticule**.
4) To work out the size of **1 division** on the **eyepiece graticule**, you need to divide 0.1 by 4.5:
   1 division on eyepiece graticule = 0.1 ÷ 4.5 = **0.022 mm**
5) So if you look at a cell under the microscope at this magnification and it's **4 eyepiece divisions** long, you know it measures:
   4 × 0.022 = **0.088 mm**.

The eyepiece graticule will need to be re-calibrated at different magnifications.

...Or You Can Use This **Formula**...

If you're given an **image** of cells under the microscope in the exam, you can calculate their **actual size** using this **formula**:

$$\text{actual size} = \frac{\text{size of image}}{\text{magnification}}$$

Example: If the image of a cell measures 5 mm and the magnification is x 100, then the actual size of the cell will be: 5 ÷ 100 = 0.05 mm.

Artefacts Can Get in the Way of Your **Observations**

The new organelle Steve had discovered looked just like his thumb print.

1) Artefacts are things that you can see down the microscope that **aren't** part of the **cell or specimen** that you're looking at.
2) They can be anything from bits of **dust**, **air bubbles** and **fingerprints**, to inaccuracies caused by **squashing** and **staining** your sample.
3) Artefacts are usually made during the **preparation** of your slides and **shouldn't** really be there at all — you'll need to prepare your root tip cells **carefully** to avoid creating artefacts.

Artefacts are especially common in electron micrographs because specimens need a lot of preparation before you can view them under an electron microscope. The first scientists to use these microscopes could only distinguish between artefacts and organelles by repeatedly preparing specimens in different ways. If an object could be seen with one preparation technique, but not another, it was more likely to be an artefact than an organelle.

Warm-Up Questions

Q1 Why do you need to squash the tissue when preparing a slide of plant root tip cells?

Exam Question

Q1 A sample of cells was prepared to observe mitosis. In total, 42 cells were observed. 32 of those had visible chromosomes. Calculate the mitotic index for this sample. Give your answer to 2 decimal places. [2 marks]

'Staining your samples' — a common problem at the start of exams...

Wow — I bet you never realised there was so much to know about using a microscope. Still, staining is pretty straightforward and so's preparing a slide. Using a graticule is tricky, but once you get your head round it you'll be fine.

Cell Membrane Structure

You might remember a bit about cell membranes from p. 27. Well now it's time to delve a little deeper...

Membranes Control What Passes Through Them

All cells are surrounded by **membranes**. In **eukaryotic cells**, many of the **organelles** are surrounded by membranes too.

1) **Cell-surface membranes** surround cells. They are a **barrier** between the **cell** and its **environment**, controlling **which substances enter and leave** the cell. They're **partially permeable** — they let some molecules through but not others. Substances can move across the cell-surface membrane by **diffusion**, **osmosis** or **active transport** (see pages 40-45).
2) The membranes around **organelles** divide the cell into different **compartments** — they act as a **barrier** between the **organelle** and the **cytoplasm**. They are also **partially permeable** and control what substances **enter** and **leave** the organelle.

Partially permeable membranes can be useful at sea.

Cell Membranes have a 'Fluid Mosaic' Structure

The basic **structure** of **all cell membranes** is pretty much the same. They're composed of **lipids** (mainly phospholipids — see page 7), **proteins** and **carbohydrates** (attached to proteins or lipids).

1) In 1972, the **fluid mosaic model** was suggested to describe the **arrangement** of **molecules** in the membrane.
2) In the model, **phospholipid molecules** form a continuous, double layer (**bilayer**).
3) This bilayer is **'fluid'** because the phospholipids are **constantly moving**.
4) **Cholesterol** molecules (see below) are present within the bilayer.
5) **Proteins** are scattered through the bilayer, like tiles in a **mosaic**. These include **channel proteins** and **carrier proteins**, which allow **large molecules** and **ions** to pass **through** the **membrane**. **Receptor proteins** on the cell-surface membrane allow the cell to **detect chemicals** released from other cells. The chemicals **signal** to the cell to **respond** in some way, e.g. the hormone insulin binds to receptor proteins on liver cells, which tells the cells to absorb glucose.
6) Some **proteins** are able to **move sideways** through the bilayer, while others are **fixed** in position.
7) Some **proteins** have a **polysaccharide** (carbohydrate) **chain** attached — these are called **glycoproteins**.
8) Some **lipids** also have a **polysaccharide chain** attached — these are called **glycolipids**.

The phospholipid bilayer is about 7 nm thick.

glycoprotein
glycolipid
phospholipids
protein
cholesterol
protein channel

The Different Components of Cell Membranes have Different Roles

Phospholipids Form a Barrier to Dissolved Substances

1) **Phospholipid molecules** have a 'head' and a 'tail'.
2) The **head** is **hydrophilic** — it **attracts water**. The **tail** is **hydrophobic** — it **repels water**.
3) The molecules automatically **arrange** themselves into a **bilayer** — the **heads face out** towards the water on either side of the membrane.

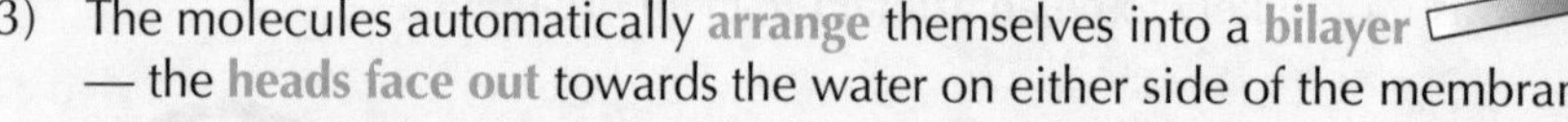

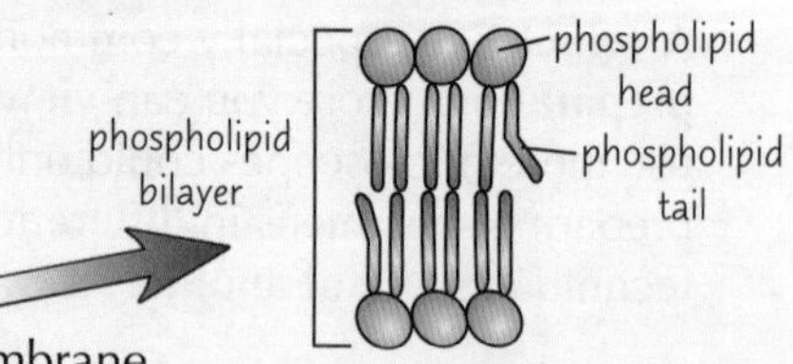

4) The **centre** of the bilayer is **hydrophobic** so the membrane **doesn't** allow **water-soluble substances** (like ions) through it — it acts as a **barrier** to these dissolved substances.

Cholesterol Gives the Membrane Stability

1) **Cholesterol** is a type of **lipid**.
2) It's present in **all** cell membranes (except bacterial cell membranes).
3) Cholesterol molecules fit **between** the phospholipids. They bind to the hydrophobic tails of the phospholipids, causing them to pack **more closely together**. This **restricts** the **movement** of the phospholipids, making the membrane **less fluid** and **more rigid**.
4) Cholesterol helps to **maintain** the **shape** of **animal cells** (which don't have cell walls). This is particularly important for cells that **aren't supported by other cells**, e.g. red blood cells, which float free in the blood.

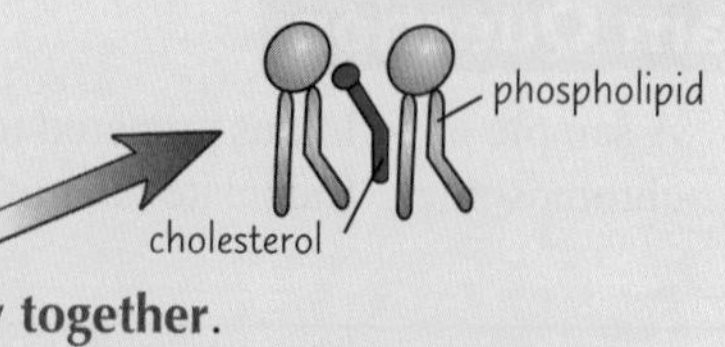

Cell Membrane Structure

The **Permeability** of the **Cell Membrane** can be **Investigated** in the **Lab**

The permeability of cell membranes is affected by **different conditions**, e.g. **temperature** and **solvent concentration**. You can investigate how these things affect permeability by doing an experiment using **beetroot**. Beetroot cells contain a **coloured pigment** that **leaks out** — the **higher** the **permeability** of the membrane, the **more pigment** leaks out of the cell. Here's how you could investigate how **temperature** affects **beetroot membrane permeability**:

1) Use a **scalpel** to carefully cut five **equal sized** pieces of beetroot. (Make sure you do your cutting on a **cutting board**.) **Rinse** the pieces to remove any pigment released during cutting.
2) Add the five pieces to five different **test tubes**, each containing **5 cm³ of water**. Use a **measuring cylinder** or **pipette** to measure the water.
3) Place each test tube in a **water bath** at a **different temperature**, e.g. 10 °C, 20 °C, 30 °C, 40 °C, 50 °C, for the **same length of time** (measured using a **stopwatch**).
4) **Remove** the pieces of beetroot from the tubes, leaving just the **coloured liquid**.
5) Now you need to use a **colorimeter** — a machine that passes **light** through the liquid and measures how much of that light is **absorbed**. The **higher** the absorbance, the **more pigment released**, so the **higher** the **permeability** of the membrane.
6) You can connect the colorimeter to a **computer** and use **software** to **collect the data** and draw a **graph** of the results.

Colorimeters need 5 minutes to stabilise before using and calibrating at zero by taking a measurement through pure water.

Increasing the **Temperature Increases Membrane Permeability**

Experiments like the one above have shown that membrane permeability **changes** with temperature:

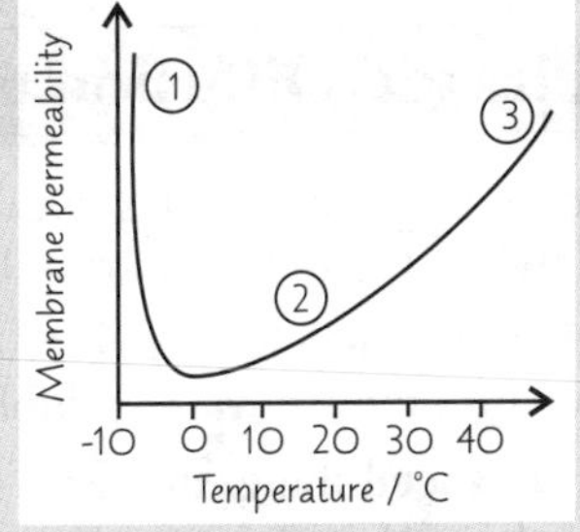

(1) **Temperatures below 0 °C** — the phospholipids don't have much energy, so they can't move very much. They're **packed closely together** and the membrane is **rigid**. But **channel proteins** and **carrier proteins** in the membrane **deform**, **increasing** the **permeability** of the membrane. **Ice crystals** may form and **pierce** the membrane making it **highly permeable** when it thaws.

(2) **Temperatures between 0 and 45 °C** — the phospholipids can **move** around and **aren't** packed as tightly together — the membrane is **partially permeable**. As the temperature **increases** the phospholipids **move more** because they have more energy — this **increases** the **permeability** of the membrane.

(3) **Temperatures above 45 °C** — the phospholipid bilayer starts to **melt** (break down) and the membrane becomes more **permeable**. **Water** inside the cell **expands**, putting pressure on the membrane. **Channel proteins** and **carrier proteins deform** so they can't control what enters or leaves the cell — this increases the **permeability** of the membrane.

You could also investigate the effect of solvents on the permeability of cell membranes. Surrounding cells in an increasing concentration of a solvent (such as alcohol or acetone) increases membrane permeability because the solvent dissolves the lipids in the cell membrane, causing it to lose its structure.

Warm-Up Questions

Q1 Give three molecules that are present in animal cell membranes.

Q2 What effect does cholesterol have on the cell membrane?

PRACTICE QUESTIONS

Exam Questions

Q1 Explain why the plasma membrane can be described as having a fluid mosaic structure. [2 marks]

Q2 The table on the right shows the results of an investigation into the effect of alcohol concentration on the permeability of beetroot cell membranes.

a) Suggest a suitable method that could have been used to obtain these results. [4 marks]

b) What conclusion can be drawn from the results? [2 marks]

Alcohol concentration / %	Absorbance
0	0.14
25	0.22
50	0.49
75	1.03
100	1.28

Fluid Mosaic Model — think I saw one being sold at a craft fair...

It's weird to think that cells are surrounded by a layer that's 'fluid' — it's a good job they are though because if cell membranes were too rigid, a cell wouldn't be able to change shape or stretch without bursting.

Exchange Across Cell Membranes — Diffusion

Ooooh it's starting to get a bit more exciting... here's how some substances can get across cell membranes without using energy. Just what you've always wanted to know, I bet.

Diffusion is the Passive Movement of Particles

1) Diffusion is the net movement of particles (molecules or ions) from an area of **higher concentration** to an area of **lower concentration**.
2) Molecules will diffuse **both ways**, but the **net movement** will be to the area of **lower concentration**. This continues until particles are **evenly distributed** throughout the liquid or gas.
3) The **concentration gradient** is the path from an area of higher concentration to an area of lower concentration. Particles diffuse **down** a concentration gradient.
4) Diffusion is a **passive process** — **no energy** is needed for it to happen.
5) Particles can diffuse **across cell membranes**, as long as they can **move freely** through the membrane.

> E.g. **oxygen** and **carbon dioxide** can diffuse easily through cell membranes because they're **small**, so they can pass through spaces between the phospholipids. They're also **non-polar**, which makes them **soluble** in **lipids**, so they can **dissolve** in the **hydrophobic bilayer**.

6) When molecules diffuse **directly** through a cell membrane, it's also known as **simple diffusion**.

Polar molecules have partial positive and negative charges (see p. 20). Non-polar molecules don't.

Facilitated Diffusion uses Carrier Proteins and Protein Channels

1) Some **larger molecules** (e.g. amino acids, glucose) would **diffuse extremely slowly** through the phospholipid bilayer because they're so **big**.
2) **Charged particles**, e.g. **ions** and **polar molecules**, would also diffuse slowly — that's because they're **water soluble**, and the **centre** of the **bilayer** is **hydrophobic** (see page 38).
3) So to **speed things up**, large or charged particles diffuse through **carrier proteins** or **channel proteins** in the membrane instead — this is called **facilitated diffusion**.
4) Like diffusion, facilitated diffusion moves particles **down** a **concentration gradient**, from a higher to a lower concentration.
5) It's also a passive process — it **doesn't** use **energy**.

Andy needed all his concentration for this particular gradient...

Carrier proteins move **large molecules** across membranes, down their concentration gradient. **Different carrier proteins** facilitate the diffusion of **different molecules**.

1) First, a large molecule **attaches** to a carrier protein in the membrane.
2) Then, the protein **changes shape**.
3) This **releases** the molecule on the **opposite side** of the membrane.

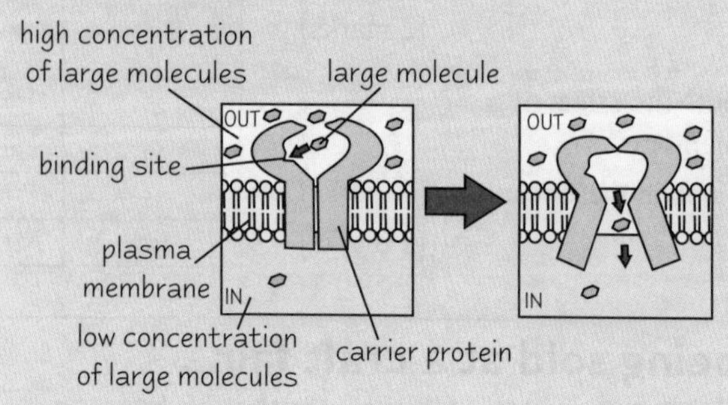

Channel proteins form **pores** in the membrane for **charged particles** to diffuse through (down their concentration gradient). **Different channel proteins** facilitate the diffusion of **different charged particles**.

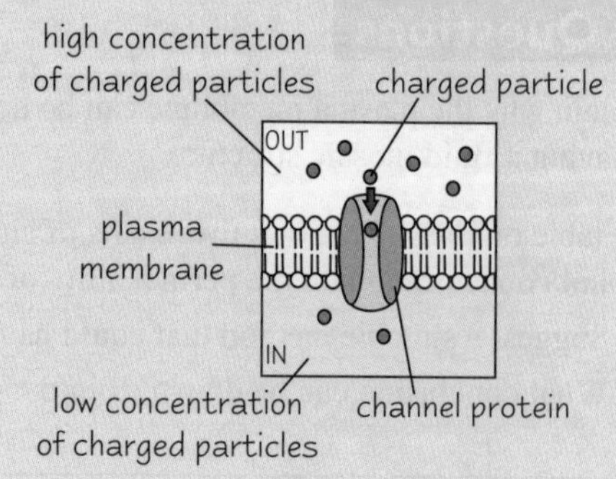

Exchange Across Cell Membranes — Diffusion

The **Rate of Diffusion** Depends on **Several Factors**

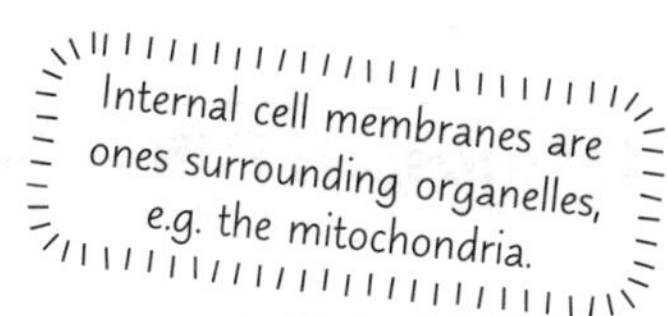

The **rate** of **diffusion** across both **external** and **internal cell membranes** can **vary**. Some cells are **adapted** for **rapid transport** across their membranes.

Simple diffusion depends on...

1) The **concentration gradient** — the **higher** it is, the **faster** the rate of diffusion. As diffusion takes place, the **difference in concentration** between the two sides of the membrane **decreases** until it reaches an **equilibrium** (i.e. the concentration on both sides is equal). This means that diffusion **slows down** over time.
2) The **thickness** of the **exchange surface** — the **thinner** the exchange surface (i.e. the **shorter** the **distance** the particles have to travel), the **faster** the rate of diffusion.
3) The **surface area** — the **larger** the surface area (e.g. of the cell-surface membrane), the **faster** the rate of diffusion.

Microvilli increase the surface area for faster diffusion

Some cells (e.g. epithelial cells in the small intestine) have **microvilli** — projections formed by the cell-surface membrane folding up on itself (see p. 29). Microvilli give the cell a **larger surface area** — in human cells, microvilli can increase the surface area by about **600 times**. A larger surface area means that **more particles** can be **exchanged** in the same amount of time — **increasing** the **rate of diffusion**.

Facilitated diffusion depends on...

1) The **concentration gradient** — the **higher** the **concentration gradient**, the **faster** the **rate** of **facilitated diffusion** (**up to a point**, see point 2 below). As **equilibrium** is reached, the **rate** of facilitated diffusion will **level off**.
2) The **number** of **channel** or **carrier proteins** — once **all** the **proteins** in a membrane are **in use**, facilitated diffusion **can't happen any faster**, even if you increase the concentration gradient. So the **greater the number** of channel or carrier proteins in the cell membrane, the **faster the rate** of facilitated diffusion.

Having more channel proteins increases the rate of facilitated diffusion

Aquaporins are special **channel proteins** that allow the **facilitated diffusion** of **water** through cell membranes. Some **kidney cells** are **adapted** to have **lots** of **aquaporins**. The aquaporins allow the cells to **reabsorb** a lot of the water that would otherwise be **excreted** by the body — about **180 litres** need re-absorbing every day.

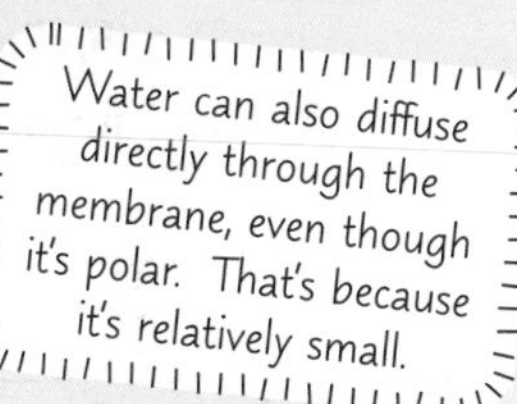

In the exams, you might be asked to calculate the **rate of diffusion** (or any other form of transport across a membrane) from a **graph**. For a straight line graph, this means finding the **gradient** of the line. For a curved graph, it means drawing a **tangent** and finding the gradient of the tangent. There's more on both of these techniques on page 228.

Warm-Up Questions

Q1 Diffusion is a passive process. What does this mean?

Q2 How do microvilli increase the rate of diffusion?

Exam Question

Q1 Chloride ions are transported into a cell across its cell-surface membrane by facilitated diffusion.

a) What type of molecule must be present in a cell membrane for the facilitated diffusion of chloride ions to take place? *[1 mark]*

b) Explain why the simple diffusion of chloride ions across a cell-surface membrane would be extremely slow. *[2 marks]*

c) The chloride ions in the cell are not immediately used up. Describe and explain what will happen to the rate of facilitated diffusion of the chloride ions into the cell over time. *[2 marks]*

All these molecules moving about — you'd think they'd get tired...

Right, I think I get it. If you're a small molecule, like oxygen, you can just cross the membrane by simple diffusion. And if you're a large or charged molecule you have a little help from a channel or carrier protein. As long as you want to go down a concentration gradient. If not, there's always active transport. Luckily that's coming up soon (page 44).

Exchange Across Cell Membranes — Osmosis

These two pages are entirely about the movement of water molecules. If you've mastered diffusion (see pages 40-41) you'll nail this lot in no time.

Osmosis is Diffusion of Water Molecules

1) Osmosis is the **diffusion** of **water molecules** across a **partially permeable membrane**, from an area of **higher water potential** (i.e. higher concentration of water molecules) to an area of **lower water potential** (i.e. lower concentration of water molecules).
2) **Water potential** is the potential (likelihood) of water molecules to diffuse out of or into a solution.
3) **Pure water** has the **highest water potential**. All solutions have a **lower** water potential than pure water.
4) If two solutions have the **same water potential**, they're said to be **isotonic**.

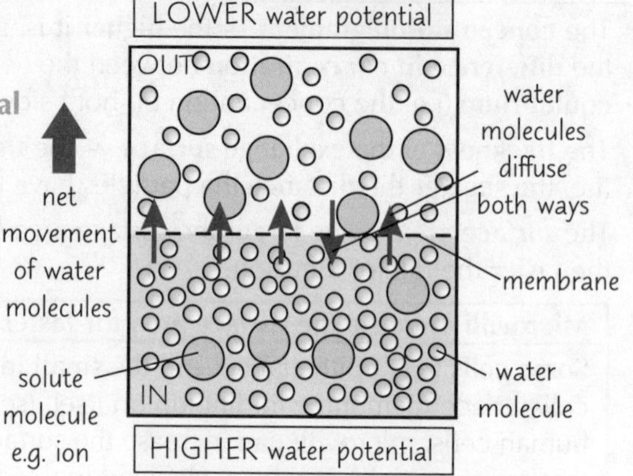

The **Rate** of **Osmosis** Depends on Several **Factors**

The factors affecting the rate of osmosis are similar to those affecting the rate of diffusion (see previous page).

1) The **water potential gradient** — the **higher** the **water potential gradient**, the **faster** the rate of osmosis. As osmosis takes place, the **difference in water potential** on either side of the membrane **decreases**, so the rate of osmosis **levels off** over time.
2) The **thickness** of the **exchange surface** — the **thinner** the **exchange surface**, the **faster** the rate of osmosis.
3) The **surface area** of the **exchange surface** — the **larger** the **surface area**, the **faster** the rate of osmosis.

You can Investigate **Water Potential** using **Serial Dilutions**

You can do a **simple experiment**, using potato cylinders, to find out the **water potential** of plant tissue (see next page). First though, you need to make up several solutions of **different, known concentrations** to test the cylinders in. You can do this using a **serial dilution** technique:

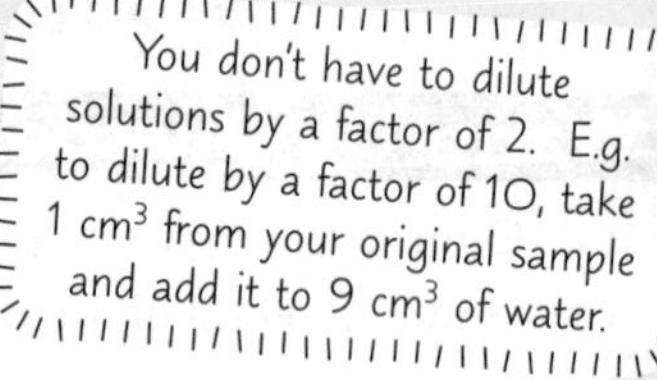

This is how you'd make **five serial dilutions** of a sucrose solution, starting with an initial sucrose concentration of **2 M** and **diluting** each solution by a **factor** of **2**...

1) Line up five **test tubes** in a rack.
2) Add **10 cm³** of the initial **2 M sucrose solution** to the first test tube and **5 cm³ of distilled water** to the other four test tubes.
3) Then, using a pipette, draw **5 cm³** of the solution from the **first** test tube, add it to the distilled water in the **second** test tube and **mix** the solution **thoroughly**. You now have **10 cm³** of solution that's **half as concentrated** as the solution in the first test tube (it's **1 M**).
4) Repeat this process **three more times** to create solutions of **0.5 M, 0.25 M** and **0.125 M**.

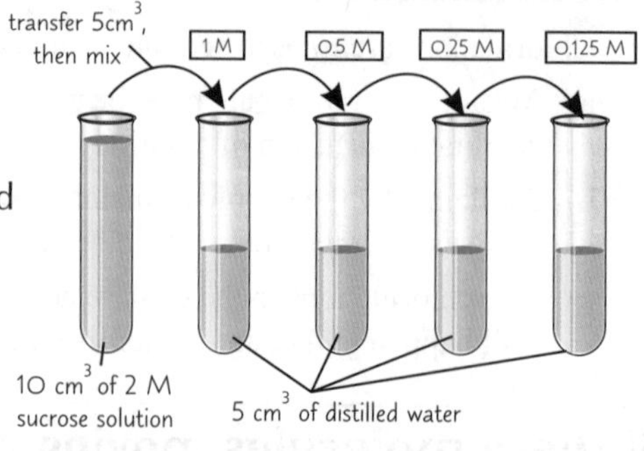

Exchange Across Cell Membranes — Osmosis

You Can Also Make **Solutions** of **Different Concentrations** By Finding the **Scale Factor**

You can make sucrose solutions of **any concentration** by finding the **scale factor**.
For example, if you want to make **15 cm³** of **0.4 M** sucrose solution...

PRACTICAL SKILLS

1) Start with a **solution** of a **known concentration**, e.g. **1 M**.
2) Find the **scale factor** by **dividing** the **concentration** of this solution by the **concentration** of the solution **you want to make**. So in this case the scale factor = 1 M ÷ 0.4 M = **2.5**.
3) This means that the solution you want to make is **2.5 times weaker** than the one you have. To make the solution 2.5 times weaker, use 2.5 times less of it, i.e. 15 cm³ ÷ 2.5 = **6 cm³**. **Transfer** this amount to a **clean test tube**.
4) **Top up** the test tube with **distilled water** to get the **volume** you want to make. In this case you want to make 15 cm³ of solution, so you need to add: 15 – 6 = **9 cm³** of distilled water.

Use Your **Solutions** To Find the **Water Potential** of **Potato Cells**

1) Use a cork borer to cut **potatoes** into **identically sized** chips, about 1 cm in diameter.
2) Divide the chips into groups of **three** and measure the **mass** of each **group** using a **mass balance**.
3) Place **one group** into **each** of your **sucrose solutions**.
4) **Leave** the chips in the solutions for **at least** 20 minutes (making sure that they all get the **same amount of time**).
5) Remove the chips and pat dry **gently** with a paper towel.
6) **Weigh** each group again and record your results.
7) Calculate the % **change in mass** for each group.
8) Use the results to make a **calibration curve**, showing % **change in mass** against **sucrose concentration**.

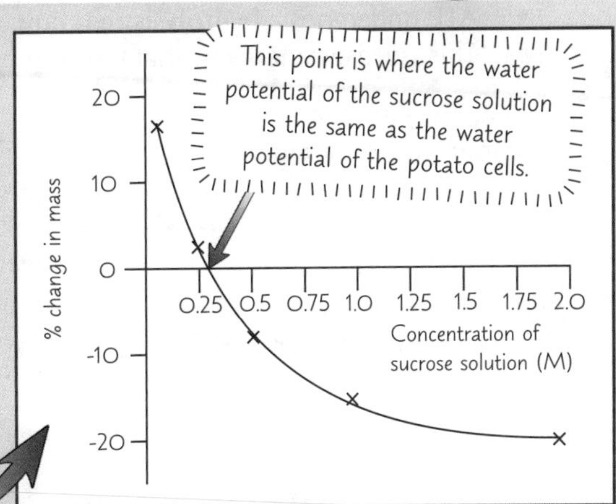

The potato chips will **gain water** (and therefore **mass**) in solutions with a **higher water potential** than the chips, and **lose water** in solutions with a **lower** water potential.

The point at which the **curve crosses** the **x-axis** (where the % change in mass is **0**) is the point at which the **water potential** of the **sucrose solution** is the **same** as the **water potential** of the **potato cells**. Find the **concentration** at this point, then **look up** the **water potential** for that concentration of sucrose solution in, e.g. a textbook.

Warm-Up Questions

Q1 Define osmosis.

Q2 Give two factors that affect the rate of osmosis.

Q3 What are serial dilutions?

Exam Question

Q1 Pieces of potato of equal mass were put into different concentrations of sucrose solution for 24 hours. The difference in mass for each is recorded in the table.

Concentration of sucrose / %	1	2	3	4
Mass difference / g	0.4	0.2	0	– 0.2

a) Explain why the pieces of potato in 1% and 2% sucrose solutions gained mass. [2 marks]

b) Suggest a reason why the mass of the piece of potato in 3% sucrose solution stayed the same. [1 mark]

c) What would you expect the mass difference for a potato in a 5% solution to be? Explain your answer. [2 marks]

I always knew that glass of water had potential...

Osmosis is just a fancy name for the diffusion of water molecules. But whether water moves in or out of a cell depends on the water potential of the surrounding solution. Water potential can be pretty confusing — if you can't make head nor tail of an exam question about it try replacing the word 'potential' with 'concentration' and it'll become clearer.

Exchange Across Cell Membranes — Active Transport

Diffusion and osmosis are passive processes — they don't require energy. So, for those of you feeling a bit more active, here's a page all about... you guessed it... active transport.

Active Transport Needs **Energy**

Active transport uses **energy** to move **molecules** and **ions** across membranes, usually **against** a **concentration gradient**.

Carrier proteins are involved in active transport. The process is pretty similar to facilitated diffusion (see p. 40) — a molecule attaches to the carrier protein, the protein changes shape and this moves the molecule across the membrane, releasing it on the other side.

There are two main differences between active transport and facilitated diffusion though:

1) Active transport usually moves solutes from a low to a high concentration — in facilitated diffusion, they always move from a high to a low concentration.
2) Active transport requires energy — facilitated diffusion does not.
   - ATP is a common source of energy in the cell. It's produced by respiration.
   - ATP undergoes a hydrolysis reaction, splitting into ADP and P_i (inorganic phosphate). This releases energy so that the solutes can be transported.

The diagram shows the active transport of calcium.

Unlike facilitated diffusion, active transport doesn't use channel proteins.

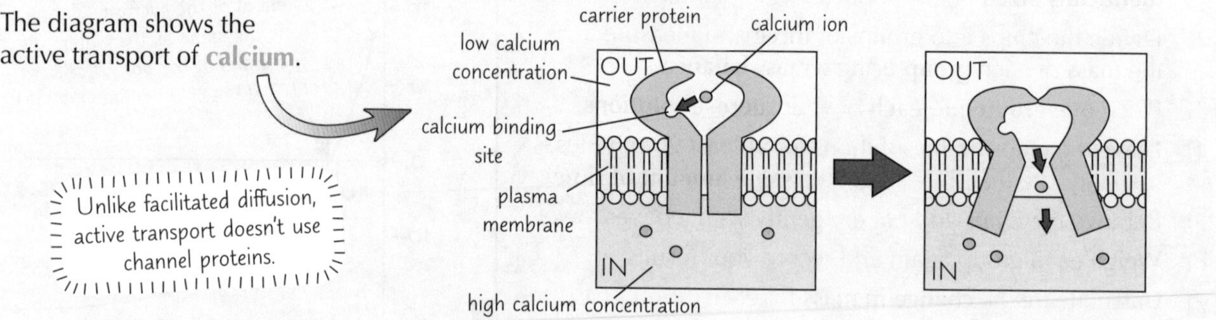

Co-transporters are a type of **carrier protein**.

1) They bind **two** molecules at a time.
2) The concentration gradient of one of the molecules is used to move the other molecule **against** its own concentration gradient.

The diagram shows the co-transport of **sodium ions** and **glucose**. Sodium ions move into the cell **down** their concentration gradient. This moves glucose into the cell too, **against** its concentration gradient.

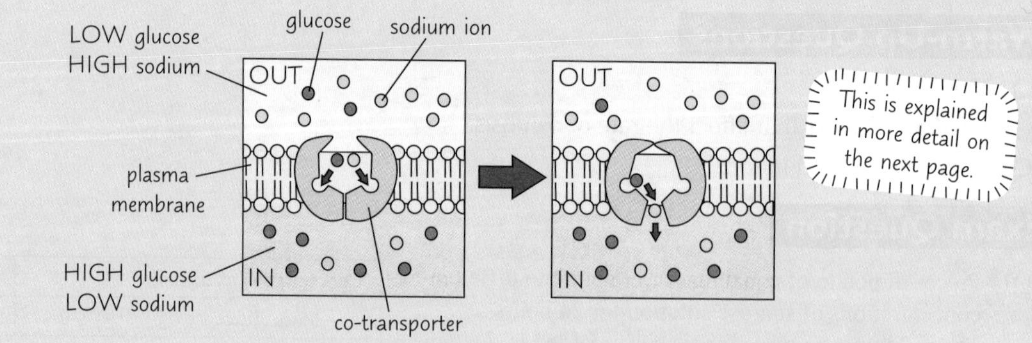

This is explained in more detail on the next page.

Learn these **Factors Affecting** the **Rate** of **Active Transport**

When **active transport** moves molecules and ions **against** their concentration gradient, a **decreasing** concentration gradient **doesn't** affect the **rate** of **active transport**. The rate of active transport is affected by:

1) The **speed** of **individual** carrier **proteins** — the **faster** they **work**, the **faster** the **rate** of active transport.
2) The **number** of **carrier proteins** present — the **more proteins** there are, the **faster** the **rate** of active transport.
3) The rate of **respiration** in the cell and the availability of **ATP**. If respiration is **inhibited**, active transport **can't** take place.

Exchange Across Cell Membranes — Active Transport

Glucose is Absorbed by Co-transport in the Mammalian Ileum

1) **Glucose** is absorbed into the **bloodstream** in the **small intestine**.
2) In the **ileum** (the final part of the small intestine) the **concentration** of glucose is **too low** for glucose to diffuse out into the blood. So glucose is absorbed from the **lumen** (middle) of the **ileum** by **co-transport**.

Glucose enters the **ileum epithelium** with **sodium ions**

1) **Sodium ions are actively transported out** of the ileum epithelial **cells**, into the **blood**, by the **sodium-potassium pump**. This creates a **concentration gradient** — there's now a higher concentration of sodium ions in the lumen of the ileum than inside the cell.
2) This causes sodium ions to **diffuse** from the lumen of the ileum **into the epithelial cell**, down their concentration gradient. They do this via the **sodium-glucose co-transporter proteins.**
3) The co-transporter carries **glucose** into the cell with the sodium. As a result the concentration of **glucose** inside the cell **increases**.
4) Glucose diffuses out of the cell, into the blood, down its concentration gradient through a protein channel, by **facilitated diffusion**.

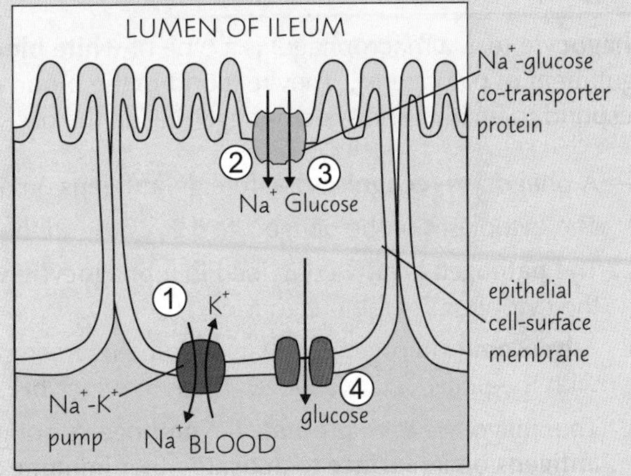

As you can see from this example, the same substance can be transported into or out of a cell in **different ways**. Sometimes **several methods of transport** are needed to move a substance from A to B.

Warm-Up Questions

Q1 Which molecule provides the energy for active transport?

Q2 Describe how carrier proteins actively transport substances across the cell membrane.

Exam Question

Q1 The graph shows the results from an experiment into the uptake of two different solutes (X and Y) by simple bacterial cells.

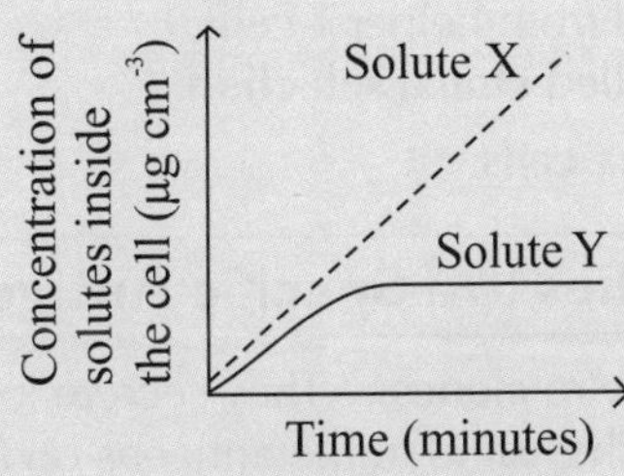

a) Which solute, X or Y, entered the cells by active transport? Give a reason for your answer. [1 mark]

b) Why is energy needed for the active transport of this solute? [1 mark]

c) Describe the process by which energy is released by the cell for active transport. [2 marks]

Revision — like working against a concentration gradient...

Don't worry if it takes you a while to learn these pages — there's quite a lot to cover. It's a good idea to learn it bit by bit. Don't move on to co-transport until you fully understand active transport in normal carrier proteins.

The Immune System

An infectious disease is one that is caused by pathogens, such as bacteria, viruses and fungi. Infectious diseases can be really nasty, but luckily there's an army of cells in the body that helps to protect us — the immune system.

Foreign Antigens Trigger an Immune Response

Antigens are **molecules** (usually proteins) that can **generate** an **immune response** when detected by the body. They are **usually** found on the **surface** of cells and are used by the **immune system** to identify: **pathogens** (organisms that cause disease), **abnormal body cells** (e.g. cancerous or pathogen-infected cells, which have abnormal antigens on their surface), **toxins** and cells from **other individuals** of the **same species** (e.g. organ transplants).
There are **four** main stages in the immune response:

1) Phagocytes Engulf Pathogens

A **phagocyte** (e.g. a macrophage) is a type of **white blood cell** that carries out **phagocytosis** (engulfment of pathogens). They're found in the **blood** and in **tissues** and are the first cells to respond to an immune system trigger inside the body. Here's how they work:

1) A phagocyte **recognises** the foreign **antigens** on a **pathogen**.
2) The cytoplasm of the phagocyte moves round the **pathogen**, **engulfing** it.
3) The **pathogen** is now contained in a **phagocytic vacuole** (a bubble) in the cytoplasm of the phagocyte.
4) A **lysosome** (an organelle that contain enzymes called **lysozymes**) **fuses** with the phagocytic vacuole. The lysozymes **break down** the pathogen.
5) The phagocyte then **presents** the **pathogen's antigens** — it sticks the **antigens** on its **surface** to **activate** other **immune system cells**.

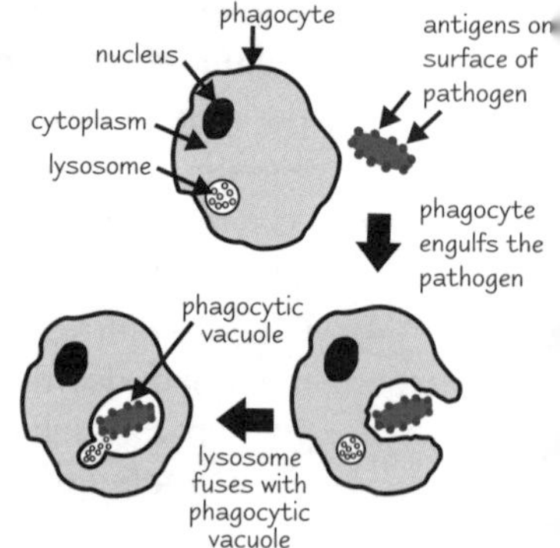

2) Phagocytes Activate T-cells

A **T-cell** (also called a T-lymphocyte) is another type of **white blood cell**. It has **receptor proteins** on its surface that **bind** to **complementary antigens** presented to it by **phagocytes**. This **activates** the T-cell. Different types of T-cells respond in different ways. For example, **helper T-cells** (T_H cells) release chemical signals that activate and **stimulate phagocytes** and **cytotoxic T-cells** (T_C cells), which kill abnormal and foreign cells. T_H cells also activate **B-cells**, which secrete antibodies (see below).

3) T-cells Activate B-cells, Which Divide into Plasma Cells

B-cells (also called B-lymphocytes) are also a type of **white blood cell**. They're covered with **antibodies** — proteins that **bind antigens** to form an **antigen-antibody complex**. Each B-cell has a **different shaped antibody** on its membrane, so different ones bind to **different shaped antigens**.

1) When the antibody on the surface of a B-cell meets a **complementary shaped** antigen, it binds to it.
2) This, together with substances released from helper T-cells, **activates** the B-cell. This process is called **clonal selection**.
3) The activated B-cell **divides** into **plasma cells**.

4) Plasma Cells Make More Antibodies to a Specific Antigen

Plasma cells are **identical** to the B-cell (they're **clones**). They secrete loads of **antibodies** specific to the antigen. These are called **monoclonal antibodies**. They bind to the antigens on the surface of the pathogen to form lots of **antigen-antibody complexes**.

An **antibody** has **two binding sites**, so can **bind** to **two pathogens** at the **same time**. This means that pathogens become **clumped** together — this is called **agglutination**. **Phagocytes** then bind to the antibodies and phagocytose **many** pathogens at once. This process leads to the **destruction** of **pathogens** carrying this **antigen** in the body.

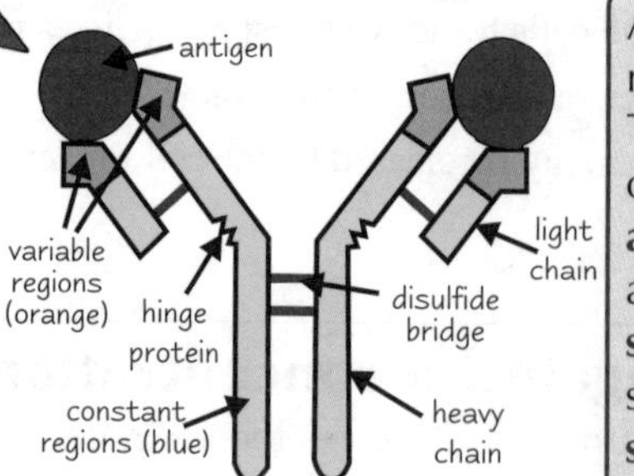

Antibodies are **proteins** — they're made up of chains of amino acids. The **specificity** of an antibody depends on its **variable regions**, which form the **antigen binding sites**. Each antibody has a variable region with a **unique tertiary structure** (due to different amino acid sequences) that's **complementary** to one **specific antigen**. All antibodies have the same **constant regions**.

The Immune System

The Immune Response Can be Split into Cellular and Humoral

Just to add to your fun, the **immune response** is split into **two** — the **cellular response** and the **humoral response**.

1) Cellular — The **T-cells** and **other** immune system **cells** that they **interact** with, e.g. phagocytes, form the cellular response.
2) Humoral — **B-cells**, **clonal selection** and the production of **monoclonal antibodies** form the **humoral response**.

Both types of response are **needed** to remove a pathogen from the body and the responses **interact** with each other, e.g. T-cells help to **activate** B-cells, and antibodies **coat** pathogens making it **easier** for **phagocytes** to **engulf** them.

The Immune Response for Antigens can be Memorised

The Primary Immune Response

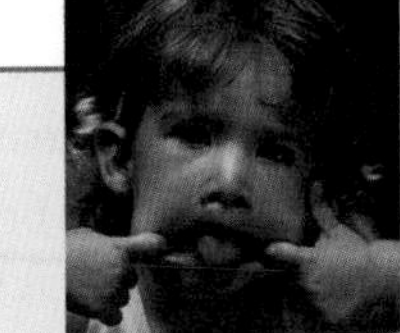

Neil's primary response — to his parents.

1) When an antigen enters the body for the **first time** it activates the immune system. This is called the **primary response.**
2) The primary response is **slow** because there **aren't many B-cells** that can make the antibody needed to bind to it.
3) Eventually the body will produce **enough** of the right antibody to overcome the infection. Meanwhile the infected person will show **symptoms** of the disease.

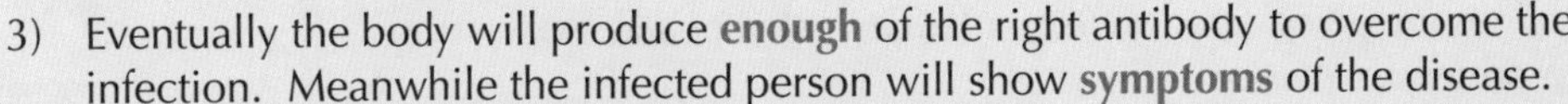

4) After being exposed to an antigen, both T- and B-cells produce **memory cells**. These memory cells **remain in the body** for a **long** time. Memory T-cells remember the **specific antigen** and will recognise it a second time round. Memory B-cells record the specific **antibodies** needed to bind the antigen.
5) The person is now **immune** — their immune system has the **ability** to respond **quickly** to a second infection.

The Secondary Immune Response

1) If the **same pathogen** enters the body again, the immune system will produce a **quicker**, **stronger** immune response — the **secondary response**.
2) **Clonal selection** happens **faster**. **Memory B-cells** are activated and divide into **plasma cells** that produce the right antibody to the antigen. **Memory T-cells** are activated and divide into the **correct type** of **T-cells** to kill the cell carrying the antigen.
3) The secondary response often gets rid of the pathogen **before** you begin to show any **symptoms** (you are **immune** to the pathogen).

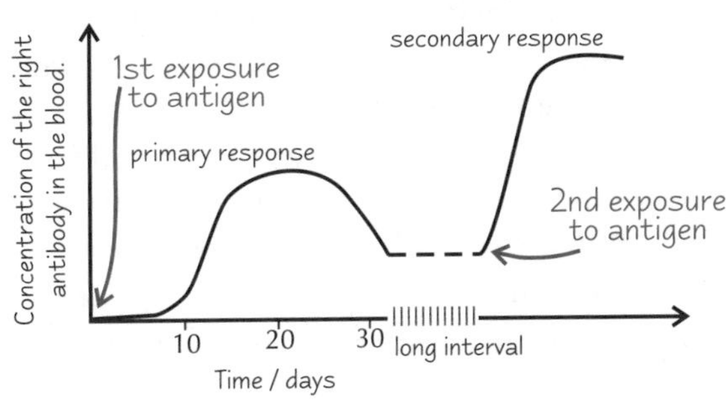

Remember, T-cells and B-cells are also called T-lymphocytes and B-lymphocytes.

Warm-Up Questions

Q1 What are antigens?

Q2 What does the humoral response involve?

Exam Questions

Q1 Describe the function of antibodies. [2 marks]

Q2 Describe and explain how a secondary immune response differs to a primary immune response. [4 marks]

Memory cells — I need a lot more to cope with these pages...

If memory cells are mentioned in the exam, remember that they are still types of T-cells and B-cells. They just hang around a lot longer than most T-cells and B-cells. When the antigen enters the body for a second time they can immediately divide into more of the specific T-cells and B-cells that can kill the pathogen or release antibodies against it.

Immunity and Vaccines

The primary response gives us immunity against a disease, but only after you've become infected. If only there was a way to stimulate memory cell production without getting the disease... well, there is — vaccination.

Vaccines can Protect Individuals and Populations Against Disease

1) While your B-cells are busy **dividing** to build up their numbers to deal with a pathogen (i.e. the **primary response** — see previous page), you **suffer** from the disease. **Vaccination** can help avoid this.
2) Vaccines **contain antigens** that cause your body to **produce memory cells** against a particular pathogen, **without** the pathogen **causing disease**. This means you become **immune** without getting any **symptoms**.
3) Vaccines protect individuals that have them and, because they reduce the **occurrence** of the disease, those **not** vaccinated are also less likely to catch the disease (because there are fewer people to catch it from). This is called **herd immunity**.
4) Vaccines always contain antigens — these may be **free** or **attached** to a **dead** or **attenuated** (weakened) **pathogen**.
5) Vaccines may be **injected** or taken **orally**. The **disadvantages** of taking a vaccine orally are that it could be **broken down** by **enzymes** in the gut or the **molecules** of the vaccine may be **too large** to be **absorbed** into the blood.
6) Sometimes **booster** vaccines are given later on (e.g. after several years) to **make sure** that memory cells are produced.

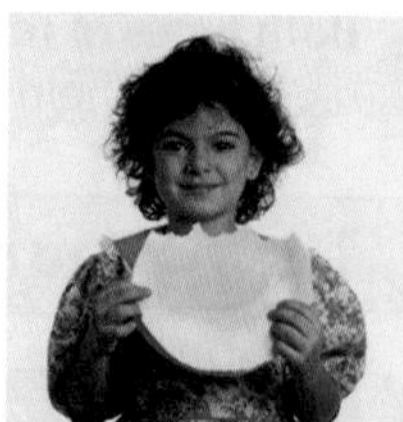

The oral vaccine was proving hard to swallow.

Antigenic Variation Helps Some Pathogens Evade the Immune System

1) **Antigens** on the surface of pathogens **activate** the **primary response**.
2) When you're **infected** a **second time** with the **same pathogen** (which has the **same antigens** on its surface) they **activate** the **secondary response** and you don't get ill.
3) However, some sneaky pathogens can **change** their surface antigens. This **antigen variability** is called **antigenic variation**. (Different antigens are formed due to changes in the **genes** of a pathogen.)
4) This means that when you're infected for a **second time**, the **memory cells** produced from the **first infection** will **not recognise** the **different antigens**. So the immune system has to start from scratch and carry out a **primary response** against these new antigens.
5) This **primary response** takes **time** to get rid of the infection, which is why you get **ill again**.
6) **Antigenic variation** also makes it **difficult** to develop **vaccines** against some pathogens for the same reason. **Examples** of pathogens that show antigenic variation include **HIV** and the **influenza virus**.
7) Here's how **antigenic variation** affects the production of **vaccines** to help prevent people catching **influenza**:

    1) The **influenza (flu) vaccine** changes every year. That's because the **antigens** on the surface of the influenza virus **change regularly**, forming **new strains** of the virus.
    2) **Memory cells** produced from **vaccination** with **one strain** of the flu will **not recognise** other strains with **different antigens**. The strains are **immunologically distinct**.
    3) Every year there are **different strains** of the influenza virus **circulating** in the **population**, so a **different vaccine** has to be made.
    4) **New vaccines** are **developed** and one is chosen **every year** that is the **most effective** against the **recently** circulating influenza viruses.
    5) Governments and health authorities then implement a **programme** of **vaccination** using the most **suitable** vaccine.

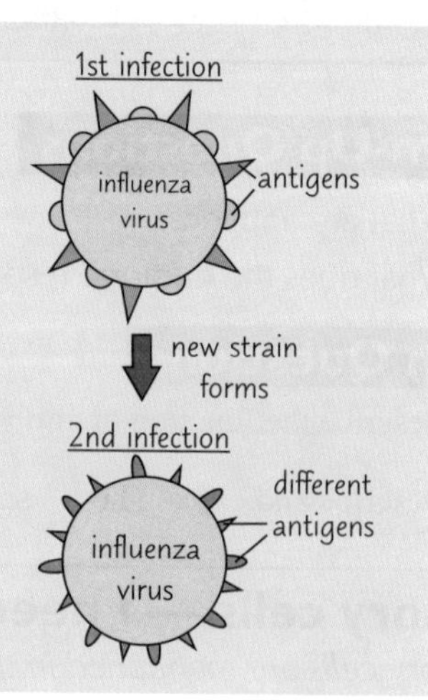

Immunity and Vaccines

Immunity can be **Active** or **Passive**

ACTIVE IMMUNITY

This is the type of immunity you get when **your immune system makes its own antibodies** after being **stimulated** by an **antigen**. There are **two** different types of active immunity:

1) **Natural** — this is when you become immune after **catching a disease**.
2) **Artificial** — this is when you become immune after you've been given a **vaccination** containing a harmless dose of antigen.

PASSIVE IMMUNITY

This is the type of immunity you get from being **given antibodies made by a different organism** — your immune system **doesn't** produce any antibodies of its own. Again, there are **two** types:

1) **Natural** — this is when a **baby** becomes immune due to the antibodies it receives from its **mother**, through the **placenta** and in **breast milk**.
2) **Artificial** — this is when you become immune after being **injected** with **antibodies** from **someone else**. E.g. If you contract tetanus you can be injected with antibodies against the tetanus toxin, collected from blood donations.

Active and **Passive** Immunity Have **Contrasting Characteristics**

In the exam you might be asked about the **differences** between these types of **immunity**:

Active immunity	Passive immunity
Requires exposure to antigen	Doesn't require exposure to antigen
It takes a while for protection to develop	Protection is immediate
Memory cells are produced	Memory cells aren't produced
Protection is long-term because the antibody is produced (after activation of memory cells) in response to complementary antigen being present in the body	Protection is short-term because the antibodies given are broken down

Warm-Up Questions

PRACTICE QUESTIONS

Q1 How do vaccines cause immunity?

Q2 Explain what antigenic variability is.

Exam Questions

Q1 Vaccines can be used to protect people against some diseases. Not all individuals in a population must receive the vaccine for a vaccination programme to be successful. Explain why this is the case. [3 marks]

Q2 The influenza virus causes the flu. Explain why it is possible to suffer from the flu more than once. [4 marks]

Q3 Immunity from a disease can be either active or passive.

a) Explain why active immunity offers long-term protection against a disease, whereas passive immunity only offers protection in the short-term. [2 marks]

b) It normally takes 14 days for immunity to develop after receiving a vaccine. Explain why vaccines do not usually offer immediate protection against a disease. [1 mark]

An injection of dead bugs — roll on my next vaccine...

The influenza virus is so clever that it would almost make you think it had a mind of its own. I mean, as soon as we catch up with it and develop a vaccine, off it goes and changes its surface antigens again. Influenza virus: one, humans: nil. This is one of the ways viruses have evolved to avoid your immune system. Well, clever them.

Antibodies in Medicine

Antibodies aren't only great for fighting off infection, they're also excellent tools for use in medical diagnosis and drug development. Let's all give three cheers for antibodies. Without them, we'd all probably be dead by now.

Monoclonal Antibodies can be used to Target Specific Substances or Cells

1) **Monoclonal antibodies** are antibodies **produced** from a **single group of genetically identical B-cells** (plasma cells). This means that they're all **identical** in **structure**.
2) As you know, antibodies are **very specific** because their binding sites have a **unique tertiary structure** (see p.46) that only one particular antigen will fit into (one with a **complementary shape**).
3) You can make monoclonal antibodies **that bind to anything** you want, e.g. a cell antigen or other substance, and they will only bind to (target) this molecule.

EXAMPLE: Targeting drugs to a particular cell type — cancer cells

1) **Different cells** in the body have **different** surface **antigens**.
2) Cancer cells have antigens called **tumour markers** that are **not** found on normal body cells.
3) **Monoclonal antibodies** can be made that will bind to the tumour markers.
4) You can also attach **anti-cancer drugs** to the antibodies.
5) When the antibodies come into **contact** with the cancer cells they will **bind** to the tumour markers.
6) This means the drug will **only accumulate** in the body where there are **cancer cells.**
7) So, the **side effects** of an antibody-based drug are lower than other drugs because they accumulate near **specific cells.**

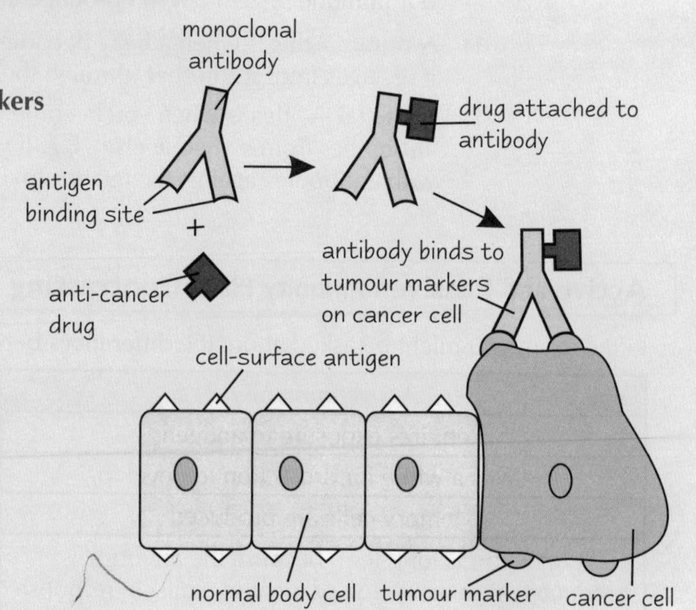

EXAMPLE: Targeting a particular substance for medical diagnosis — pregnancy testing

Pregnancy tests detect the hormone **human chorionic gonadotropin (hCG)** that's found in the **urine** of pregnant women:

1) The application area contains **antibodies for hCG** bound to a **coloured bead** (**blue**).
2) When urine is applied to the application area any hCG will **bind** to the antibody on the beads, forming an **antigen-antibody complex**.
3) The urine **moves** up the stick to the **test strip**, **carrying** any **beads** with it.
4) The test strip contains **antibodies to hCG** that are stuck in place (**immobilised**).
5) If there **is hCG present** the test strip turns **blue** because the **immobilised** antibody binds to any **hCG** — concentrating the hCG-antibody complex with the **blue beads** attached. If **no hCG** is present, the beads will **pass through** the test area **without** binding to anything, and so it **won't** go blue.

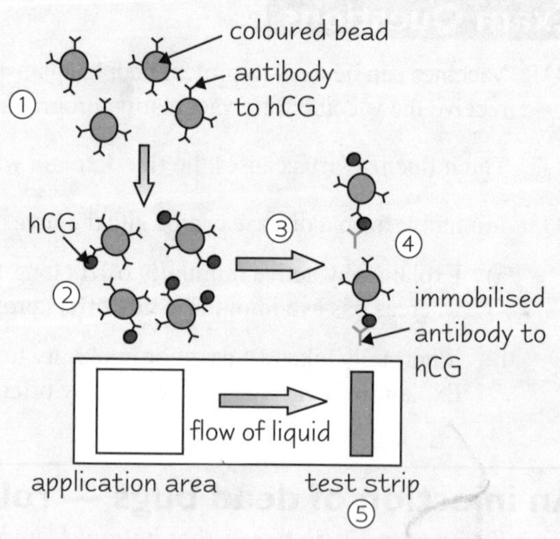

Antibodies in Medicine

The ELISA Test is a Medical Diagnostic Test that Uses Antibodies

1) The **enzyme-linked immunosorbent assay** (ELISA) allows you to see if a patient has any **antibodies** to a certain **antigen** (see example below) or any **antigen** to a certain **antibody**.
2) It can be used to test for **pathogenic infections**, for **allergies** (e.g. to nuts or lactose) and for just about **anything** you can make an **antibody** for.
3) In an ELISA test, an antibody is used which has an **enzyme attached** to it. This enzyme can **react** with a substrate to produce a **coloured product**. This causes the solution in the reaction vessel to **change colour**.
4) If there's a **colour change**, it demonstrates that the **antigen** or **antibody** of interest is **present** in the sample being tested (e.g. blood plasma). In some types of ELISA, the **quantity** of this antigen/antibody can be worked out from the **intensity** of the colour change.
5) There are several **different types** of ELISA. **Direct ELISA** uses a **single** antibody that is complementary to the antigen you're testing for. **Indirect ELISA** is different because it uses **two** different antibodies. This method is outlined below:

EXAMPLE: Using an ELISA as a HIV (Human Immunodeficiency Virus) Test

An **indirect ELISA test** can be used to see if a patient possesses **antibodies** to the HIV virus:

① **HIV antigen** is **bound** to the bottom of a **well** in a **well plate** (a plastic tray with loads of little circular pits in it).

② A sample of the **patient's blood plasma**, which might contain several different antibodies, is **added** to the **well**. If there are any **HIV-specific antibodies** (i.e. antibodies against HIV) these will **bind** to the **HIV antigen** stuck to the bottom of the **well**. The well is then **washed out** to remove any **unbound antibodies**.

③ A **secondary antibody**, that has a specific **enzyme** attached to it, is added to the **well**. This secondary antibody can bind to the **HIV-specific antibody** (which is also called the **primary antibody**). The well is **washed out** again to remove any **unbound secondary antibody**. If there's no primary antibody in the sample, all of the secondary antibody will be **washed away**.

④ A **solution** is added to the **well**. This solution contains a **substrate**, which is able to react with the **enzyme** attached to the secondary antibody and produce a **coloured product**. If the solution **changes colour**, it indicates that the patient has **HIV-specific antibodies** in their blood and is **infected** with HIV.

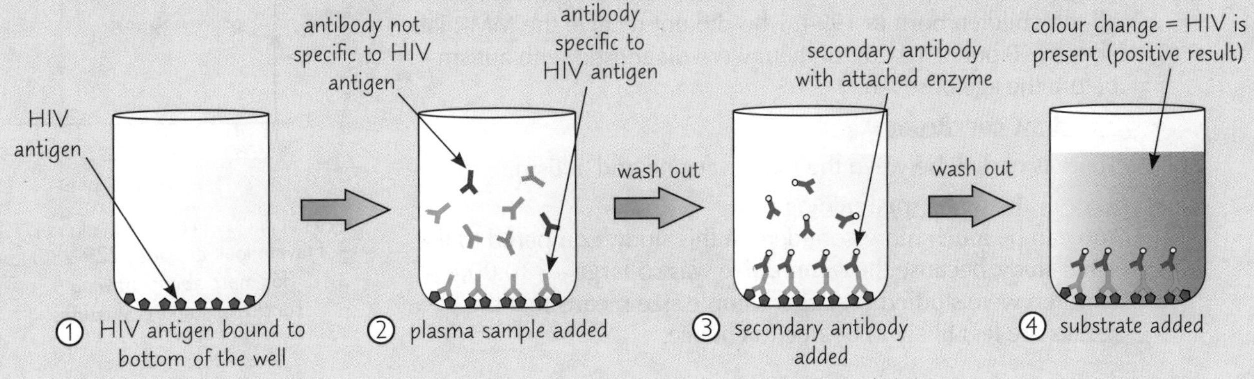

The washing steps are important to make sure unbound antibodies aren't left in the well which could affect the results. E.g. unbound secondary antibodies could cause the test to appear positive when there are no HIV antibodies present.

If the ELISA result was negative, there would be no colour change because there would be no HIV-specific antibodies for the secondary antibodies to bind to.

Warm-Up Questions

Q1 What are monoclonal antibodies?

PRACTICE QUESTIONS

Exam Question

Q1 Describe how monoclonal antibodies can be used to target a drug to cancer cells. [4 marks]

Antibodies — the multi-tool of the immune system...

Monoclonal antibodies are really useful — they can even be made against other antibodies. For example, people with asthma produce too many of a type of antibody that causes inflammation in the lungs. Monoclonal antibodies can be made to bind this type of antibody, so it can no longer cause inflammation, which can reduce the asthma symptoms.

Interpreting Vaccine and Antibody Data

If someone claims anything about a vaccine or antibody, the claim has to be validated (confirmed) before it's accepted. To do this, you need to evaluate the data used to support the claim and the methodology behind it.

New Knowledge About Vaccines and Antibodies is Validated by Scientists

When a **study** presents evidence for a **new theory** (e.g. a vaccine has a dangerous side effect) it's important that other scientists come up with **more evidence** to **validate** (confirm) the theory. Other scientists may **repeat** the study and try to **reproduce** the results, or **conduct other studies** to try to prove the same theory.

EXAMPLE 1: The MMR Vaccine

1) In 1998, a study was published about the **safety** of the **measles**, **mumps** and **rubella (MMR) vaccine**. The study was based on **12 children** with **autism** (a life-long developmental disability) and concluded that there may be a **link** between the MMR vaccine and autism.
2) Not everyone was convinced by this study because it had a **very small sample size** of 12 children, which increased the likelihood of the results being due to **chance**. The study may have been **biased** because one of the scientists was helping to gain evidence for a **lawsuit** against the MMR vaccine manufacturer. Also, studies carried out by different scientists found no link between autism and the MMR vaccine.
3) There have been **further scientific studies** to sort out the **conflicting** evidence. In **2005**, a **Japanese** study was published about the incidence of autism in Yokohama (an area of Japan). They looked at the medical records of **30 000 children** born between **1988 and 1996** and counted the number of children that developed **autism** before the age of seven. The **MMR jab** was first **introduced in Japan in 1989** and was **stopped in 1993**. During this time the MMR vaccine was administered to children at **12 months old**. The graph shows the results of the study.
4) In the exam you could be asked to **evaluate evidence** like this.
   - <u>You might be asked to **describe the data**...</u>
     The graph shows that the number of children diagnosed with autism continued to **rise** after the MMR vaccine was **stopped**. For example, from all the children born in 1992, who did receive the MMR jab, about 60 out of 10 000 were diagnosed with autism before the age of seven. However, from all the children born in 1994, who did not receive the MMR jab, about 160 out of 10 000 of them were diagnosed with autism before the age of seven.
   - <u>...or **draw conclusions**</u>
     There is **no link** between the MMR vaccine and autism.
   - <u>... or **evaluate the methodology**</u>
     You can be much more confident in this study, compared to the 1998 study, because the **sample size** was so **large** — 30 000 children were studied. A larger sample size means that the results are less likely to be due to **chance**.

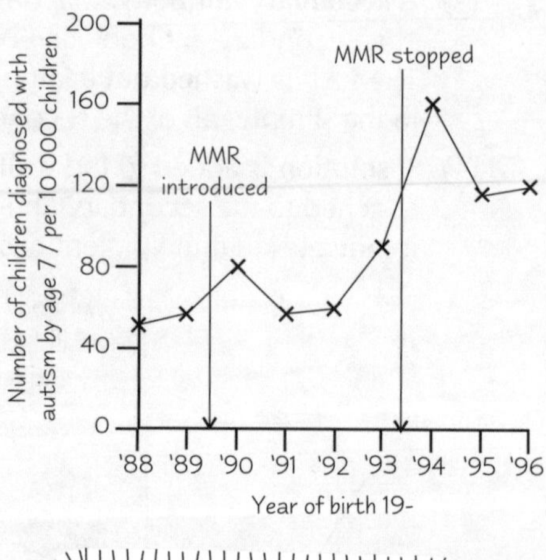

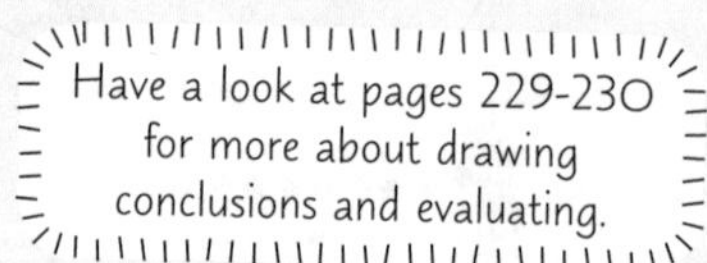

EXAMPLE 2: Herceptin® — Monoclonal Antibodies

About **20%** of **women with breast cancer** have tumours that produce more than the usual amount of a **receptor** called **HER2**. **Herceptin®** is a **drug** used to treat this type of breast cancer — it contains **monoclonal antibodies** that **bind the HER2 receptor** on a **tumour cell** and **prevent** the cells from growing and dividing.

In **2005**, a study **tested** Herceptin® on women who had already undergone **chemotherapy** for HER2-type **breast cancer**. **1694** women took the **drug** for a **year** after chemotherapy and another **1694** women were **observed** for the **same time** (the control group). The results are shown in the graph on the right.

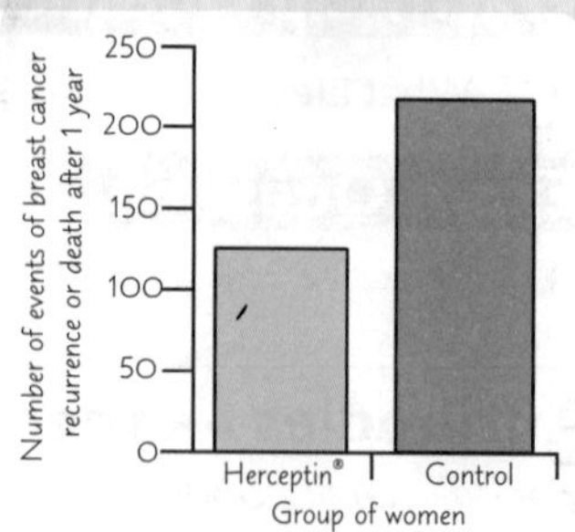

<u>**Describe the data:**</u> Almost **twice as many** women in the **control group** developed breast cancer again or died **compared** to the group taking Herceptin®.

<u>**Draw conclusions:**</u> A **one-year treatment** with Herceptin®, after chemotherapy, **increases** the disease-free survival rate for women with HER2-type breast cancer.

Interpreting Vaccine and Antibody Data

Use of Vaccines and Antibodies Raises Ethical Issues

Ethical issues surrounding vaccines include:

1) All vaccines are **tested on animals** before being tested on humans — some people **disagree** with animal testing. Also, **animal based substances** may be used to **produce** a vaccine, which some people disagree with.
2) **Testing** vaccines on **humans** can be **tricky**, e.g. volunteers may put themselves at **unnecessary risk** of contracting the disease because they think they're fully protected (e.g. they might have unprotected sex because they have had a new HIV vaccine and think they're protected — and the vaccine might not work).
3) Some people **don't** want to take the vaccine due to the **risk** of **side effects**, but they are **still protected** because of **herd immunity** (see p. 48) — other people think this is **unfair**.
4) If there was an **epidemic** of a **new disease** (e.g. a new influenza virus) there would be a rush to **receive** a vaccine and **difficult decisions** would have to be made about **who** would be the **first** to receive it.

Ethical issues surrounding monoclonal antibody therapy often involve animal rights issues. **Animals** are used to **produce the cells** from which the monoclonal antibodies are produced. Some people **disagree** with the use of animals in this way.

Warm-Up Questions

Q1 Suggest one ethical issue surrounding vaccines.

Q2 Suggest one ethical issue surrounding monoclonal antibodies.

Exam Question

Q1 The graph below shows the number of laboratory reports of *Haemophilus influenzae* type b (Hib), in England and Wales, from 1990 to 2004. Hib affects children and can lead to meningitis and pneumonia.

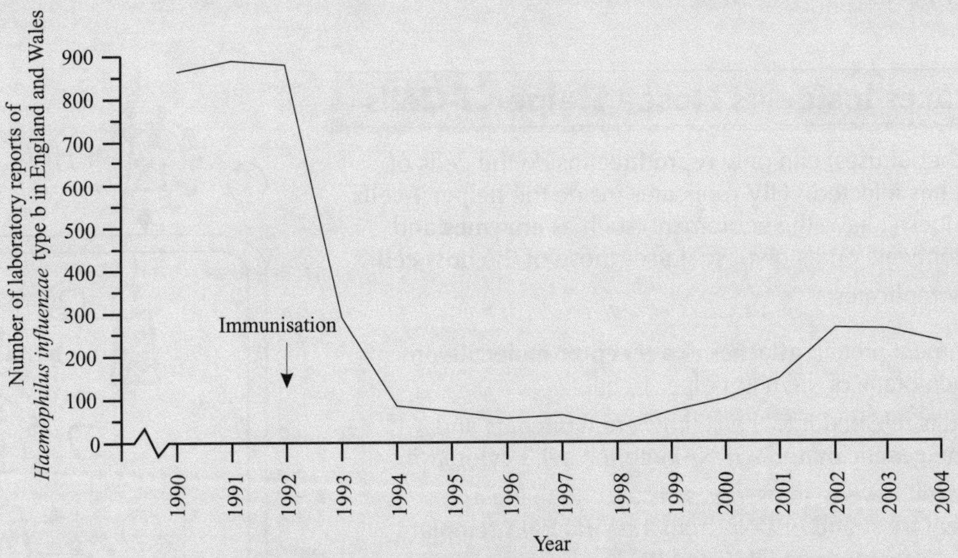

a) Explain how immunisation could have caused the sharp decrease in Hib cases after 1992. [2 marks]

b) Suggest a possible explanation for the increase in Hib cases after 1998. [1 mark]

Some scientists must have to validate the taste of chocolate — nice job...

After the 1998 study, some parents were worried about giving their kids the MMR vaccine, so the number of children given the vaccine fell. With fewer children in each community protected by the vaccine, herd immunity decreased. This meant that more people were vulnerable to measles, mumps and rubella, so the number of cases of went up.

HIV and Viruses

Viruses aren't cells like bacteria. They're not even living things — they can only reproduce inside the cells of another organism (called the host). All viruses cause disease, and you need to know all about one particularly nasty blighter...

HIV is the Virus That Causes AIDS

1) **HIV (Human Immunodeficiency Virus)** is a virus that affects the **immune system**. It eventually leads to **acquired immune deficiency syndrome (AIDS)**.
2) **AIDS** is a condition where the immune system **deteriorates** and eventually **fails**. This makes someone with AIDS more **vulnerable** to **other infections**, like pneumonia (see next page).
3) **HIV** infects (and eventually kills) **helper T-cells**, which act as the **host cells** (see p. 30) for the virus. Remember, helper T-cells send chemical signals that **activate phagocytes**, **cytotoxic T-cells** and **B-cells** (see p. 46) so they're **hugely important cells** in the **immune response**. Without enough helper T-cells, the immune system is **unable** to mount an **effective** response to **infections** because other immune system cells **don't behave** how they **should**.
4) People infected with HIV develop **AIDS** when the **helper T-cell numbers** in their body reach a critically **low** level.

HIV has a Spherical Structure

You might get asked about the structure of HIV in your exam.

Attachment proteins can also be called envelope proteins.

1) A **core** that contains the **genetic material** (RNA) and some **proteins** (including the enzyme **reverse transcriptase**, which is needed for virus replication).
2) An **outer coating** of protein called a **capsid**.
3) An **extra outer layer** called an **envelope**. This is made of **membrane** stolen from the cell membrane of a previous host cell.
4) Sticking out from the envelope are **loads of copies** of an **attachment protein** that help HIV **attach** to the **host helper T-cell**.

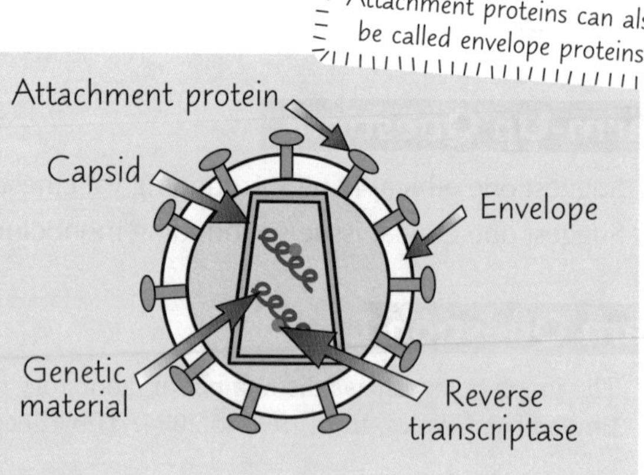

HIV Replicates Inside its Host's Helper T-Cells

HIV (and all other viruses) can only **reproduce** inside the cells of the organism it has infected. HIV replicates inside the **helper T-cells** of the host. It doesn't have the equipment (such as **enzymes** and **ribosomes**) to replicate on its own, so it uses those of the **host cell**.

Here's how **HIV** replicates:

1) The attachment protein **attaches** to a **receptor molecule** on the cell membrane of the host helper T-cell.
2) The capsid is released into the cell, where it **uncoats** and releases the **genetic material** (RNA) into the cell's cytoplasm.
3) Inside the cell, reverse transcriptase is used to make a **complementary strand** of DNA from the **viral RNA template** (see p. 16-17 for more on DNA and RNA).
4) From this, **double-stranded DNA** is made and **inserted** into the human DNA.
5) Host cell enzymes are used to make **viral proteins** from the **viral DNA** found within the human DNA.
6) The viral proteins are **assembled** into **new viruses**, which **bud** from the cell and go on to infect other cells.

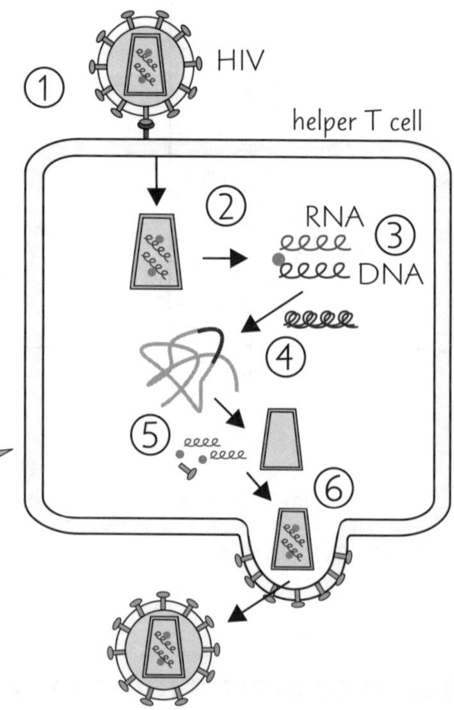

During the initial infection period, HIV replicates rapidly and the infected person may experience severe flu-like symptoms. After this period, HIV replication drops to a lower level. This is the **latency period**. During the latency period (which can last for years), the infected person **won't experience** any **symptoms**.

HIV and Viruses

People with AIDS are Susceptible to a Range of Illnesses

People with HIV are classed as having AIDS when **symptoms** of their **failing immune system** start to **appear** or their **helper T-cell count drops** below a certain level. People with AIDS generally develop diseases that **wouldn't** cause serious problems in people with a **healthy** immune system. The length of time between **infection** with HIV and the **development** of AIDS **varies** between individuals but without treatment it's usually around **10 years**.

1) The **initial symptoms** of AIDS include **minor infections** of mucous membranes (e.g. the inside of the nose, ears and genitals), and recurring respiratory infections.
2) As AIDS **progresses** the number of **immune system cells decreases** further. Patients become susceptible to **more serious infections** including chronic diarrhoea, severe bacterial infections and tuberculosis.
3) During the **late stages** of AIDS patients have a very **low number** of immune system cells and can develop a **range of serious infections** such as toxoplasmosis of the brain (a parasite infection) and candidiasis of the respiratory system (fungal infection). It's these serious infections that kill AIDS patients, not HIV itself.

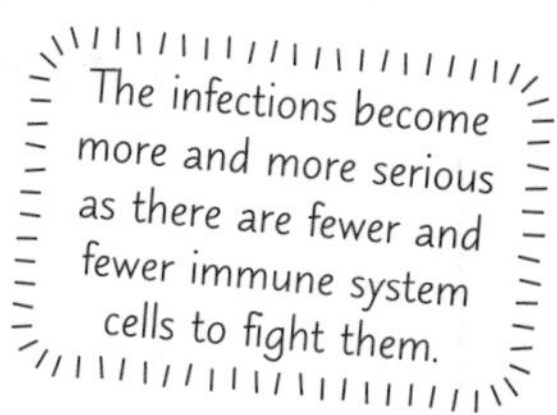

The length of time that people survive with AIDS varies a lot. Factors that affect progression of HIV to AIDS and survival time with AIDS include **existing infections**, the **strain of HIV** they're infected with, **age** and access to **healthcare**.

Antibiotics Don't Work Against Viruses

1) Antibiotics kill **bacteria** by **interfering** with their metabolic reactions. They target the **bacterial enzymes** and **ribosomes** used in these reactions.
2) Bacterial enzymes and ribosomes are **different** from **human** enzymes and ribosomes. Antibiotics are designed to **only target** the bacterial ones so they don't damage human cells. Makes sense.
3) Viruses **don't have their own** enzymes and ribosomes — they use the ones in the host's cells. So because human viruses use human enzymes and ribosomes to replicate, antibiotics **can't** inhibit them because they **don't** target human processes.
4) Most **antiviral drugs** are designed to target the few **virus-specific enzymes** (enzymes that only the virus uses) that exist. For example, HIV uses **reverse transcriptase** to replicate (see previous page). Human cells **don't** use this enzyme so drugs can be designed to inhibit it **without affecting** the host cell. These drugs are called reverse-transcriptase inhibitors.

There's No Cure for HIV

1) There's currently **no cure** or **vaccine** for HIV but **antiviral** drugs can be used to **slow down** the **progression** of HIV infection and AIDS in an infected person.
2) The best way to control HIV infection in a population is by **reducing** its **spread**. HIV can be **spread** via **unprotected sexual intercourse**, through **infected bodily fluids** (e.g. like blood from sharing contaminated needles) and from a HIV-positive **mother** to her **fetus**. Not all babies from HIV-positive mothers are born infected with HIV and taking antiviral drugs during pregnancy can reduce the chance of the baby being HIV-positive.

HIV testing, based on HIV antibody detection, before a baby is 18 months old can be inaccurate. This is because the baby of an HIV-positive mother may have some HIV antibodies in their blood regardless of whether or not they're infected.

Warm-Up Questions

Q1 What type of cell does HIV replicate in?
Q2 Why can't antibiotics be used to treat HIV?

PRACTICE QUESTIONS

Exam Question

Q1 HIV is the virus that eventually causes AIDS. Describe the structure of HIV. [4 marks]

Viruses can be dangerous and hard to treat — they're just not funny...

Well, apart from rhinoviruses, which cause colds, but they're only funny because of the name. It's actually quite a logical name — rhino is from the Greek for nose. They're literally nose viruses. If I was a virus I'd choose somewhere better to infect. Anyway, you need to learn this stuff. Scribble everything down and see what you remember.

Extra Exam Practice

That's Topic 2 nearly finished — which is good 'cos I'm starting to feel a bit woozy. All that's left is for you to have a go at these extra exam practice questions, which mix content from the whole of Topic 2. Good luck...

- Have a look at this fantastic example of how to answer a tricky exam question.
- Then check how much you've understood from Topic 2 by having a go at the questions on the next page.

1 **Figure 1** shows how the number of copies of HIV and antibodies complementary to HIV vary over time in the blood of a person who has been infected with HIV.

Figure 1

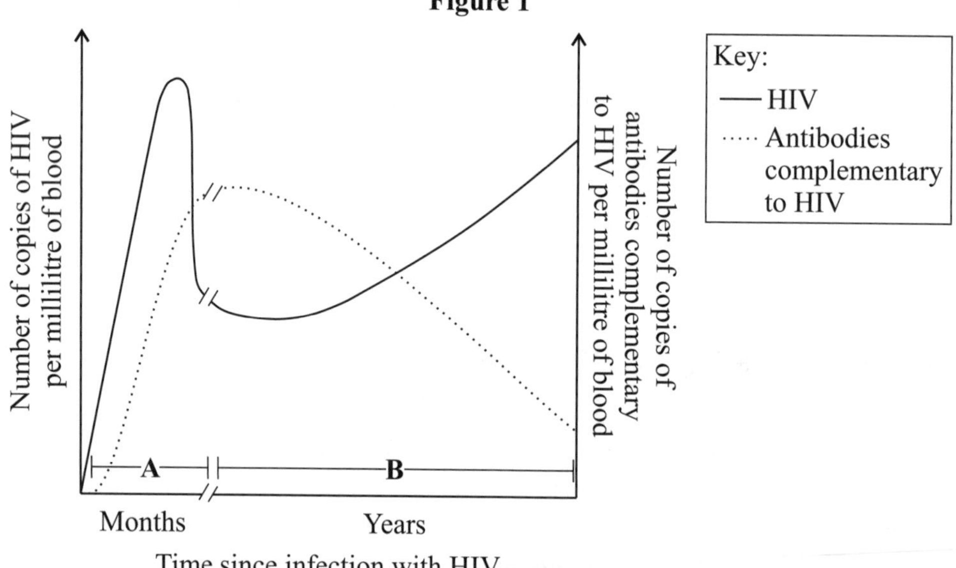

1.1 Describe the process which results in the rise in the number of copies of antibodies complementary to HIV, during time period **A** in **Figure 1**.

(6 marks)

1.2 The number of copies of antibodies complementary to HIV falls during time period **B** in **Figure 1**. Explain why.

(4 marks)

You might be tempted to write everything you know about the immune response, but it's important to focus on what you're actually being asked — e.g. don't get carried away and write about cytotoxic T-cells here — they're not relevant to the question.

1.1

Phagocytes recognise the foreign antigens on HIV and engulf the virus. The phagocytes present the HIV antigens on their surface, which receptor proteins on helper T-cells can recognise and bind to. When helper T-cells bind to these antigens they become activated and release chemical signals that activate B-cells, in a process called clonal selection. The activated B-cells divide into plasma cells that secrete lots of antibodies specific to the HIV antigen, causing the rise shown in Figure 1.

You've been asked to describe the process so make sure you clearly describe each step in the correct order.

1.2

HIV infects, and eventually kills, helper T-cells, which act as the host cells for the virus. Over time, this reduces the number of helper T-cells in the blood of a person infected with HIV. Helper T-cells activate B-cells, so if their numbers fall then fewer B-cells are activated, so fewer plasma cells are produced. As plasma cells secrete antibodies, the number of copies of complementary antibodies to HIV in the blood also falls.

The graph shows that as the level of antibodies complementary to HIV in the blood falls, the HIV level rises. You just need to use your knowledge of HIV to figure out why this happens.

Extra Exam Practice

2 A student investigated the effect of a chemical called lectin on the growth of roots of onion plants. She grew some onion plants in distilled water and others in a lectin solution. After leaving the plants to grow for two days, she removed the tips of the roots and observed them under a microscope to determine how many root tip cells were undergoing mitosis.

2.1 The cells measured 75 μm in size. The student used a magnification of × 200 to view them. Calculate the size of the cells in the image. Give your answer in millimetres.

(2 marks)

Lectins can bind to carbohydrates. Some lectins can bind to molecules on cell-surface membranes, and cause signals to be passed to the cells that result in the duration of interphase being reduced.

2.2 Suggest which type of molecule on the cell-surface membrane these lectins bind to. Explain your answer.

(2 marks)

2.3 Predict the effect that the lectin solution would have had on the mitotic index in the root tip cells of the onion plant, compared to the mitotic index of the root tip cells from the onion plant grown in distilled water. Explain your answer.

(2 marks)

3 Beetroot cells contain a red pigment called betalain. A student investigated the permeability of beetroot cell-surface membranes in different concentrations of ethanol. Ethanol can denature proteins and dissolve lipids. The student placed beetroot discs in different concentrations of ethanol for 20 minutes, then removed the discs and recorded the absorbance value of the solution. The absorbance value illustrates the proportion of light absorbed by the betalain in the solution. **Figure 2** shows the results of the investigation.

Figure 2

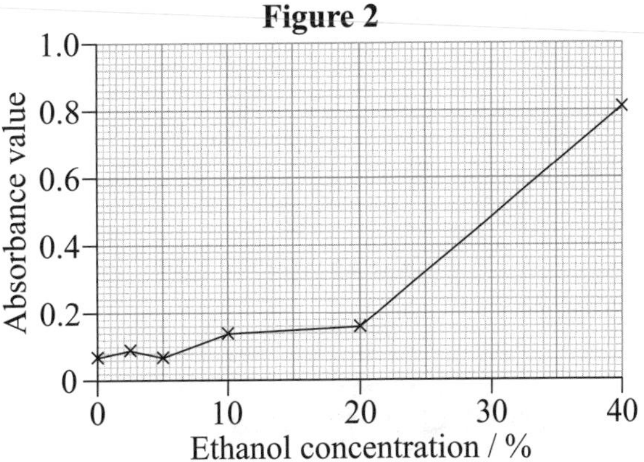

3.1 The student used serial dilutions to produce the different concentrations of ethanol used in the investigation. The student started with 20 cm^3 of 40% ethanol solution. Calculate the volume of the original solution and distilled water that he needs to make 20 cm^3 of 30% ethanol solution.

(2 marks)

3.2 A student suggested that, as it leaves the cells, the betalain may accumulate around the beetroot discs. He suggested that the results would be more accurate if the ethanol solutions containing the beetroot discs had been mixed regularly throughout the investigation. Explain why regularly mixing the solutions may increase the accuracy of the results.

(2 marks)

3.3 Sitosterol is a molecule that has a role in plant cell-surface membranes similar to the role of cholesterol in cell-surface membranes. Suggest how the addition of sitosterol to the beetroot discs may affect the results shown in **Figure 2**. Explain your answer.

(3 marks)

Size and Surface Area

Exchanging things with the environment is pretty easy if you're a single-celled organism, but if you're multicellular it all gets a bit more complicated... and it's all down to this 'surface area to volume ratio' malarkey.

Organisms Need to **Exchange Substances** with their **Environment**

Every organism, whatever its size, needs to exchange things with its environment. Otherwise there'd be no such thing as poop scoops...

Raj was glad he'd exchanged his canoe for a bigger boat.

1) Cells need to take in **oxygen** (for aerobic respiration) and **nutrients**.
2) They also need to excrete **waste products** like **carbon dioxide** and **urea**.
3) Most organisms need to stay at roughly the **same temperature**, so **heat** needs to be exchanged too.

How easy the exchange of substances is depends on the organism's **surface area to volume ratio**.

Smaller Animals have **Higher Surface Area : Volume Ratios**

A mouse has a bigger surface area **relative to its volume** than a hippo. This can be hard to imagine, but you can prove it mathematically. Imagine these animals as cubes:

The hippo could be represented by a block measuring 2 cm × 4 cm × 4 cm.

Its **volume** is 2 × 4 × 4 = **32 cm³**

Its **surface area** is 2 × 4 × 4 = 32 cm² (top and bottom surfaces of cube)
+ 4 × 2 × 4 = 32 cm² (four sides of the cube)

Total surface area = **64 cm²**

So the hippo has a **surface area : volume ratio** of 64 : 32 or **2 : 1**.

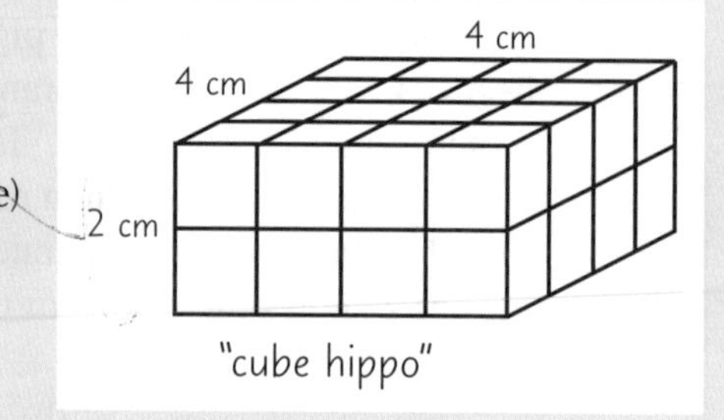

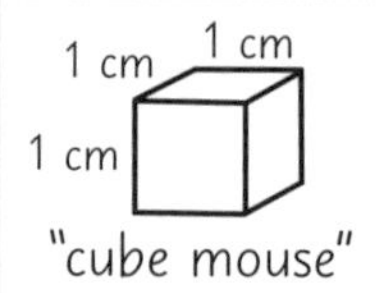

Compare this to a cube mouse measuring 1 cm × 1 cm × 1 cm.

Its **volume** is 1 x 1 x 1 = **1 cm³**

Its **surface area** is 6 x 1 x 1 = **6 cm²**

So the mouse has a **surface area : volume ratio** of **6 : 1**.

The cube mouse's surface area is six times its volume, but the cube hippo's surface area is only twice its volume. Smaller animals have a bigger surface area compared to their volume.

Multicellular Organisms need **Exchange Organs** and **Mass Transport Systems**

An organism needs to supply **every one of its cells** with substances like **glucose** and **oxygen** (for respiration). It also needs to **remove waste products** from every cell to avoid damaging itself.

1) In **single-celled** organisms, these substances can **diffuse directly** into (or out of) the cell across the cell-surface membrane. The diffusion rate is quick because of the small distances the substances have to travel (see p. 41).
2) In **multicellular** animals, diffusion across the outer membrane is **too slow**, for two reasons:
   - Some cells are **deep within the body** — there's a big distance between them and the **outside environment**.
   - Larger animals have a **low surface area to volume ratio** — it's difficult to exchange **enough** substances to supply a **large volume of animal** through a relatively **small outer surface**.

So rather than using straightforward diffusion to absorb and excrete substances, multicellular animals need specialised **exchange organs** (like lungs — see p. 62).

They also need an efficient system to carry substances to and from their individual cells — this is **mass transport**. In mammals, 'mass transport' normally refers to the **circulatory system** (see p. 74), which uses **blood** to carry glucose and oxygen around the body. It also carries **hormones**, **antibodies** (p. 46) and **waste** like CO_2. Mass transport in **plants** involves the transport of **water** and **solutes** in the **xylem** and **phloem** (see pages 82 and 84).

Size and Surface Area

Body Size and Shape Affect Heat Exchange

As well as creating **waste products** that need to be transported away, the metabolic activity inside cells creates **heat**. Staying at the right temperature is difficult, and it's pretty heavily influenced by your **size** and **shape**...

Size

The **rate of heat loss** from an organism depends on its **surface area**. If an organism has a large volume, e.g. a hippo, its surface area is relatively **small**. This makes it **harder** for it to lose heat from its body. If an organism is small, e.g. a mouse, its relative surface area is **large**, so heat is lost more **easily**. This means **smaller** organisms need a relatively **high metabolic rate**, in order to **generate** enough **heat** to stay warm.

Shape

1) Animals with a **compact** shape have a **small surface area** relative to their volume — **minimising heat loss** from their surface.
2) Animals with a **less compact** shape (those that are a bit **gangly** or have **sticky outy** bits) have a **larger surface area** relative to their volume — this **increases heat loss** from their surface.
3) Whether an animal is compact or not depends on the **temperature** of its **environment**. Here's an example:

Arctic fox
Body temperature 37 °C
Average outside temperature 0 °C

The Arctic fox has **small ears** and a **round head** to **reduce** its SA : V ratio and heat loss.

African bat-eared fox
Body temperature 37 °C
Average outside temperature 25 °C

The African bat-eared fox has **large ears** and a more **pointed nose** to **increase** its SA : V ratio and heat loss.

European fox
Body temperature 37 °C
Average outside temperature 12 °C

The European fox is **intermediate** between the two, matching the temperature of its environment.

Organisms have Behavioural and Physiological Adaptations to Aid Exchange

Not all organisms have a body size or shape to suit their climate — some have **other adaptations** instead...

1) Animals with a high SA : volume ratio tend to **lose more water** as it evaporates from their surface. Some **small desert mammals** have **kidney structure adaptations** so that they produce **less urine** to compensate.
2) To support their **high metabolic rates**, small mammals living in **cold regions** need to eat large amounts of **high energy foods** such as seeds and nuts.
3) Smaller mammals may have thick layers of **fur** or **hibernate** when the weather gets really cold.
4) **Larger organisms** living in **hot regions**, e.g. elephants and hippos, find it hard to keep cool as their heat loss is relatively slow. **Elephants** have developed **large flat ears** to **increase** their **surface area**, allowing them to lose more heat. **Hippos** spend much of the day in the **water** — a **behavioural adaptation** to help them lose heat.

Warm-Up Questions

PRACTICE QUESTIONS

Q1 Give four things that organisms need to exchange with their environment.

Q2 Describe how body shape affects heat exchange.

Exam Question

Q1 Explain why a small mammal needs a relatively high metabolic rate compared to a large mammal. [3 marks]

Cube animals indeed — it's all gone a bit Picasso...

You need to know how size and surface area to volume ratio are related, as well as what adaptations multicellular organisms have to help with exchange and transport. Don't panic, there are more adaptations coming up next...

Gas Exchange

Lots of organisms have developed adaptations to improve their rate of gas exchange. It's a tricky business if you're an insect or a plant though — you've got to exchange enough gas but avoid losing all your water and drying to a crisp...

Gas Exchange Surfaces have **Two** Major **Adaptations**

Most gas exchange surfaces have two things in common:

1) They have a **large surface area**.
2) They're **thin** (often just one layer of epithelial cells) — this provides a **short diffusion pathway** across the gas exchange surface.

The organism also maintains a **steep concentration gradient** of gases across the exchange surface.

All these features **increase** the **rate of diffusion** — see page 41.

Single-celled Organisms Exchange Gases across their **Body Surface**

1) Single-celled organisms absorb and release gases by **diffusion** through their **outer surface.**
2) They have a relatively **large surface area**, a **thin surface** and a **short diffusion pathway** (oxygen can take part in **biochemical reactions** as soon as it **diffuses** into the cell) — so there's **no need** for a gas exchange system.

Fish Use a **Counter-Current System** for Gas Exchange

There's a **lower concentration** of oxygen in water than in air. So **fish** have special **adaptations** to get enough of it.

1) Water, containing oxygen, enters the fish through its **mouth** and passes out through the gills.
2) Each gill is made of lots of **thin plates** called **gill filaments**, which give a **big surface area** for **exchange** of **gases.**
3) The gill filaments are covered in lots of tiny structures called **lamellae**, which **increase** the **surface area** even more.
4) The lamellae have lots of **blood capillaries** and a thin surface layer of cells to speed up diffusion.

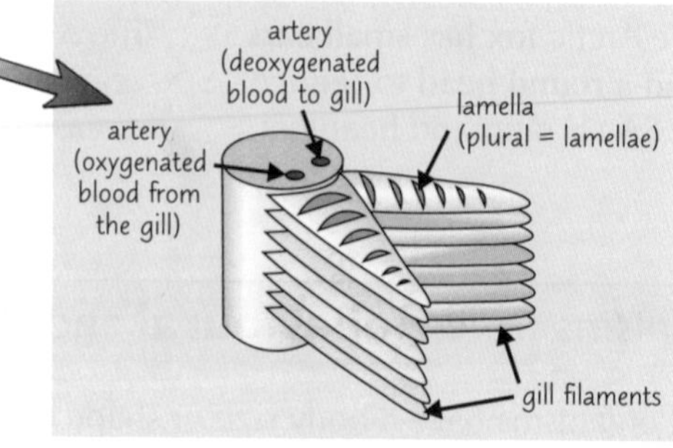

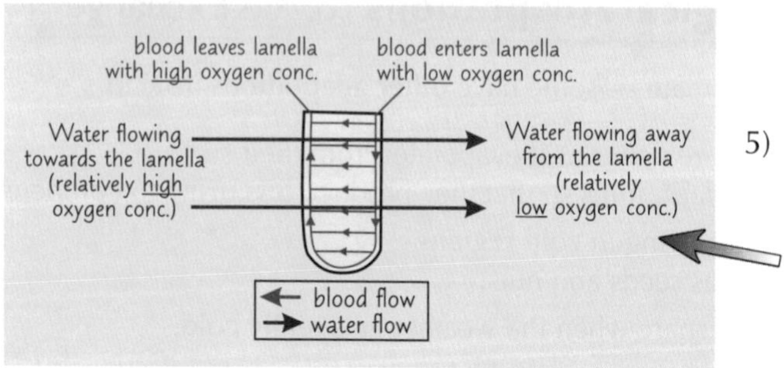

5) **Blood** flows through the lamellae in one direction and **water** flows over in the opposite direction. This is called a **counter-current system.** It maintains a **large concentration gradient** between the water and the blood. The **concentration of oxygen** in the **water** is always **higher** than that in the **blood**, so as much oxygen as possible diffuses from the water into the blood.

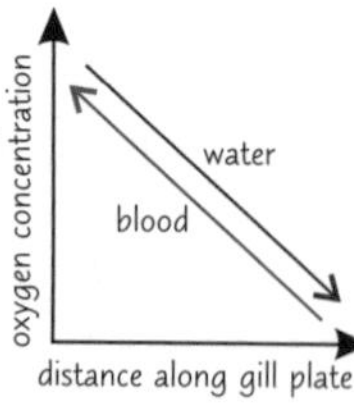

Insects use **Tracheae** to **Exchange Gases**

1) Insects have microscopic air-filled pipes called **tracheae** which they use for gas exchange.
2) Air moves into the tracheae through pores on the surface called **spiracles.**
3) **Oxygen** travels down the **concentration gradient** towards the **cells.**
4) The tracheae branch off into smaller **tracheoles** which have **thin, permeable walls** and go to individual cells. This means that oxygen diffuses directly into the respiring cells — the insect's circulatory system doesn't transport O_2.
5) **Carbon dioxide** from the cells moves down its own concentration gradient towards the **spiracles** to be **released** into the atmosphere.
6) Insects use **rhythmic abdominal movements** to move air in and out of the spiracles.

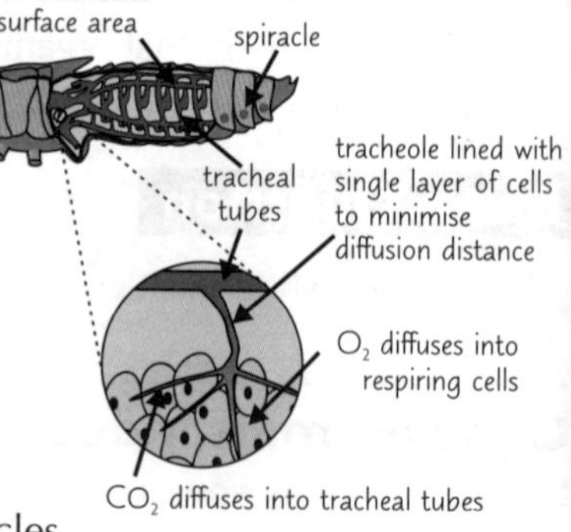

Gas Exchange

Dicotyledonous Plants Exchange Gases at the Surface of the Mesophyll Cells

1) Plants need CO_2 for **photosynthesis**, which produces O_2 as a waste gas. They need O_2 for **respiration**, which produces CO_2 as a waste gas.
2) The main gas exchange surface is the **surface of the mesophyll cells** in the leaf. They're well adapted for their function — they have a **large surface area**.
3) The mesophyll cells are inside the leaf. Gases move in and out through special pores in the **epidermis** called **stomata** (singular = stoma).
4) The stomata can **open** to allow exchange of gases, and **close** if the plant is losing too much water. **Guard cells** control the opening and closing of stomata.

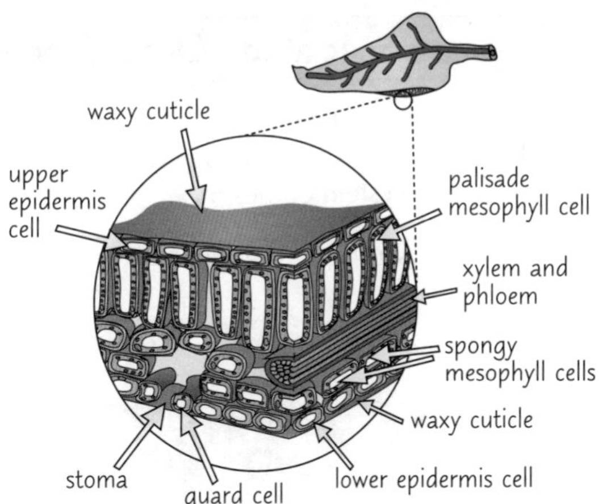

Insects and Plants can Control Water Loss

Exchanging gases tends to make you **lose water** — there's a sort of **trade-off** between the two. Luckily for plants and insects though, they've evolved **adaptations** to **minimise water loss** without reducing gas exchange too much.

1) If **insects** are losing too much water, they **close** their **spiracles** using muscles. They also have a **waterproof, waxy cuticle** all over their body and **tiny hairs** around their spiracles, both of which **reduce evaporation**.
2) Plants' stomata are usually kept **open** during the day to allow **gaseous exchange**. Water enters the guard cells, making them **turgid**, which **opens** the stomatal pore. If the plant starts to get **dehydrated**, the guard cells lose water and become **flaccid**, which **closes** the pore.
3) Some plants are specially adapted for life in **warm**, **dry** or **windy** habitats, where **water loss** is a problem. These plants are called **xerophytes**.

See p. 82 for more on water loss in plants.

Examples of **xerophytic adaptations** include:

- **Stomata** sunk in **pits** that trap moist air, reducing the concentration gradient of water between the leaf and the air. This reduces the amount of water diffusing out of the leaf and evaporating away.
- A layer of '**hairs**' on the epidermis — again to trap moist air round the stomata.
- **Curled leaves** with the **stomata inside**, protecting them from wind (windy conditions increase the rate of diffusion and evaporation).
- A **reduced number of stomata**, so there are fewer places for water to escape.
- **Waxy**, **waterproof cuticles** on leaves and stems to reduce evaporation.

sunken stomata
upper epidermis with thick waxy cuticle
lower epidermis
epidermal hairs trap humid air
curled leaf traps humid air and lowers exposed surface area

Warm-Up Questions

Q1 How are single-celled organisms adapted for efficient gas exchange?

Q2 What is the advantage to fish of having a counter-current system in their gills?

Q3 What are an insect's spiracles?

Q4 Through which pores are gases exchanged in plants?

Exam Questions

Q1 Describe, using an example, one way that gas exchange organs are adapted to their function. [2 marks]

Q2 Explain why plants that live in the desert often have sunken stomata or stomata surrounded by hairs. [2 marks]

Keep revising and you'll be on the right trachea...

There's a pretty strong theme on these pages — whatever organism it is, to exchange gases efficiently it needs exchange organs with a large surface area, a thin exchange surface and a high concentration gradient. Just don't you forget that.

Gas Exchange in Humans

In humans, gas exchange takes place in the lungs. You need to know the structure of the lungs as well as how they're ventilated... take a deep breath...

Lungs are Specialised Organs for **Gas Exchange**

Humans need to get **oxygen** into the blood (for respiration) and they need to **get rid** of **carbon dioxide** (made by respiring cells). This is where **breathing** (or **ventilation** as it's sometimes called) and the **gas exchange system** comes in.

Human Gas Exchange System

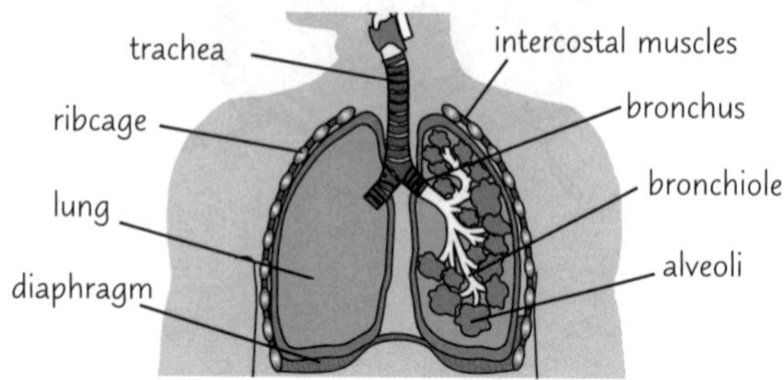

1) As you breathe in, air enters the **trachea** (windpipe).
2) The trachea splits into two **bronchi** — one **bronchus** leading to each lung.
3) Each bronchus then branches off into smaller tubes called **bronchioles**.
4) The bronchioles end in small 'air sacs' called **alveoli** (this is where gases are exchanged — see next page).
5) The **ribcage**, **intercostal muscles** and **diaphragm** all work together to move air in and out (see below).

There are actually three layers of intercostal muscles. You need to know about two of them (the internal and external intercostal muscles — see below) for your exam. We've only shown one layer here for simplicity.

Ventilation is **Breathing In** and **Out**

Ventilation consists of **inspiration** (breathing in) and **expiration** (breathing out).
It's controlled by the movements of the **diaphragm**, **internal** and **external intercostal muscles** and **ribcage**.

Inspiration

1) The **external intercostal** and **diaphragm muscles contract**.
2) This causes the **ribcage** to move **upwards and outwards** and the **diaphragm** to **flatten**, **increasing the volume** of the **thoracic cavity** (the space where the lungs are).
3) As the volume of the thoracic cavity increases, the **lung pressure decreases** (to below atmospheric pressure).
4) Air will always flow from an area of **higher pressure** to an area of **lower pressure** (i.e. down a pressure gradient) so air flows down the trachea and **into the lungs**.
5) Inspiration is an **active process** — it requires **energy**.

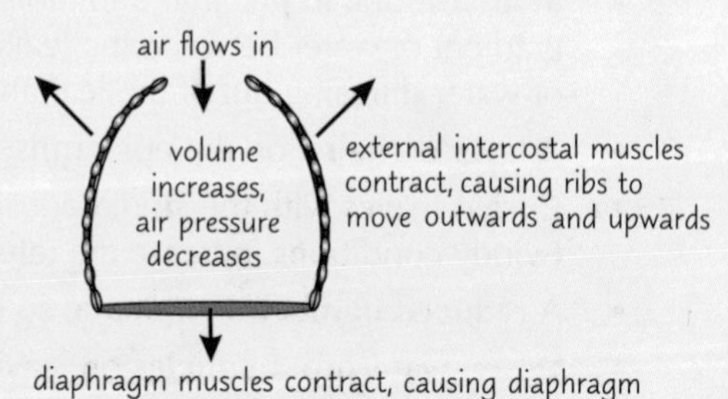

Expiration

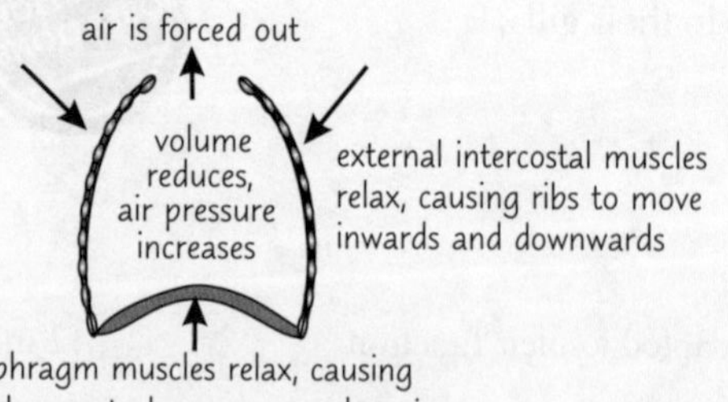

1) The **external intercostal** and **diaphragm muscles relax**.
2) The **ribcage** moves **downwards and inwards** and the **diaphragm** becomes **curved** again.
3) The **volume** of the thoracic cavity **decreases**, causing the **air pressure** to **increase** (to above atmospheric pressure).
4) Air is forced down the pressure gradient and **out of the lungs**.
5) Normal expiration is a **passive process** — it **doesn't** require energy.
6) Expiration can be **forced** though (e.g. if you want to blow out the candles on your birthday cake).
7) During forced expiration, the external intercostal muscles relax and **internal** intercostal muscles **contract**, **pulling** the **ribcage further down** and **in**. During this time, the movement of the two sets of intercostal muscles is said to be **antagonistic** (opposing).

Gas Exchange in Humans

In Humans Gaseous Exchange Happens in the Alveoli

Epithelial tissue is pretty common in the body. It's usually found on exchange surfaces.

Lungs contain millions of microscopic air sacs where gas exchange occurs — called **alveoli**. Each **alveolus** is made from a **single layer** of **thin**, **flat cells** called **alveolar epithelium**.

1) There's a huge number of alveoli in the lungs, which means there's a **big surface area** for exchanging oxygen (O_2) and carbon dioxide (CO_2).
2) The alveoli are surrounded by a network of **capillaries**.

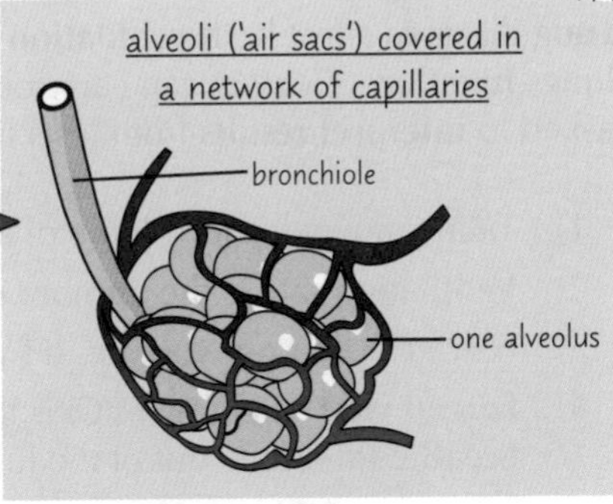

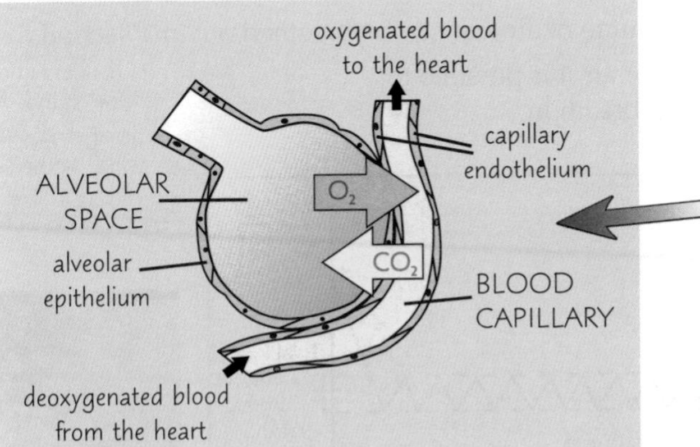

3) O_2 diffuses **out of** the alveoli, across the **alveolar epithelium** and the **capillary endothelium** (a type of epithelium that forms the capillary wall), and into **haemoglobin** (see p. 72) in the **blood**.
4) CO_2 diffuses **into** the alveoli from the blood, and is breathed out.

So, **in summary**: **oxygen** from the **air** moves down the **trachea**, **bronchi** and **bronchioles** into the **alveoli**. This movement happens **down** a **pressure gradient**. Once in the alveoli, the oxygen **diffuses** across the **alveolar epithelium**, then the **capillary endothelium**, ending up in the capillary itself. This movement happens **down** a **diffusion gradient**.

The Alveoli are Adapted for Gas Exchange

Alveoli have features that **speed up** the **rate of diffusion** so gases can be exchanged quickly:

1) **A thin exchange surface** — the **alveolar epithelium** is only **one cell thick**. This means there's a **short diffusion pathway** (which speeds up diffusion).
2) **A large surface area** — the **large number** of alveoli means there's a large surface area for gas exchange.

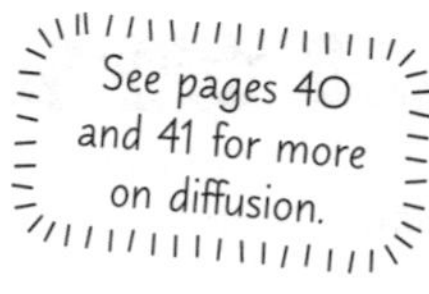

There's also a **steep concentration gradient** of oxygen and carbon dioxide between the alveoli and the capillaries, which increases the rate of diffusion. This is constantly maintained by the **flow of blood** and **ventilation**.

Warm-Up Questions

Q1 Describe the structure of the human gas exchange system.

Q2 How is normal expiration different to forced expiration?

Q3 Describe the movement of carbon dioxide and oxygen across the alveolar epithelium.

Exam Questions

Q1 Describe two ways in which lungs are adapted for efficient gas exchange. [2 marks]

Q2 Describe the process of inspiration. [4 marks]

Alveoli — useful things... always make me think about pasta...

A mammal's lungs act as an interface with the environment — they take in air and give out waste gases. Ventilation moves these gases into and out of the lungs, but the alveoli have the task of getting them in and out of the bloodstream. Luckily, like many other biological structures, they're well adapted for doing their job.

The Effects of Lung Disease

It's all very well when your lungs are working perfectly, but some pathogens (and even your lifestyle) can muck them up good and proper. This can make it more difficult to breathe and reduce the rate of gas exchange.

Measures of Lung Function Can Help to Diagnose Lung Diseases

Lung diseases affect both **ventilation** (breathing) and **gas exchange** in the lungs — in other words, how well the lungs **function**. Doctors can carry out **tests** to investigate lung function and diagnose a lung disease. You might be asked to **interpret results** from tests like these in your exams. Here are some **terms** you might come across:

1) **Tidal volume** is the volume of air in **each breath** — usually between **0.4 dm³** and **0.5 dm³** for adults.
2) **Ventilation rate** is the **number of breaths per minute**. For a healthy person at rest it's about **15 breaths**.
3) **Forced expiratory volume$_1$** (FEV_1) is the maximum volume of air that can be breathed out in **1 second**.
4) **Forced vital capacity** (FVC) is the **maximum volume of air** it is possible to **breathe forcefully out** of the lungs after a really deep breath in.

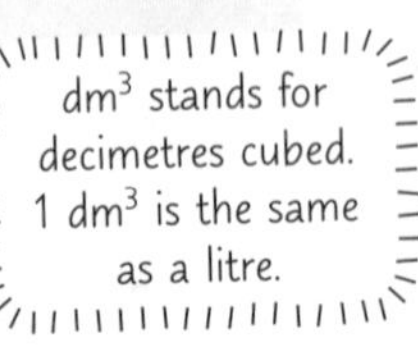

You can figure out **tidal volume**, **ventilation rate** and other measures of breathing from the graph produced from a **spirometer** (a fancy machine that scientists and doctors use to measure the volume of air breathed in and out):

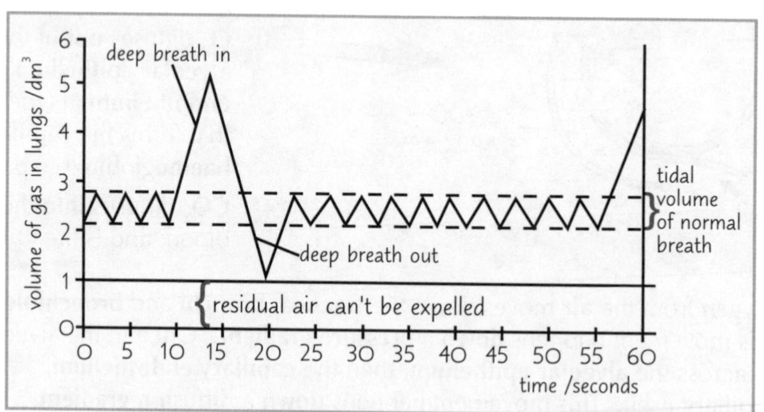

Measuring tidal volume is one of the hardest jobs in the world.

Different Diseases Affect the Lungs in Different Ways

Here are some examples of different **lung diseases** and how they affect breathing.

Pulmonary Tuberculosis (TB)

1) When someone becomes infected with **tuberculosis bacteria**, immune system cells build a **wall** around the bacteria in the **lungs**. This forms small, hard lumps known as **tubercles**.
2) Infected tissue within the tubercles **dies** and the **gaseous exchange surface** is **damaged**, so **tidal volume** is **decreased**.
3) Tuberculosis also causes **fibrosis** (see below), which further **reduces** the **tidal volume**.
4) A reduced tidal volume means **less air** can be **inhaled** with each breath. In order to take in enough oxygen, patients have to **breathe faster**, i.e. **ventilation rate** is **increased**.
5) Common symptoms include a persistent **cough**, coughing up **blood** and **mucus**, **chest pains**, **shortness of breath** and **fatigue**.

Fibrosis

1) Fibrosis is the formation of **scar tissue** in the lungs. This can be the result of an **infection** or exposure to substances like **asbestos** or **dust**.
2) Scar tissue is **thicker** and **less elastic** than normal lung tissue.
3) This means that the lungs are **less able to expand** and so **can't hold as much air** as normal — **tidal volume** is **reduced**, and so is **FVC** (i.e. a smaller volume of air can be forcefully breathed out).
4) There's a **reduction** in the rate of **gaseous exchange** — **diffusion** is **slower** across a **thicker** scarred membrane.
5) Symptoms of fibrosis include **shortness of breath**, a **dry cough**, **chest pain**, **fatigue** and **weakness**.
6) Fibrosis sufferers have a **faster ventilation rate** than normal — to get enough air into their lungs to **oxygenate** their blood.

The Effects of Lung Disease

Asthma

1) Asthma is a respiratory condition where the airways become **inflamed** and **irritated**. The causes vary from case to case but it's usually because of an **allergic reaction** to substances such as **pollen** and **dust**.
2) During an asthma attack, the **smooth muscle** lining the **bronchioles contracts** and a large amount of **mucus** is produced.
3) This causes **constriction** of the airways, making it difficult for the sufferer to **breathe properly**. **Air flow** in and out of the lungs is **severely reduced**, so less oxygen enters the alveoli and moves into the blood. Reduced air flow means that **FEV_1** is severely **reduced** (i.e. less air can be breathed out in 1 second).
4) Symptoms include **wheezing**, a **tight chest** and **shortness of breath**. During an attack the symptoms come on very suddenly. They can be relieved by **drugs** (often in **inhalers**) which cause the muscle in the bronchioles to **relax**, opening up the airways.

Emphysema

1) Emphysema is a lung disease caused by **smoking** or long-term exposure to **air pollution** — foreign particles in the smoke (or air) become **trapped** in the alveoli.
2) This causes **inflammation**, which attracts **phagocytes** to the area. The phagocytes produce an **enzyme** that breaks down **elastin** (a protein found in the **walls** of the **alveoli**).

See p. 46 for more on phagocytes.

3) Elastin is **elastic** — it helps the alveoli to **return** to their **normal shape** after inhaling and exhaling air.
4) Loss of elastin means the alveoli **can't recoil** to **expel air** as well (it remains **trapped** in the alveoli).
5) It also leads to **destruction** of the **alveoli walls**, which **reduces** the **surface area** of the alveoli, so the rate of **gaseous exchange** decreases.
6) Symptoms of emphysema include **shortness of breath** and **wheezing**. People with emphysema have an **increased ventilation rate** as they try to increase the amount of air (containing oxygen) reaching their lungs.

TB, fibrosis, asthma and emphysema all **reduce the rate of gas exchange** in the alveoli. Less oxygen is able to diffuse into the bloodstream, the body cells **receive less oxygen** and the rate of **aerobic respiration** is **reduced**. This means **less energy is released** and sufferers often feel **tired** and **weak**.

Warm-Up Questions

PRACTICE QUESTIONS

Q1 What is tidal volume?

Q2 What happens to the lung tissue of someone with TB?

Q3 What happens to the alveoli of someone who suffers from emphysema?

Exam Question

Q1 FVC (forced vital capacity) is the maximum amount of air it is possible to expel from the lungs after a deep breath in. A hospital patient has emphysema. The patient has a lower FVC than normal.

a) Explain how emphysema could reduce FVC. [2 marks]

FEV_1 is the maximum volume of air that can be breathed out in 1 second. FEV_1 is around 80% of FVC in a healthy adult. The emphysema patient has an FVC of 3.2 dm^3 and a FEV_1 of 1.7 dm^3.

b) Calculate FEV_1 as a percentage of FVC in the emphysema patient. [1 mark]

c) In a fibrosis patient, FEV_1 is close to 80% of FVC even though FVC is reduced. Suggest an explanation for this. [1 mark]

Spirometers — they're not machines for measuring spirals...

The examiners like to mix things up a bit, so you could get asked questions about a lung disease you've not come across before or a measure of lung function you've not heard of. If so, take a deep breath and don't panic — the question should give you all the information you need, then it's just a case of applying what you already know to answer it.

Interpreting Lung Disease Data

It's very possible that you could be asked to interpret some data on lung disease in the exam. So being my usual nice self, I've given you some examples to show you how to do it. I know it looks a bit dull but believe me, it'll really help

You Need to be Able to **Interpret Data** on **Risk Factors** and **Lung Disease**

1) All diseases have factors that will **increase** a person's **chance** of getting that disease. These are called **risk factors**. For example, it's widely known that if you **smoke** you're more likely to get **lung cancer** (smoking is a risk factor for lung cancer).
2) This is an example of a **correlation** — a link between two things (see page 229). However, a correlation doesn't always mean that one thing **causes** the other. Smokers have an **increased risk** of getting cancer but that doesn't necessarily mean smoking **causes** the disease — there are lots of other factors to take into consideration.
3) You need to be able to describe and analyse data given to you in your exams. Here's an example of the sort of thing you might get:

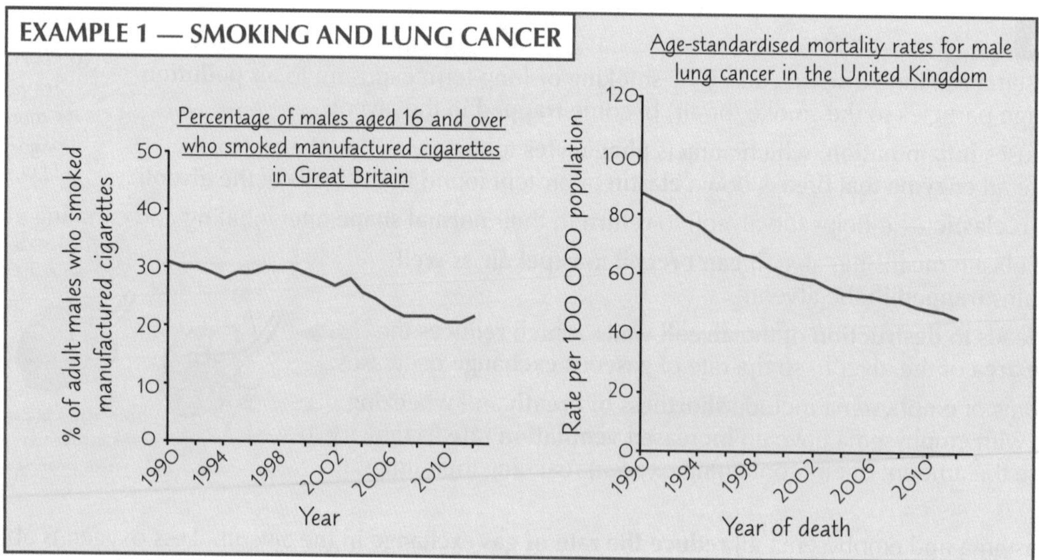

You might be asked to:

1) <u>**Describe the data**</u> — The graph on the left shows that the **number** of adult males in Great Britain who **smoke decreased** between 1990 and 2012. The graph on the right shows that the male lung cancer **mortality (death) rate decreased** between 1990 and 2012 in the United Kingdom.

2) <u>**Draw conclusions**</u> — You need to be careful what you say here. There's a **correlation** (link) between the **number** of males **who smoked** and the **mortality rate** for male lung cancer. But you **can't** say that one **caused** the other. There could be **other reasons** for the trend, e.g. deaths due to lung cancer may have decreased because less asbestos was being used in homes (not because fewer people were smoking).

<u>**Other points to consider**</u> — The graph on the right shows mortality (**death**) rates. The rate of **cases** of lung cancer **may have been increasing** but medical advances may mean more people were **surviving** (so only mortality was decreasing).

You might also need to **evaluate** the way in which **scientific data** has led to **government restrictions** on the **sources** of **risk factors**. E.g.

Responses to data

Medical studies in 1950s and 1960s documented the **link** between **smoking** and various forms of **cancer**, particularly lung cancer. The evidence prompted the first **voluntary agreement** between the UK government and tobacco companies in 1971, which stated that tobacco products and adverts should carry a **health warning label**. As of October 2008, **picture health warnings** were made **compulsory** on all UK boxes of cigarettes after studies suggested they were more effective than written warnings alone.

Interpreting Lung Disease Data

EXAMPLE 2 — AIR POLLUTION AND ASTHMA

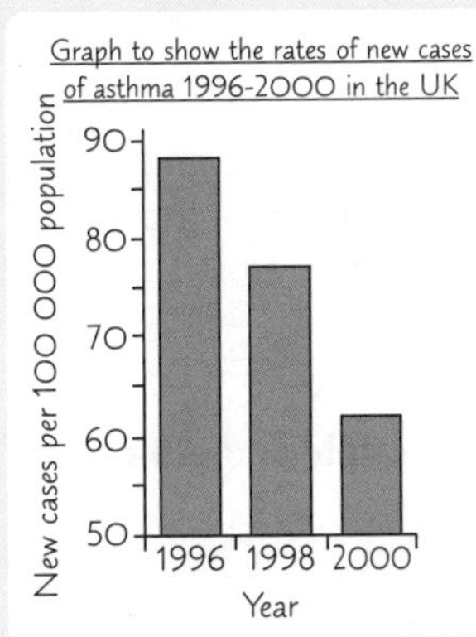

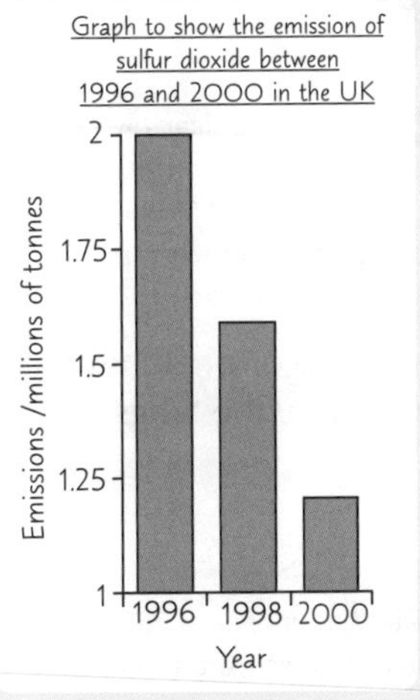

The **top graph** shows the number of **new cases of asthma** per 100 000 of the population diagnosed in the UK from 1996 to 2000. The **bottom graph** shows the **emissions** (in millions of tonnes) of **sulfur dioxide** (an **air pollutant**) from 1996 to 2000 in the UK.

You might be asked to describe the data...

1) The **top graph** shows that the number of **new cases of asthma** in the UK **fell** between 1996 and 2000, from 87 to 62 per 100 000 people.
2) The **bottom graph** shows that the **emissions of sulfur dioxide** in the UK **fell** between 1996 and 2000, from 2 to 1.2 million tonnes.

... **or draw conclusions**

1) Be careful what you say when drawing conclusions. Here there's a **link** between the **number** of new cases of **asthma** and **emissions** of **sulfur dioxide** in the **UK** — the rate of new cases of asthma has **fallen** as sulfur dioxide emissions have **fallen**. You **can't** say that one **causes** the other though because there could be **other reasons** for the trend, e.g. the number of new cases of asthma could be falling due to the **decrease** in the number of people **smoking**.
2) You can't say the **reduction** in asthma cases is **linked** to a **reduction in air pollution** (in general) either as **only** sulfur dioxide levels were studied.

Other points to consider:

1) The top graph shows **new cases** of asthma. The rate of new cases may be **decreasing** but existing cases may be becoming **more severe**.
2) The emissions were for the whole of the UK but air pollution **varies from area to area**, e.g. **cities** tend to be **more polluted**.
3) The asthma data doesn't take into account any **other factors** that may **increase** the risk of developing asthma, e.g. allergies, smoking, etc.

Responses to data

In response to **studies** connecting **air pollution** to various **diseases**, the EU adopted the **National Emissions Ceilings Directive**. This set **upper limits** on the total emissions of **four major pollutants** in the **atmosphere**, to be achieved by **2010**. **New limits** are being agreed on for **2020**. The EU also introduced the **Clean Power for Transport Package** to promote **cleaner fuels** for vehicles, and the UK **taxes car owners** according to their car's **emissions**.

Warm-Up Question

Q1 Give an example of where scientific data has led to restrictions on the source of a risk factor in lung disease.

PRACTICE QUESTIONS

Exam Question

Q1 In early December 1952, a dense layer of cold air trapped pollutants close to ground level in London. The graph opposite shows daily deaths and levels of sulfur dioxide and smoke between 1st and 15th December.

a) Describe the changes in the daily death rate and the levels of pollutants over the days shown. [3 marks]

b) What conclusion can be drawn from this graph? [1 mark]

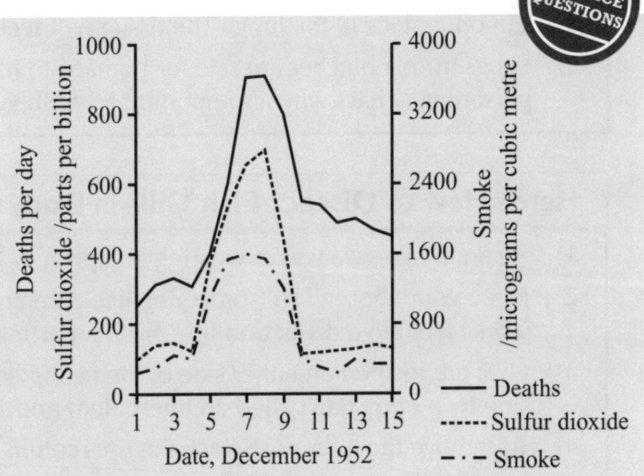

Drawing conclusions — you'll need your wax crayons and some paper...

These pages give examples to help you deal with what the examiners are sure to hurl at you — and boy, do they like throwing data around. There's some important advice here (even if I say so myself) — it's easy to leap to a conclusion that isn't really there — stick to your guns about the difference between correlation and cause and you'll be just fine.

Dissecting Gas Exchange Systems

After learning all about how different organisms are adapted for efficient gas exchange, you might be wondering what those structures really look like. Well, here are some lovely dissections that you might do. How exciting.

You Need to be Able to **Carry Out Dissections**

1) As part of your AS or A level in Biology, you're expected to carry out at least one **dissection**. It could be a dissection of a **gaseous exchange system** or a **mass transport system** (or an organ within one of those systems) in either an **animal** or a **plant**. You could also be asked about dissections in your exams.
2) There's an example of a **plant dissection** that you could do on page 83. These two pages cover some **animal dissections** that you could do as well or instead.
3) Whether it's a plant or animal dissection, you're expected to know how to carry it out **safely** and **ethically**. You might also need to **record** your **observations** using **labelled diagrams**.

You Can **Dissect** the **Gaseous Exchange Systems** of **Animals**

Lungs Can be **Dissected** To Show the Main **Structures**

You can learn more about the lungs on pages 62-63.

1) First up, lung dissection is **messy**, so make sure you're wearing a **lab coat**. Your **dissecting tools** (e.g. scalpels, dissecting scissors) should all be **clean**, **sharp** and **free from rust** — blunt tools **don't cut well** and can be **dangerous**.
2) Lay the **lungs** your teacher has given you on a **cutting board**. They'll probably be sheep or pig lungs from a butcher's shop. You should be able to see the **trachea** and two **bronchi** going into the lungs.
3) To see the **lungs inflate**, attach a piece of **rubber tubing** to the **trachea** and pump air into the lungs using a **foot** or **bicycle pump**. The lungs will **deflate** by themselves because of the **elastin** in the walls of the **alveoli** (see p. 65).

   Never blow down the tube to inflate the lungs — you could end up sucking up stale air from inside the lungs into your mouth. Pop the lungs in a clear plastic bag before you start to stop bacteria inside the lungs from being released into the room.
4) Once you've seen the lungs inflate, you can examine the different **tissue types** in the lungs.
5) The **trachea** is supported by **C-shaped rings** of **cartilage**. A **cross-section** of the trachea looks like this:

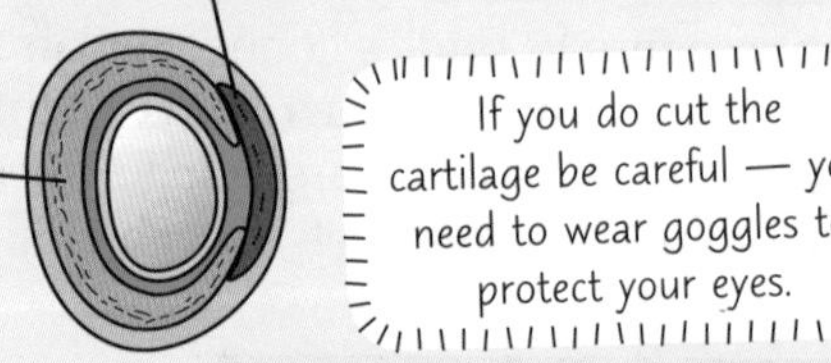

6) **Cartilage** is **tough**, so if you want to **open up** the **trachea**, it's best to cut it **lengthways**, down the **gap** in the **C-shaped rings**. Use **dissecting scissors** or a **scalpel** to make the cut. If using a scalpel, **cut downwards** (not towards you) and **don't apply** too much **pressure** to the blade.

   If you do cut the cartilage be careful — you need to wear goggles to protect your eyes.
7) Continue cutting down one of the **bronchi**. You should be able to see the **bronchioles** branching off.
8) Cut off a piece of the lung. The tissue will **feel spongy** because of the air trapped in all the **alveoli**.
9) Lungs from a butcher are safe for humans to handle, but they could still contain **bacteria** that cause **food poisoning**. That's why it's important to **wash your hands** after the dissection and **disinfect work surfaces**.

Here's How to **Dissect Fish Gills** in **Bony Fish**

1) Again, make sure you're wearing an **apron** or **lab coat**.
2) Place your chosen fish (something like a perch or salmon works well) in a **dissection tray** or on a **cutting board**.
3) Gills are located on either side of the fish's head. They're protected on each side by a **bony flap** called an **operculum** and supported by **gill arches**.
4) To remove the gills, **push back** the **operculum** and use **scissors** to carefully **remove the gills**. Cut each **gill arch** through the bone at the **top** and **bottom**. They should look a bit like this:
5) If you look closely, you should be able to see the **gill filaments**.

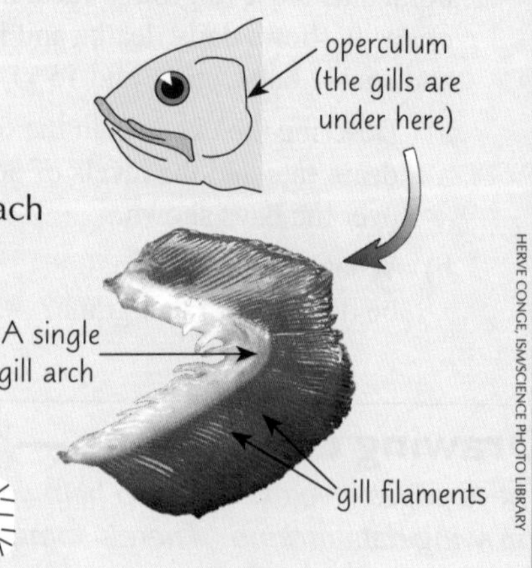

For more information about the structure and function of fish gills, see page 60.

Dissecting Gas Exchange Systems

You Can **Dissect** the **Gaseous Exchange System** in **Insects** too

You can find more information about the gas exchange systems of insects on page 60.

Big insects like **grasshoppers** or **cockroaches** are usually **best** for dissecting because they're easier to handle. For dissection, you'll need to use an insect that's been humanely killed **fairly recently**.

1) First fix the insect to a **dissecting board**. You can put **dissecting pins** through its legs to hold it in place.
2) To examine the **tracheae**, you'll need to carefully **cut** and **remove** a piece of **exoskeleton** (the insect's hard outer shell) from along the length of the insect's **abdomen**.
3) Use a syringe to fill the abdomen with **saline solution**. You should be able to see a network of **very thin**, **silvery-grey tubes** — these are the **tracheae**. They look silver because they're **filled** with **air**.
4) You can examine the tracheae under an **optical microscope** using a **temporary mount slide** (see p. 32). Again, the tracheae will appear silver or grey. You should also be able to see **rings** of **chitin** in the walls of the **tracheae** — these are there for **support** (like the rings of cartilage in a human trachea).

Some live insects, e.g. grasshoppers, can cause allergic reactions in some people. They need to be handled very carefully.

There are Some **Ethical Issues** Involved in **Dissecting Animals**

Dissecting animals (including **fish** and **insects**) can give you a **better understanding** of their anatomy. However, there are some **ethical issues** involved. Here are some points to think about:

1) Some people argue that it is **morally wrong** to **kill** animals **just for dissections**, as it is **unnecessary** killing. However many dissections that are carried out in schools involve animals that have **already been killed** for their **meat**, e.g. the sheep's lung dissection on the previous page. (Some people disagree with killing animals altogether though.)
2) There are concerns that the animals used for dissections are **not** always **raised in a humane way** — they may be subject to **overcrowding**, **extremes of temperature** or **lack of food** — and they may **not** be **killed humanely** either. If animals (e.g. insects) are raised in school for dissection, it's important to make sure they are **looked after properly** and killed humanely to **minimise** any **suffering** or **distress**.

Warm-Up Questions

Q1 Why is it important that dissecting tools are sharp?

Q2 Describe how to remove the gills in a bony fish.

Exam Questions

Q1 A student is examining grasshopper tracheae under the microscope. The tracheae were taken from a preserved grasshopper specimen. The grasshopper was killed some time ago and kept in a liquid preservative. The tracheae do not appear silver under the microscope. Instead they are a dark grey. Suggest why this is the case. [1 mark]

Q2 A student is performing a dissection of a pig's lungs.

a) The student cuts off a piece of lung tissue and drops it into a beaker of water. The lung tissue floats in the water. Explain why it floats. [1 mark]

b) Give one safety precaution the student should take when carrying out this dissection. [1 mark]

Dissection tools should be like your mind — clean and sharp...

Well, that was all a bit grim. But don't look at it like that — dissections are all about cutting open organisms, so you can see everything you've been learning about and hopefully understand it better. You should be able to apply your knowledge of gas exchange systems (see pages 60-63) to any dissections you get asked about in the exams.

Digestion and Absorption

The whole point of digestion is to break down the food you eat into small molecules that your cells can absorb. As you might imagine, this involves loads of different chemical reactions and our old friends, enzymes.

Food is Broken Down into Smaller Molecules During Digestion

1) The **large biological molecules** (e.g. starch, proteins) in food are **too big** to cross **cell membranes**. This means they **can't** be **absorbed** from the gut into the blood.
2) During digestion, these large molecules are **broken down** into **smaller molecules** (e.g. glucose, amino acids), which **can** move across cell membranes. This means they can be **easily absorbed** from the gut into the blood, to be **transported** around the body for use by the body cells.
3) You might remember from Topic 1A, that most **large biological molecules** are **polymers**, which can be **broken down** into **smaller** molecules (**monomers**) using **hydrolysis reactions**. Hydrolysis reactions **break bonds** by **adding water**.
4) During hydrolysis, **carbohydrates** are broken down into **disaccharides** and then **monosaccharides**. **Fats** are broken down into **fatty acids** and **monoglycerides**. **Proteins** are broken down into **amino acids**.

Digestive Enzymes are Used to Break Down Biological Molecules in Food

1) A variety of different **digestive enzymes** are produced by **specialised cells** in the **digestive systems** of mammals. These enzymes are then released into the gut to mix with food.
2) Since enzymes only work with **specific substrates** (see page 11), **different enzymes** are needed to **catalyse** the breakdown of **different food molecules**.

Carbohydrates are Broken Down by Amylase and Membrane-Bound Disaccharidases

1) **Amylase** is a digestive enzyme that catalyses the conversion of **starch** (a polysaccharide) into the smaller sugar **maltose** (a disaccharide). This involves the **hydrolysis** of the **glycosidic bonds** in starch.
2) Amylase is produced by the **salivary glands** (which release amylase into the **mouth**) and also by the **pancreas** (which releases amylase into the **small intestine**).
3) **Membrane-bound disaccharidases** are enzymes that are attached to the **cell membranes** of **epithelial cells** lining the **ileum** (the final part of the small intestine). They help to break down **disaccharides** (e.g. maltose, sucrose and lactose) into **monosaccharides** (e.g. glucose, fructose and galactose). Again, this involves the hydrolysis of glycosidic bonds.

Disaccharide	Disaccharidase	Monosaccharide Products
maltose	maltase	glucose + glucose
sucrose	sucrase	glucose + fructose
lactose	lactase	glucose + galactose

There's more on polysaccharides, disaccharides and monosaccharides on pages 2-5.

4) **Monosaccharides** can be transported across the cell membranes of the ileum epithelial cells via specific **transporter proteins** (see next page).

Lipids are Broken Down by Lipase (with the Help of Bile Salts)

1) **Lipase** enzymes catalyse the breakdown of **lipids** into **monoglycerides** and **fatty acids**. This involves the **hydrolysis** of the **ester bonds** in lipids.
2) Lipases are made in the **pancreas**. They work in the **small intestine**.
3) **Bile salts** are produced by the **liver** and **emulsify** lipids — this means they cause the lipids to form **small droplets**.
4) Bile salts are really important in the process of lipid digestion. **Several small lipid droplets** have a **bigger surface area** than a **single large droplet** (for the same volume of lipid). So the formation of small droplets greatly increases the surface area of lipid that's available for **lipases** to work on.

A monoglyceride is a glycerol molecule with one fatty acid attached.

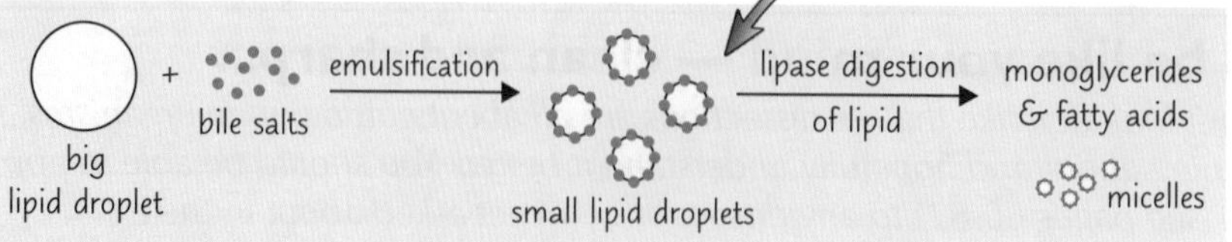

5) Once the lipid has been broken down, the **monoglycerides** and **fatty acids** stick with the **bile salts** to form tiny structures called **micelles**.

Digestion and Absorption

Proteins are Broken Down by **Endopeptidases** and **Exopeptidases**

Proteins are broken down by a combination of different **proteases** (or **peptidases**). These are enzymes that catalyse the conversion of **proteins** into **amino acids** by **hydrolysing** the **peptide bonds** between amino acids. You need to know about **endopeptidases** and **exopeptidases** (including **dipeptidases**):

Endopeptidases

- Endopeptidases act to hydrolyse peptide bonds **within** a protein.
- **Trypsin** and **chymotrypsin** are two examples of endopeptidases. They're synthesised in the **pancreas** and secreted into the **small intestine**.
- **Pepsin** is another endopeptidase. It's released into the **stomach** by cells in the **stomach lining**. Pepsin only works in **acidic conditions** — these are provided by **hydrochloric acid** in the **stomach**.

Remember: endopeptidases break bonds inside the protein.

Exopeptidases

- Exopeptidases act to hydrolyse peptide bonds **at the ends** of protein molecules. They remove **single amino acids** from proteins.
- **Dipeptidases** are exopeptidases that work specifically on **dipeptides**. They act to separate the two amino acids that make up a dipeptide by **hydrolysing** the **peptide bond** between them.
- Dipeptidases are often located in the **cell-surface membrane** of **epithelial cells** in the **small intestine**.

The **Products** of **Digestion** are **Absorbed Across Cell Membranes**

The products of digestion are absorbed across the **ileum epithelium** into the bloodstream.

Monosaccharides

- **Glucose** is absorbed by **active transport** with **sodium ions** via a **co-transporter protein** (see page 45). **Galactose** is absorbed in the same way using the same co-transporter protein.
- **Fructose** is absorbed via **facilitated diffusion** through a different transporter protein.

Monoglycerides and fatty acids

Micelles (see previous page) help to **move** monoglycerides and fatty acids **towards** the **epithelium**. Because micelles constantly break up and reform they can '**release**' monoglycerides and fatty acids, allowing them to be absorbed — whole micelles are **not** taken up across the epithelium. **Monoglycerides** and **fatty acids** are **lipid-soluble**, so can **diffuse** directly across the epithelial cell membrane.

Amino acids

Amino acids are absorbed via **co-transport**, in a similar way to glucose and galactose. **Sodium ions** are **actively transported** out of the ileum epithelial cells into the blood. This creates a sodium ion **concentration gradient**. Sodium ions can then **diffuse** from the lumen of the ileum into the epithelial cells through **sodium-dependent transporter proteins**, carrying the amino acids with them.

Warm-Up Questions

PRACTICE QUESTIONS

Q1 What is the function of amylase in digestion?

Q2 Describe the role of bile salts in lipid digestion.

Exam Question

Q1 Some people suffer from lactose intolerance.
This can be caused by an inability to break down lactose in the upper small intestine.

a) Suggest which disaccharidase enzyme is deficient or missing in people who are lactose-intolerant. [1 mark]

b) How are the digestion products of lactose absorbed across the epithelial cells of the ileum? [2 marks]

Crikey, this all looks a bit tricky to digest... belch...

Don't panic. There's a lot to take in here but as long as you break it down a bit (ha, just like digestion) then it's not too bad. You can't escape learning what all the enzymes act on, but helpfully their names are usually linked to what they do — maltase breaks down maltose, dipeptidases break down dipeptides. See, it's not as bad as it looks...

Haemoglobin

Haemoglobin's a protein that carries oxygen around the body. Different species have different versions of it depending on where each species lives. All of which adds up to two pages of no-holds-barred fun...

Oxygen is **Carried** Round the Body by **Haemoglobin**

1) **Red blood cells** contain **haemoglobin** (Hb).
2) Haemoglobin is a large **protein** with a **quaternary** structure (see p. 8 for more) — it's made up of **more than one** polypeptide chain (**four** of them in fact).
3) Each chain has a **haem group**, which contains an **iron ion** (see page 23) and gives haemoglobin its **red** colour.
4) Haemoglobin has a **high affinity for oxygen** — each molecule can carry **four oxygen molecules**.
5) In the lungs, oxygen **joins** to haemoglobin in red blood cells to form **oxyhaemoglobin**.
6) This is a **reversible reaction** — when oxygen leaves oxyhaemoglobin (**dissociates** from it) near the body cells, it turns back to haemoglobin.

'Affinity' for oxygen means 'tendency to combine with' oxygen.

$$\text{Hb} + 4\text{O}_2 \rightleftharpoons \text{HbO}_8$$
$$\text{haemoglobin} + \text{oxygen} \rightleftharpoons \text{oxyhaemoglobin}$$

There are many **chemically similar** types of haemoglobin found in many different organisms, all of which carry out the **same function**. As well as being found in all vertebrates, haemoglobin is found in earthworms, starfish, some insects, some plants and even in some bacteria.

Haemoglobin Saturation Depends on the **Partial Pressure** of **Oxygen**

1) The **partial pressure** of **oxygen** (**pO_2**) is a measure of **oxygen concentration**. The **greater** the concentration of dissolved oxygen in cells, the **higher** the partial pressure.
2) Similarly, the **partial pressure** of **carbon dioxide** (**pCO_2**) is a measure of the concentration of CO_2 in a cell.
3) Haemoglobin's **affinity** for oxygen **varies** depending on the **partial pressure** of **oxygen**:

> Oxygen **loads onto** haemoglobin to form oxyhaemoglobin where there's a **high pO_2**.
> Oxyhaemoglobin **unloads** its oxygen where there's a **lower pO_2**.

4) Oxygen enters blood capillaries at the **alveoli** in the **lungs**. Alveoli have a **high pO_2** so oxygen **loads onto** haemoglobin to form oxyhaemoglobin.
5) When **cells respire**, they use up oxygen — this **lowers the pO_2**. Red blood cells deliver oxyhaemoglobin to respiring tissues, where it unloads its oxygen.
6) The haemoglobin then returns to the lungs to pick up more oxygen.

There was no use pretending — the partial pressure of CH_4 had just increased, and Keith knew who was to blame.

Dissociation Curves Show How **Affinity for Oxygen** Varies

A **dissociation curve** shows how **saturated** the haemoglobin is with oxygen at any given partial pressure.

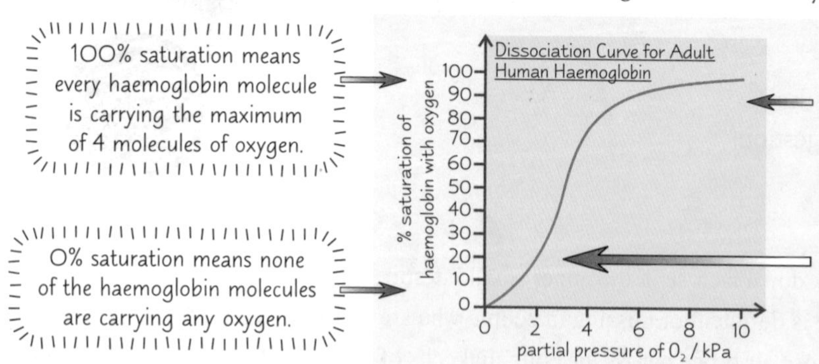

Where **pO_2 is high** (e.g. in the lungs), haemoglobin has a **high affinity** for oxygen (i.e. it will **readily combine** with oxygen), so it has a **high saturation** of oxygen.

Where **pO_2 is low** (e.g. in respiring tissues), haemoglobin has a **low affinity** for oxygen, which means it **releases oxygen** rather than combines with it. That's why it has a **low saturation** of oxygen.

The graph is '**S-shaped**' because when haemoglobin (Hb) combines with the **first O_2 molecule**, its **shape alters** in a way that makes it **easier** for other molecules to join too. But as the Hb starts to become saturated, it gets **harder** for more oxygen molecules to join. As a result, the curve has a **steep** bit in the middle where it's really easy for oxygen molecules to join, and **shallow** bits at each end where it's harder. When the curve is steep, a **small change in pO_2** causes a **big change** in the **amount of oxygen** carried by the Hb.

Haemoglobin

Carbon Dioxide Concentration Affects Oxygen Unloading

To complicate matters, haemoglobin gives up its oxygen **more readily** at **higher partial pressures of carbon dioxide** (pCO_2). It's a cunning way of getting more oxygen to cells during activity.

1) When cells respire they produce carbon dioxide, which **raises the pCO_2**.
2) This increases the rate of **oxygen unloading** (i.e. the rate at which oxyhaemoglobin **dissociates** to form haemoglobin and oxygen) — so the dissociation curve '**shifts**' right. The saturation of blood with oxygen is **lower** for a given pO_2, meaning that **more oxygen** is being **released**.
3) This is called the **Bohr effect**.

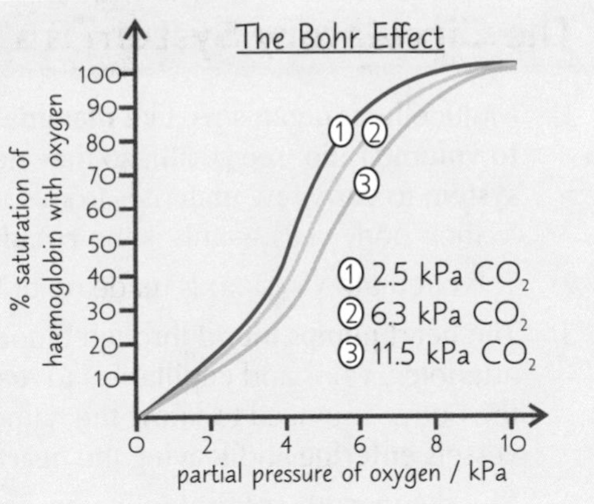

Haemoglobin is Different in Different Organisms

Different organisms have different **types** of haemoglobin with different **oxygen transporting capacities**. Having a particular type of haemoglobin is an **adaptation** that helps the organism to **survive** in a **particular environment**.

1) Organisms that live in environments with a **low concentration of oxygen** have haemoglobin with a **higher affinity** for oxygen than human haemoglobin — the dissociation curve is to the **left** of ours.
2) Organisms that are very **active** and have a **high oxygen demand** have haemoglobin with a **lower affinity** for oxygen than human haemoglobin — the curve is to the **right** of the human one.

A = animal living in depleted oxygen environment, e.g. a lugworm.
B = animal living at high altitude where the partial pressure of oxygen is lower, e.g. a llama in the Andes.
C = human dissociation curve.
D = active animal with a high respiratory rate living where there's plenty of available oxygen, e.g. a hawk.

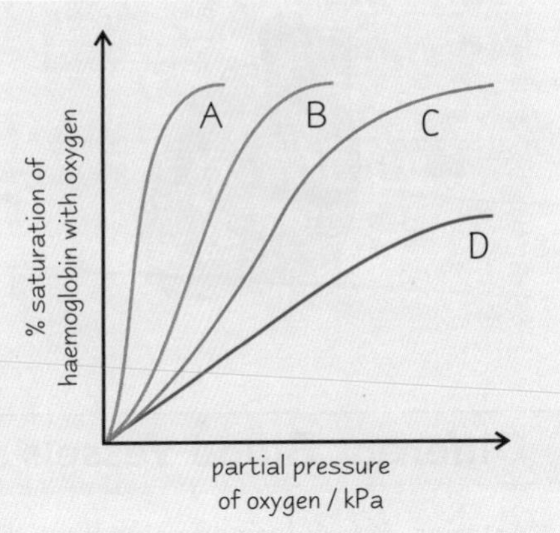

Warm-Up Questions

Q1 How many oxygen molecules can each haemoglobin molecule carry?
Q2 Where in the body would you find a low partial pressure of oxygen?
Q3 Why are oxygen dissociation curves S-shaped?

PRACTICE QUESTIONS

Exam Question

Q1 a) Haemoglobin is a protein with a quaternary structure. Explain what this means. [1 mark]

b) The graph shows the normal oxygen dissociation curve for human haemoglobin.

i) On the graph, sketch the curve you would expect to see for a human in a high carbon dioxide environment. Explain the position of your sketched curve. [3 marks]

ii) Earthworms live in a low oxygen environment. On the graph, sketch the curve you would expect to see for an earthworm. [1 mark]

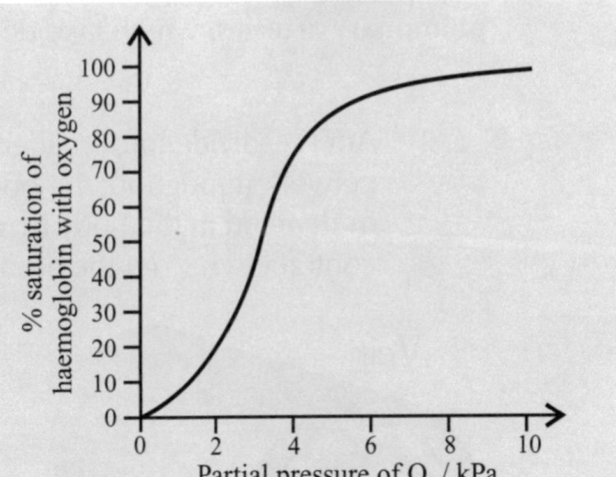

There's more than partial pressure on you to learn this stuff...

Well, I don't know about you but after these pages I need a sit down. Most people get their knickers in a twist over partial pressure — it's not the easiest thing to understand. Whenever you see it written down just pretend it says concentration instead (cross it out and write concentration if you like) and everything should become clearer. Honest.

The Circulatory System

As the name suggests, the circulatory system is responsible for circulating stuff around the body — blood, to be specific. Most multicellular organisms (mammals, insects, fish) have a circulatory system of some type.

The **Circulatory System** is a **Mass Transport System**

1) Multicellular organisms, like **mammals**, have a **low surface area to volume ratio** (see p. 58), so they need a specialised **transport system** to carry raw materials from specialised **exchange organs** to their body cells — this is the **circulatory system**.
2) The circulatory system is made up of the **heart** and **blood vessels**.
3) The heart **pumps blood** through blood vessels (arteries, arterioles, veins and capillaries) to reach different parts of the body. You need to **know** the names of **all** the blood vessels **entering** and **leaving** the **heart**, **lungs** and **kidneys**.
4) Blood transports **respiratory gases**, products of **digestion**, **metabolic wastes** and **hormones** round the body.

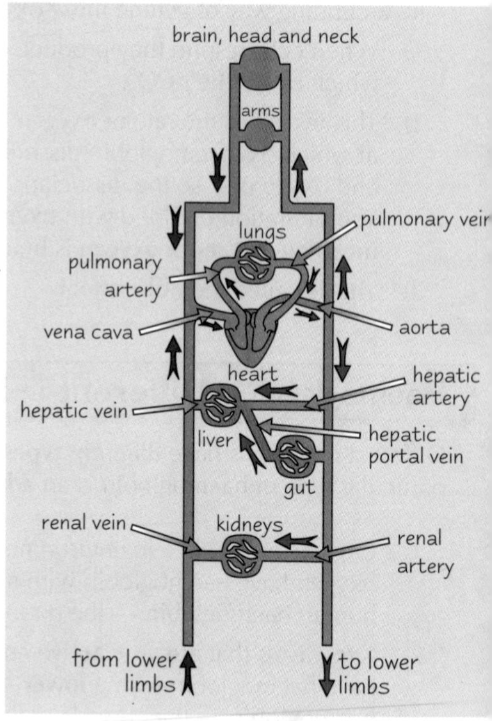

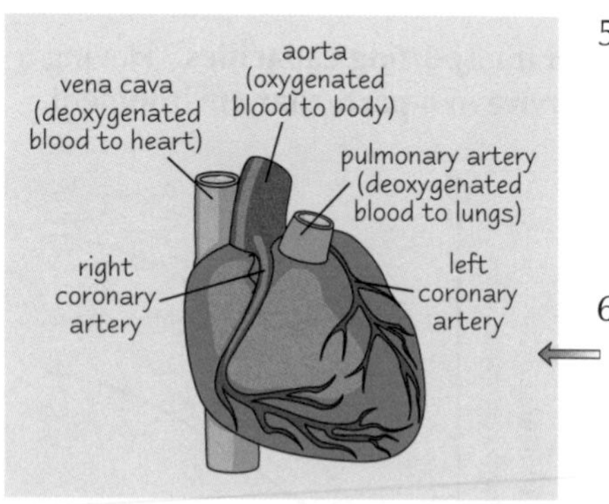

5) There are **two circuits**. One circuit takes blood from the **heart** to the **lungs**, then **back to the heart**. The other loop takes blood around the **rest of the body**.
6) The heart has its own blood supply — the left and right **coronary arteries**.

Different Blood Vessels are Adapted for **Different Functions**

Arteries, **arterioles** and **veins** have different **characteristics**, and you need to know **why**...

1) **Arteries** carry blood **from** the heart **to** the rest of the body. Their walls are thick and **muscular** and have elastic tissue to **stretch** and **recoil** as the heart beats, which helps **maintain** the **high pressure**. The inner lining (**endothelium**) is **folded**, allowing the artery to **stretch** — this also helps it to maintain high pressure. All arteries carry **oxygenated** blood except for the **pulmonary arteries**, which take deoxygenated blood to the lungs.

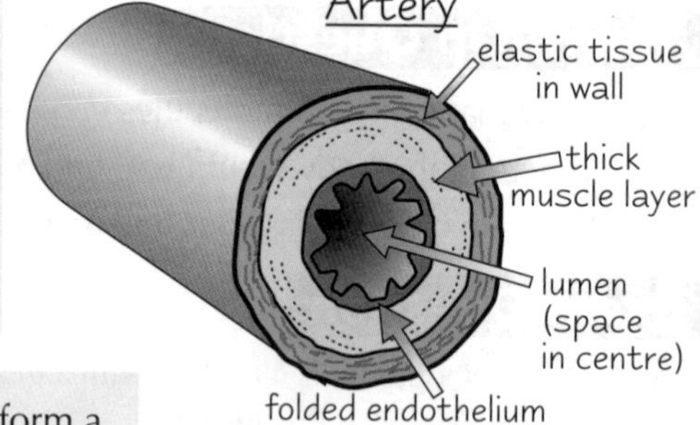

2) Arteries divide into smaller vessels called **arterioles**. These form a network throughout the body. Blood is directed to different **areas of demand** in the body by **muscles** inside the arterioles, which contract to restrict the blood flow or relax to allow full blood flow.

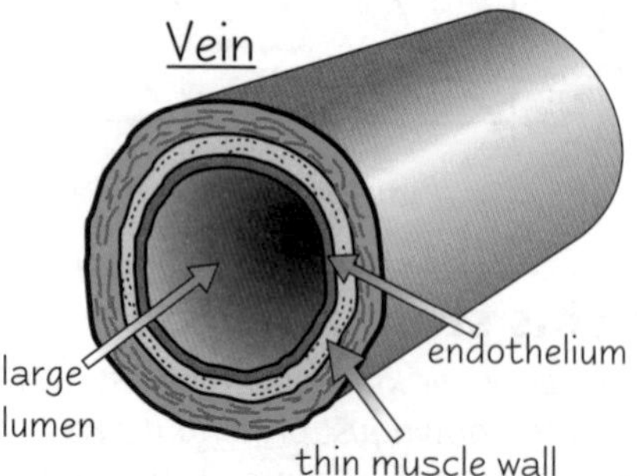

3) **Veins** take blood back **to the heart** under **low pressure**. They have a **wider lumen** than equivalent arteries, with **very little elastic** or **muscle tissue**. Veins contain **valves** to stop the blood flowing backwards. Blood flow through the veins is helped by contraction of the **body muscles** surrounding them. All veins carry **deoxygenated** blood (because oxygen has been used up by body cells), except for the **pulmonary veins**, which carry oxygenated blood to the heart from the lungs.

The Circulatory System

Substances are Exchanged between Blood and Body Tissues at Capillaries

Arterioles branch into **capillaries**, which are the **smallest** of the blood vessels. Substances (e.g. glucose and oxygen) are **exchanged** between cells and capillaries, so they're adapted for **efficient diffusion**.

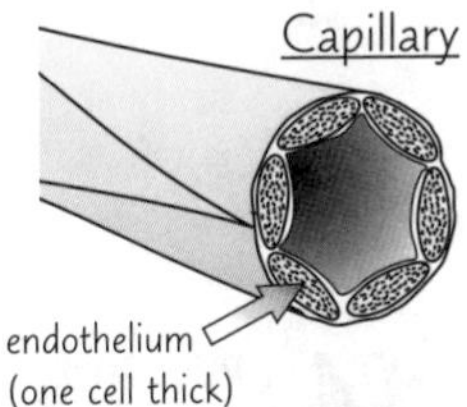

1) They're always found very **near cells in exchange tissues** (e.g. alveoli in the lungs), so there's a **short diffusion pathway**.
2) Their walls are only **one cell thick**, which also shortens the diffusion pathway.
3) There are a large number of capillaries, to **increase surface area** for exchange. Networks of capillaries in tissue are called **capillary beds**.

Tissue Fluid is Formed from Blood

Tissue fluid is the fluid that **surrounds cells** in tissues. It's made from **small molecules** that leave the blood plasma, e.g. oxygen, water and nutrients. (Unlike blood, tissue fluid **doesn't** contain **red blood cells** or **big proteins**, because they're **too large** to be pushed out through the capillary walls.) Cells take in oxygen and nutrients from the tissue fluid, and release metabolic waste into it. In a **capillary bed**, substances move out of the capillaries, into the tissue fluid, by **pressure filtration**:

1) At the **start** of the capillary bed, nearest the arteries, the **hydrostatic (liquid) pressure** inside the capillaries is **greater** than the hydrostatic pressure in the tissue fluid.
2) This **difference** in hydrostatic pressure means an overall outward pressure **forces fluid out** of the **capillaries** and into the **spaces** around the cells, forming **tissue fluid**.
3) As fluid leaves, the hydrostatic pressure **reduces** in the **capillaries** — so the hydrostatic pressure is much **lower** at the **venule end** of the capillary bed (the end that's nearest to the veins).
4) Due to the fluid loss, and an increasing concentration of plasma proteins (which don't leave the capillaries), the **water potential** at the **venule end** of the capillary bed is **lower** than the water potential in the **tissue fluid**.
5) This means that some **water re-enters** the capillaries from the tissue fluid at the venule end by **osmosis** (see p. 42 for more on osmosis).

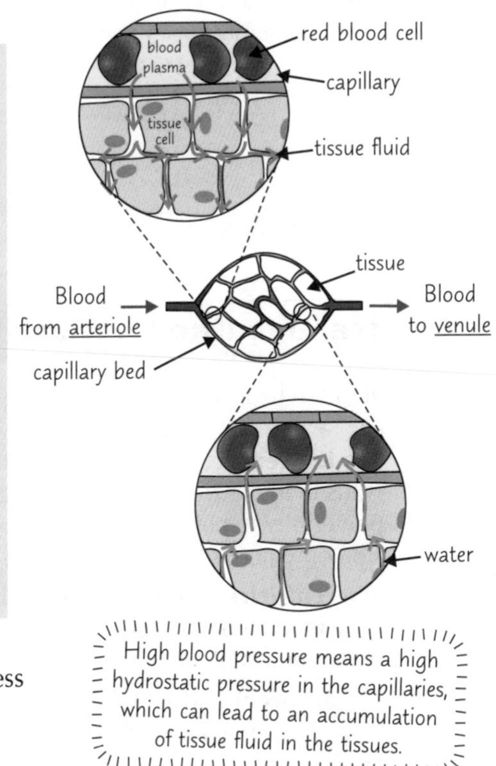

Any **excess** tissue fluid is drained into the **lymphatic system** (a network of tubes that acts a bit like a drain), which transports this excess fluid from the tissues and dumps it back into the circulatory system.

High blood pressure means a high hydrostatic pressure in the capillaries, which can lead to an accumulation of tissue fluid in the tissues.

Warm-Up Questions

Q1 Name all the blood vessels entering and leaving the heart.
Q2 List four types of blood vessel.
Q3 Explain why water returns to the capillary at the venule end of the capillary bed.

PRACTICE QUESTIONS

Exam Questions

Q1 Describe two structural features of an artery and explain how each feature relates to its function. [4 marks]

Q2 At the arteriole end of a capillary bed the hydrostatic pressure is 5.1 kPa in a capillary and 0.13 kPa in the space around the cells. Explain the effect this has on the movement of fluid between the capillary and cell space. [2 marks]

If blood can handle transport this efficiently, the trains have no excuse...

Four hours I was waiting at the train station this weekend. Four hours! Anyway, you may have noticed that biologists are obsessed with the relationship between structure and function, so whenever you're learning the structure of something, make sure you know how this relates to its function. Veins, arteries and capillaries are good examples.

The Heart

As I'm sure you know already, your heart is the 'pump' that gets oxygenated blood to your cells. It's very important, so unsurprisingly, you need to know how it works. You'll find that these pages definitely get to the heart of it... groan

The **Heart** Consists of **Two Muscular Pumps**

1) The diagram on the right shows the **internal structure** of the heart.
2) The **right side** pumps **deoxygenated blood** to the **lungs** and the **left side** pumps **oxygenated blood** to the **whole body**.
3) Note — the **left and right sides** are **reversed** on the diagram, cos it's the left and right of the person that the heart belongs to.

The diagram is a good reminder that veins carry blood into the heart (vena cava and pulmonary vein) and arteries carry blood away from it (pulmonary artery and aorta).

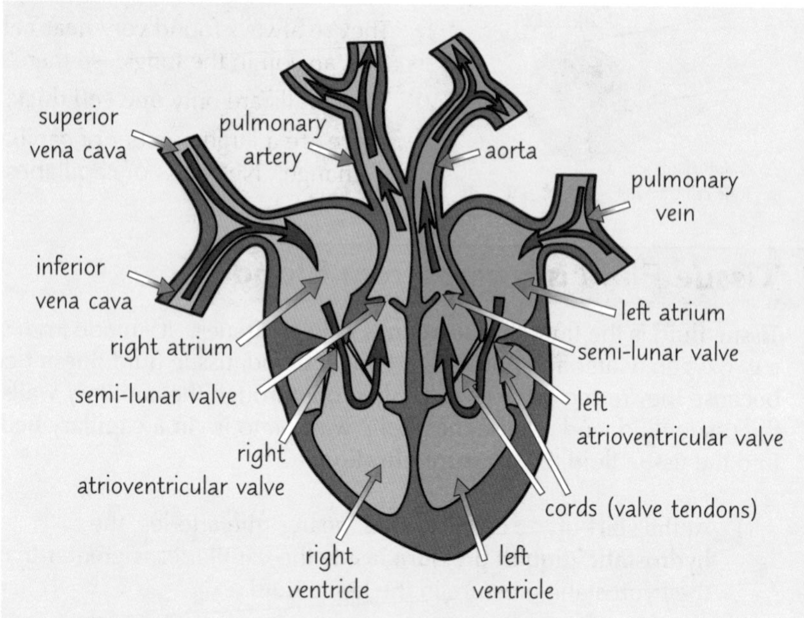

You Need to **Know What** the **Different Parts** of the **Heart Do**

Each bit of the heart is adapted to do its job effectively.

1) The **left ventricle** of the heart has **thicker**, more muscular walls than the **right ventricle**, because it needs to contract powerfully to pump blood all the way round the body. The right side only needs to get blood to the lungs, which are nearby.
2) The **ventricles** have **thicker walls** than the **atria**, because they have to push blood out of the heart whereas the atria just need to push blood a short distance into the ventricles.
3) The **atrioventricular (AV) valves** link the atria to the ventricles and **stop blood flowing back** into the atria when the ventricles contract.
4) The **semi-lunar (SL) valves** link the ventricles to the pulmonary artery and aorta, and **stop blood flowing back** into the heart after the ventricles contract.
5) The **cords** attach the atrioventricular valves to the ventricles to stop them being forced up into the atria when the ventricles contract.

Captain Jeff reckoned the lock gates were just like a big heart valve — with enough pressure he would be able to force his way through.

The **valves** only **open one way** — whether they're open or closed depends on the relative **pressure** of the heart chambers. If there's higher pressure **behind** a valve, it's forced **open**, but if pressure is higher **in front** of the valve it's forced **shut**. This means blood only flows in **one direction** through the heart.

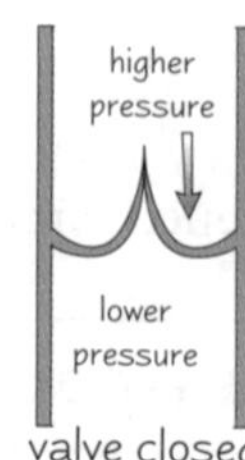

valve closed

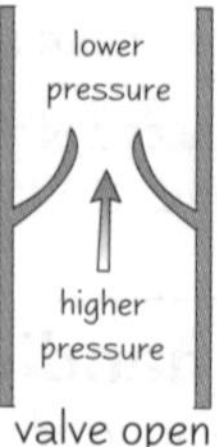

valve open

The Heart

The **Cardiac Cycle** Pumps Blood Round the Body

The cardiac cycle is an ongoing sequence of **contraction** and **relaxation** of the atria and ventricles that keeps blood **continuously** circulating round the body. The **volume** of the atria and ventricles **changes** as they contract and relax. **Pressure** changes also occur, due to the changes in chamber volume (e.g. decreasing the volume of a chamber by contraction will increase the pressure in a chamber). The cardiac cycle can be simplified into three stages:

1 Ventricles relax, atria contract

The **ventricles are relaxed.** The **atria contract**, decreasing the volume of the chambers and **increasing** the **pressure** inside the chambers. This **pushes** the blood into the **ventricles**. There's a slight **increase** in **ventricular pressure** and **chamber volume** as the **ventricles receive the ejected blood** from the contracting atria.

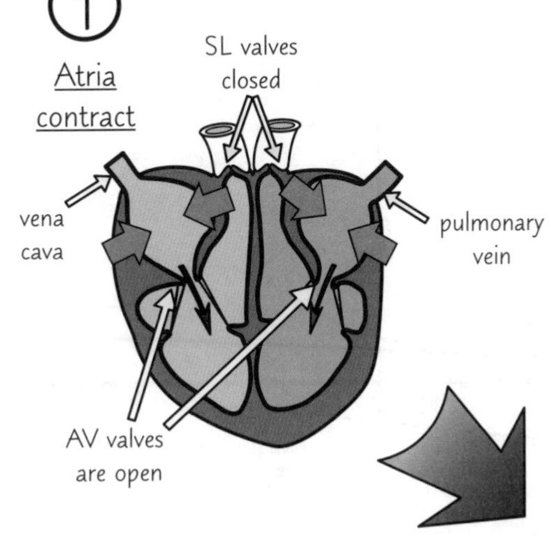

Cardiac contraction is also called systole and relaxation is called diastole.

2 Ventricles contract, atria relax

The **atria relax**. The **ventricles contract** (decreasing their volume), **increasing** their **pressure**. The pressure becomes **higher** in the ventricles than the atria, which forces the **AV valves shut** to prevent back-flow. The **pressure** in the **ventricles** is also **higher** than in the **aorta** and **pulmonary artery**, which forces **open** the **SL valves** and blood is forced out into these arteries.

3 Ventricles relax, atria relax

The **ventricles and the atria both relax.** The higher pressure in the pulmonary artery and aorta closes the SL valves to prevent back-flow into the ventricles. Blood returns to the heart and the **atria fill again** due to the higher pressure in the vena cava and pulmonary vein. In turn this starts to **increase** the **pressure** of the atria. As the ventricles continue to **relax**, their **pressure falls below the pressure of the atria** and so the **AV valves open**. This allows blood to flow **passively** (without being pushed by atrial contraction) into the ventricles from the atria. The atria contract, and the whole process begins again.

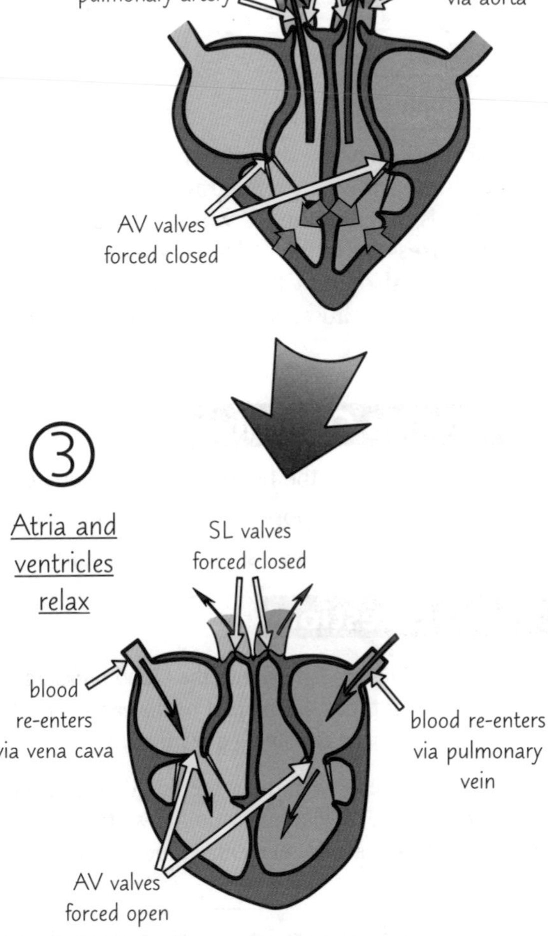

There's a bit about interpreting cardiac cycle data on the next page. So turn over now — it's well exciting...

The Heart

You Might be Asked to **Interpret Data** on the **Cardiac Cycle**

You may well be asked to analyse or interpret **data** about the changes in **pressure** and **volume** during the cardiac cycle. Here are two examples of the kind of things you might get:

Example 1

1 Ventricles relaxed Atria contract
2 Ventricles contract Atria relax
3 Ventricles relax Atria relaxed

KEY
— ventricles
— atria

Pressure / mmHg
pressure increase due to contraction
slight increase due to passive filling
A
pressure increase due to contraction
pressure decrease as atria relax
pressure decrease as ventricles relax
pressure increase as atria fill
pressure decrease as some blood passively moves from atria into ventricle
pressure increase as atria continue to fill
C
pressure increase as ventricles fill

Volume / ml
ventricles stretch while filling
volume decrease due to contraction
B
atria expand as they relax and fill with blood
some blood passively moves from atria to ventricle as AV valves open
ventricles expand as they relax and fill with blood
Time / s

If you get a graph you could be asked **questions** like this:

1) **When** does blood start flowing into the **aorta**?
At **point A**, the ventricles are **contracting** (and the AV valves are shut), forcing blood into the aorta.
2) Why is **ventricular volume decreasing** at **point B**?
The ventricles are **contracting**, **reducing** the volume of the chamber.
3) Are the **semi-lunar valves** open or closed at **point C**?
Closed. The ventricles are **relaxed** and **refilling**, so the pressure is **higher** in the **pulmonary artery** and **aorta**, forcing the SL valves **closed**.

Example 2

You may have to describe the changes in pressure and volume shown by a **diagram**, like the one on the right. In this diagram the **AV valves** are **open**. So you know that the **pressure** in the **atria** is **higher** than in the **ventricles**. So you also know that the **atria are contracting** because that's what causes the **increase** in **pressure**.

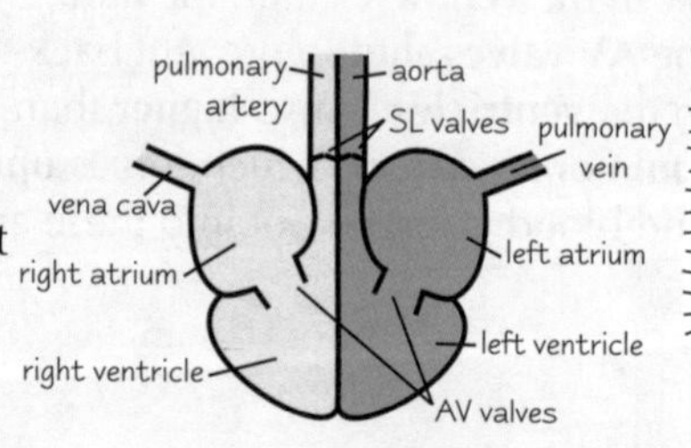

The left ventricle has a thicker wall than the right ventricle and so it contracts more forcefully. This means the pressure is higher in the left ventricle (and in the aorta compared to the pulmonary artery).

PRACTICE QUESTIONS

Warm-Up Questions

Q1 Which side of the heart carries oxygenated blood?
Q2 Explain the purpose of the semi-lunar valves.
Q3 Name the blood vessel that carries blood from the lungs to the heart.

Exam Question

Q1 The table opposite shows the blood pressure in two heart chambers at different times during part of the cardiac cycle. Use the data in the table to answer the following questions.

	Blood pressure / kPa	
Time / s	Left atrium	Left ventricle
0.0	0.6	0.5
0.1	1.3	0.8
0.2	0.4	6.9
0.3	0.5	16.5
0.4	0.9	7.0

a) Between what times are the AV valves shut? [1 mark]
b) Between what times do the ventricles start to relax? [1 mark]
c) Calculate the percentage increase in left ventricle blood pressure between 0.0 s and 0.3 s. [1 mark]

The cardiac cycle — a bewilderingly complicated pump-action bicycle...

Three whole pages to learn here, all full of really important stuff. If you understand all the pressure and volume changes then whether you get a diagram, graph or something else in the exam, you'll be able to interpret it, no probs.

Cardiovascular Disease

Diseases associated with your heart and blood vessels are called cardiovascular diseases (cardio = heart, vascular = blood vessels — geddit?). There are certain factors that increase the risk of developing cardiovascular disease.

Most **Cardiovascular Disease** Starts with **Atheroma** Formation

1) The wall of an artery is made up of **several layers** (see p. 74).
2) The **endothelium** (inner lining) is usually smooth and unbroken.
3) If **damage** occurs to the endothelium (e.g. by high blood pressure) **white blood cells** (mostly macrophages) and **lipids** (fat) from the blood, clump together under the lining to form **fatty streaks**.
4) Over time, more white blood cells, lipids and **connective tissue** build up and harden to form a **fibrous plaque** called an **atheroma**.
5) This plaque **partially blocks** the lumen of the **artery** and **restricts blood flow**, which causes **blood pressure** to **increase**.
6) **Coronary heart disease** (**CHD**) is a type of cardiovascular disease. It occurs when the coronary arteries have **lots of atheromas** in them, which **restricts blood flow** to the **heart muscle**. It can lead to **myocardial infarction** (see below).

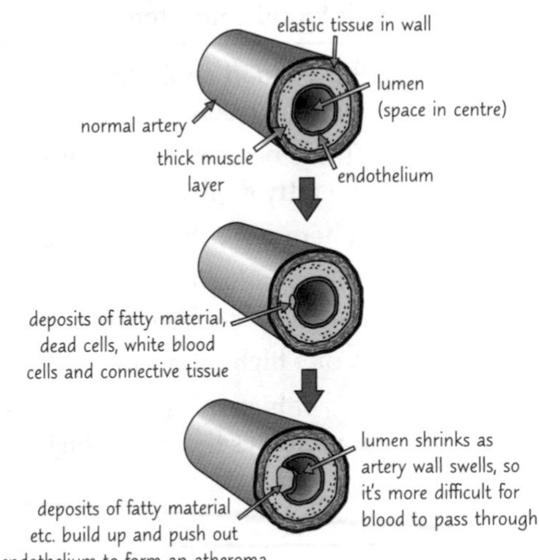

Atheromas Increase the **Risk** of **Aneurysm** and **Thrombosis**

Two types of **disease** that affect the **arteries** are:

Aneurysm — a **balloon-like swelling** of the artery.

1) Atheroma plaques **damage** and **weaken arteries**. They also **narrow** arteries, **increasing blood pressure**.
2) When **blood** travels through a weakened artery at **high pressure**, it may **push** the **inner layers** of the artery **through the outer elastic layer** to form a **balloon-like swelling** — an **aneurysm**.
3) This aneurysm may **burst**, causing a **haemorrhage** (bleeding).

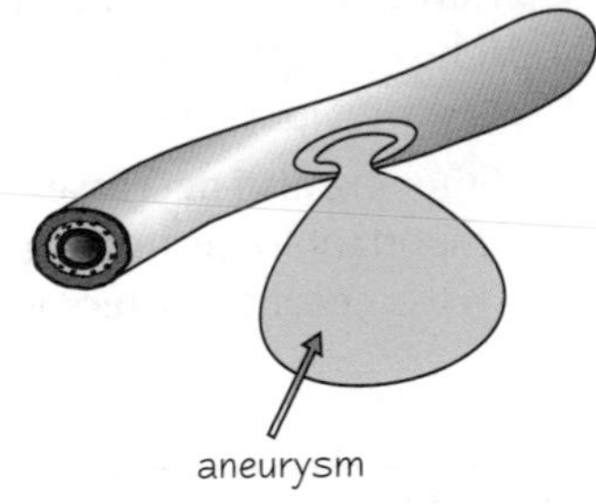

Thrombosis — formation of a **blood clot**.

1) An atheroma plaque can **rupture** (burst through) the **endothelium** (inner lining) of an artery.
2) This **damages** the artery wall and leaves a **rough** surface.
3) **Platelets** and **fibrin** (a protein) accumulate at the site of damage and form a **blood clot** (a thrombus).
4) This blood clot can cause a complete **blockage** of the artery, or it can become **dislodged** and block a blood vessel elsewhere in the body.
5) **Debris** from the rupture can cause another blood clot to form further down the artery.

Interrupted Blood Flow to the **Heart** can Cause a **Myocardial Infarction**

1) The **heart muscle** is supplied with **blood** by the **coronary arteries**.
2) This blood contains the **oxygen** needed by heart muscle cells to carry out **respiration**.
3) If a coronary artery becomes **completely blocked** (e.g. by a **blood clot**) an area of the heart muscle will be totally **cut off** from its blood supply, receiving **no oxygen**.
4) This causes a **myocardial infarction** — more commonly known as a **heart attack**.
5) A heart attack can cause **damage** and **death** of the **heart muscle**.
6) **Symptoms** include **pain** in the chest and upper body, **shortness of breath** and **sweating**.
7) If **large areas** of the heart are affected complete **heart failure** can occur, which is often **fatal**.

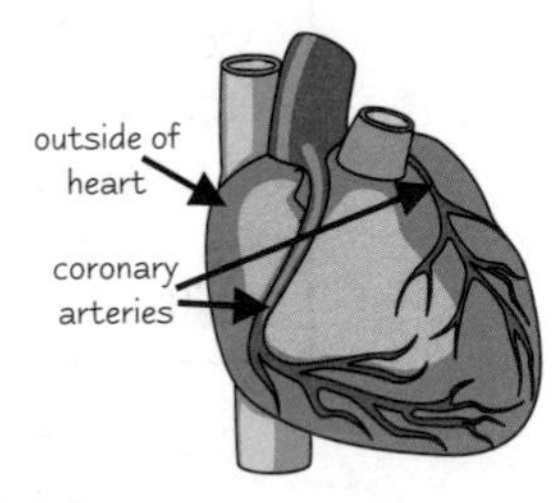

Cardiovascular Disease

Some **Factors Increase** the **Risk** of **Cardiovascular Disease**

Some of the most common risk factors for cardiovascular disease are:

1 High blood cholesterol and poor diet

1) If the **blood cholesterol level** is **high** (above 240 mg per 100 cm^3) then the risk of cardiovascular disease is increased.
2) This is because **cholesterol** is one of the main constituents of the **fatty deposits** that form **atheromas** (see previous page).
3) Atheromas can lead to **increased blood pressure** and **blood clots**.
4) This could **block** the flow of blood to **coronary arteries**, which could cause a **myocardial infarction** (see previous page for details).
5) A diet **high in saturated fat** is associated with high blood cholesterol levels.
6) A diet **high in salt** also **increases** the **risk** of cardiovascular disease because it increases the risk of **high blood pressure** (see below).

John decided to live on the edge and ordered a fry-up.

2 Cigarette smoking

1) Both **nicotine** and **carbon monoxide**, found in **cigarette smoke**, increase the risk of cardiovascular disease.
2) **Nicotine** increases the risk of **high blood pressure** (see below).
3) **Carbon monoxide** combines with **haemoglobin** and **reduces** the amount of **oxygen transported** in the **blood**, and so reduces the amount of oxygen available to tissues. If heart muscle doesn't receive enough oxygen it can lead to a **heart attack** (see previous page).
4) Smoking also **decreases** the **amount** of **antioxidants** in the blood — these are important for **protecting cells** from damage. Fewer antioxidants means **cell damage** in the **coronary artery walls** is more likely, and this can lead to **atheroma formation**.

3 High blood pressure

1) High blood pressure **increases** the **risk** of **damage** to the **artery walls**.
2) Damaged walls have an **increased risk** of **atheroma** formation, causing a further increase in blood pressure.
3) Atheromas can also cause **blood clots** to form (see previous page).
4) A blood clot could **block flow** of **blood** to the heart muscle, possibly resulting in **myocardial infarction**.
5) So **anything** that **increases** blood pressure also increases the risk of **cardiovascular disease**, e.g. being **overweight**, **not exercising** and excessive **alcohol** consumption.

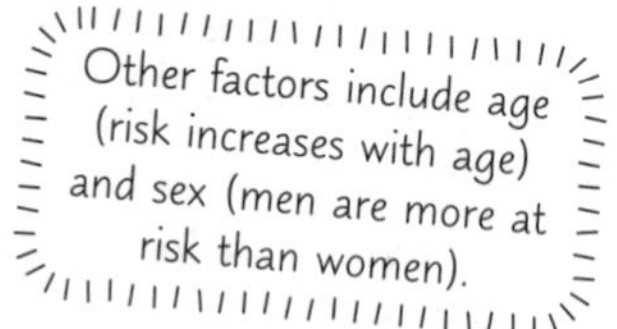

Most of these factors are within our **control** — a person can **choose** to smoke, eat fatty foods, etc. However, some risk factors can't be controlled, such as having a **genetic predisposition** to coronary heart disease or having high blood pressure as a result of **another condition**, e.g. some forms of diabetes. Even so, the risk of developing cardiovascular disease can be reduced by removing as many **risk factors** as you possibly can.

Cardiovascular Disease

You May Have to Interpret Data on Risk Factors and Cardiovascular Disease

Example: The graph shows the results of a study involving **27 939 American women**. The **LDL cholesterol level** was **measured** for each woman. These women were then **followed** for an average of **8 years** and the **occurrence** of **cardiovascular events** (e.g. heart attack, surgery on coronary arteries) or **death** from cardiovascular diseases was **recorded**. The **relative risk** of a cardiovascular event, **adjusted** for **other factors** that can affect cardiovascular disease, was then calculated.

Relative risk of cardiovascular event
1.6
1.4
1.2
1.0
0.8
0.6
0
>97.6 - 115.4
>115.4 - 132.2
>132.2 - 153.9
>153.9
Level of LDL cholesterol in blood / mg dl^{-1}

Here are some of the things you might be asked to do:

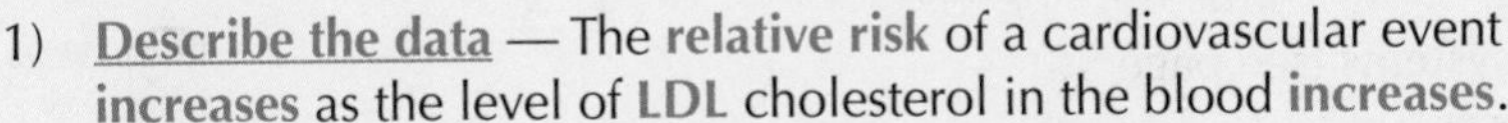

1) **Describe the data** — The **relative risk** of a cardiovascular event **increases** as the level of **LDL** cholesterol in the blood **increases**.
2) **Draw conclusions** — The graph shows a **positive correlation** between the **relative risk** of a cardiovascular event and the level of **LDL cholesterol** in the blood.
3) **Check any conclusions are valid** — Make sure any conclusions **match** the data, e.g.
   - This data only looked at **women** — no males were involved, so you can't say that this trend is true for **everyone**.
   - You can't say that a high LDL cholesterol level is **correlated with** an increased risk of **heart attacks**, because the data shows **all** first cardiovascular events, including surgery on coronary arteries.
   - Also, you can't conclude that a high LDL cholesterol level **caused** the increased relative risk of a cardiovascular event — there may be other reasons for the trend.
4) **Other things to think about** — A **large sample size** was used (27 939). Data based on large samples is **better** than data based on small samples. This is because a large sample is **more representative** of the whole population (i.e. it shares more of the various **characteristics** of the population).

There's more on correlation and cause on page 229.

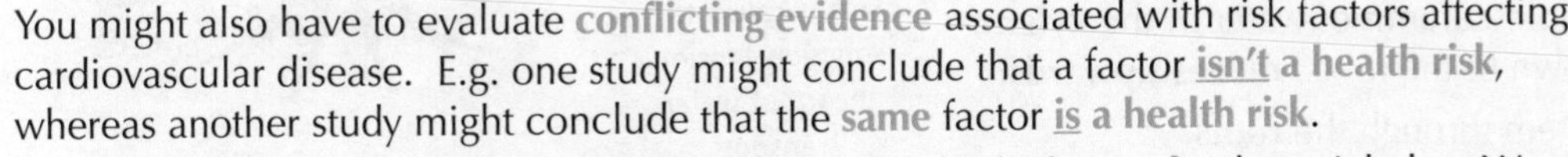

You might also have to evaluate **conflicting evidence** associated with risk factors affecting cardiovascular disease. E.g. one study might conclude that a factor **isn't a health risk**, whereas another study might conclude that the **same** factor **is a health risk**.

1) If two studies have produced conflicting results, think about **why** that might be. Was it to do with **study design**? Was one study based on a **small sample size**? Did both studies take into account **other risk factors** (variables) that could have affected the results? Knowing whether both studies used **similar groups** can be helpful, e.g. same age, gender, etc.
2) Sometimes, the only way to **resolve** the problem of conflicting evidence is to **carry out more studies** and **collect more results**. Results need to be **reproduced** by **other scientists** before they're accepted.

The way in which information is collected can also be important. Some studies rely on the results of questionnaires (e.g. asking people how many cigarettes they smoke). Questionnaires can be unreliable as people can tell fibs or give inaccurate information.

Warm-Up Questions

Q1 Give three factors that can increase the risk of developing cardiovascular disease.

PRACTICE QUESTIONS

Exam Question

Q1 The results of a study involving 168 000 people in 63 countries have shown a strong correlation between waist measurement and risk of cardiovascular disease. Analysis of the results has shown that waist circumference is independently associated with cardiovascular disease.

a) Give two reasons why the study provides strong evidence for a link between waist measurement and risk of cardiovascular disease. [2 marks]

b) Suggest why waist measurement might be related to risk of cardiovascular disease. [3 marks]

Revision — increasing my risk of headache, stress, boredom...

There's a lot to take in on these pages... but make sure you understand the link between atheromas, thrombosis and heart attacks — basically an atheroma forms, which can cause thrombosis, which can lead to a heart attack. Anything that increases the chance of an atheroma forming (high blood pressure, smoking, fatty diet) is bad news for your heart...

Transport in Plants — Xylem

Transport in plants isn't really roads and railways, but I guess it's a bit like a flowing river that's carrying stuff around in a network of tubes. When you consider the movement of water, it's all about the xylem. It's pretty exciting stuff...

Two Types of Tissue are Involved in Transport in Plants

1) **Xylem tissue** transports **water** and **mineral ions** in solution. These substances move **up** the plant from the roots to the leaves.
2) **Phloem tissue** transports organic substances like **sugars** (also in solution) both **up and down** the plant — there's more about the phloem on pages 84-85.
3) Xylem and phloem are **mass transport systems** (see page 58) — they move substances over **large distances**.

Xylem vessels are the part of the xylem tissue that actually transports the water and ions. Xylem vessels are very **long, tube-like** structures formed from dead cells (vessel elements) joined end to end. There are **no end walls** on these cells, making an **uninterrupted tube** that allows water to pass up through the middle easily.

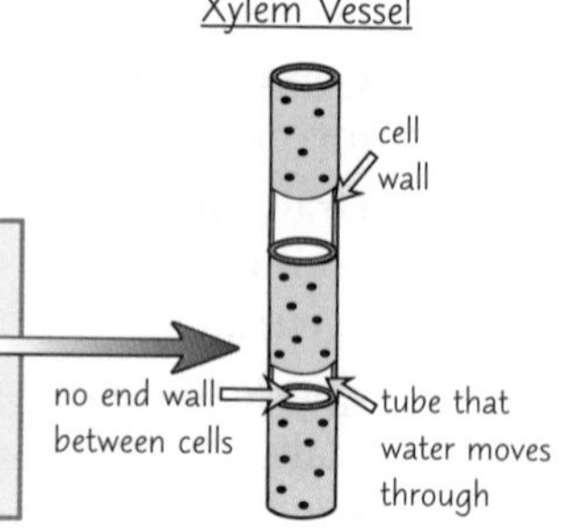

Water Moves Up a Plant Against the Force of Gravity

Cohesion and **tension** help water move up plants, from roots to leaves, against the force of gravity.

1) Water **evaporates** from the **leaves** at the 'top' of the xylem (this is transpiration — see below).
2) This creates **tension** (**suction**), which pulls more water into the leaf.
3) Water molecules are **cohesive** (they **stick together** — see page 21) so when some are pulled into the leaf others follow. This means the whole **column** of water in the **xylem**, from the leaves down to the roots, **moves upwards**.
4) **Water** enters the stem through the **roots**.

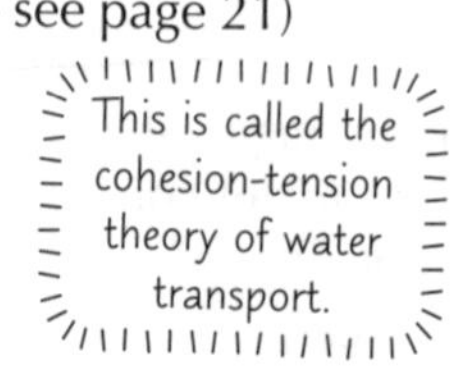

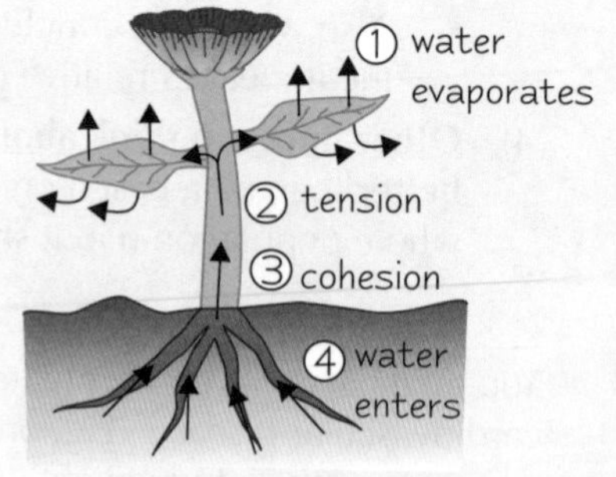

Transpiration is Loss of Water from a Plant's Surface

Transpiration is the **evaporation** of **water** from a plant's surface, especially the **leaves**.

1) Water **evaporates** from the moist cell walls and accumulates in the spaces between cells in the leaf.
2) When the **stomata** open (see page 61), it moves out of the leaf down the **concentration gradient** (there's more water inside the leaf than in the air outside).

Transpiration's really a side effect of photosynthesis — the plant needs to open its stomata to let in CO_2 so that it can produce glucose, but this also lets water out.

Four Main Factors Affect Transpiration Rate

1) **Light** — the **lighter** it is the **faster** the **transpiration rate** (i.e. there's a **positive correlation** between light intensity and transpiration rate). This is because the **stomata open** when it gets **light** to let in CO_2 for **photosynthesis**. When it's **dark** the stomata are usually **closed**, so there's little transpiration.
2) **Temperature** — the **higher the temperature** the **faster** the **transpiration rate**. Warmer water molecules have more energy so they **evaporate** from the cells inside the leaf **faster**. This **increases** the **concentration gradient** between the inside and outside of the leaf, making water **diffuse out** of the leaf **faster**.
3) **Humidity** — the **lower** the **humidity**, the **faster** the **transpiration rate** (i.e. there's a **negative correlation** between humidity and transpiration rate). If the air around the plant is **dry**, the **concentration gradient** between the leaf and the air is **increased**, which increases transpiration.
4) **Wind** — the **windier** it is, the **faster** the **transpiration rate**. Lots of air movement **blows away** water molecules from around the stomata. This **increases** the **concentration gradient**, which increases the rate of transpiration.

Transport in Plants — Xylem

A Potometer can be Used to Estimate Transpiration Rate

A **potometer** is a special piece of apparatus used to **estimate transpiration rates**. It actually measures **water uptake** by a plant, but it's **assumed** that water uptake by the plant is **directly related** to **water loss** by the **leaves**. You can use it to estimate how different factors **affect** the transpiration rate.

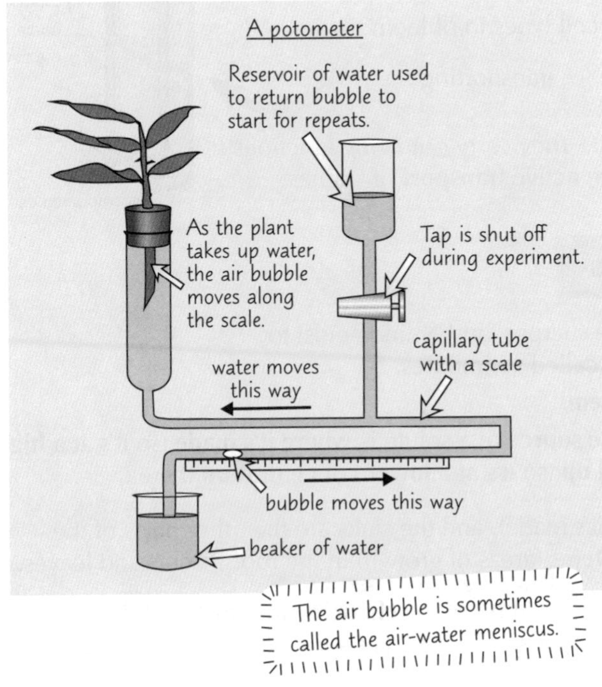

The air bubble is sometimes called the air-water meniscus.

Here's what you'd do:

1) **Cut** a **shoot underwater** to prevent air from entering the xylem. Cut it at a **slant** to increase the surface area available for water uptake.
2) Assemble the potometer **in water** and insert the shoot **underwater**, so no air can enter.
3) Remove the apparatus from the water but keep the **end of the capillary tube submerged** in a beaker of water.
4) Check that the apparatus is **watertight** and **airtight**.
5) **Dry** the leaves, allow time for the shoot to **acclimatise**, and then **shut the tap**.
6) Remove the end of the capillary tube from the beaker of water until **one air bubble** has formed, then put the end of the tube back into the water.
7) Record the **starting position** of the **air bubble**.
8) Start a **stopwatch** and record the **distance** moved by the bubble **per unit time**, e.g. per hour. The **rate** of **air bubble movement** is an estimate of the **transpiration rate**.
9) Remember, only change **one variable** (e.g. temperature) at a time. All other **conditions** (e.g. light, humidity) must be kept **constant**.

You Might Have to Dissect Plants

You can **look at xylem** or **phloem** in **plant tissue** (e.g. part of a plant stem) under a **microscope**, and then **draw** them. But first you need to **dissect** the plant and **prepare** a section of the tissue. You can do this using the following method:

1) Use a **scalpel** (or razor blade) to cut a **cross-section** of the stem. Cut the sections as **thinly** as possible — thin sections are better for viewing under a microscope.
2) Use **tweezers** to gently place the cut sections in **water** until you come to use them. This stops them from **drying out**.
3) Transfer each section to a dish containing a **stain**, e.g. **toluidine blue O (TBO)**, and leave for one minute. TBO stains the **lignin** in the walls of the xylem vessels **blue-green**. This will let you see the **position** of the xylem vessels and examine their **structure**.
4) **Rinse off** the sections in water and **mount** each one onto a slide (see page 32).

You can use different stains to highlight different parts of the cells.

Warm-Up Questions

PRACTICE QUESTIONS

Q1 What is the function of xylem tissue?
Q2 Give four factors that affect transpiration rate.
Q3 Name a piece of apparatus used to measure transpiration rate.

Exam Question

Q1 a) What is meant by the term transpiration? [1 mark]
b) Describe how the cohesion-tension theory helps explain water movement in plants. [3 marks]

Xylem — not to be confused with Wylam, a small village in Northumberland...

So, one of the key things you need to take away from these pages is that xylem is the plant tissue that water is transported through. Water can make its way from the root to the leaves in the lovely xylem tissue and then some of it will end up being lost into the air from the plant's surfaces (particularly the leaves) via good ol' transpiration.

Transport in Plants — Phloem

Next up, it's time to look at phloem. I know, I almost can't contain my excitement. When it comes to the phloem it's all about moving dissolved organic substances around the plant, so it's a bit different to xylem.

Phloem Tissue is Adapted for **Transporting Solutes**

1) **Solutes** are **dissolved substances**. Phloem tissue **transports solutes** (mainly sugars like sucrose) round plants. Like xylem, phloem is formed from cells arranged in **tubes**.
2) **Sieve tube elements** and **companion cells** are important cell types in phloem tissue:

- **Sieve tube elements** are living cells that form the **tube** for transporting solutes. They have no nucleus and few organelles, so...
- ...there's a **companion cell** for each sieve tube element. They carry out living functions for sieve cells, e.g. providing the **energy** needed for the **active transport** of solutes.

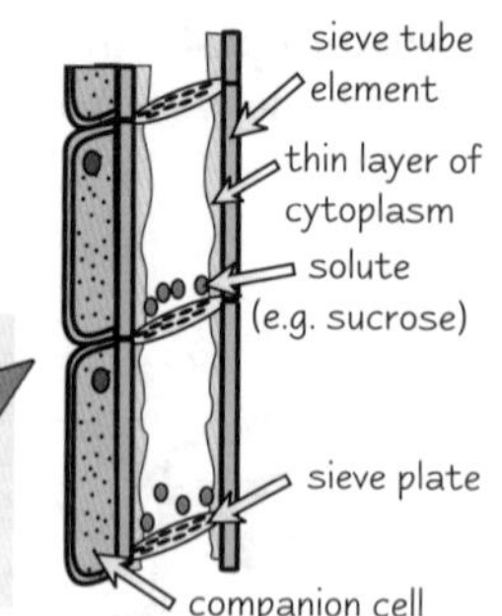

Translocation is the **Movement** of **Solutes**

1) **Translocation** is the **movement** of solutes (e.g. sugars like sucrose, and amino acids) to **where they're needed** in a plant. Solutes are sometimes called **assimilates**.
2) It's an **energy-requiring** process that happens in the **phloem**.
3) Translocation moves solutes from '**sources**' to '**sinks**'. The **source** of a solute is **where it's made** (so it's at a **high concentration** there). The **sink** is the area where it's **used up** (so it's at a **lower concentration** there).

E.g. the **source** for **sucrose** is usually the **leaves** (where it's made), and the **sinks** are the **other parts** of the plant, especially the **food storage organs** and the **meristems** (areas of growth) in the roots, stems and leaves.

4) **Enzymes** maintain a **concentration gradient** from the source to the sink by **changing** the solutes at the **sink** (e.g. by breaking them down or making them into something else). This makes sure there's always a **lower concentration** at the sink than at the source.

E.g. in **potatoes**, **sucrose** is converted to **starch** in the **sink** areas, so there's always a **lower concentration** of sucrose **at the sink** than inside the phloem. This makes sure a **constant supply** of new sucrose reaches the sink from the phloem.

The **Mass Flow Hypothesis** Best Explains **Phloem Transport**

Scientists still aren't certain **exactly how** the solutes are transported from source to sink by **translocation**. The best supported theory is the **mass flow hypothesis**:

(1)
1) Active transport (see p. 44) is used to **actively load** the solutes (e.g. sucrose from photosynthesis) from **companion cells** into the **sieve tubes** of the phloem at the **source** (e.g. the **leaves**).
2) This **lowers the water potential** inside the sieve tubes, so water **enters** the tubes by **osmosis** from the **xylem** and **companion cells**.
3) This creates a **high pressure** inside the sieve tubes at the **source end** of the phloem.

(2)
1) At the **sink** end, **solutes** are removed from the phloem to be used up.
2) This **increases** the **water potential** inside the sieve tubes, so water also **leaves** the tubes by **osmosis**.
3) This **lowers the pressure** inside the sieve tubes.

(3)
1) The result is a **pressure gradient** from the **source** end to the **sink** end.
2) This gradient pushes solutes along the sieve tubes **towards the sink**.
3) When they reach the sink the solutes will be **used** (e.g. in respiration) or **stored** (e.g. as starch).

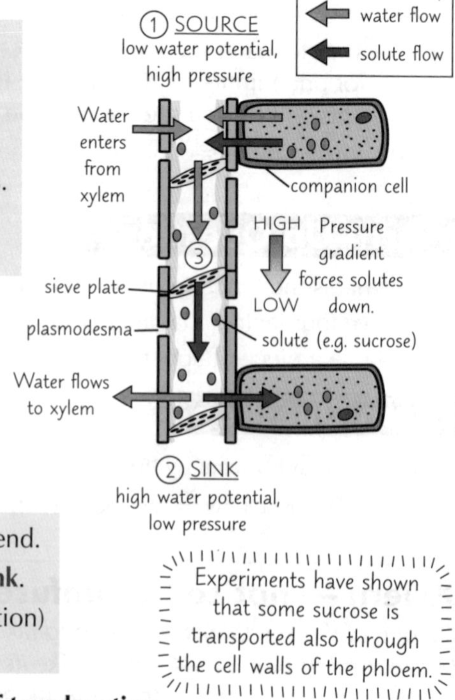

Experiments have shown that some sucrose is transported also through the cell walls of the phloem.

The higher the **concentration of sucrose** at the source, the higher the **rate of translocation**.

Transport in Plants — Phloem

You Need to be Able to **Evaluate Evidence** For and Against **Mass Flow**

Supporting evidence

1) If a **ring of bark** (which includes the phloem, but not the xylem) is removed from a woody stem, a **bulge forms above** the ring. The fluid from the bulge has a **higher concentration** of **sugars** than the fluid from below the ring — this is evidence that there's a **downward flow** of sugars.
2) A **radioactive tracer** such as radioactive carbon (^{14}C) can be used to **track** the movement of organic substances in a plant (see below).
3) Pressure in the phloem can be investigated using **aphids** (they pierce the phloem, then their bodies are removed leaving the mouthparts behind, which allows the sap to flow out... gruesome). The sap flows out **quicker nearer the leaves** than further down the stem — this is evidence that there's a **pressure gradient**.
4) If a **metabolic inhibitor** (which stops ATP production) is put into the **phloem**, then **translocation stops** — this is evidence that **active transport** is involved.

Objections

1) Sugar travels to **many different sinks**, not just to the one with the **highest water potential**, as the model would suggest.
2) The **sieve plates** would create a **barrier** to mass flow. A **lot of pressure** would be needed for the solutes to get through at a reasonable rate.

You could get asked about correlations and casual relationships in data relating to mass transport in plants. There's loads on correlation and cause on p. 229.

The **Translocation** of **Solutes** Can be **Demonstrated Experimentally**

Translocation of solutes in plants can be modelled in an experiment using **radioactive tracers**.

1) This can be done by supplying part of a plant (often a **leaf**) with an **organic substance** that has a **radioactive label**. One example is **carbon dioxide** containing the radioactive isotope ^{14}**C**. This radioactively-labelled CO_2 can be supplied to a single leaf by being pumped into a container which completely surrounds the leaf.
2) The radioactive carbon will then be **incorporated** into organic substances produced by the leaf (e.g. sugars produced by **photosynthesis**), which will be moved around the plant by **translocation**.
3) The movement of these substances can be tracked using a technique called **autoradiography**. To reveal where the radioactive tracer has **spread to** in a plant, the plant is killed (e.g. by freezing it using liquid nitrogen) and then the whole plant (or sections of it) is placed on **photographic film** — the radioactive substance is present wherever the film turns **black**.
4) The results demonstrate the translocation of substances from **source** to **sink** over time — for example, autoradiographs of plants **killed** at **different times** show an overall movement of solutes (e.g. products of photosynthesis) from the leaves **towards the roots**.

Photosynthesis produces glucose. This is converted to sucrose for transport around the plant.

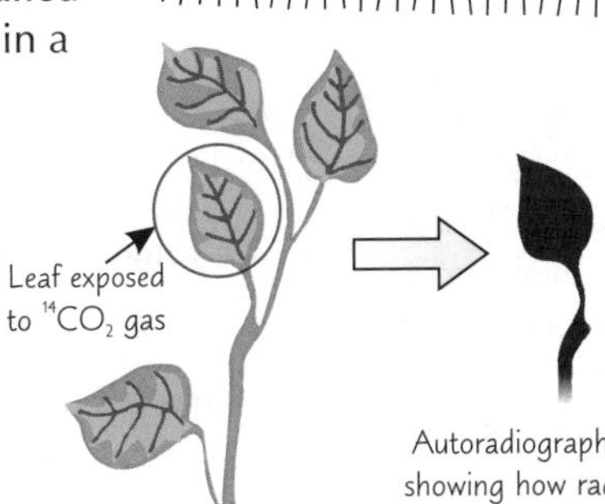

Autoradiograph of plant showing how radioactivity has spread from the leaf.

Warm-Up Questions

Q1 According to the mass flow hypothesis, how is a pressure gradient set up in the phloem?

PRACTICE QUESTIONS

Exam Question

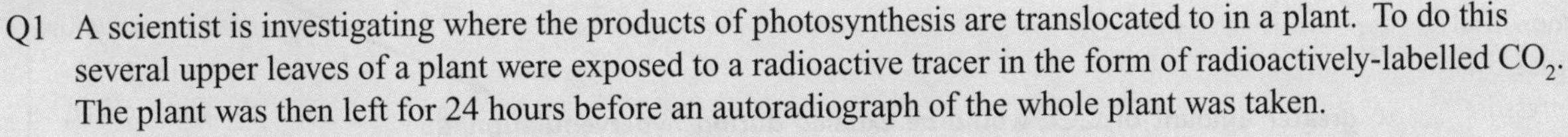

Q1 A scientist is investigating where the products of photosynthesis are translocated to in a plant. To do this several upper leaves of a plant were exposed to a radioactive tracer in the form of radioactively-labelled CO_2. The plant was then left for 24 hours before an autoradiograph of the whole plant was taken.

a) Explain how the leaves of the plant can act as a source in translocation. [1 mark]

b) The autoradiograph showed radioactivity in the roots and fruits. Explain why radioactivity was seen in the fruits. [1 mark]

Human mass flow — running out of the hall at the end of an exam...

The mass flow hypothesis is just the best theory that scientists have come up with so far. If other evidence came along, a different theory could be developed based on the new findings. It could happen tomorrow, you never know...

Extra Exam Practice

You should probably make sure that everything from Topic 3 is circulating around in there before you move on. Handily, these questions pull some of the ideas from the topic together...

- Have a look at this example of how to answer a tricky exam question.
- Then check how much you've understood from Topic 3 by having a go at the questions on the next page.

1 Hyperventilation is very fast and deep breathing. **Figure 1** shows the volume of air in the lungs of a person at rest, who begins hyperventilating after 10 seconds.

Figure 1

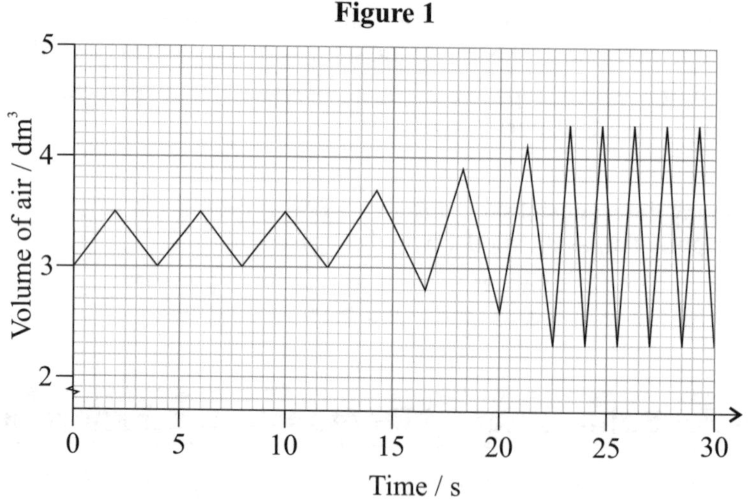

1.1 The pulmonary ventilation rate (PVR) is the volume of air inspired or expired in one minute. Calculate the maximum PVR of the person while they were hyperventilating, in $dm^3 min^{-1}$. Show your working.

(3 marks)

1.2 Using **Figure 1** and your knowledge of the Bohr effect, suggest how hyperventilation when at rest can result in body cells being unable to obtain sufficient oxygen from red blood cells.

(4 marks)

Start by working out what you need to calculate and think about how you can use the information from the graph to do it.

1.1

PVR = **volume of air in each breath × number of breaths per minute**

Volume of air in each breath when hyperventilating = 4.3 – 2.3 = 2 dm^3

Time taken for one breath = 1.5 s

Number of breaths per minute = 60 ÷ 1.5 = 40 breaths per minute

PVR = 2 × 40 = **80 $dm^3 min^{-1}$**

Here you'd get 3 marks for the correct answer. But, don't worry, if you got the answer wrong you'd still get 1 mark for working out the volume of air in each breath and 1 mark for calculating the number of breaths per minute.

It's a good idea to use a ruler to draw lines to the axes when you're reading values from a graph — this helps to make sure the values you read off are accurate.

From your knowledge of the Bohr effect you should know that an increase the partial pressure of carbon dioxide reduces the affinity of haemoglobin for oxygen — here you need to think what would happen in the opposite situation.

You've been asked to use your knowledge of the Bohr effect, so you need to think about how hyperventilation could affect the partial pressure of carbon dioxide.

1.2

A **greater amount of CO_2 would be exhaled during hyperventilation**, due to the increased breathing rate and volume of air in each breath. Because the person is at rest, the respiration rate of the body cells is not increased, so there is no increase in the amount of CO_2 entering the blood. This causes a reduction in the partial pressure of CO_2, which **increases the affinity of haemoglobin for oxygen**. This reduces the rate of oxygen unloading from red blood cells, meaning the body cells may be unable to obtain sufficient oxygen.

Extra Exam Practice

2 Insects are very important organisms in many different ecosystems.

Figure 2 shows the circulatory system of an insect. The arrows indicate the direction of blood flow. Blood is pumped through each segment of the insect's heart in a wave of contractions. The blood then enters the body cavity and flows around the insect's organs before re-entering the heart through a series of valves.

Figure 2

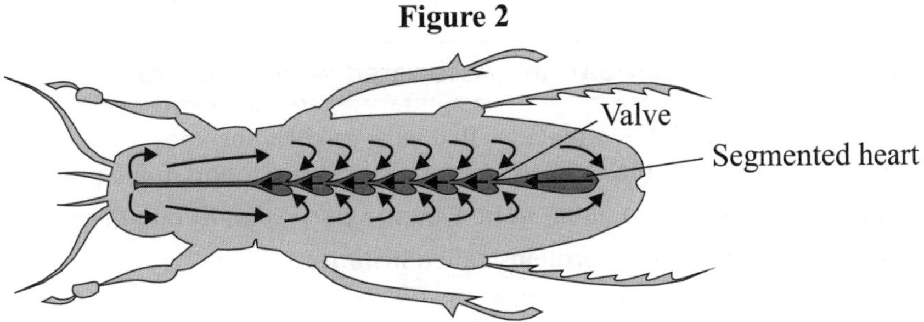

An insect's circulatory system has similarities and differences to that of a mammal.

2.1 Using your knowledge of valves in the mammalian heart, suggest why blood from the body cavity does not re-enter segments of the insect heart while they are contracting.

(2 marks)

2.2 Haemoglobin has an important role in the mammalian circulatory system. Suggest why most insects do not require haemoglobin.

(2 marks)

Some insects feed on the phloem sap of plants. They do this using a long, sharp mouthpiece called a stylet, which is used to pierce the stem of the plant and then draw up the phloem sap. A scientist investigated the mass flow hypothesis in a plant using insects that feed with a stylet. She selected insects that were feeding at three different sites on a plant, as shown in **Figure 3**. She removed each insect from the plant but left the stylets in place. She then collected the sap excreted through each stylet over 30 minutes.

Figure 3

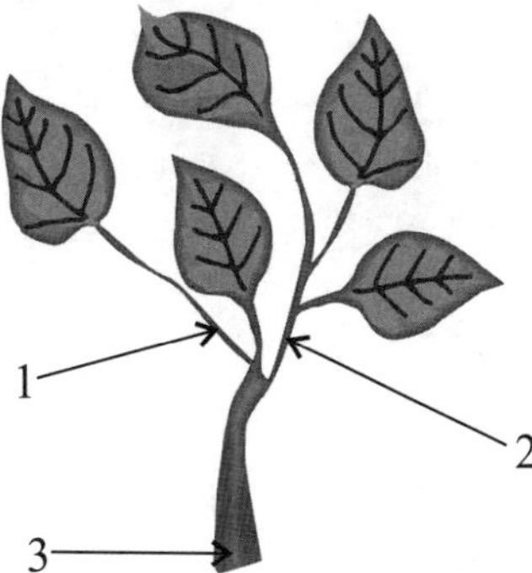

2.3 Explain how the water potential inside the sieve tubes changes between sites **2** and **3**.

(4 marks)

2.4 The volume of sap collected at site **2** was greater than at site **1**. Suggest an explanation for this.

(4 marks)

Plant growth rate can be reduced by insects feeding on plants.

2.5 Some plants produce a compound that inhibits alpha-amylase in insects. Suggest how this compound helps to defend the plant against insect pests.

(3 marks)

DNA, Genes and Chromosomes

DNA can be cruel — it gave me two feet, but made me bad at football... OK, maybe that's not completely DNA's fault. These pages give you plenty of info on how DNA is packaged, what genes are and how they code for stuff.

DNA is **Stored Differently** in **Different Organisms**

Although the **structure** of DNA is the same in all organisms, **eukaryotic** and **prokaryotic** cells store DNA in slightly different ways. (For a recap on the differences between prokaryotic and eukaryotic cells see pages 26 and 30.)

Nuclear Eukaryotic DNA is **Linear** and Associated with **Proteins**

1) Eukaryotic cells contain **linear** DNA molecules that exist as **chromosomes** — thread-like structures, each made up of **one long molecule** of DNA. Chromosomes are found in the **nucleus**.
2) The DNA molecule is **really long** so, it has to be **wound up** so it can **fit** into the nucleus.
3) The DNA molecule is wound around **proteins** called **histones**.
4) Histone proteins also help to **support** the DNA.
5) The DNA (and protein) is then coiled up **very tightly** to make a **compact chromosome**.

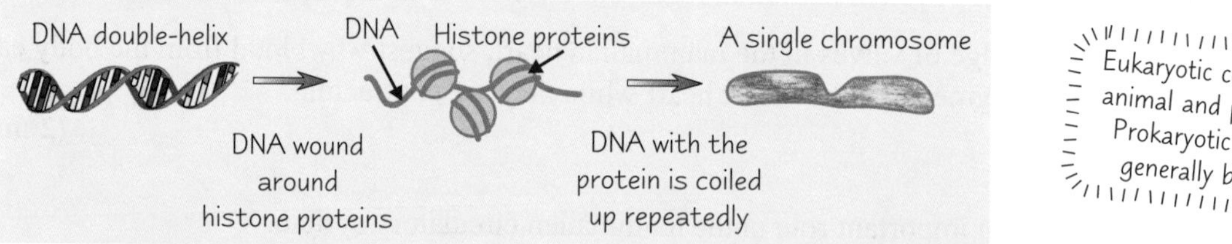

Eukaryotic cells include animal and plant cells. Prokaryotic cells are generally bacteria.

6) The **mitochondria** and **chloroplasts** in eukaryotic cells also have their **own** DNA. This is pretty similar to prokaryotic DNA (see below) because it's **circular** and **shorter** than DNA molecules in the nucleus. It's **not associated** with **histone proteins**.

DNA Molecules are **Shorter** and **Circular** in **Prokaryotes**

1) Prokaryotes also carry DNA as **chromosomes** — but the DNA molecules are **shorter** and **circular**.
2) The DNA **isn't** wound around histones — it condenses to fit in the cell by **supercoiling**.

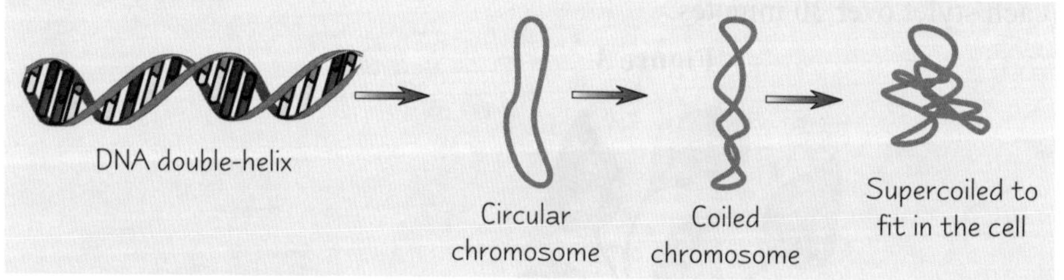

If one more person confused Clifford with supercoiled DNA, he'd have 'em.

DNA Contains **Genes**

1) A **gene** is a **sequence** of **DNA bases** (see p. 16) that codes for either a **polypeptide** or **functional RNA** (see below).
2) The sequence of **amino acids** in a polypeptide forms the **primary structure** of a **protein** (see p. 8).
3) Different polypeptides have a **different number** and **order** of amino acids. It's the **order** of **bases** in a gene that determines the **order of amino acids** in a particular **polypeptide**.
4) Each amino acid is coded for by a sequence of **three bases** in a gene called a **triplet**.
5) To make a **polypeptide**, DNA is first copied into **messenger RNA** (mRNA). This is the first stage of **protein synthesis** (see p. 90).
6) Genes that don't code for a polypeptide code for **functional RNA** instead. Functional RNA is RNA molecules other than mRNA, which perform **special tasks** during protein synthesis, e.g. **tRNA** (see p. 90) and **ribosomal RNA** (rRNA), which forms part of ribosomes.

Bases on DNA
G T C T G A
DNA triplet = one amino acid

A cell's **GENOME** is the **complete** set of **genes** in the cell.
A cell's **PROTEOME** is the **full range** of **proteins** that the cell is able to produce.

DNA, Genes and Chromosomes

Most DNA in Eukaryotic Cells Doesn't Code for Polypeptides

1) **Some** genes don't code for **polypeptides** at all — they code for **functional RNA** (see previous page).
2) In **eukaryotic** DNA, genes that do code for **polypeptides** contain **sections** that **don't code** for **amino acids**.
3) These sections of DNA are called **introns**. There can be several introns within a gene.
4) All the bits of a gene that do code for amino acids are called **exons**.
5) **Introns** are **removed** during **protein synthesis** — so they **don't** affect the amino acid **order**. Their purpose isn't known for sure. (Prokaryotic DNA **doesn't** have introns.)
6) Eukaryotic DNA also contains regions of **multiple repeats** outside of genes.
7) These are DNA sequences that **repeat** over and over. For example: CCTTCCTTCCTT.
8) These areas **don't code** for amino acids either, so they're called **non-coding repeats**.

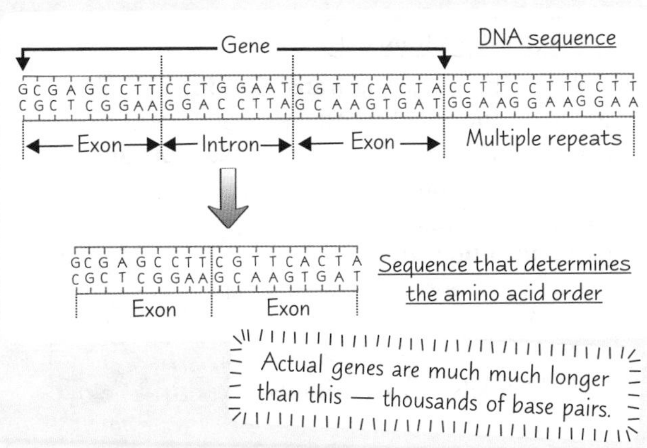

Genes Can Exist in Different Forms Called Alleles

1) A gene can exist in more than one form. These forms are called **alleles**.
2) The order of bases in each allele is slightly different, so they code for **slightly different versions** of the **same polypeptide**. For example, the gene that determines **blood type** exists as one of three alleles — one determines type O, another type A and the other type B.

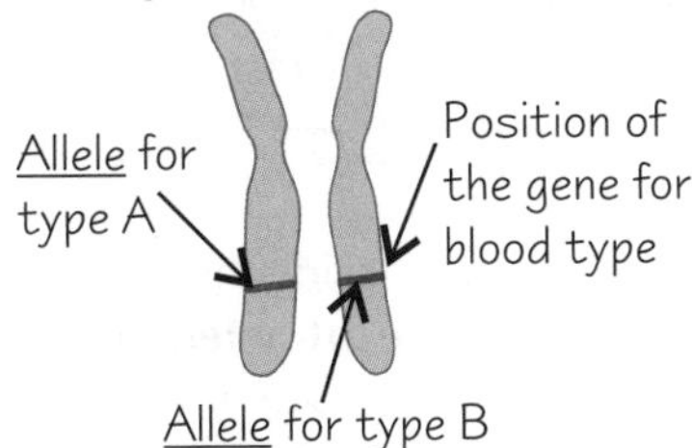

In a **eukaryotic** cell nucleus, DNA is stored as **chromosomes**. Humans have **23 pairs** of chromosomes, 46 in total — two number 1s, two number 2s, two number 3s, etc. Pairs of matching chromosomes (e.g. the 1s) are called **homologous pairs**. In a homologous pair, both chromosomes are the same size and have the **same genes**, although they could have **different alleles**. Alleles coding for the same characteristic will be found at the same **fixed position** (**locus**) on each chromosome in a homologous pair.

Warm-Up Questions

Q1 What is a DNA triplet?
Q2 What is an intron?
Q3 What are non-coding repeats?
Q4 What is a locus?

PRACTICE QUESTIONS

Exam Questions

Q1 Describe how DNA is stored in eukaryotic cells. [5 marks]

Q2 A scientist is studying a DNA sequence that is made up of 3800 nucleotide pairs. Exons account for 672 of the nucleotide pairs. Introns account for 3128 of the nucleotide pairs. The sequence codes for a section of a polypeptide. How many amino acids will make up this section of the polypeptide? [2 marks]

Exons stay in, introns go out, in, out, in, out, and shake it all about...

Quite a few terms to learn here, I'm afraid. Some are a bit confusing too. Just try to remember which way round they go. Introns are the non-coding regions, but exons are extremely important — they actually code for the polypeptide.

RNA and Protein Synthesis

Protein synthesis involves two stages — transcription and translation. They both involve RNA.

There's **More Than One** Type of **RNA**

Remember, RNA is a **single** polynucleotide strand and it contains **uracil** (**U**) as a base instead of thymine (see p. 16). Uracil **always pairs** with **adenine** during protein synthesis. RNA isn't all the same though. You need to know about:

Messenger RNA (mRNA)

mRNA is made during **transcription** (see below). It **carries the genetic code** from the DNA to the ribosomes, where it's used to make a **protein** during **translation** (see next page).
mRNA is a **single polynucleotide strand.**
In mRNA, groups of three adjacent bases are usually called **codons** (they're sometimes called **triplets** or **base triplets**).

Base
A
C
U
codon
Phosphate
Ribose sugar

Transfer RNA (tRNA)

tRNA is involved in **translation**. It **carries** the amino acids that are used to make **proteins** to the **ribosomes.** tRNA is a **single polynucleotide strand** that's folded into a **clover shape**. **Hydrogen bonds** between **specific base pairs** hold the molecule in this shape.
Every tRNA molecule has a **specific sequence** of **three bases** at one end called an **anticodon**.
They also have an **amino acid binding site** at the other end.

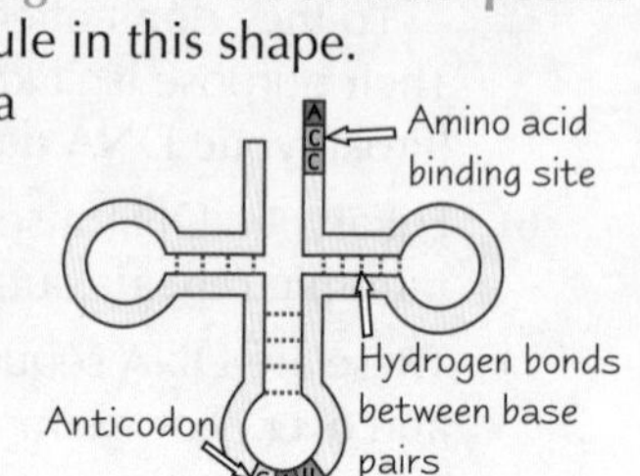

First Stage of **Protein Synthesis — Transcription**

During transcription, an **mRNA copy** of a gene is made from **DNA**.
In **eukaryotic** cells, transcription takes place in the **nucleus**.
(Prokaryotes don't have a nucleus, so transcription takes place in the cytoplasm.)

1) Transcription starts when **RNA polymerase** (an **enzyme**) **attaches** to the **DNA** double-helix at the **beginning** of a **gene**.
2) The **hydrogen bonds** between the two DNA strands in the gene **break**, **separating** the strands, and the DNA molecule **uncoils** at that point, **exposing** some of the **bases**.
3) **One** of the strands is then used as a **template** to make an **mRNA copy**.

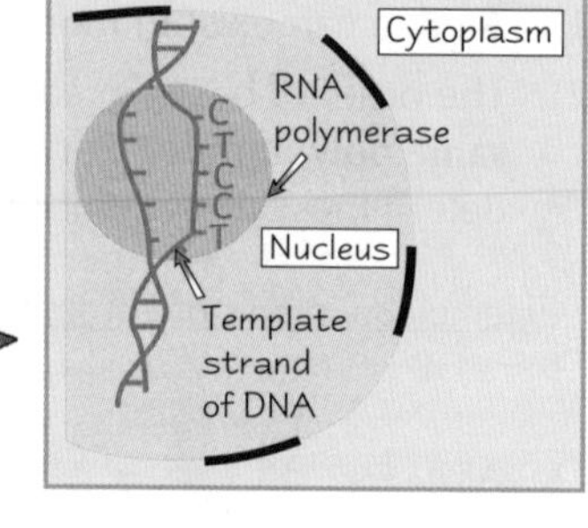

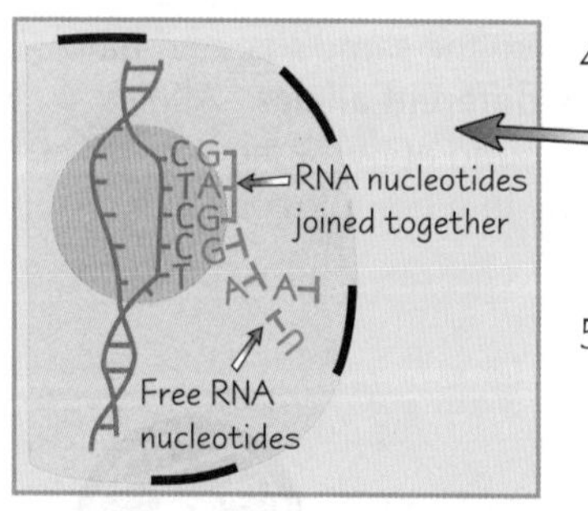

4) The RNA polymerase lines up **free RNA nucleotides** alongside the exposed bases on the template strand. The free bases are **attracted** to the exposed bases. Specific, **complementary base pairing** (see p. 17) means that the mRNA strand ends up being a **complementary copy** of the DNA template strand (except the base **T** is replaced by **U** in **RNA**).
5) Once the RNA nucleotides have **paired up** with their **specific bases** on the DNA strand, they're **joined together** by **RNA polymerase**, forming an **mRNA** molecule.

DNA triplet: A T C
codon on mRNA: U A G

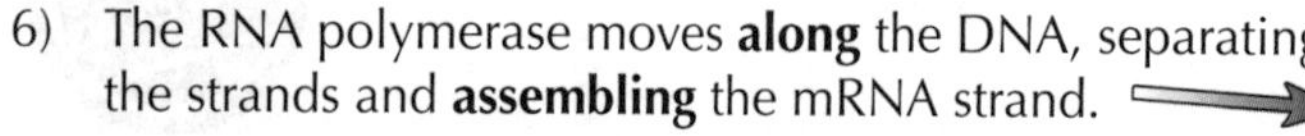

6) The RNA polymerase moves **along** the DNA, separating the strands and **assembling** the mRNA strand.
7) The **hydrogen bonds** between the uncoiled strands of DNA **re-form** once the RNA polymerase has passed by and the strands **coil back into a double-helix**.

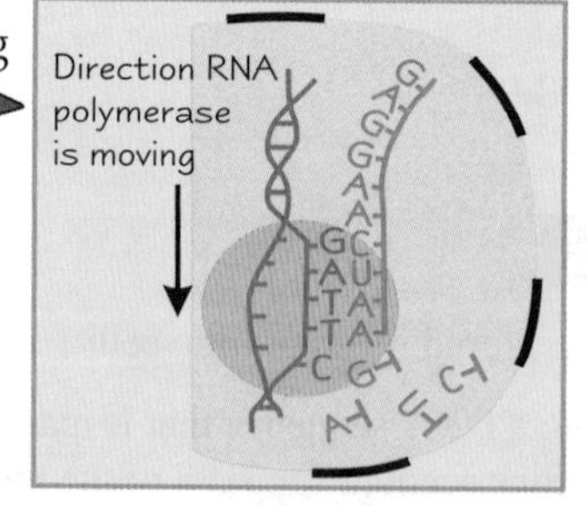

In eukaryotes, it's actually a complex of proteins including a DNA helicase that separates the strands. RNA polymerase just assembles the mRNA strand.

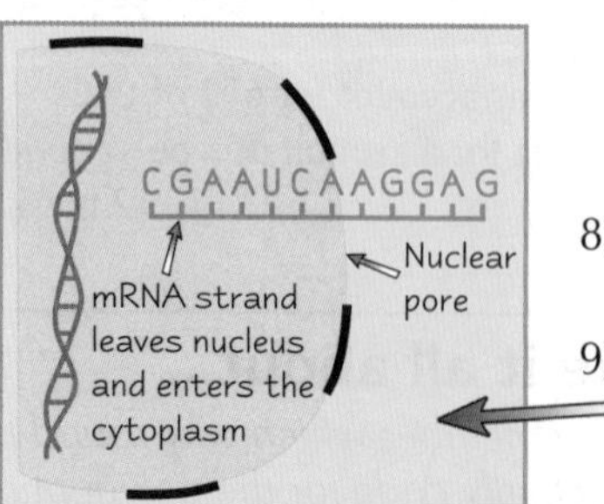

8) When RNA polymerase reaches a particular sequence of DNA called a **stop signal**, it stops making mRNA and **detaches** from the DNA.
9) In eukaryotes, **mRNA** moves **out** of the **nucleus** through a nuclear pore and attaches to a **ribosome** in the cytoplasm, where the next stage of protein synthesis takes place (see next page).

RNA and Protein Synthesis

Transcription Makes Different Products in Eukaryotes and Prokaryotes

1) In **eukaryotes**, the **introns** and **exons** are **both copied** into mRNA during transcription. mRNA strands containing introns and exons are called **pre-mRNA**. A process called **splicing** then occurs — **introns** are removed and the **exons** joined together — forming **mRNA** strands. This takes place in the **nucleus**. The mRNA then **leaves** the nucleus for the next stage of protein synthesis (**translation**).
2) In **prokaryotes**, mRNA is produced **directly** from the DNA — **without** splicing taking place. (There's no need for splicing because there are no introns in prokaryotic DNA.)

Turn to page 89 for more on introns and exons.

Second Stage of Protein Synthesis — Translation

In both eukaryotes and prokaryotes, translation occurs at the **ribosomes** in the **cytoplasm**. During **translation**, **amino acids** are **joined together** to make a **polypeptide chain** (protein), following the sequence of **codons** (triplets) carried by the mRNA.

1) The **mRNA attaches** itself to a **ribosome** and **transfer RNA** (**tRNA**) molecules **carry amino acids** to it. **ATP** provides the energy needed for the **bond** between the **amino acid** and the **tRNA** molecule to form.
2) A **tRNA** molecule (carrying an amino acid), with an **anticodon** that's **complementary** to the **first codon** on the mRNA, attaches itself to the mRNA by **specific base pairing**.

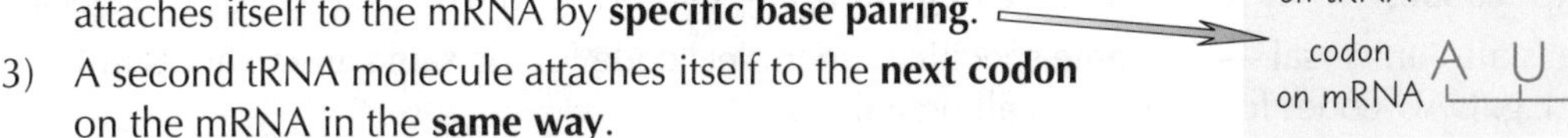

3) A second tRNA molecule attaches itself to the **next codon** on the mRNA in the **same way**.
4) The two amino acids attached to the tRNA molecules are **joined** by a **peptide bond**. The first tRNA molecule **moves away**, leaving its amino acid behind.
5) A third tRNA molecule binds to the **next codon** on the mRNA. Its amino acid **binds** to the first two and the second tRNA molecule **moves away**.
6) This process continues, producing a chain of linked amino acids (a **polypeptide chain**), until there's a **stop signal** on the mRNA molecule.
7) The polypeptide chain **moves away** from the ribosome and translation is complete.

Protein synthesis is also called polypeptide synthesis.

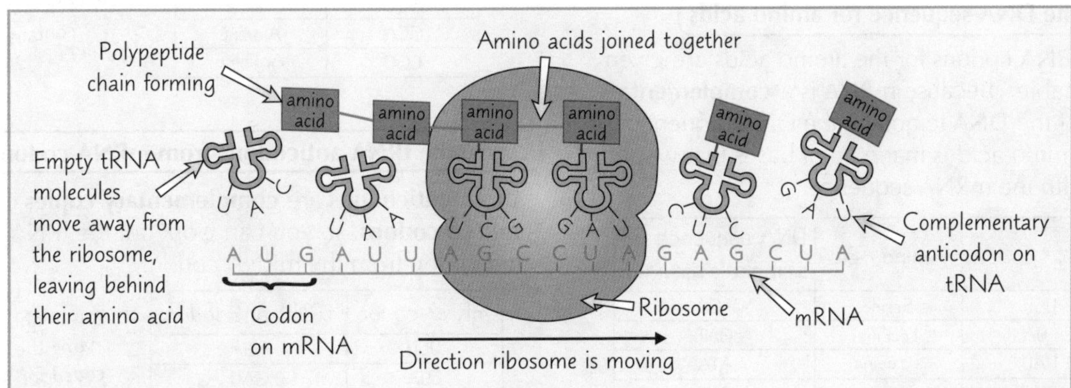

Warm-Up Questions

Q1 Describe the structure of tRNA.

Q2 Where does transcription take place in eukaryotes?

PRACTICE QUESTIONS

Exam Question

Q1 A drug that inhibits cell growth is found to be able to bind to DNA, preventing RNA polymerase from binding. Explain how this drug will affect protein synthesis. [2 marks]

The only translation I'm interested in is a translation of this page into English

So you start off with DNA, lots of cleverness happens and bingo... you've got a protein. Only problem is, you need to know the cleverness in quite a bit of detail. So scribble it down, recite it to yourself, explain it to your best mate or do whatever else helps you remember the joys of protein synthesis. And then think how clever you must be to know it all.

The Genetic Code and Nucleic Acids

The genetic code is exactly as it sounds — a code found in your genes that tells your body how to make proteins. It can be interpreted, just like any other code, which is exactly what you might have to do in your exam...

The Genetic Code is **Non-Overlapping, Degenerate** and **Universal**

1) The genetic code is the **sequence of base triplets** (codons) in **mRNA** which **code** for specific **amino acids**.
2) In the genetic code, each base triplet is **read** in sequence, **separate** from the triplet **before** it and **after** it. Base triplets **don't share** their **bases** — the code is **non-overlapping**.

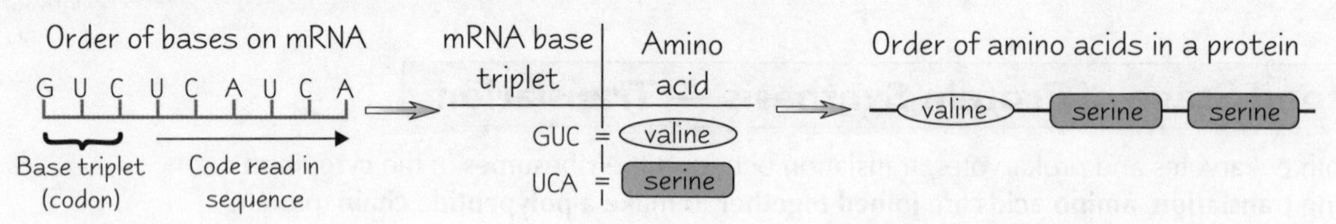

3) The genetic code is also **degenerate** — there are **more** possible combinations of **triplets** than there are amino acids (20 amino acids but 64 possible triplets). This means that some **amino acids** are coded for by **more than one** base triplet, e.g. tyrosine can be coded for by UAU or UAC.
4) Some triplets are used to tell the cell when to **start** and **stop** production of the protein — these are called **start** and **stop** signals (or **codons**). They're found at the **beginning** and **end** of the mRNA. E.g. UAG is a stop signal.
5) The genetic code is also **universal** — the **same** specific base triplets code for the **same** amino acids in **all living things**. E.g. UAU codes for tyrosine in all organisms.

You need to be able to **Interpret Data** about **Nucleic Acids**

The table on the right shows the **mRNA codons** (triplets) for some amino acids. You might have to **interpret** information like this in the exam. For example, using the table, you could be asked to...

mRNA codon	Amino Acid
UCU	Serine
CUA	Leucine
UAU	Tyrosine
GUG	Valine
GCA	Alanine
CGC	Arginine

When interpreting data on nucleic acids remember that DNA contains T and RNA contains U.

...give the DNA sequence for amino acids

The mRNA codons for the amino acids are given in the table. Because **mRNA** is a **complementary copy** of the **DNA** template, the DNA sequence for each amino acid is made up of bases that would **pair** with the mRNA sequence:

mRNA codon	Amino Acid	DNA sequence (of template strand)
UCU	Serine	AGA
CUA	Leucine	GAT
UAU	Tyrosine	ATA
GUG	Valine	CAC
GCA	Alanine	CGT
CGC	Arginine	GCG

You could also be asked to work out the amino acids from a given DNA sequence and a table.

...give the tRNA anticodons from mRNA codons

tRNA anticodons are **complementary copies** of **mRNA codons**, so you can work out the tRNA anticodon from the mRNA codon:

mRNA codon	tRNA anticodon
UCU	AGA
CUA	GAU
UAU	AUA
GUG	CAC
GCA	CGU
CGC	GCG

You might be asked to name the amino acid coded for by a tRNA anticodon using a table like the one above.

...write the amino acid sequence for a section of mRNA

To **work out** the sequence of **amino acids** from some mRNA, you need to break the genetic code into **codons** and then use the information in the table to work out what **amino acid** they code for.

You might have to work out the sequence of some mRNA from a sequence of amino acids and a table.

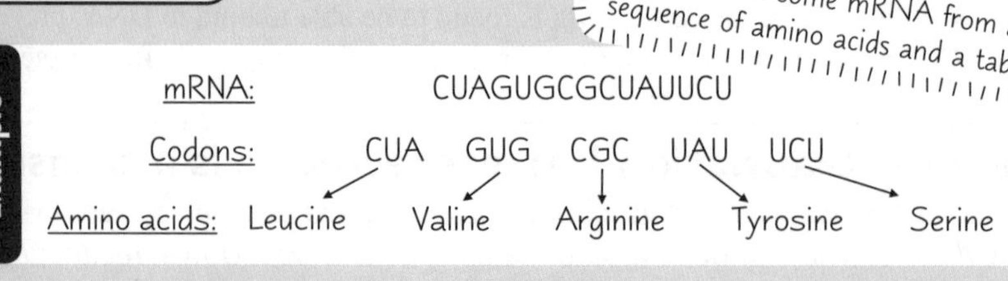

The Genetic Code and Nucleic Acids

You Might Have to **Interpret Data** About The **Role** of **Nucleic Acids**

In the exam you might have to **interpret data** from experiments done to **investigate nucleic acids** and their **role** in **protein synthesis**. Here's an example (you **don't** need to **learn** it):

Investigating the effect of new drugs on nucleic acids

1) To investigate **how** two new drugs affect **nucleic acids** and their **role** in protein synthesis, **bacteria** were **grown** in **normal conditions** for a few generations, then moved to media containing the drugs.
2) After a short period of time, the **concentration** of **protein** and **complete strands** of **mRNA** in the bacteria were analysed. The results are shown in the **bar graph**.
3) Both mRNA **and** protein concentration were **lower** in the presence of **drug 1 compared** to the **no-drug control**. This suggests that drug 1 **affects the production** of **full length mRNA**, so there's no mRNA for protein synthesis during **translation**.
4) **mRNA production** in the presence of **drug 2** was **unaffected**, but **less protein** was produced — **3** mg cm^{-3} compared to **8** mg cm^{-3}. This suggests that drug 2 **interferes** with **translation**. **mRNA was produced**, but **less protein** was **translated** from it.
5) **Further tests** to establish the **nature** of the two drugs were carried out.
6) **Drug 1** was found to be a **ribonuclease** (an enzyme that **digests RNA**). This could **explain** the results of the first experiment — **any strands** of **mRNA** produced by the cell would be **digested** by drug 1, so **couldn't be used** in **translation** to make proteins.
7) **Drug 2** was found to be a **single-stranded**, **clover-shaped** molecule capable of binding to the **ribosome**. Again, this helps to **explain** the **results** from the first experiment — drug 2 could work by **binding** to the ribosome, **blocking tRNAs** from binding to it and so **preventing translation**.

Bar chart to show mRNA and protein concentration in the presence and absence of drugs

Protein / mg cm^{-3} (0–10); mRNA / µg cm^{-3} (0–5)

■ Protein ■ mRNA

No drug | Drug 1 | Drug 2

A molecule capable of binding to mRNA would have had a similar effect to drug 1, since it would have prevented mRNA being read by the ribosomes and stopped it being translated.

Transcription and translation are on pages 90-91.

Warm-Up Questions

Q1 What is the genetic code?

Q2 Why is the genetic code described as degenerate?

Q3 Why is the genetic code described as universal?

mRNA codon	amino acid
UGU	Cysteine
CGC	Arginine
GGG	Glycine
GUG	Valine
GCA	Alanine
UUG	Leucine
UUU	Phenylalanine

Exam Questions

Q1 The table shows the mRNA codons for some amino acids. Show your working for the following questions.

a) Give the amino acid sequence for the mRNA sequence: GUGUGUCGCGCA. [2 marks]

b) Give the DNA template strand sequence that codes for the amino acid sequence: valine, arginine, alanine. [3 marks]

Q2 An artificial mRNA was synthesised to code for a particular polypeptide. Part of the mRNA sequence was: UUGUGUGGGUUUGCAGCA. This produced the following sequence of amino acids: Leucine–Cysteine–Glycine–Phenylalanine–Alanine–Alanine. Use the table above to help you answer the following questions.

a) Explain how the result suggests that the genetic code is based on triplets of nucleotides in mRNA. [2 marks]

b) Explain how the result suggests that the genetic code is non-overlapping. [2 marks]

—•• ——— ••• ——— —— • •—— ——— •—• —•—

Yay — a page with slightly fewer confusing terms and a lot less to remember. The key to the genetic code is to be able to interpret it, so if you know how DNA, mRNA and tRNA work together to make a protein you should be able to handle any data they can throw at you. Remember, C pairs with G, A pairs with T. Unless it's RNA — then it's U.

Meiosis and Genetic Variation

Ahh, now on to some really exciting stuff — the production of gametes (sex cells to you and me). This is how we end up different from our parents and our siblings — and yet, in some ways, strangely alike...

DNA from One Generation is Passed to the Next by Gametes

1) **Gametes** are the **sperm** cells in males and **egg** cells in females. They join together at **fertilisation** to form a **zygote**, which divides and develops into a **new organism**.
2) Normal **body cells** have the **diploid number** (**2n**) of chromosomes — meaning each cell contains **two** of each chromosome, one from the mum and one from the dad.
3) **Gametes** have a **haploid** (**n**) number of chromosomes — there's only one copy of each chromosome.
4) At **fertilisation**, a **haploid sperm** fuses with a **haploid egg**, making a cell with the normal diploid number of chromosomes. Half these chromosomes are from the father (the sperm) and half are from the mother (the egg).

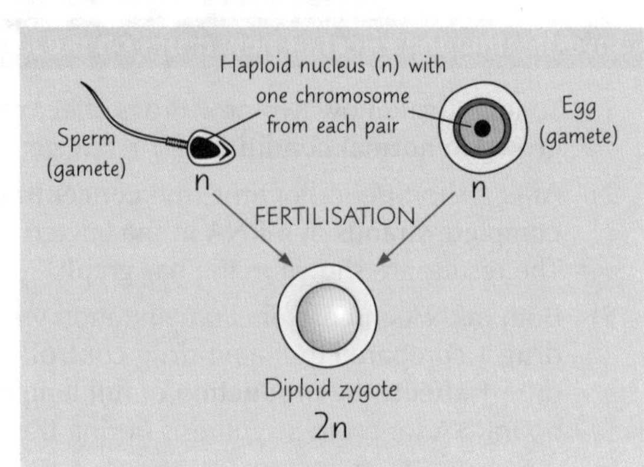

5) During sexual reproduction, any sperm can fertilise any egg — **fertilisation** is **random**. Random fertilisation produces zygotes with **different combinations** of **chromosomes** to both parents. This **mixing** of **genetic material** in sexual reproduction **increases genetic diversity** within a **species** (there's more on genetic diversity on page 98).

Gametes are Formed by Meiosis

Meiosis is a type of cell division. It takes place in the **reproductive organs**. Cells that divide by meiosis are **diploid** to start with, but the cells that are formed from meiosis are **haploid** — the chromosome number **halves**. Without meiosis, you'd get **double** the number of chromosomes when the gametes fused. Not good.

1) Before meiosis starts, the DNA unravels and **replicates** so there are **two** copies of **each** chromosome, called **chromatids**.
2) The DNA condenses to form double-armed chromosomes, each made from **two sister chromatids**. The sister chromatids are joined in the middle by a **centromere**.
3) **Meiosis I** (first division) — the chromosomes arrange themselves into **homologous pairs**.
4) These homologous **pairs** are then **separated**, **halving** the chromosome number.
5) **Meiosis II** (second division) — the pairs of sister **chromatids** that make up each chromosome are **separated** (the **centromere** is divided).
6) **Four haploid cells** (gametes) that are **genetically different** from each other are produced.

A Note About Homologous Pairs: Humans have 46 chromosomes in total — 23 pairs. One chromosome in each pair came from mum and one from dad, e.g. there are two number 1's (one from mum and one from dad), two number 2's etc. The chromosomes that make up each pair are the same size and have the same genes, although they could have different versions of those genes (called alleles). These pairs of chromosomes are called homologous pairs.

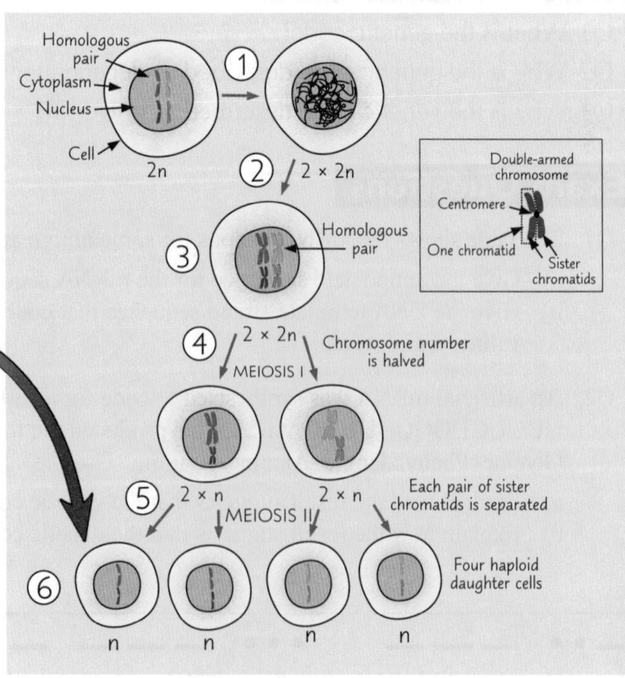

In the exams, you might need to spot **when meiosis happens** in an organism with a **life cycle you haven't seen before**, e.g. an insect or plant. Just remember that in any organism, **meiosis** is needed for **sexual reproduction** because it produces daughter cells (gametes) with **half** the number of **chromosomes** of the parent cell.

You might also be told **how many** chromosomes are in a parent cell, then asked to **complete diagrams** showing how many chromosomes will be in the daughter cells after the **first** and **second divisions** of meiosis. Remember that the **chromosome number** is **halved** during the **first division**.

Meiosis and Genetic Variation

Chromatids **Cross Over** in **Meiosis I**

During meiosis I, **homologous pairs** of chromosomes come together and pair up. The chromatids twist around each other and bits of **chromatids** swap over. The chromatids still contain the **same genes** but now have a different combination of **alleles**.

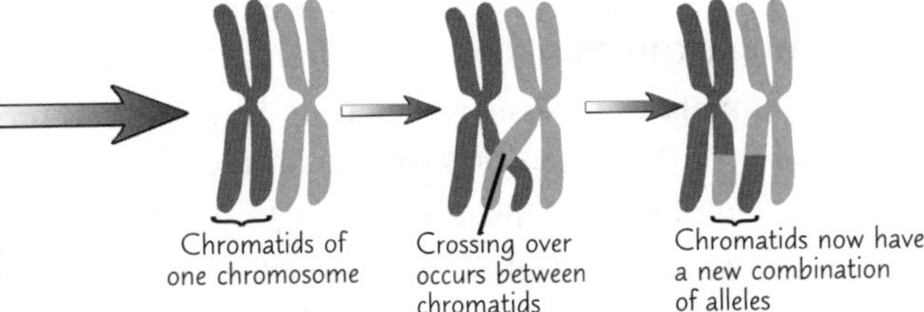

Meiosis Produces Cells that are **Genetically Different**

There are two main events during meiosis that lead to **genetic variation**:

1 Crossing over of chromatids

The **crossing over** of chromatids in meiosis I means that each of the **four daughter cells** formed from meiosis contains chromatids with **different alleles**:

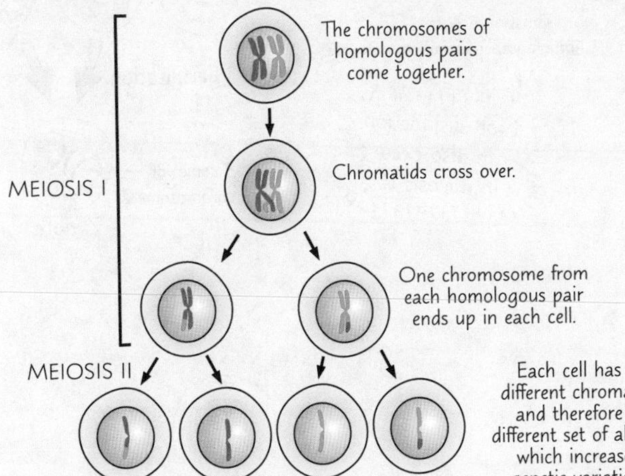

2 Independent segregation of chromosomes

1) Each **homologous pair** of chromosomes in your cells is made up of **one chromosome** from your mum (**maternal**) and **one chromosome** from your dad (**paternal**).
2) When the homologous pairs are **separated** in **meiosis I**, it's completely **random** which chromosome from each pair ends up in which daughter cell.
3) So the **four daughter cells** produced by meiosis have completely **different combinations** of those **maternal** and **paternal chromosomes**.
4) This is called **independent segregation** (separation) of the chromosomes.
5) This '**shuffling**' of chromosomes leads to **genetic variation** in any **potential** offspring.

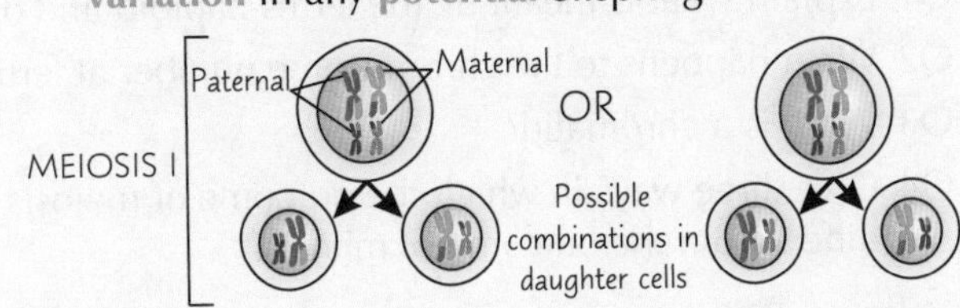

Meiosis Has a **Different Outcome** to **Mitosis**

You may remember **mitosis** from page 34. **Mitosis** and **meiosis** have **different outcomes**:

	Outcomes:		
Mitosis	Produces cells with the **same number** of chromosomes as the parent cell.	Daughter cells are **genetically identical** to each other and to the parent cell.	Produces **two** daughter cells.
Meiosis	Produces cells with **half** the number of chromosomes as the parent cell.	Daughter cells are **genetically different** from one another and the parent cell.	Produces **four** daughter cells.

You need to be able to **explain** the different outcomes of mitosis and meiosis. They're different because **mitosis** only involves **one division** (which separates the sister chromatids) whereas **meiosis** has **two divisions** (which separate the homologous pairs and then the sister chromatids). There's **no pairing** or **separating** of **homologous chromosomes** in **mitosis**, and so **no crossing over** or **independent segregation of chromosomes**. This produces **genetically identical** daughter cells — unlike **meiosis**.

Meiosis and Genetic Variation

Chromosome Mutations are Caused by Errors in Cell Division

1) In humans, when meiosis **works properly**, all four daughter cells will end up with **23 whole chromosomes** — one from each homologous pair (1 to 23).
2) But sometimes meiosis **goes wrong** and the cells produced contain **variations** in the numbers of whole chromosomes or **parts** of chromosomes.
3) For example, two cells might have 23 whole chromosomes, one each of 1 to 23, but the other two might get a bit muddled up, one having two chromosome 6's and the other no chromosome 6.
4) This is called **chromosome mutation** and is caused by **errors** during meiosis.
5) Chromosome mutations lead to **inherited conditions** because the errors are present in the **gametes** (the hereditary cells).

One type of chromosome mutation is called **non-disjunction** — it's a **failure** of the **chromosomes** to **separate** properly. In humans, non-disjunction of **chromosome 21** during **meiosis** can lead to **Down's Syndrome**.

1) **Down's syndrome** is caused by a person having an **extra copy** of **chromosome 21** (or sometimes an extra copy of part of chromosome 21).
2) Non-disjunction means that chromosome 21 **fails to separate properly** during **meiosis**, so one cell gets an **extra copy** of 21 and another gets **none**.
3) When the gamete with the **extra copy** fuses to another gamete at **fertilisation**, the resulting zygote will have **three** copies of chromosome 21.

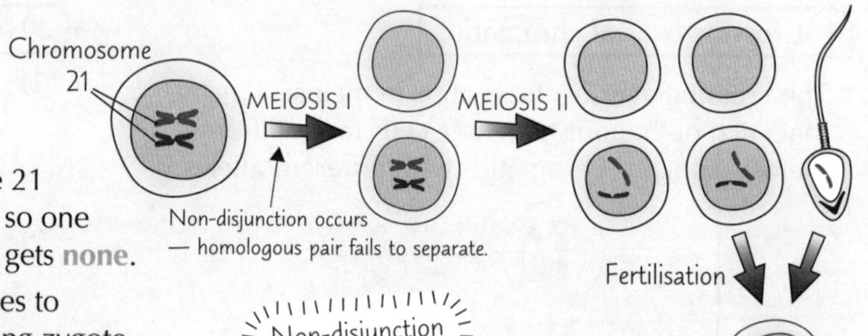

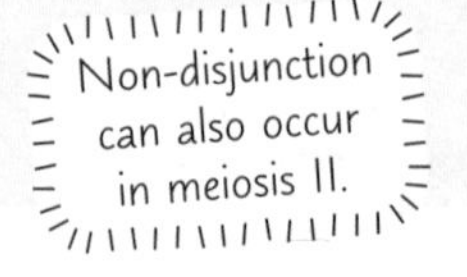

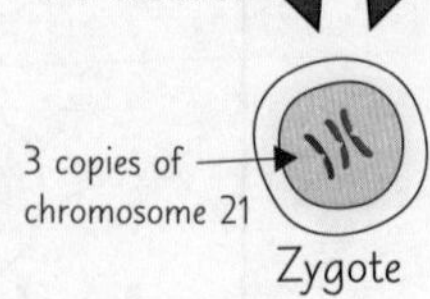

Warm-Up Questions

Q1 Explain what is meant by the terms haploid and diploid.

Q2 What happens to the chromosome number at fertilisation?

Q3 What is a chromatid?

Q4 Give three ways in which the outcome of meiosis differs from the outcome of mitosis.

Exam Questions

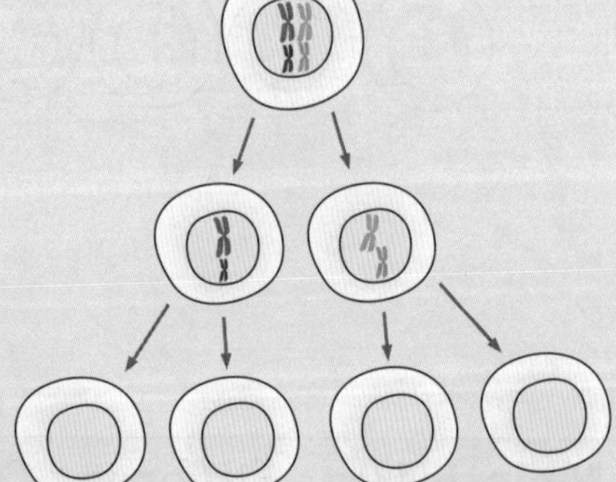

Q1 *Drosophila* (fruit flies) only have four chromosomes. The diagram on the right summarises meiosis in *Drosophila*.

a) Complete the diagram to show the chromosomes in the four daughter cells. [1 mark]

b) Crossing over does not occur very frequently in male *Drosophila*. Explain what crossing over is and how it leads to genetic variation. [4 marks]

c) Explain how independent segregation leads to genetic variation. [2 marks]

Q2 Turner syndrome is a genetic condition affecting females. It is caused by non-disjunction of the sex chromosomes. Females usually have two X chromosomes. Some females with Turner syndrome have only one X chromosome.

Suggest and explain how chromosome non-disjunction could cause Turner syndrome. [3 marks]

Reproduction isn't as exciting as some people would have you believe...

These pages are quite tricky, so use the diagrams to help you understand — they might look evil, but they really do help. The key thing to understand is that meiosis produces four genetically different haploid (n) daughter cells. And that the genetic variation in the daughter cells occurs because of two processes — crossing over and independent segregation.

Mutations

Aside from chromosome mutations, other types of genetic mutations can also occur — some useful, some not so.

Mutations are Changes to the Base Sequence of DNA

Gene **mutations** involve a **change** in the **DNA base sequence** of chromosomes.

1) The **types** of errors that can occur include:

> Errors can also be caused by insertion, duplication, addition and translocation of bases.

Substitution — one base is substituted with another, e.g. ATGCCT becomes ATTCCT (G is swapped for T).
Deletion — one base is deleted, e.g. ATGCCT becomes ATCCT (G is deleted).

2) The **order** of **DNA bases** in a gene determines the **order of amino acids** in a particular **protein** (see p. 88). If a mutation occurs in a gene, the **sequence** of **amino acids** it codes for (and the protein formed) could be **altered**:

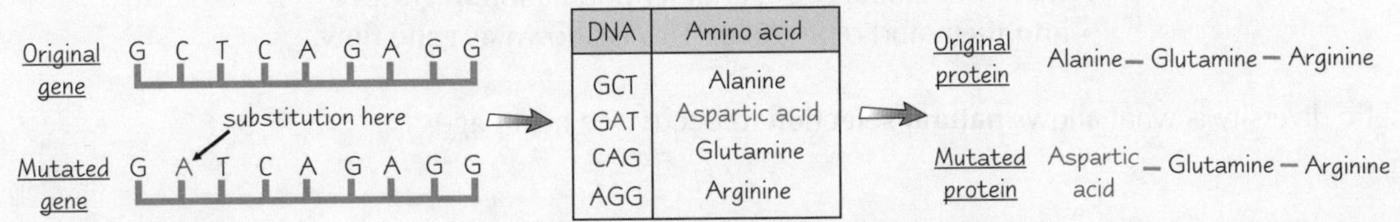

Not All Mutations Affect the Order of Amino Acids

The **degenerate nature** of the genetic code (see page 92) means that some amino acids are coded for by **more than one DNA triplet** (e.g. tyrosine can be coded for by TAT or TAC in DNA). This means that **not all** substitution mutations will result in a change to the amino acid sequence of the protein — some substitutions will still **code** for the **same amino acid**. For example:

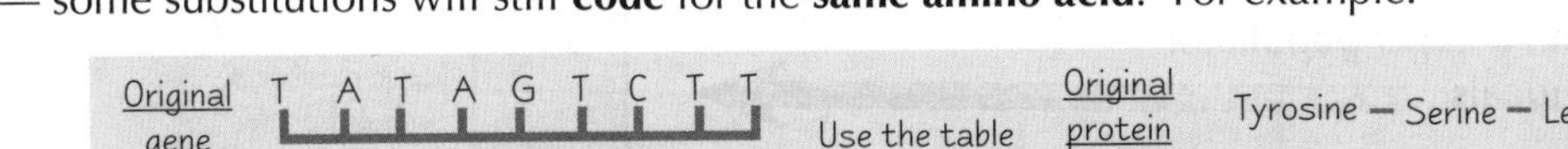

DNA	Amino acid
TAT	Tyrosine
TAC	Tyrosine
AGT	Serine
CTT	Leucine
GTC	Valine

Substitution mutations **won't always** lead to changes in the amino acid sequence, but **deletions will** — the deletion of a base will change the **number** of bases present, which will cause a **shift** in all the base triplets after it.

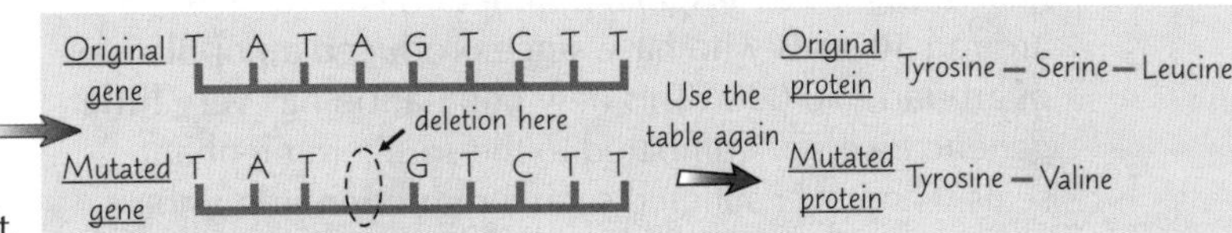

Mutagenic Agents Increase the Rate of Mutation

Mutations occur **spontaneously**, e.g. when DNA is **misread** during **replication**. But some things can cause an **increase** in the **rate** of **mutations** — these are called **mutagenic agents**. **Ultraviolet radiation**, **ionising radiation**, some **chemicals** and some **viruses** are examples of mutagenic agents.

Warm-Up Questions

PRACTICE QUESTIONS

Q1 What are mutagenic agents?

Exam Question

DNA	Amino acid
AGT	Serine
TAT	Tyrosine
CTT	Leucine
AGG	Arginine

Original gene T A T A G T C T T

Mutated gene T A T A G G C T T

Q1 A mutation occurred during DNA replication. The diagram on the right shows part of the original gene and the mutated gene.

a) What type of mutation has occurred? [1 mark]

b) Using the table provided, explain the effects that this mutation would have on the amino acid sequence. [2 marks]

What do you get if you cross James Bond with the Hulk™?*

Mutations affect the sequence of amino acids produced — learn what happens for substitutions and deletions.

*A mutagenic agent

Genetic Diversity and Natural Selection

Genetic diversity describes the number of alleles in a species or population, and natural selection acts to increase the proportion of advantageous alleles. It's all about the most well-adapted organisms getting on with some reproductio

Lots of Different Alleles Means a High Genetic Diversity

1) Remember, there can be **different versions** of a single **gene** — these are called **alleles** (see page 89).
2) **Genetic diversity** is the number of **different alleles** of genes in a species or population.
3) Genetic diversity **within** a **population** is increased by:

- **Mutations** in the DNA — forming **new alleles**.
- **Different alleles** being **introduced** into a population when individuals from another population **migrate into them** and reproduce. This is known as **gene flow**.

A population is a group of organisms of one species living in a particular habitat.

4) Genetic diversity is what allows **natural selection** to occur (see next page).

Genetic Bottlenecks Reduce Genetic Diversity

1) A **genetic bottleneck** is an event that causes a big **reduction** in a population, e.g. when a large number of organisms within a population **die** before reproducing.
2) This reduces the number of **different alleles** in the **gene pool** and so reduces **genetic diversity**.
3) The survivors **reproduce** and a larger population is created from a few individuals.

The gene pool is the complete range of alleles in a population.

Example — Northern Elephant Seals

Northern elephant seals were hunted by humans in the late 1800s. Their **original population** was reduced to around **50 seals** who have since produced a population of around 170 000. This new population has **very little** genetic diversity compared to the southern elephant seals who never suffered such a **reduction** in numbers.

The Founder Effect is a Type of Genetic Bottleneck

1) The **founder effect** describes what happens when just a **few** organisms from a population start a **new colony** and there are only a **small number** of **different alleles** in the **initial gene pool**.
2) The **frequency** of each allele in the **new colony** might be **very different** to the frequency of those alleles in the original population — for example, an allele that was **rare** in the original population might be **more common** in the new colony. This may lead to a **higher incidence** of **genetic disease**.
3) The founder effect can occur as a result of **migration** leading to geographical **separation** or if a new colony is separated from the original population for **another reason**, such as **religion**.

Example — The Amish

The **Amish population** of North America are all descended from a **small** number of Swiss who **migrated** there. The population shows **little genetic diversity**. They have remained **isolated** from the surrounding population due to their **religious beliefs**, so **few new alleles** have been introduced. The population has an unusually high incidence of certain **genetic disorders**.

Genetic Diversity and Natural Selection

Natural Selection Increases Advantageous Alleles in a Population

Randomly-occurring **mutations** sometimes result in a **new allele** being formed. This can be **harmful**, which usually means that the **mutated allele** quickly **dies** out. However, **some mutations** can produce **alleles** that are **beneficial** to an organism (e.g. a protein is produced that works better than the original), helping the organism to **survive** in certain environments. When the allele codes for a characteristic that **increases** the **chances** of an organism **surviving**, its **frequency** within the population can **increase**. This process is known as **natural selection**. Here's how it works:

1) Not all individuals are as likely to **reproduce** as each other. There's **differential reproductive success** in a population — individuals that have an allele that **increases** their **chance of survival** are **more likely** to **survive**, **reproduce** and **pass on** their genes (including the **beneficial** allele), than individuals with different alleles.
2) This means that a **greater proportion** of the next generation **inherits** the **beneficial allele**.
3) They, in turn, are **more likely** to **survive**, **reproduce** and **pass on** their genes.
4) So the **frequency** of the beneficial allele **increases** from generation to generation.
5) Over **generations** this leads to **evolution** as the **advantageous alleles** become **more common** in the population.

Adaptation and **selection** are both key factors in **evolution** — the **gradual change** in **species** over **time**. Evolution has led to the **huge diversity** of **living organisms** on Earth.

Natural Selection Leads to Populations Becoming Better Adapted

Adaptations help organisms to **survive** in their **environment**. They can be **behavioural**, **physiological** or **anatomical**. Here are some examples:

Bob and Sue were well adapted to hiding in candyfloss shops.

1) Behavioural adaptations

Ways an organism **acts** that increase its chance of survival and reproduction. For example, **possums** sometimes **'play dead'** if they're being threatened by a **predator** to **escape attack**.

2) Physiological adaptations

Processes inside an organism's body that increase its chance of survival. For example, **brown bears hibernate** over **winter**. They **lower their rate of metabolism** (all the chemical reactions taking place in their body). This **conserves energy**, so they don't need to look for **food** in the months when it's scarce.

3) Anatomical adaptations

Structural features of an organism's body that increase its chance of survival. For example, **whales** have a **thick layer of blubber** (fat) which helps them keep **warm** in the cold sea.

Warm-Up Questions

Q1 What is genetic diversity?

Q2 Explain how a genetic bottleneck reduces genetic diversity.

Q3 Give an example of a behavioural adaptation.

Exam Question

Q1 Tawny owls show variation in colour. There are light grey owls and darker brown owls. Before the 1970s there were more grey owls than brown owls in Finland. Since then, climate change has been causing a decrease in the amount of snowfall in Finland. During this period, the darker brown owls have become more common.

a) Suggest why the brown owls are better adapted to living in an area with less snowfall than the grey owls. [2 marks]

b) Explain how the brown owls have become more common. [3 marks]

I'm perfectly adapted — for staying in bed...

Just remember that any mutation that increases the chances of an organism surviving (e.g. thicker blubber for keeping warm) or reproducing will increase in the population due to the process of natural selection.

Investigating Selection

Now you get to apply what you know about natural selection to bacteria and babies (amongst other things). Natural selection affects different populations in different ways, as you'll soon discover...

Different Types of Natural Selection Lead to Different Frequency Patterns

You might remember from the previous page that **natural selection** alters **allele frequency** in a population. **Stabilising selection** and **directional selection** are **types** of **natural selection** that affect **allele frequency** in different ways. You need to learn these examples:

Antibiotic Resistance Shows Directional Selection

Directional selection is where individuals with alleles for characteristics of an **extreme type** are more likely to **survive** and **reproduce**. This could be in response to an **environmental change**. **Bacteria** evolving **antibiotic resistance** is an example of **directional selection**. Here's how it works:

1) Some individuals in a population have alleles that give them **resistance** to an **antibiotic**.
2) The population is **exposed** to the antibiotic, **killing** bacteria **without** the resistant allele.
3) The **resistant bacteria survive** and **reproduce** without competition, passing on the **allele** that gives antibiotic resistance to their offspring.
4) After some time, **most** organisms in the population will carry the **antibiotic resistance allele**.

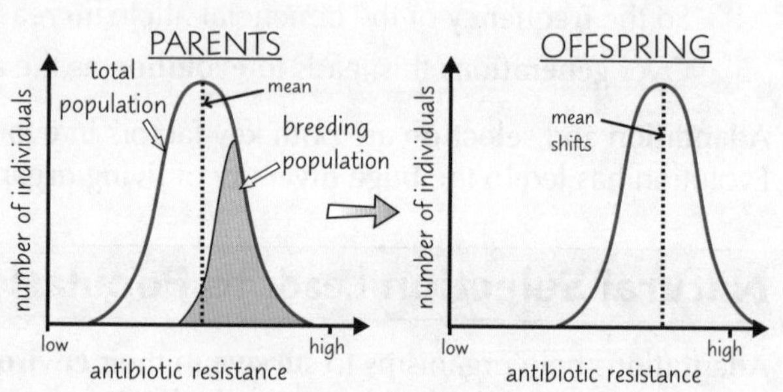

Human Birth Weight Shows Stabilising Selection

Stabilising selection is where individuals with alleles for characteristics towards the **middle** of the range are more likely to **survive** and **reproduce**. It occurs when the environment **isn't changing**, and it **reduces the range** of possible **characteristics**. An example of **stabilising selection** is **human birth weight**.

1) Humans have a **range** of **birth weights**.
2) Very **small babies** are **less likely** to **survive** — partly because they find it **hard** to **maintain** their **body temperature**.
3) Giving birth to **large babies** can be difficult, so large babies are **less likely** to **survive** too.
4) Conditions are **most favourable** for **medium-sized babies** — so weight of human babies tends to **shift towards** the **middle** of the range.

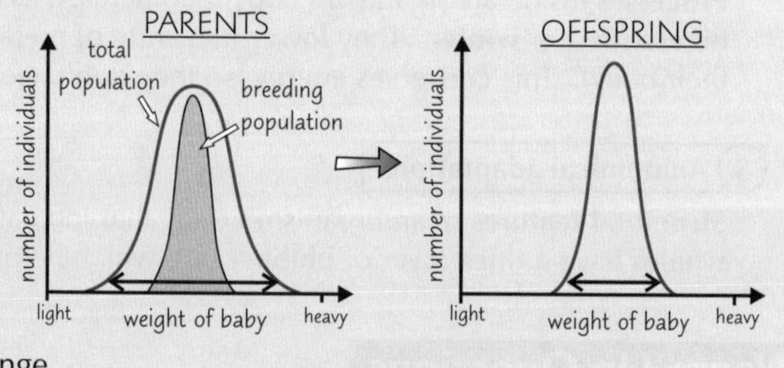

You Need to be Able to Interpret Data on the Effects of Selection

You might be asked to **interpret** information about an **unfamiliar species** in the exam. For example:
There is a population of **rabbits** with **varying fur length**. **Longer fur** helps to keep the rabbits **warmer**. The graph shows how the **average fur length** of the rabbits **changed** over a period of six years, which had particularly **cold winters**. The **bars** span the **difference** between the **shortest** and **longest fur lengths** recorded.

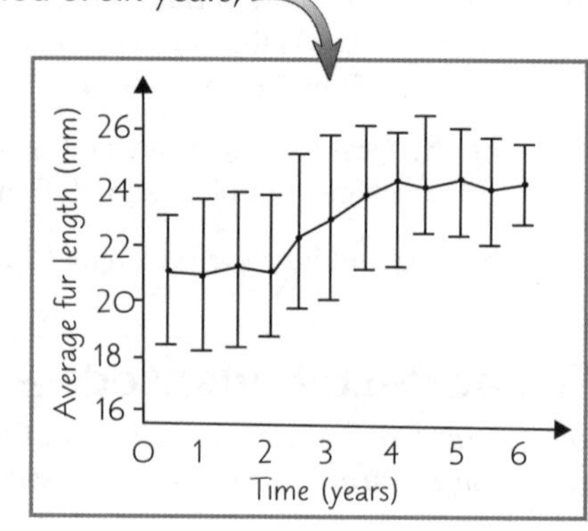

Describe what the data shows:
Over the first two years the average fur length is about 21 mm. However, the average length gradually increases from 21 mm to 24 mm. This shows directional selection.

Suggest a possible cause:
The rabbits with the longer fur are more likely to survive the cold winters than the short-furred rabbits. This makes them more likely to reproduce and, when they do, they pass on the allele for longer fur to the next generation. Over time, the allele for longer fur becomes more common in the population and the average fur length of the rabbits increases.

Investigating Selection

You Can **Investigate** the **Effects** of **Antibiotics** on **Bacterial Growth**

PRACTICAL SKILLS

You need to know how to **investigate** the effect of **antimicrobial substances** (e.g. antibiotics, antiseptics or disinfectants) on **microbial growth**, using **aseptic techniques**.

Test the Effects of **Antibiotics** Using **Agar Plates**

1) The bacteria you will use are likely to have been grown in a **liquid broth** (a mixture of distilled water, bacterial culture and nutrients).
2) Use a **sterile pipette** to **transfer** the bacteria from the broth to an **agar plate** (a **Petri dish** containing **agar jelly**). Spread the bacteria over the plate using a sterile plastic spreader.
3) Use **sterile forceps** to place paper discs **soaked** with different **antibiotics** spaced apart on the plate. Make sure you add a **negative control** disc soaked only in **sterile water**.
4) **Lightly tape** a **lid** on, invert, and **incubate** the plate at about **25 °C** for **48 hours**. This allows the bacteria to **grow** (forming a '**lawn**'). Anywhere the bacteria **can't grow** can be seen as a **clear patch** in the lawn of bacteria. This is called an **inhibition zone**.
5) The size of an **inhibition zone** tells you how well an antibiotic works. The **larger** the zone, the **more** the bacteria were inhibited from growing.
6) A **similar technique** can be used to test the effects of **antiseptics** or **disinfectants** on microbial growth.

An agar plate after incubation with discs of **meticillin**, **tetracycline** and **streptomycin** at different concentrations:

125 mg
250 mg
Inhibition zone
Agar plate
Meticillin
Tetracycline
Streptomycin
Negative control (soaked in water)
Disc
Lawn of bacteria

- The **tetracycline** discs have **no** inhibition zones, so the bacteria are **resistant** to tetracycline up to 250 mg.
- The **streptomycin** discs have **small** inhibition zones, with the zone at 250 mg slightly larger than the one at 125 mg. So streptomycin has **some effect** on the bacteria.
- The **meticillin** discs have the **largest** inhibition zones, so meticillin has the **strongest effect** on these bacteria.

Always Use **Aseptic Techniques** to **Prevent Contamination** of Microbial Cultures

Aseptic techniques are used to **prevent contamination** of cultures by **unwanted** microorganisms. This is important because contamination can affect the **growth** of the microorganism that you're **working** with. It's also important to avoid contamination with **disease-causing microbes** that could make you **ill**.
When carrying out the investigation above, you need to use the following **aseptic techniques**:

- Regularly **disinfect work surfaces** to minimise contamination. Don't put any **utensils** on the work surface. **Contaminated** utensils should be placed in a beaker of **disinfectant**.
- Use sterile equipment and discard safely after use. E.g. glassware can be sterilised before and after use in an autoclave (which steams equipment at high pressure). Pre-sterilised plastics instruments are used once, then discarded.
- Work **near** a **Bunsen flame**. **Hot air rises**, so any microbes in the air should be drawn away from your culture.
- **Minimise** the **time** spent with the lid off the agar plate, to reduce the chance of **airborne microorganisms** contaminating the culture.
- Briefly **flame** the neck of the glass **container of broth** just after it's **opened** and just before it's **closed** — this causes air to move out of the container, preventing **unwanted** organisms from **falling in**.

You should also take steps to protect yourself, e.g. wash your hands thoroughly before and after handling cultures.

Warm-Up Questions

PRACTICE QUESTIONS

Q1 Describe how you could investigate the effects of antibiotics on bacterial growth.

Exam Question

Q1 A group of scientists monitored how the colour of oyster shells on a beach changed over time. The graph shows the colour of the oyster shells in the scientists' initial sample and in their final sample. The oysters were mainly found on the sand, which was a mid-brown colour.

a) What type of selection is shown in the graph? Explain your answer. [3 marks]

b) Suggest how the changes shown in the graph might have taken place. [4 marks]

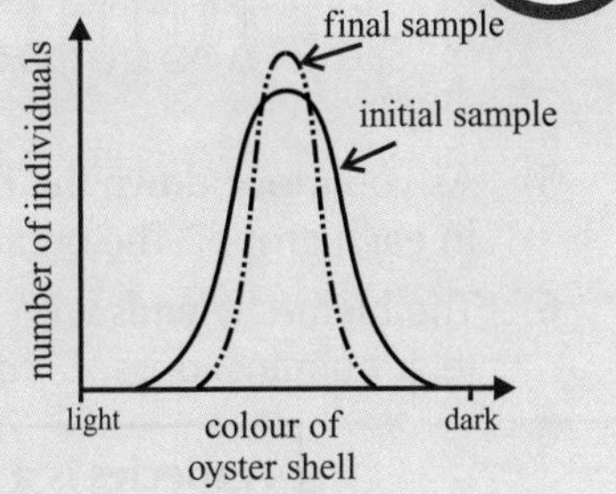

One Directional selection — the shift in the population of teenage girls...

Quite a bit to learn here — maybe try the whole cover, scribble, check thing to make sure you remember the details.

Classification of Organisms

For hundreds of years people have been putting organisms into groups to make it easier to recognise and name them. For example, my brother is a member of the species Idioto bigearian (Latin for idiots with big ears).

Phylogeny Tells Us About the **Evolutionary History** of Organisms

1) **Phylogeny** is the study of the **evolutionary history** of groups of **organisms**. Phylogeny tells us **who's related** to whom and how **closely related** they are.
2) All organisms have **evolved** from shared **common ancestors** (relatives). This can be shown on a **phylogenetic tree**, like this one.

Orangutan
Human
Chimpanzee
Bonobo
Gorilla
First branch point

3) This tree shows the **relationship** between members of the Hominidae family (great apes and humans). The **first branch point** represents a **common ancestor** of **all** the family members. This ancestor is now **extinct**. **Orangutans** were the first group to **diverge** (evolve to become a different species) from this common ancestor.
4) Each of the following branch points represents **another common ancestor** from which a **different group diverged**. Gorillas diverged next, then humans, closely followed by bonobos and chimpanzees.
5) Closely related species **diverged** away from each other **most recently**. E.g. humans and **chimpanzees** are **closely** related, as they diverged very **recently**. You can see this because their branches are **close** together.

Classification is All About **Grouping Together Related Organisms**

Taxonomy is the science of classification. It involves **naming** organisms and **organising them** into **groups**. This makes it **easier** to **identify** and **study** them. Scientists now take into account **phylogeny** when classifying organisms, and group organisms according to their **evolutionary relationships**.

1) There are **eight** levels of groups used to classify organisms. These groups are called **taxa**. Each group is called a **taxon**.
2) The groups are arranged in a **hierarchy**, with the **largest groups** at the **top** and the smallest groups at the bottom. Organisms can only belong to **one group** at **each level** in the hierarchy — there's **no overlap**.
3) Organisms are first sorted into **three** large groups (or taxa) called **domains** — the **Eukarya**, **Bacteria** and **Archaea**.
4) **Related organisms** in a domain are then sorted into **slightly smaller groups** called **kingdoms**, e.g. all animals are in the animal kingdom. **More closely related** organisms from that kingdom are then grouped into a **phylum**, then grouped into a **class**, and **so on** down the eight levels of the hierarchy.

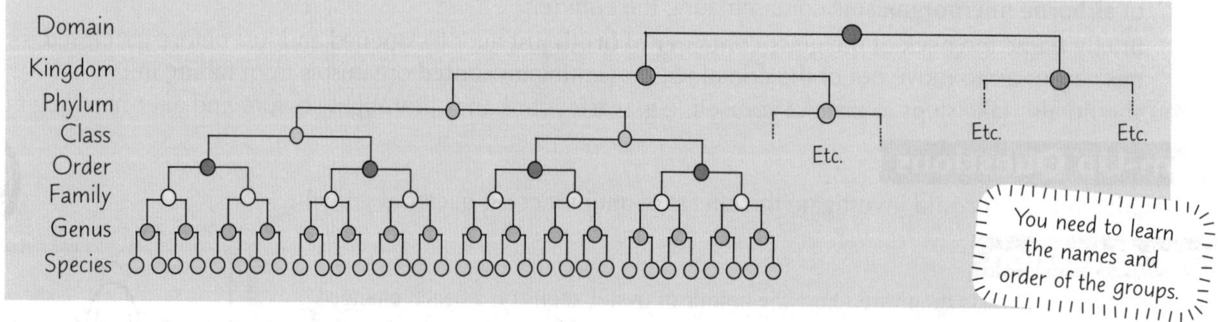

5) As you move **down** the hierarchy, there are **more groups** at each level but **fewer organisms** in each group. The organisms in each group also become **more closely related**.
6) The hierarchy **ends** with **species** — the groups that contain only **one type** of organism (e.g. humans, dogs, *E. coli*). You need to **learn** the definition of a **species**:

A species is a group of similar organisms able to reproduce to give fertile offspring.

7) Scientists constantly **update** classification systems because of **discoveries** about new species and new **evidence** about known organisms (e.g. **DNA sequence** data — see page 104).

Classification of Organisms

The **Binomial Naming System** is Used in **Classification**

1) The **nomenclature (naming system)** used for classification is called the **binomial system** — all organisms are given **one** internationally accepted scientific **name** in **Latin** that has **two parts**.
2) The **first part** of the name is the **genus** name and has a capital letter. The **second part** is the **species** name and begins with a lower case letter. E.g. using the binomial system humans are ***Homo sapiens***. Names are always written in ***italics*** (or they're **underlined** if they're **handwritten**).
3) The binomial system helps to avoid the **confusion** of using **common names**. E.g. over 100 different plant species are called **raspberries** and one species of buttercup has over 90 different common names.

Courtship Behaviour can be Used to Help **Classify Species**

1) **Courtship behaviour** is carried out by organisms to **attract** a mate of the **right species**.
2) It can be fairly simple, e.g. **releasing chemicals**, or quite complex, e.g. a series of **displays**.
3) Courtship behaviour is **species specific** — only members of the same species will do and respond to that courtship behaviour. This allows members of the **same species** to **recognise** each other, preventing **interbreeding** and making reproduction **more successful** (as mating with the wrong species won't produce **fertile** offspring).
4) Because of this specificity, courtship behaviour can be used to **classify organisms**.
5) The more **closely related** species are, the **more similar** their courtship behaviour.

Geoff's jive never failed to attract a mate.

Some examples of courtship behaviour include:

1) **Fireflies** give off **pulses of light**. The pattern of flashes is specific to each species.
2) **Crickets** make **sounds** that are similar to Morse code, the code being different for different species.
3) **Male peacocks** show off their **colourful tails**. This tail pattern is only found in peacocks.
4) **Male butterflies** use **chemicals** to attract females. Only those of the correct species respond.

Warm-Up Questions

Q1 What is phylogeny?

Q2 What is a taxon?

Q3 List the groups of the phylogenetic hierarchy in order, starting with domain.

Q4 How does courtship behaviour help to prevent interbreeding?

Q5 How is courtship behaviour used in classification?

Exam Question

Q1 The brown trout is a species of fish and is part of the Salmonidae family. Its Latin name is *Salmo trutta*.

a) Complete the table below for the classification of the brown trout. [2 marks]

Domain		Phylum				Genus	Species
Eukarya	Animalia	Chordata	Actinopterygii	Salmoniformes			

b) The brook trout is another member of the Salmonidae family. Rarely, a brook trout and a brown trout are able to mate to produce offspring known as tiger trout. Tiger trout are unable to reproduce. Explain how you know that a brook trout and a brown trout are different species. [1 mark]

Phylum — I thought that was the snot you get with a cold...

Learning the order of the levels in the phylogenetic hierarchy is about as easy as licking your elbow... try making up a mnemonic to help (like 'Dopey King Prawns Can't Order Fried Green Sausages' for Domain, Kingdom, Phylum, Class, Order, etc). Don't be put off if you get funny Latin names in the exam — just apply what you know. Right, onwards...

DNA Technology, Classification and Diversity

Advances in DNA and molecular technology have led to advances in many other fields.
For example, scientists have been able to use the technology to help classify organisms more accurately...

Advances in Techniques Can Clarify Evolutionary Relationships

New or **improved technologies** can result in **new discoveries** being made and the **relationships** between organisms being **clarified**. This can lead to **classification systems** being **updated**. Technologies that have been useful for clarifying evolutionary relationships include:

Genome sequencing — Advances in genome sequencing have meant that the **entire base sequence** of an organism's **DNA** can be determined. The DNA base sequence of one organism can then be **compared** to the DNA base sequence of another organism, to see how closely related they are. **Closely related** species will have a **higher percentage** of similarity in their DNA **base order**, e.g. humans and chimps share around 94%, humans and mice share about 86%.

Genome sequencing has clarified the relationship between **skunks** and members of the **Mustelidae family** (e.g. weasels and badgers). Skunks **were** classified in the Mustelidae family until their **DNA sequence** was revealed to be **significantly different** to other members of that family. So they were reclassified into the family **Mephitidae**.

Comparing amino acid sequence — Proteins are made of **amino acids**. The **sequence** of amino acids in a protein is coded for by the **base sequence** in DNA (see p. 88). **Related organisms** have similar DNA sequences and so **similar amino acid sequences** in their proteins. E.g. **cytochrome C** is a short protein found in many species. The more **similar** the **amino acid sequence** of cytochrome C in two different species, the **more closely related** the species are likely to be.

Immunological comparisons — **Similar proteins** will also bind the same **antibodies** (see p. 46). E.g. if antibodies to a **human version** of a protein are added to isolated samples from some other **species**, any protein that's like the human version will also be **recognised** (bound) by that antibody.

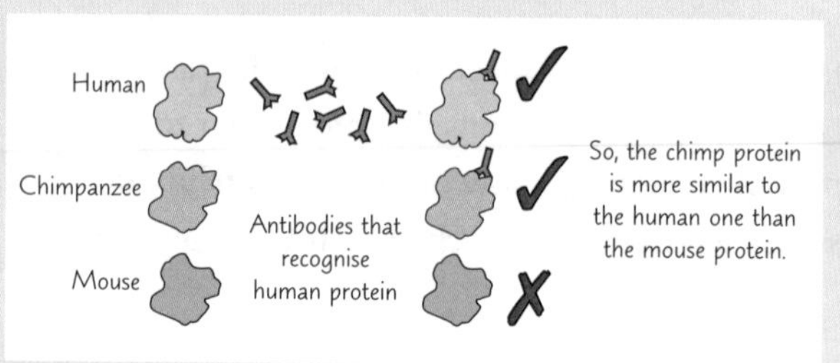

You Need to be Able to Interpret Data on DNA and Protein Similarities

Here are two examples of the kind of thing you might get:

	Species A	Species B	Species C	Species D
Species A	100%	86%	42%	44%
Species B	86%	100%	51%	53%
Species C	42%	51%	100%	91%
Species D	44%	53%	91%	100%

The table on the left shows the **% similarity of DNA** using DNA sequence analysis between several species of bacteria.

The data shows that species **A** and **B** are **more closely related** to each other than they are to either C or D. Species **C** and **D** are also **more closely related** to each other than they are to either A or B.

You can also use DNA base sequences to see how closely related two members of the same species are.

The diagram on the right shows the **amino acid sequences** of a certain protein from three different species.

You can see that the amino acid sequences from species **A** and **B** are **very similar**. The sequence from species **C** is **very different** to any of the other sequences. This would suggest that species **A** and **B** are **more closely related**.

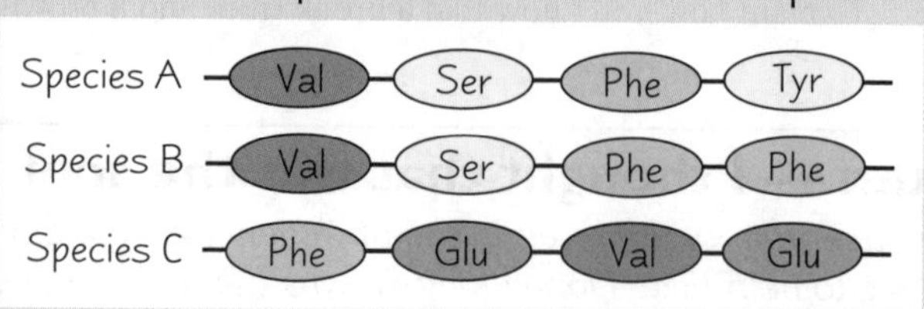

DNA Technology, Classification and Diversity

Gene Technologies Have Changed the Way Genetic Diversity is Assessed

You might remember from page 98, that **genetic diversity** is the **number** of **different alleles** in a population.

1) Early **estimates** of genetic diversity were made by looking at the **frequency** of **measurable** or **observable characteristics** in a population, e.g. the number of different eye colours in a population and the number of people with each particular eye colour.
2) Since different **alleles** determine different **characteristics** (see page 89) a **wide variety** of **each characteristic** in a population indicates a **high number** of different **alleles** — and so a **high genetic diversity**.
3) However gene technologies have now been developed that allow us to **measure genetic diversity directly**:

There weren't many people with Sid's observable characteristics.

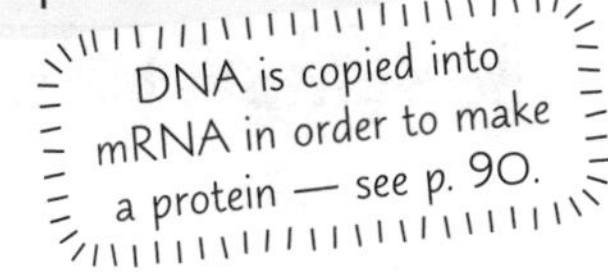

For example:

- **Different alleles** of the same gene will have **slightly different DNA base sequences**. Comparing the DNA base sequences of the same gene in **different organisms** in a population allows scientists to find out **how many alleles** of that gene there are in that population.
- Different alleles will also produce slightly different **mRNA base sequences**, and may produce **proteins** with slightly different **amino acid sequences**, so these can also be compared.

DNA is copied into mRNA in order to make a protein — see p. 90.

4) These **new technologies** can all be used to give more **accurate estimates** of genetic diversity within a population or species. They also allow the genetic diversity of **different species** to be **compared** more easily.

Warm-Up Questions

Q1 Give one technological advance that has helped to clarify evolutionary relationships.

Q2 Suggest two techniques that could be used to assess genetic diversity within species.

Q3 Why can observable characteristics be used as a measure of genetic diversity?

Q4 How has the way in which genetic diversity is assessed changed over time?

Exam Question

Q1 The amino acid sequence of a specific protein was used to make comparisons between four species of animal. The results are shown below.

Species	Amino acid 1	Amino acid 2	Amino acid 3	Amino acid 4
Rabbit	His	Ala	Asp	Lys
Mouse	Thr	Ala	Asp	Val
Chicken	Ala	Thr	Arg	Arg
Rat	Thr	Ala	Asp	Phy

a) Which two species are the most closely related? [1 mark]

b) Which species is the most distantly related to the other three? Explain your answer. [2 marks]

These pages have a PG classification — Protein Guidance...

...on evolutionary relationships. It's the latest release. It's important that you understand that the more similar the DNA and proteins, the more closely related two species are. This is because relatives have similar DNA, which codes for similar proteins, made of similar sequences of amino acids. DNA really is the key to everything, eh?

Investigating Variation

It's a lot of work studying variation in an entire population (imagine studying all the ants in one nest) — so instead you can take a random sample and use this to give you a good idea of what's going on in the entire population.

Variation Can be **Caused** by **Genes**, the **Environment**, or **Both**

1) **Variation** is the **differences** that exists between individuals. There's variation **between** species and **within** species.
2) Variation can be caused by **genetic factors**. **Different** species have **different genes**, which causes variation **between** species. Individuals of the **same** species have the **same genes**, but **different alleles** (versions of genes) — this causes variation **within** a species.
3) Variation **within** a species can also be caused by differences in the **environment**, e.g. climate, food, lifestyle.
4) Most variation **within** a species is caused by a **combination** of **genetic** and **environmental** factors. E.g. **genes** determine how tall an organism **can grow**, but **nutrient availability** affects how tall the organism **actually grows**.

To **Study** Variation You Have to **Sample** a **Population**

When investigating variation you usually only look at a **sample** of the population, **not** the **whole thing**. For most species it would be too **time-consuming** or **impossible** to catch all the individuals in the group. So samples are used as **models** for the **whole population**.

The **Sample** has to be **Random**

Because sample data will be used to **draw conclusions** about the **whole population**, it's important that it **accurately represents** the whole population and that any patterns observed are tested to make sure they're not due to chance.

1) To make sure the sample isn't **biased**, it should be **random**. For example, if you were looking at plant species in a field you could pick random sample sites by dividing the field into a **grid** and using a **random number generator** to select coordinates.
2) To ensure any variation observed in the sample isn't just due to **chance**, it's important to analyse the results **statistically**. This allows you to be more **confident** that the results are true and therefore will reflect what's going on in the **whole population**.

You Can Use the **Mean** to Look for **Variation Between Samples**

1) The **mean** is an **average** of the values collected in a sample. Find it using this **formula**:

$$\text{mean} = \frac{\text{total of all the values in your data}}{\text{the number of values in your data}}$$

Example:
The heights of different seedlings in a group are: 6 cm, 4 cm, 7 cm, 6 cm, 5 cm, 8 cm, 7 cm, 5 cm, 7 cm and 9 cm.
To calculate the mean, add all of the heights together and divide by the number of seedlings:
Mean height = (6 + 4 + 7 + 6 + 5 + 8 + 7 + 5 + 7 + 9) ÷ 10 = 64 ÷ 10 = **6.4 cm**

2) The mean can be used to tell if there is **variation between samples**. For example:

    The **mean height** of a species of **tree** in woodland A = **26 m**, in woodland B = **32 m** and in woodland C = **35 m**. So the **mean height varies**.

3) Most samples will include values **either side** of the **mean**, so you end up with a **bell-shaped graph** — this is called a **normal distribution**. A normal distribution is **symmetrical** about the mean.

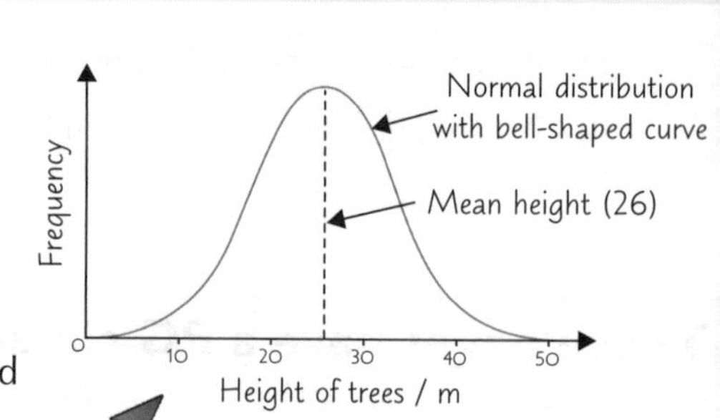

Investigating Variation

The **Standard Deviation** Tells You About **Variation Within a Sample**

1) The **standard deviation** tells you how much the **values** in a **single sample vary**. It's a measure of the **spread** of **values** about the **mean**.
2) Sometimes you'll see the mean written as, e.g. **9 ± 3**. This means that the **mean** is **9** and the **standard deviation** is **3**, so most of the **values** are spread between **6** to **12**.
3) A **large standard deviation** means the values in the sample **vary a lot**. A **small standard deviation** tells you that most of the sample data is around the mean value, so **varies little**.

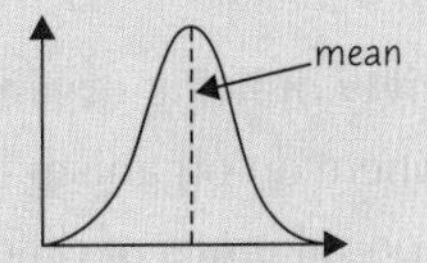

Here, all the values are similar and close to the mean, so the graph is steep and the standard deviation is small.

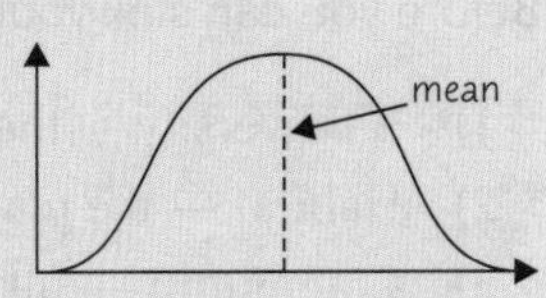

Here, the values vary a lot, so the graph is fatter and the standard deviation is large.

You won't be asked to calculate standard deviation in the exams, but you might be asked to interpret data that includes standard deviations.

You Can Use the **Standard Deviation** to Draw **Error Bars**

1) **Standard deviations** can be **plotted** on a graph or chart of **mean values** using **error bars**, e.g.
2) Error bars extend **one standard deviation above** and **one standard deviation below** the mean (so the total **length** of an error bar is **twice the standard deviation**).
3) The **longer** the **bar**, the **larger** the **standard deviation** and the **more spread out** the sample data is from the mean.

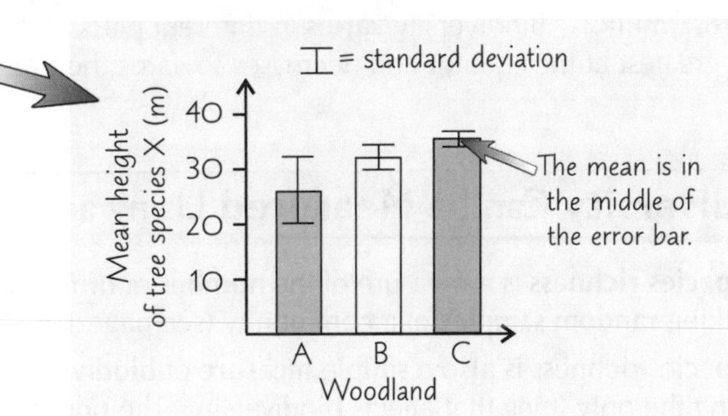

Warm-Up Questions

Q1 Why do scientists look at a sample of a population, rather than the whole population?

Q2 Why does a population sample have to be chosen at random?

Q3 What does the standard deviation of a data set tell us?

Exam Question

Q1 A student was investigating the variation in development time for two beetle species. The student recorded the development time for six beetle larvae from species A and six from species B. The results are shown in the table.

Development Time (Days)	
Species A	Species B
8	12
11	10
9	6
10	12
7	15
9	11

a) Calculate the mean development time for each species. [2 marks]

b) The standard deviation for species A is 1.3 and for species B is 2.7 (to 1 decimal place). What conclusions can you draw from this information? [2 marks]

Sex and drugs and rock and roll — it's all just standard deviation...

Bet you thought you'd finished with maths — 'fraid not. Luckily, calculating a mean is probably one of the easiest bits of maths you could be asked to do — so make sure you can. Also, make sure you understand what both the mean and the standard deviation can tell you about a bit of data. Interpreting data is an all-time favourite with the examiners.

Biodiversity

Bet you've noticed how there are loads of different living things in the world — well that's biodiversity in a nutshell.

Biodiversity is the Variety of Organisms

Before you can sink your teeth into the real meat of biodiversity, there are a few definitions you need to know:

1) **Biodiversity** — the **variety** of **living organisms** in an **area**.
2) **Habitat** — the **place** where an organism **lives**, e.g. a rocky shore or field.
3) **Community** — all the **populations** of different **species** in a **habitat**.

A species is a group of similar organisms able to reproduce to give fertile offspring (see page 102).

Areas with a **high** biodiversity are those with lots of **different species**.

Biodiversity Can be Considered at Different Levels

Biodiversity can be considered at a range of scales from the **local** to the **global**.

1) **Local biodiversity** — you could consider the **variety** of different **species** living in a **small habitat** that's **local** to you — e.g. a pond or meadow, or even your back garden. Some habitats will be more **biodiverse** than others.
2) **Global biodiversity** — you could also consider the variety of **species** on **Earth**. Recent estimates put the **total number** of species on Earth at about 8.7 million. Biodiversity **varies** in different parts of the world — it is greatest at the equator and decreases towards the poles.

Pete wasn't sure that the company's new increased biodiversity policy would be good for productivity.

Biodiversity Can be Measured Using an Index of Diversity

1) **Species richness** is a measure of the **number** of **different species** in a **community**. It can be worked out by taking **random samples** of a community (see page 106) and **counting** the number of different species.
2) Species richness is also a simple **measure** of **biodiversity**. But the number of different species in a community isn't the only thing that affects biodiversity. The **population sizes** of those species do too. Species that are in a community in very **small numbers** shouldn't be treated the same as those with bigger populations.
3) An **index of diversity** is another way of measuring biodiversity. It's calculated using an equation that takes **both** the **number of species** in a community (species richness) and the **abundance of each species** (population sizes) into account.
4) You can **calculate** an index of diversity (d) using this formula:

The number of species in a community and the abundance of each species is also known as the species diversity.

$$d = \frac{N(N-1)}{\sum n(n-1)}$$

Where...
N = **Total number** of organisms of **all** species
n = **Total number** of organisms of **one** species
$\sum$ = **'Sum of'** (i.e. added together)

The **higher** the **number**, the **more diverse** the area is. If all the individuals are of the **same species** (i.e. no biodiversity) the **index is 1**. Here's an example:

There are 3 different species of flower in this field — a red species, a white and a blue.

There are 11 organisms altogether, so N = 11.

There are 3 of the red species, 5 of the white and 3 of the blue.

So the species diversity index of this field is:

$$d = \frac{11\,(11-1)}{3\,(3-1) + 5\,(5-1) + 3\,(3-1)} = \frac{110}{6 + 20 + 6} = 3.44$$

When calculating the bottom half of the equation you need to work out the n(n–1) bit for each different species then add them all together.

Biodiversity

Agricultural Practices Can Reduce Biodiversity

Farmers try to **maximise** the **amount of food** that they can produce from a given area of land. But many of the **methods** they use **reduce biodiversity**. For example:

1) **Woodland clearance** — this is done to **increase** the **area** of farmland. It directly reduces the **number** of **trees** and sometimes the **number** of **different tree species**. It also **destroys habitats**, so some species could lose their **shelter** and **food source**. This means that species will **die** or be forced to **migrate** to another suitable area, further **reducing** biodiversity.
2) **Hedgerow removal** — this is also done to **increase** the **area** of farmland by turning **lots of small fields** into **fewer large fields**. This **reduces** biodiversity for the same reasons as woodland clearance.
3) **Pesticides** — these are chemicals that **kill** organisms (**pests**) that feed on **crops**. This **reduces** diversity by **directly killing** the pests. Also, any species that feed on the pests will **lose** a food source, so their numbers could **decrease** too.
4) **Herbicides** — these are chemicals that kill **unwanted plants** (**weeds**). This **reduces** plant diversity and could **reduce** the number of organisms that feed on the weeds.
5) **Monoculture** — this is when farmers have fields containing only **one type of plant**. A **single type** of plant **reduces** biodiversity **directly** and will **support fewer organisms** (e.g. as a habitat or food source), which **further reduces** biodiversity.

Whilst **agriculture** is **important**, we don't want to **lose** too much **biodiversity**. So there has to be a **balance** between agriculture and conservation. Conservationists try to **protect** biodiversity. Some **examples** of **conservation** schemes are:

- Giving **legal protection** to **endangered species**.
- Creating **protected areas** such as SSSIs (Sites of Special Scientific Interest) and AONBs (Areas of Outstanding Natural Beauty). These **restrict** further **development**, including **agricultural** development.
- The **Environmental Stewardship Scheme** which encourages **farmers** to **conserve biodiversity,** e.g. by replanting hedgerows and leaving margins around fields for wild flowers to grow.

PRACTICE QUESTIONS

Warm-Up Questions

Q1 What is biodiversity?

Q2 What is species richness?

Q3 Give three ways in which agriculture can reduce biodiversity.

Exam Question

Site 1 — No Field Margins		Site 2 — Enhanced Field Margins	
Bombus lucorum	15	*Bombus lucorum*	35
Bombus lapidarius	12	*Bombus lapidarius*	25
Bombus pascuorum	24	*Bombus pascuorum*	34
		Bombus ruderatus	12
		Bombus terrestris	26

Q1 A study was conducted to investigate the impact of introducing enhanced field margins on the diversity of bumblebees. Enhanced field margins are thick bands of land around the edges of fields that are not farmed, but instead are planted with plants that are good for wildlife. Scientists studied two wheat fields, one where the farmer sowed crops right to the edge of the field and another where the farmer created enhanced field margins.
The scientists counted the number of bees of different species at each site. Their results are shown in the table above.

a) What two things does an index of diversity take into account when measuring biodiversity? [2 marks]

b) Use the data in the table and the formula below to calculate the index of diversity for each site.

$$d = \frac{N(N-1)}{\sum n(n-1)}$$

[4 marks]

c) What conclusions can be drawn from the findings of this study? [2 marks]

Species richness — goldfish and money spiders top the list...

Agricultural practices can threaten biodiversity — I never knew a field of corn could cause so much bother. Make sure you know the definition of species richness and that population size is important in biodiversity measures too. As for the formula for the index of diversity — be prepared to use it and to say what the numbers it churns out actually mean.

Extra Exam Practice

That's Topic 4 almost over. I hope you enjoyed it, although let's be honest, just enjoying biology isn't going to get you through the exams. You need to make sure what you're learning sticks in your noggin. Do these practice exam questions to check how well you know Topic 4.

- Have a look at this example of how to answer a tricky exam question.
- Then check how much you've understood from Topic 4 by having a go at the questions on the next page.

Revising topic by topic is great, but make sure you try your hand at the synoptic questions on p.233-241 too. They'll test your knowledge of the whole course.

1 Ecologists conducted a survey to assess whether the biodiversity of insect larvae in an area of a park changed between 2016 and 2017. The data they collected is shown in **Table 1**.

Table 1

Species	No. of individuals found in July 2016	No. of individuals found in July 2017
Hemerobius micans	18	2
Hemerobius stigma	15	3
Hemerobius alpestris	11	9
Forficula auricularia	22	19
Index of diversity		2.5

1.1 With reference to the data in **Table 1**, explain the difference between species richness and an index of diversity and use these measures to make a conclusion for the study. You should use the following formula in your answer and show your working:

$$d = \frac{N(N-1)}{\Sigma n(n-1)}$$

where N = total number of organisms of all species and n = total number of organisms of one species

(4 marks)

1.2 The scientists planned to investigate the evolutionary relationships between the *Hemerobius* species by looking at amino acid sequences within each of the three species. They predicted that two of the species, *H. micans* and *H. stigma*, would be more closely related than the third, *H. alpestris*. Explain how the scientists could test their prediction.

(3 marks)

Make sure your working is nice and clear — if your answer's wrong you might still get a mark for the method as long as the examiner can see what you were trying to do.

1.1

N = 18 + 15 + 11 + 22 = 66

$$d = \frac{66(66-1)}{18(18-1) + 15(15-1) + 11(11-1) + 22(22-1)} = \frac{4290}{1088}$$

= 3.9 (to 2 s.f.)

Species richness is a measure of the number of different species in a community, whereas an index of diversity is a measure of the number of individuals of each species present in a community. Although the species richness remained at **4** in this study, the index of diversity decreased from **3.9 to 2.5**. Based on these findings, it could be concluded that the biodiversity of insect larvae in the area of the park studied fell between 2016 and 2017.

Double-check the formula to make sure you've put the right numbers in the right places.

You've been asked to use data from Table 1 — so make sure you include relevant figures to support your answer.

Here you need to work out how the amino acid sequences could be used to show how closely related the species are.

1.2

They could **compare** the amino acid sequence for specific proteins that are common to all three species. If they find a greater degree of similarity in the sequences for *H. micans* and *H. stigma* than for *H. alpestris* and *H. micans* or *H. stigma* then this would suggest that the **DNA base sequence** for *H. micans* and *H. stigma* is **more similar** than for *H. alpestris* and *H. micans* or *H. stigma*. This would indicate that *H. micans* and *H. stigma* are more closely related.

Make sure you explain how the amino acid sequences show that species are genetically similar.

Extra Exam Practice

2 Protein synthesis is an essential process in all living organisms.

2.1 Mutations can affect the proteins that are made by cells.
Explain how the degenerate nature of the genetic code reduces the effects of mutations on proteins.
(3 marks)

2.2 It is likely that the mRNA needed to produce one polypeptide is created more quickly in prokaryotic cells than in eukaryotic cells.
Using your knowledge of the structure of chromosomes and genes, suggest why.
(3 marks)

Tetracycline is an antibiotic that kills many species of bacteria.
During protein synthesis, tetracycline attaches to the prokaryotic ribosome as shown in **Figure 1**.

Figure 1

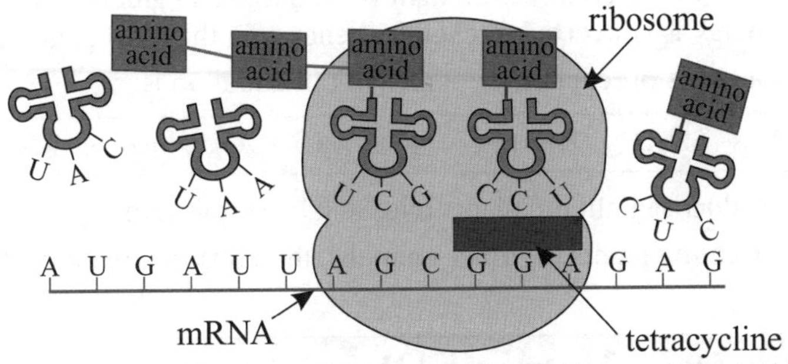

2.3 Using **Figure 1** and your own knowledge, describe how tetracycline prevents protein synthesis in prokaryotic cells.
(3 marks)

2.4 Some bacterial cells have a natural resistance to tetracycline. Suggest how being exposed to tetracycline would affect genetic diversity within a bacterial population. Explain your answer.
(2 marks)

3 **Figure 2** shows part of a phylogenetic tree for some aquatic organisms.

Figure 2

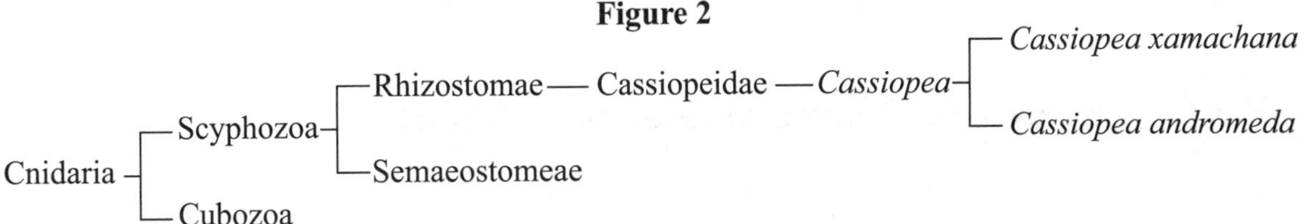

3.1 Using **Figure 2**, name the order that *Cassiopea andromeda* belongs to.
(1 mark)

3.2 It is possible that there are undiscovered species that are closely related to the species shown in **Figure 2**. If scientists discover one of these species, explain how antibodies could be used to help position it on the tree.
(3 marks)

3.3 *Cassiopea andromeda* can reproduce via sexual reproduction. Explain how chromosome non-disjunction could result in two offspring of *Cassiopea andromeda* having different genomes.
(2 marks)

Photosynthesis, Respiration and ATP

All organisms need energy for life processes (and you'll need some for revising), so it's pretty important stuff. Annoyingly, it's pretty complicated stuff too, but 'cos I'm feeling nice today we'll take it slowly, one bit at a time...

Biological Processes Need Energy

Plant and animal cells **need energy** for biological processes to occur:

- **Plants** need energy for things like **photosynthesis**, **active transport** (e.g. to take in minerals via their roots), **DNA replication**, **cell division** and **protein synthesis**.
- **Animals** need energy for things like **muscle contraction**, maintenance of **body temperature**, **active transport**, **DNA replication**, **cell division** and **protein synthesis**.

Without energy, these biological processes would stop and the plant or animal would die.

Photosynthesis Stores Energy in Glucose

1) **Photosynthesis** is the process where **energy** from **light** is used to **make glucose** from H_2O and CO_2 (the light energy is **converted** to **chemical energy** in the form of glucose).
2) Photosynthesis occurs in a **series** of **reactions**, but the overall equation is:

$$6CO_2 + 6H_2O + \text{Energy} \longrightarrow C_6H_{12}O_6 \text{ (glucose)} + 6O_2$$

3) So, energy is **stored** in the **glucose** until the plants **release** it by **respiration**.
4) Animals obtain glucose by **eating plants** (or **other animals**), then respire the glucose to release energy.

Cells Release Energy from Glucose by Respiration

1) **Plant** and **animal** cells **release energy** from **glucose** — this process is called **respiration**.
2) This energy is used to power all the **biological processes** in a cell.
3) There are two types of respiration:
   - **Aerobic respiration** — respiration **using oxygen**.
   - **Anaerobic respiration** — respiration **without oxygen**.
4) **Aerobic** respiration produces **carbon dioxide** and **water**, and releases **energy**. The overall equation is:

$$C_6H_{12}O_6 \text{ (glucose)} + 6O_2 \longrightarrow 6CO_2 + 6H_2O + \text{Energy}$$

5) **Anaerobic** respiration in **plants** and **yeast** produces **ethanol** and **carbon dioxide** and releases energy. In **humans**, anaerobic respiration produces **lactate** and releases energy.

ATP is the Immediate Source of Energy in a Cell

You should remember most of this stuff from **Topic 1**. Here's a quick recap:

1) A cell **can't** get its energy **directly** from glucose.
2) So, in respiration, the **energy released** from glucose is used to **make ATP** (adenosine triphosphate). ATP **carries energy** around the cell to where it's **needed**.

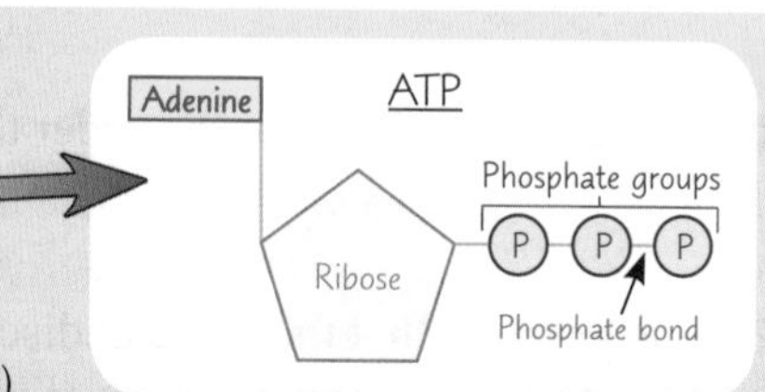

3) **ATP** is **synthesised** via a **condensation reaction** between **ADP** (adenosine diphosphate) and **inorganic phosphate** (P_i) using energy from an **energy-releasing** reaction, e.g. the **breakdown** of **glucose** in **respiration**. The energy is stored as **chemical energy** in the **phosphate bond**. The enzyme **ATP synthase** catalyses this reaction.
4) ATP **diffuses** to the part of the cell that **needs** energy.
5) Here, it's **hydrolysed** back into **ADP** and **inorganic phosphate** (P_i). Chemical **energy** is **released** from the phosphate bond and used by the cell. **ATP hydrolase** catalyses this reaction.
6) The ADP and inorganic phosphate are **recycled** and the process starts again.

Photosynthesis, Respiration and ATP

ATP has **Specific Properties** that Make it a **Good Energy Source**

1) ATP stores or releases only a **small**, **manageable amount** of energy at a time, so **no** energy is **wasted** as **heat**.
2) It's a **small**, **soluble** molecule so it can be **easily transported** around the cell.
3) It's **easily broken down**, so energy can be **easily released instantaneously**.
4) It can be **quickly re-made**.
5) It can make **other molecules** more **reactive** by **transferring** one of its **phosphate groups** to them (**phosphorylation**).
6) ATP **can't pass out** of the **cell**, so the cell **always** has an immediate supply of energy.

You Need to **Know Some Basics** Before You Start

There are some pretty confusing technical terms in this section that you need to get your head around:

- **Metabolic pathway** — a **series** of **small reactions** controlled by **enzymes**, e.g. **respiration** and **photosynthesis**.
- **Phosphorylation** — **adding phosphate** to a molecule, e.g. **ADP** is phosphorylated to **ATP** (see previous page).
- **Photophosphorylation** — **adding phosphate** to a molecule using **light**.
- **Photolysis** — the **splitting** (lysis) of a molecule using **light** (photo) energy.
- **Photoionisation** — when **light energy 'excites' electrons** in an **atom** or **molecule**, giving them **more energy** and causing them to be **released**. The release of electrons causes the atom or molecule to become a **positively-charged ion**.
- **Hydrolysis** — the **splitting** (lysis) of a molecule using **water** (hydro).
- **Decarboxylation** — the **removal** of **carbon dioxide** from a molecule.
- **Dehydrogenation** — the **removal** of **hydrogen** from a molecule.
- **Redox reactions** — reactions that involve **oxidation** and **reduction**.

Remember redox reactions:

1) If something is **reduced** it has **gained electrons** (e^-), and may have **gained hydrogen** or lost oxygen.
2) If something is **oxidised** it has **lost electrons**, and may have **lost hydrogen** or gained oxygen.
3) Oxidation of one molecule **always** involves reduction of another molecule.

One way to remember electron and hydrogen movement is OILRIG. Oxidation Is Loss, Reduction Is Gain.

Photosynthesis and **Respiration** Involve **Coenzymes**

1) A **coenzyme** is a molecule that **aids** the **function** of an **enzyme**.
2) They work by **transferring** a **chemical group** from one molecule to another.
3) A coenzyme used in **photosynthesis** is **NADP**. NADP transfers **hydrogen** from one molecule to another — this means it can **reduce** (give hydrogen to) or **oxidise** (take hydrogen from) a molecule.
4) Examples of coenzymes used in **respiration** are: **NAD**, **coenzyme A** and **FAD**.
    - NAD and FAD transfer **hydrogen** from one molecule to another — this means they can **reduce** (give hydrogen to) or **oxidise** (take hydrogen from) a molecule.
    - **Coenzyme A** transfers **acetate** between molecules (see page 123).

When hydrogen is transferred between molecules, electrons are transferred too.

Warm-Up Questions

Q1 What is photoionisation?

Q2 Give the name of a coenzyme involved in photosynthesis.

Exam Question

Q1 ATP is the immediate source of energy inside a cell.
Describe how the synthesis and breakdown of ATP meets the energy needs of a cell. [6 marks]

Oh dear, I've used up all my energy on these two pages...

Well, I won't beat about the bush, this stuff is pretty tricky... nearly as hard as a cross between Hugh Jackman and concrete. With a little patience and perseverance (and plenty of [chocolate] [coffee] [marshmallows] — delete as you wish), you'll get there. Once you've got these pages straight in your head, the next ones will be easier to understand.

Photosynthesis

Right, pen at the ready. Check. Brain switched on. Check. Cuppa piping hot. Check. Sweets on standby. Check. Okay, I think you're all sorted to start photosynthesis. Finally, take a deep breath and here we go...

Photosynthesis Takes Place in the Chloroplasts of Plant Cells

1) **Chloroplasts** are **flattened organelles** surrounded by a **double membrane**. They are found in **plant cells**.
2) **Thylakoids** (fluid-filled sacs) are **stacked up** in the chloroplast into structures called **grana** (singular = **granum**). The grana are **linked** together by bits of thylakoid membrane called **lamellae** (singular = **lamella**).
3) Chloroplasts contain **photosynthetic pigments** (e.g. **chlorophyll a**, **chlorophyll b** and **carotene**). These are **coloured substances** that **absorb** the **light energy** needed for photosynthesis. The pigments are found in the **thylakoid membranes** — they're attached to **proteins**. The protein and pigment is called a **photosystem**.
4) There are **two** photosystems used by plants to capture light energy. **Photosystem I** (or PSI) absorbs light best at a wavelength of **700 nm** and **photosystem II** (PSII) absorbs light best at **680 nm**.
5) Contained within the inner membrane of the chloroplast and **surrounding** the thylakoids is a gel-like substance called the **stroma**. It contains **enzymes**, **sugars** and **organic acids**.
6) Carbohydrates produced by photosynthesis and not used straight away are stored as **starch grains** in the **stroma**.

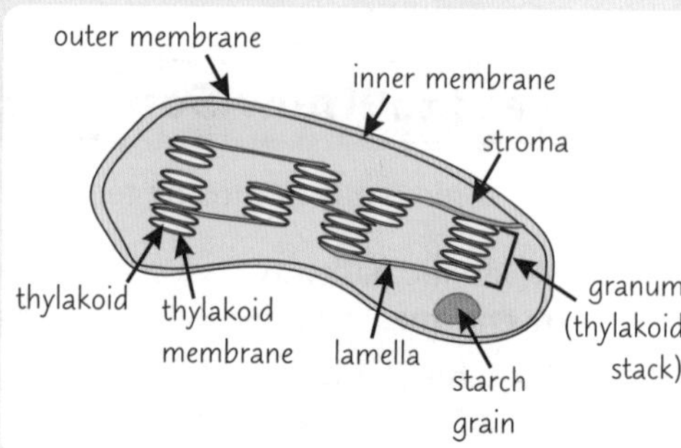

Photosynthesis can be Split into Two Stages

See p. 116 for loads more information on the Calvin cycle.

There are actually **two stages** that make up **photosynthesis**:

1 The Light-Dependent Reaction

1) As the name suggests, this reaction **needs light energy**.
2) It takes place in the **thylakoid membranes** of the chloroplasts.
3) Here, light energy is absorbed by **chlorophyll** (and other photosynthetic pigments) in the **photosystems**. The light energy **excites** the **electrons** in the **chlorophyll**, leading to their eventual **release** from the molecule. The **chlorophyll** has been **photoionised**.
4) Some of the **energy** from the **released electrons** is used to add a phosphate group to ADP to form **ATP**, and some is used to reduce NADP to form **reduced NADP**. ATP **transfers energy** and reduced NADP **transfers hydrogen** to the light-independent reaction.
5) During the process H_2O is **oxidised** to O_2.

2 The Light-Independent Reaction

1) This is also called the **Calvin cycle** and as the name suggests it **doesn't use light energy** directly. (But it does **rely** on the **products** of the light-dependent reaction.)
2) It takes place in the **stroma** of the chloroplast.
3) Here, the **ATP** and **reduced NADP** from the light-dependent reaction supply the **energy** and **hydrogen** to make **simple sugars** from CO_2.

This diagram shows how the two reactions link together in the chloroplast.

In the Light-Dependent Reaction ATP is Made by Photophosphorylation

In the light-dependent reaction, the **energy** resulting from the **photoionisation** of **chlorophyll** is used for three things:

1) Making **ATP** from **ADP** and **inorganic phosphate**. This reaction is called **photophosphorylation** (see p. 113).
2) Making **reduced NADP** from **NADP**.
3) Splitting **water** into **protons** (H^+ ions), **electrons** and **oxygen**. This is called **photolysis** (see p. 113).

The light-dependent reaction actually includes **two types** of **photophosphorylation** — **non-cyclic** and **cyclic**. Each of these processes has **different products** (see next page).

Photosynthesis

Non-cyclic Photophosphorylation Produces ATP, Reduced NADP and O_2

To understand the process you need to know that the photosystems (in the thylakoid membranes) are **linked** by **electron carriers**. Electron carriers are **proteins** that **transfer electrons**. The photosystems and electron carriers form an **electron transport chain** — a **chain** of **proteins** through which **excited electrons flow**. All the processes in the diagrams are happening together — I've just split them up to make it easier to understand.

1) Light energy excites electrons in chlorophyll

- **Light energy** is absorbed by **PSII**.
- The light energy **excites electrons** in **chlorophyll**.
- The electrons move to a **higher energy level** (i.e. they have more energy).
- These **high-energy electrons** are **released** from the **chlorophyll** and **move down** the **electron transport chain** to **PSI**.

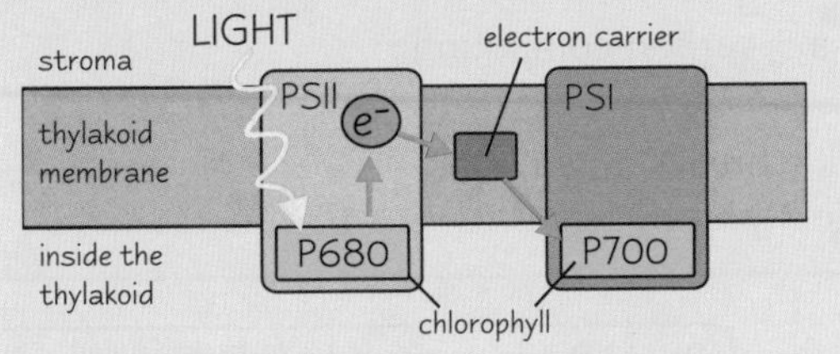

2) Photolysis of water produces protons (H^+ ions), electrons and O_2

- As the excited electrons **from chlorophyll leave PSII** to **move down** the electron transport chain, they must be **replaced**.
- **Light** energy splits **water** into **protons** (H^+ ions), **electrons** and **oxygen** — **photolysis**. (So the O_2 in photosynthesis comes from water and is made in the light-dependent reaction.)
- The reaction is: $H_2O \rightarrow 2H^+ + \frac{1}{2}O_2$

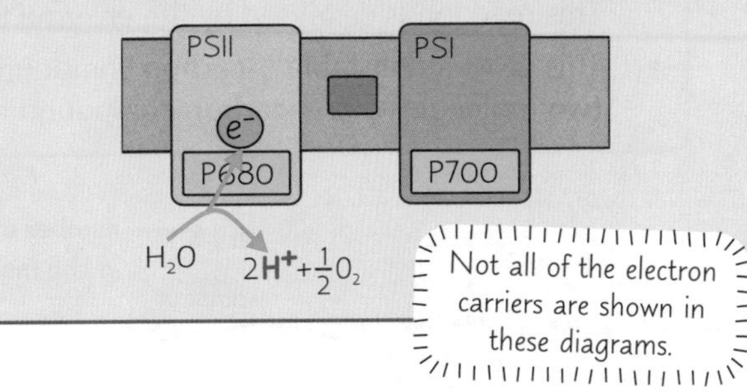

Not all of the electron carriers are shown in these diagrams.

3) Energy from the excited electrons makes ATP...

- The excited electrons **lose energy** as they **move down** the **electron transport chain**.
- This energy is used to **transport protons into** the **thylakoid**, so that the thylakoid has a **higher concentration** of protons than the stroma. This forms a **proton gradient** across the thylakoid membrane.
- Protons move **down** their concentration gradient, into the stroma, **via** the enzyme **ATP synthase**, which is **embedded** in the **thylakoid membrane**. The energy from this movement combines **ADP** and **inorganic phosphate** (P_i) to form **ATP**.

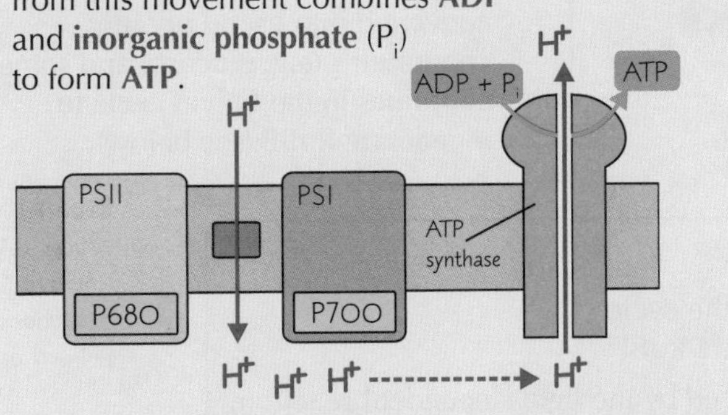

4) ...and generates reduced NADP.

- Light energy is **absorbed** by PSI, which excites the electrons again to an **even higher** energy level.
- Finally, the electrons are **transferred** to **NADP**, along with a **proton** (H^+ ion) from the **stroma**, to form **reduced NADP**.

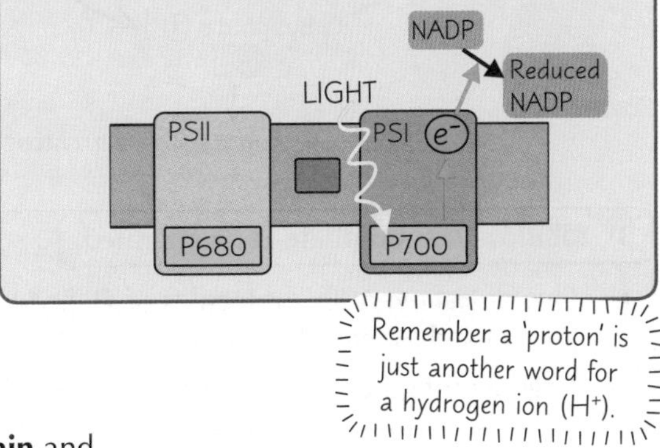

Remember a 'proton' is just another word for a hydrogen ion (H^+).

The process of **electrons** flowing down the **electron transport chain** and creating a **proton gradient** across the **membrane** to drive **ATP synthesis** is called **chemiosmosis**. It's described by the **chemiosmotic theory**.

Cyclic Photophosphorylation Only Produces ATP

Cyclic photophosphorylation **only uses PSI**. It's called 'cyclic' because the electrons from the chlorophyll molecule **aren't** passed onto NADP, but are **passed back** to PSI via electron carriers. This means the electrons are **recycled** and can repeatedly flow through PSI. This process doesn't produce any reduced NADP or O_2 — it **only produces** small amounts of **ATP**.

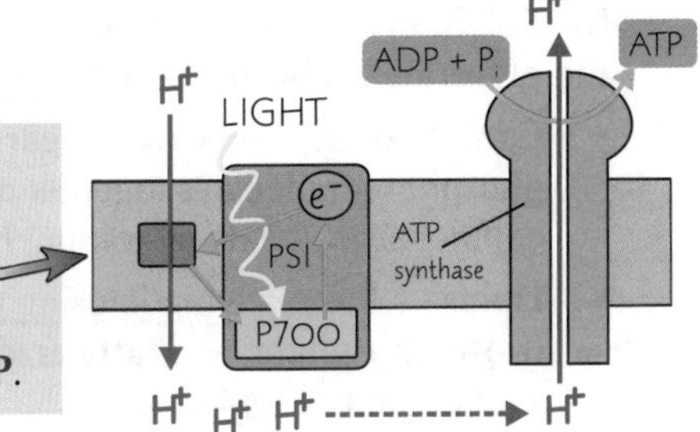

Photosynthesis

Don't worry, you're over the worst of photosynthesis now. Instead of electrons flying around, there's a nice cycle of reactions to learn. What more could you want from life? Money, fast cars and nice clothes have nothing on this...

The **Light-Independent** Reaction is also called the **Calvin Cycle**

1) The **Calvin cycle** takes place in the **stroma** of the chloroplasts.
2) It makes a molecule called **triose phosphate** from CO_2 and **ribulose bisphosphate** (a 5-carbon compound). Triose phosphate can be used to make **glucose** and other **useful organic substances** (see below).
3) There are a few steps in the cycle, and it needs **ATP** and **H^+ ions** to keep it going.
4) The reactions are linked in a **cycle**, which means the starting compound, **ribulose bisphosphate**, is **regenerated**.

The Calvin cycle is also known as carbon dioxide fixation because carbon from CO_2 is 'fixed' into an organic molecule.

Here's what happens at each stage in the cycle:

1) Carbon dioxide is combined with ribulose bisphosphate to form two molecules of glycerate 3-phosphate

- CO_2 enters the leaf through the stomata and diffuses into the stroma of the chloroplast.
- Here, it's combined with ribulose bisphosphate (RuBP), a 5-carbon compound. This reaction is catalysed by the enzyme rubisco.
- This gives an unstable 6-carbon compound, which quickly breaks down into two molecules of a 3-carbon compound called glycerate 3-phosphate (GP).

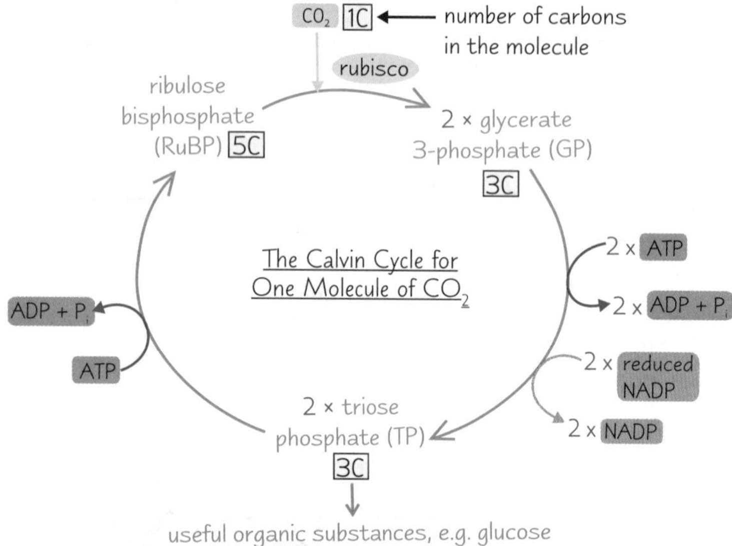

2) ATP and reduced NADP are required for the reduction of GP to triose phosphate

- The hydrolysis of ATP (from the light-dependent reaction) provides energy to turn the 3-carbon compound, GP, into a different 3-carbon compound called triose phosphate (TP).
- This reaction also requires H^+ ions, which come from reduced NADP (also from the light-dependent reaction). Reduced NADP is recycled to NADP.
- Some triose phosphate is then converted into useful organic compounds (e.g. glucose) and some continues in the Calvin cycle to regenerate RuBP (see below).

3) Ribulose bisphosphate is regenerated

- Five out of every six molecules of TP produced in the cycle aren't used to make hexose sugars, but to regenerate RuBP.
- Regenerating RuBP uses the rest of the ATP produced by the light-dependent reaction.

Reduced NADP reduces GP to TP — reduction reactions are explained on p. 113.

TP and **GP** are **Converted** into **Useful Organic Substances** like **Glucose**

The **Calvin cycle** is the starting point for making **all** the organic substances a plant needs. **Triose phosphate** (TP) and **glycerate 3-phosphate** (GP) molecules are used to make **carbohydrates**, **lipids** and **amino acids**:

- **Carbohydrates** — **hexose sugars** (e.g. glucose) are made by joining **two triose phosphate molecules** together and **larger** carbohydrates (e.g. sucrose, starch, cellulose) are made by joining **hexose sugars** together in **different ways**.
- **Lipids** — these are made using **glycerol**, which is synthesised from **triose phosphate**, and **fatty acids**, which are synthesised from **glycerate 3-phosphate**.
- **Amino acids** — some **amino acids** are made from **glycerate 3-phosphate**.

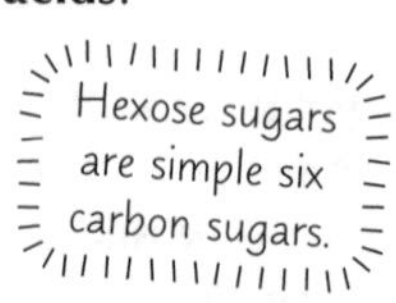

Photosynthesis

The Calvin Cycle Needs to Turn Six Times to Make One Hexose Sugar

Here's the reason why:

1) **Three turns** of the cycle produces **six** molecules of **triose phosphate** (TP), because two molecules of TP are made for every one CO_2 molecule used.
2) **Five out of six** of these TP molecules are used to **regenerate ribulose bisphosphate** (RuBP).
3) This means that for **three turns** of the cycle only **one TP** is produced that's used to make a **hexose sugar**.
4) A hexose sugar has **six carbons** though, so **two TP** molecules are needed to form one hexose sugar.
5) This means the cycle must turn **six times** to produce **two molecules** of **TP** that can be used to make **one hexose sugar**.
6) Six turns of the cycle need **18 ATP** and **12 reduced NADP** from the light-dependent reaction.

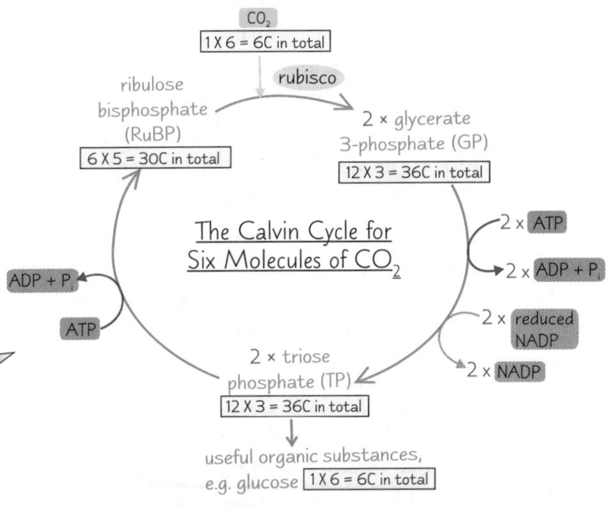

This might seem a bit inefficient, but it keeps the cycle going and makes sure there's always **enough RuBP** ready to combine with CO_2 taken in from the atmosphere.

Morag had to turn one million times to make a sock... two million for a scarf.

Warm-Up Questions

PRACTICE QUESTIONS

Q1 Name two photosynthetic pigments in the chloroplasts of plants.

Q2 At what wavelength does photosystem I absorb light best?

Q3 What three substances does non-cyclic photophosphorylation produce?

Q4 Which photosystem is involved in cyclic photophosphorylation?

Q5 Where in the chloroplasts does the light-independent reaction occur?

Q6 How many carbon atoms are there in a molecule of TP?

Q7 Name two organic substances made from triose phosphate.

Q8 How many CO_2 molecules need to enter the Calvin cycle to make one hexose sugar?

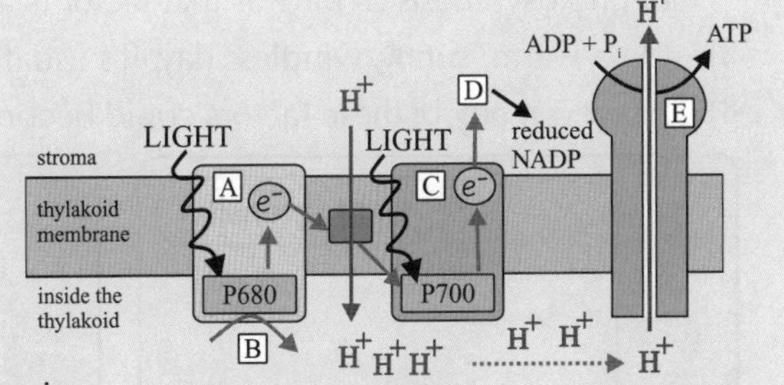

Exam Questions

Q1 The diagram above shows the light-dependent reaction of photosynthesis.
- a) What does object A represent? [1 mark]
- b) Describe process B and explain its purpose. [3 marks]
- c) Explain how reactant D is made into reduced NADP. [2 marks]

Q2 Rubisco is an enzyme that catalyses the first reaction of the Calvin cycle. CA1P is an inhibitor of rubisco.
- a) Describe how triose phosphate is produced in the Calvin cycle. [5 marks]
- b) Briefly explain how ribulose bisphosphate (RuBP) is regenerated in the Calvin cycle. [2 marks]
- c) Explain the effect that CA1P would have on glucose production. [3 marks]

Calvin cycles — bikes made by people who normally make pants...

Next thing we know there'll be people swanning about in their pants riding highly fashionable bikes. Sounds awful I know, but let's face it, anything would look better than cycling shorts. Anyway, it would be a good idea to go over these pages a couple of times — I promise you, there's still room left in your head for more information.

Limiting Factors in Photosynthesis

I'd love to tell you that you'd finished photosynthesis... but I'd be lying.

There are **Optimum Conditions** for **Photosynthesis**

The **ideal conditions** for photosynthesis vary from one plant species to another, but the conditions below would be ideal for **most** plant species in temperate climates like the UK.

1. High light intensity of a certain wavelength

- Light is needed to provide the **energy** for the **light-dependent reaction** — the **higher** the **intensity** of the light, the **more energy** it provides.
- Only certain **wavelengths** of light are used for photosynthesis. The photosynthetic pigments chlorophyll a, chlorophyll b and carotene only **absorb** the **red** and **blue** light in sunlight. (**Green** light is **reflected**, which is why plants look green.)

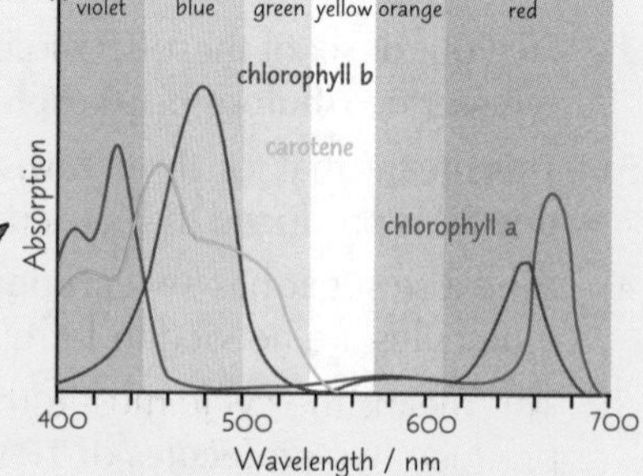

2. Temperature around 25 °C

- Photosynthesis involves **enzymes** (e.g. ATP synthase, rubisco). If the temperature falls **below 10 °C** the enzymes become **inactive**, but if the temperature is **more than 45 °C** they may start to **denature**.
- Also, at **high** temperatures **stomata close** to avoid losing too much water. This causes photosynthesis to slow down because **less CO_2** enters the leaf when the stomata are closed.

3. Carbon dioxide at 0.4%

- Carbon dioxide makes up **0.04%** of the gases in the atmosphere.
- Increasing this to **0.4%** gives a **higher rate** of photosynthesis, but any higher and the stomata start to **close**.

Plants also need a **constant supply** of **water** — **too little** and photosynthesis has to **stop** but **too much** and the soil becomes **waterlogged** (**reducing** the uptake of **minerals** such as **magnesium**, which is needed to make **chlorophyll a**).

Light, **Temperature** and **CO_2** can all **Limit Photosynthesis**

1) **All three** of these things need to be at the **right level** to allow a plant to photosynthesise as quickly as possible.
2) If any **one** of these factors is **too low** or **too high**, it will **limit photosynthesis** (slow it down). Even if the other two factors are at the perfect level, it won't make **any difference** to the speed of photosynthesis as long as that factor is at the wrong level.
3) On a warm, sunny, windless day, it's usually **CO_2** that's the limiting factor, and at night it's the **light intensity**.
4) However, **any** of these factors could become the limiting factor, depending on the **environmental conditions**.

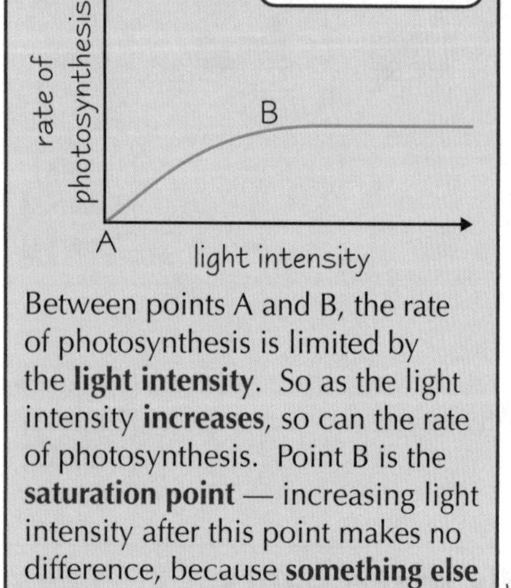

Between points A and B, the rate of photosynthesis is limited by the **light intensity**. So as the light intensity **increases**, so can the rate of photosynthesis. Point B is the **saturation point** — increasing light intensity after this point makes no difference, because **something else** has become the limiting factor. The graph now **levels off**.

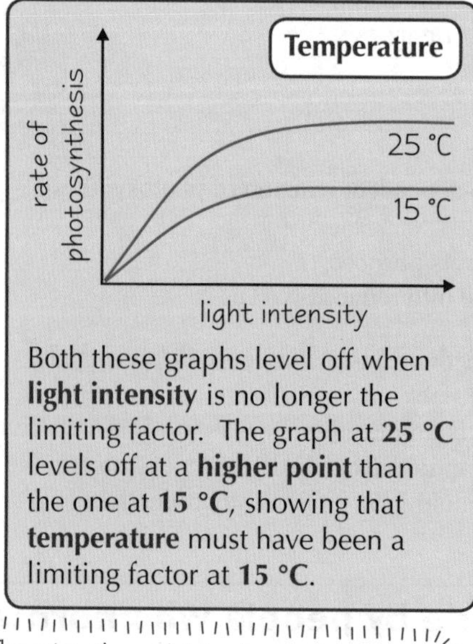

Both these graphs level off when **light intensity** is no longer the limiting factor. The graph at **25 °C** levels off at a **higher point** than the one at **15 °C**, showing that **temperature** must have been a limiting factor at **15 °C**.

The saturation point is where a factor is no longer limiting the reaction — something else has become the limiting factor.

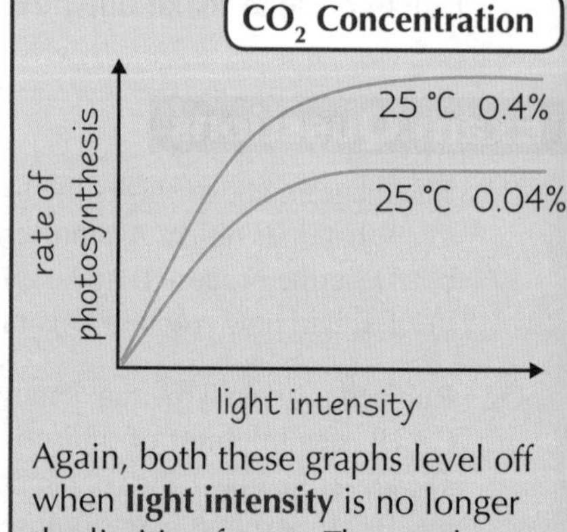

Again, both these graphs level off when **light intensity** is no longer the limiting factor. The graph at **0.4% CO_2** levels off at a **higher point** than the one at **0.04%**, so **CO_2 concentration** must have been a limiting factor at **0.04% CO_2**. The limiting factor here **isn't temperature** because it's the **same** for both graphs (25 °C).

Limiting Factors in Photosynthesis

Growers Use Information About Limiting Factors to Increase Plant Growth

Agricultural growers (e.g. farmers) know the **factors** that **limit photosynthesis** and therefore limit **plant growth**. This means they try to create an **environment** where plants get the **right amount** of everything that they need, which **increases growth** and so **increases yield**.

Growers create optimum conditions in **glasshouses** in the following ways:

Limiting Factor	Management in Glasshouse
Carbon dioxide concentration	CO_2 is added to the air, e.g. by burning a small amount of propane in a CO_2 generator.
Light	Light can get in through the glass. Lamps provide light at night-time.
Temperature	Glasshouses trap heat energy from sunlight, which warms the air. Heaters and cooling systems can also be used to keep a constant optimum temperature, and air circulation systems make sure the temperature is even throughout the glasshouse.

Similar techniques can also be used in **polytunnels** (tunnels made of polythene, under which plants can be grown).

You Need to be Able to Interpret Data on Limiting Factors

Here are some **examples** of the kind of **data** you might get in the exam:

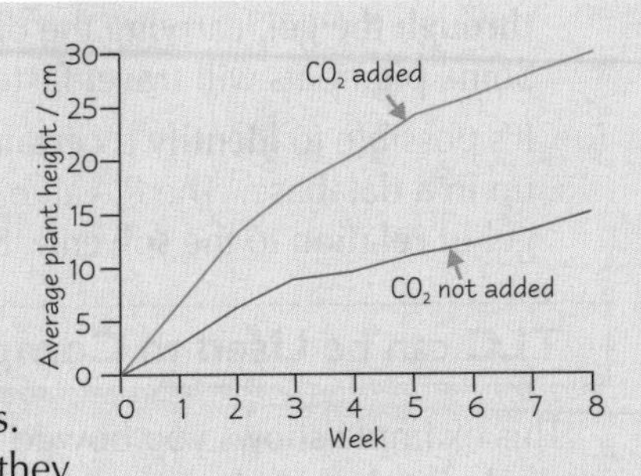

The graph on the **right** shows the effect on plant growth of **adding carbon dioxide** to a greenhouse.

1) In the greenhouse **with added CO_2** plant **growth** was **faster** (the line is steeper) and on average the plants were **larger** after 8 weeks than they were in the control greenhouse (30 cm compared to only 15 cm in the greenhouse where no CO_2 was added).
2) This is because the plants use CO_2 to produce **glucose** by photosynthesis. The more CO_2 they have, the more glucose they can produce, meaning they can **respire more** and so have **more ATP** for **DNA replication**, **cell division** and **protein synthesis**, i.e. growth.

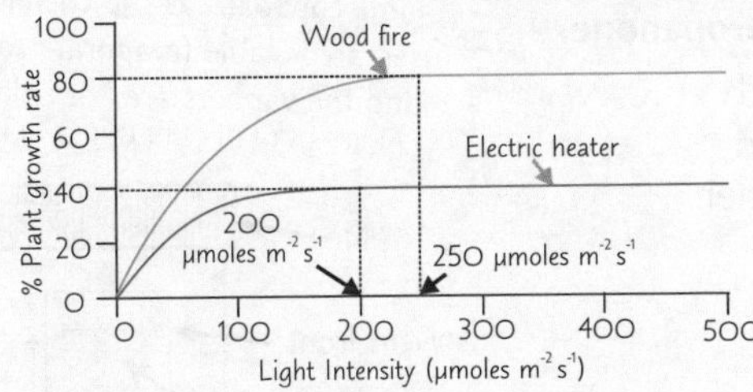

The graph on the **left** shows the effect of **light intensity** on plant growth, and the effect of two **different types** of **heater**.

1) At the start of the graph, the **greater** the **light intensity** the **greater** the **plant growth**.
2) At **200 μmoles m^{-2} s^{-1}** (micromoles per metre^2 per second) of light the **bottom** graph flattens out, showing that **CO_2 concentration** or **temperature** is **limiting growth** in these plants.
3) At **250 μmoles m^{-2} s^{-1}** of light the **top** graph flattens out. The difference between the two graphs could be because the **wood fire increases** the **temperature more** than the electric heater or because it's **increasing** the **concentration** of **CO_2** in the air (an electric heater **doesn't** release CO_2).

Warm-Up Questions

Q1 Name two factors that can limit plant growth.

Exam Question

Crop	Yield in glasshouse / kg	Yield grown outdoors / kg
Tomato	1000	200
Lettuce	750	230
Potato	850	680
Wheat	780	550

Q1 The table above shows the yields of various crops when they are grown in glasshouses and when grown outdoors.

a) Yields are usually higher overall in glasshouses. Describe two ways in which conditions can be controlled in glasshouses to increase yields. [2 marks]

b) Glasshouses are not always financially viable for all crops. Which crop above benefits the least from being grown in glasshouses? Explain your answer. [2 marks]

I'm a whizz at the factors that limit revision...

... watching Hollyoaks, making tea, watching EastEnders, walking the dog... not to mention staring into space (one of my favourites). Anyway, an interpreting data question could well come up in the exams — it could be any kind of data, but don't panic if it's not like the graphs above — as long as you understand limiting factors you'll be able to interpret it.

Photosynthesis Experiments

Everyone loves a good experiment — especially when they involve bright colours. Here are two really colourful photosynthesis experiments for you to enjoy. Let's chop up some plants and marvel at the beauty of Biology...

You Can **Investigate** the **Pigments** in **Leaves** Using **Chromatography**

1) All plants contain several **different photosynthetic pigments** in their leaves. Each pigment absorbs a **different wavelength** of light, so having **more than one** type of pigment **increases** the **range** of **wavelengths** of light that a plant can **absorb**.
2) In addition to photosynthetic pigments, some plants also have **other pigments** in their leaves, which play other essential roles, e.g. protecting the leaves from excessive UV radiation. **Different species** of plants contain **different proportions** and **mixtures** of **pigments**.
3) You can use **thin layer chromatography** (**TLC**) to determine what **pigments** are present in the leaves of a plant. Like all chromatography, TLC involves:
   - A **mobile phase** — where molecules can move. In TLC, this is a **liquid solvent**.
   - A **stationary phase** — where molecules can't move. In TLC, this consists of a **solid** (e.g. glass) **plate** with a **thin layer of gel** (e.g. silica gel) on top.
4) A **sample** of pigments can be **extracted** from the plant and put on the TLC plate. When the plate is placed vertically in the **solvent**, the solvent moves upwards **through** the gel, carrying the dissolved pigments with it. Some pigments will **travel faster** or **further** through the gel than others, which **separates** them out.
5) It's possible to **identify** a **certain pigment** by calculating its **R_f value** and looking it up in a database. The R_f value is the **distance** a substance has moved through the gel in **relation** to the **solvent**. Each pigment has a specific R_f value.

TLC can be **Used** to **Compare** the **Pigments** in **Different Plants**

This example shows you how to use **TLC** to **compare** the **pigments** present in **shade-tolerant plants** and **shade-intolerant plants**. Make sure you're wearing a lab coat, eye protection and gloves before you start. Many of the chemicals involved are toxic and highly flammable.

It's best to do steps 2 and 5 in a fume cupboard as the chemicals used are volatile (evaporate easily) and the vapours are hazardous.

1) **Grind** up several leaves from the shade-tolerant plant you're investigating with some **anhydrous sodium sulfate**, then add a few drops of **propanone**.
2) **Transfer** the **liquid** to a test tube, add some **petroleum ether** and gently shake the tube. **Two distinct layers** will form in the liquid — the **top layer** is the **pigments** mixed in with the petroleum ether.
3) Transfer some of the liquid from the **top layer** into a second test tube with some **anhydrous sodium sulfate**.
4) Draw a horizontal **pencil line** near the bottom of a **TLC plate**. Build up a single **concentrated spot** of the liquid from step 3 on the line by applying several drops and ensuring each one is **dry** before the next is added. This is the **point of origin**.
5) Once the point of origin is completely dry, put the plate into a small glass container with some prepared **solvent** (e.g. a mixture of **propanone**, **cyclohexane** and **petroleum ether**) — just enough so that the **point of origin** is a little bit **above** the solvent. Put a **lid** on the container and leave the plate to develop. As the solvent spreads up the plate, the different **pigments** move with it, but at **different rates** — so they **separate**.

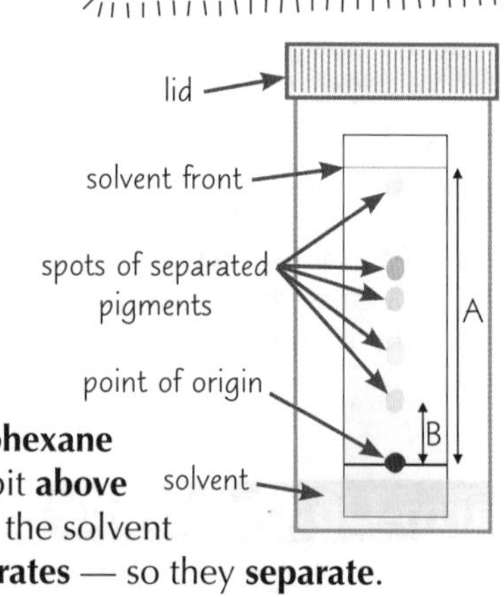

6) When the solvent has **nearly** reached the top, take the plate out and **mark** the **solvent front** (the furthest point the solvent has reached) with a **pencil** and leave the plate to dry in a well-ventilated place.
7) There should be **several** new coloured spots on the chromatography plate between the **point of origin** and the **solvent front**. These are the separated **pigments**. You can calculate their **R_f values** and look them up in a database to **identify** what the pigments are.
8) **Repeat** the process for the **shade-intolerant** plant you're investigating and **compare** the **pigments** present in their leaves.

$$R_f \text{ value} = \frac{B}{A} = \frac{\text{distance travelled by spot}}{\text{distance travelled by solvent}}$$

You may find that the mixture of pigments in the leaves of the shade-tolerant plant is quite different compared to the shade-intolerant plant. One way that shade-tolerant plants can adapt to the light conditions in their environment is by possessing a different proportion of photosynthetic pigments, which allows the plant to make the best use of the light available to it. The mixture of non-photosynthetic pigments is also likely to be different. For example, the chloroplasts of shade-tolerant plants are adapted for photosynthesis in low light conditions, but really sensitive to higher levels of light. These plants sometimes produce dark red and purple pigments called anthocyanins, which are thought to protect their chloroplasts from brief exposure to higher light levels.

Photosynthesis Experiments

You Can **Investigate** the **Activity** of **Dehydrogenase** in **Chloroplasts**

1) In **photosystem I**, during the **light-dependent** stage of photosynthesis, **NADP** acts as an **electron acceptor** and is **reduced** (see page 115). The reaction is **catalysed** by a **dehydrogenase enzyme**.
2) The **activity** of this **enzyme** can be investigated by adding a **redox indicator dye** to extracts of **chloroplasts**. Like **NADP**, the dye acts as an **electron acceptor** and gets **reduced** by the dehydrogenase in the chloroplasts. As the dye gets reduced, you'll see a **colour change**. For example, the dye **DCPIP** changes from **blue** to **colourless** when it gets reduced.
3) You can measure the rate of the dehydrogenase activity by measuring the **rate** at which DCPIP **loses** its **blue colour**. To do this, you need a **colorimeter**. A colorimeter measures how much light a solution **absorbs** when a light source is **shone** directly **through** it. A **coloured solution** absorbs **more light** than a colourless solution.

The experiment below shows you how to **investigate** the **effect** of **light intensity** on **dehydrogenase activity** in extracts of chloroplasts. It uses a **bench lamp** as a **light source** and involves placing tubes of chloroplast extract mixed with DCPIP at a range of **different distances** from the light source. **Light intensity** should **decrease** with **increasing distance** from the lamp. You'll need to **choose** the **distances** you're going to investigate (e.g. 15 cm, 30 cm and 45 cm) before you start.

You can use a similar method to investigate the effects of other factors on dehydrogenase activity in chloroplasts.

1) **Cut** a few **leaves** (spinach works well) into pieces. Remove any tough stalks.
2) Using a pestle and mortar, **grind** up the **leaf pieces** with some **chilled isolation solution** (a solution of **sucrose**, **potassium chloride** and **phosphate buffer** at **pH 7**). **Filter** the **liquid** you make into a **beaker** through a **funnel** lined with **muslin cloth**.
3) Transfer the liquid to **centrifuge tubes** and centrifuge them at **high speed** for **10 minutes**. This will make the **chloroplasts** gather at the **bottom** of each tube in a **'pellet'**.
4) **Get rid** of the **liquid** from the top of the tubes, **leaving** the **pellets** in the bottom.
5) **Re-suspend** the pellets in **fresh**, chilled **isolation solution**. This is your **chloroplast extract**. **Store** it on **ice** for the rest of the experiment.
6) Set up a **colorimeter** with a **red filter** and **zero** it using a cuvette (a cuboid-shaped vessel used in colorimeters) containing the **chloroplast extract** and **distilled water**.
7) Set up a **test tube rack** at a **set distance** from a **bench lamp**. Switch the lamp on.
8) Put a test tube in the rack, add a **set volume** of **chloroplast extract** to the tube and a **set volume** of **DCPIP**. **Mix** the contents of the tube together.
9) **Immediately** take a sample of the mixture from the tube and add it to a **clean cuvette**. Then place the cuvette in your colorimeter and **record** the **absorbance**. Do this every **2 minutes** for the next **ten minutes**.
10) **Repeat** steps **7** to **9** for **each distance** under investigation.

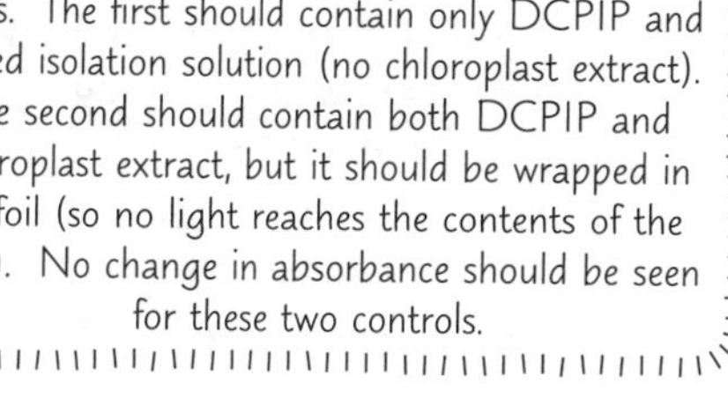

If dehydrogenase activity is taking place, the **absorbance** will **decrease** as the DCPIP gets **reduced** and **loses** its **blue colour**. The **faster** the absorbance decreases, the **faster** the **rate** of **dehydrogenase activity**. You can plot a **graph** of **absorbance against time** for each distance from the light source. Then **compare** your results to **determine** how light intensity affects the **rate** of the dehydrogenase enzyme.

You should also check whether the absorbance changes at each distance in two negative control tubes. The first should contain only DCPIP and chilled isolation solution (no chloroplast extract). The second should contain both DCPIP and chloroplast extract, but it should be wrapped in tin foil (so no light reaches the contents of the tube). No change in absorbance should be seen for these two controls.

Warm-Up Questions

Q1 What is a chromatography plate?

PRACTICE QUESTIONS

Exam Question

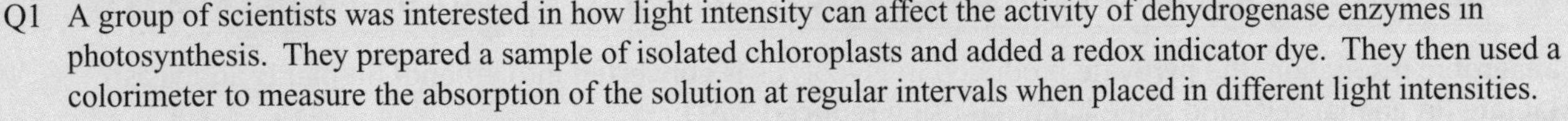

Q1 A group of scientists was interested in how light intensity can affect the activity of dehydrogenase enzymes in photosynthesis. They prepared a sample of isolated chloroplasts and added a redox indicator dye. They then used a colorimeter to measure the absorption of the solution at regular intervals when placed in different light intensities.

a) What role do dehydrogenase enzymes play in photosystem I of photosynthesis? [1 mark]

b) Explain how a redox indicator dye is able to indicate dehydrogenase activity in photosystem I. [3 marks]

No animals were harmed in the making of these experiments...

... but I did ruin my garden. Make sure that the plant's sacrifice wasn't in vain and learn how these experiments work. You might get a question in your exams that involves experiments pretty similar to these, so they're worth remembering.

Respiration

Roses are red, violets are blue, I love respiring and I bet you do too. Now you've enjoyed that poem, it's time to concentrate. I hope you like remembering reactions involved in respiration, because these pages have several.

There are Two Types of Respiration

1) The two types of respiration are **aerobic** (requires oxygen) and **anaerobic** (doesn't require oxygen).
2) Both **produce ATP** (see p. 112), although **anaerobic respiration** produces **less**.
3) Both **start** with the process of **glycolysis** (see below). The stages **after** glycolysis **differ**.

Glycolysis Makes Pyruvate from Glucose

1) Glycolysis involves splitting **one molecule** of glucose (with 6 carbons — 6C) into **two** smaller molecules of **pyruvate** (3C).
2) The process happens in the **cytoplasm** of cells.
3) Glycolysis is the **first stage** of both aerobic and anaerobic respiration and **doesn't need oxygen** to take place — so it's an **anaerobic** process.

There are Two Stages in Glycolysis — Phosphorylation and Oxidation

First, **ATP** is **used** to **phosphorylate glucose** to triose phosphate. Then **triose phosphate** is **oxidised**, **releasing ATP**. Overall there's a **net gain** of **2 ATP** and **2 reduced NAD**.

1 Stage One — Phosphorylation

1) Glucose is **phosphorylated** using a **phosphate** from a molecule of **ATP**. This creates **1** molecule of **glucose phosphate** and **1** molecule of **ADP**.
2) **ATP** is then used to add another **phosphate**, forming **hexose bisphosphate**.
3) **Hexose bisphosphate** is then **split** into 2 molecules of **triose phosphate**.

2 Stage Two — Oxidation

1) **Triose phosphate** is **oxidised** (loses hydrogen), forming 2 molecules of **pyruvate**.
2) **NAD** collects the hydrogen ions, forming **2 reduced NAD**.
3) **4 ATP** are **produced**, but 2 were used up in stage one, so there's a **net gain** of **2 ATP**.

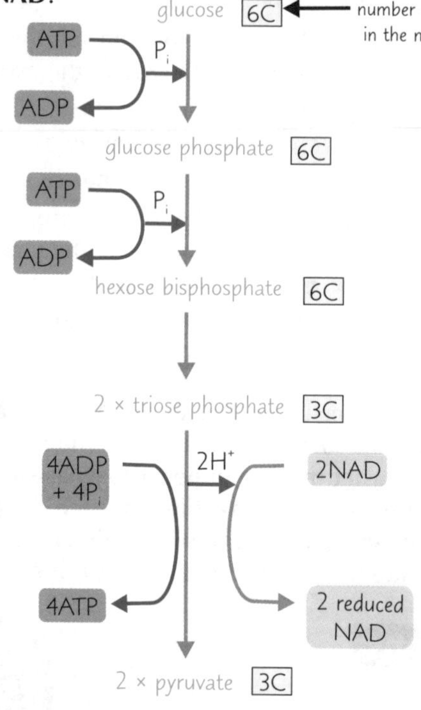

In **aerobic** respiration...

1) The **two** molecules of **reduced NAD** go to **oxidative phosphorylation** — see page 124.
2) The **two pyruvate** molecules are **actively transported** into the **matrix** of the **mitochondria** for the **link reaction** (see the next page).

In Anaerobic Respiration Pyruvate is Converted to Ethanol or Lactate

In **anaerobic** respiration, the **pyruvate** produced in glycolysis is **converted** into **ethanol** (in **plants** and **yeast**) or **lactate** (in **animal** cells and some **bacteria**) using **reduced NAD**:

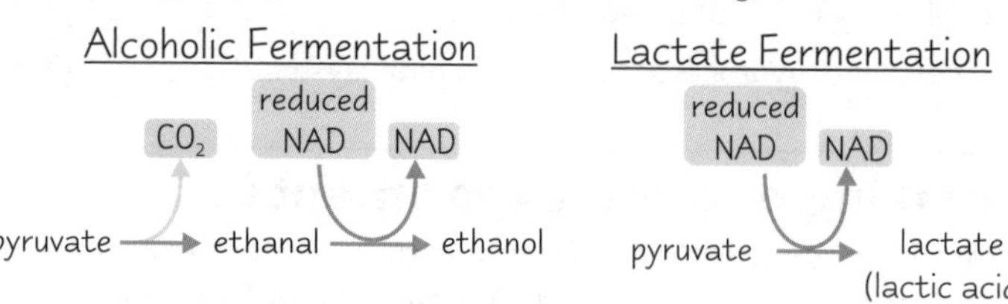

The production of ethanol or lactate **regenerates oxidised NAD**. This means **glycolysis** can **continue** even when there **isn't** much oxygen around, so a **small amount of ATP** can still be **produced** to keep some biological processes going... clever.

Aerobic Respiration

So, as you know from the previous page, in aerobic respiration, the two molecules of pyruvate from glycolysis enter the mitochondrial matrix for the link reaction. Here's what happens next...

The **Link Reaction** converts **Pyruvate** to **Acetyl Coenzyme A**

1) **Pyruvate** is **decarboxylated** (one carbon atom is **removed** from pyruvate in the form of CO_2).
2) **Pyruvate** is **oxidised** to form **acetate** and **NAD** is reduced to form **reduced** NAD.
3) **Acetate** is combined with **coenzyme A** (CoA) to form **acetyl coenzyme A** (**acetyl CoA**).
4) **No ATP** is produced in this reaction.

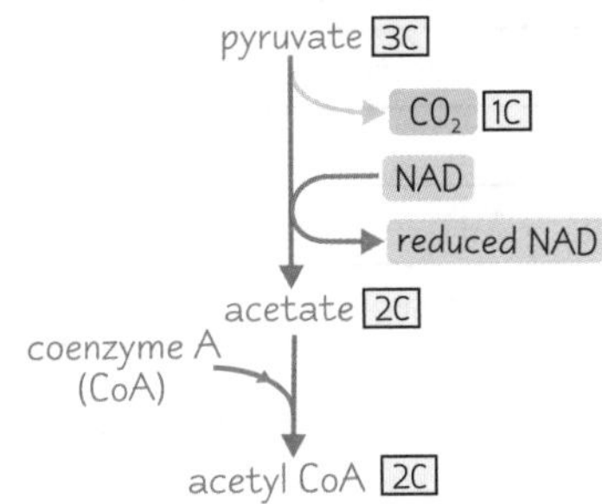

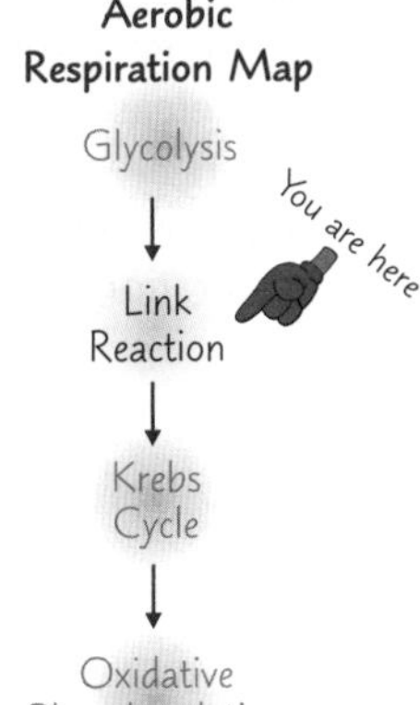

The **Link Reaction** Occurs **Twice** for Every **Glucose Molecule**

Two pyruvate molecules are made for **every glucose molecule** that enters glycolysis. This means the **link reaction** and the third stage (the **Krebs cycle**) happen **twice** for every glucose molecule. So for each glucose molecule:

- **Two** molecules of **acetyl coenzyme A** go into the Krebs cycle (see below).
- **Two CO_2 molecules** are released as a waste product of respiration.
- **Two** molecules of **reduced NAD** are formed and go to the last stage (oxidative phosphorylation, see page 124).

The **Krebs Cycle** Produces **Reduced Coenzymes** and **ATP**

The Krebs cycle involves a series of **oxidation-reduction reactions**, which take place in the **matrix** of the **mitochondria**. The cycle happens **once** for **every pyruvate** molecule, so it goes round **twice** for **every glucose** molecule.

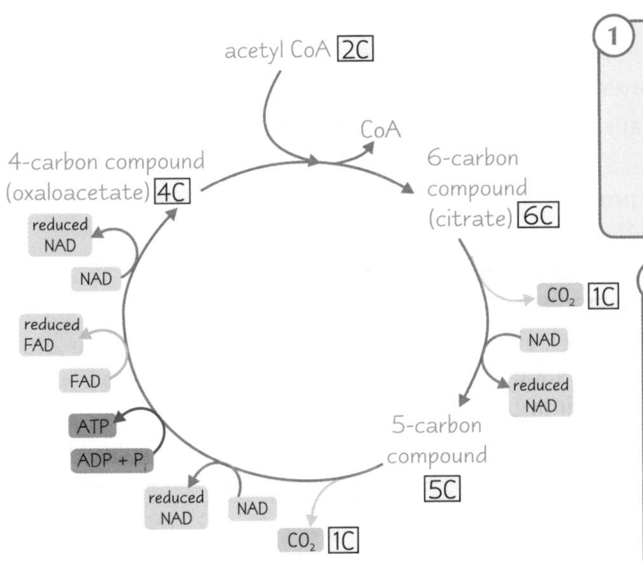

1
- **Acetyl CoA** from the link reaction combines with a **four-carbon molecule** (oxaloacetate) to form a **six-carbon molecule** (citrate).
- **Coenzyme A** goes back to the **link reaction** to be used again.

2
- The **6C citrate molecule** is converted to a **5C molecule**.
- **Decarboxylation** occurs, where CO_2 is **removed**.
- **Dehydrogenation** also occurs, where **hydrogen** is **removed**.
- The hydrogen is used to **produce reduced NAD** from NAD.

Aerobic
Respiration Map
Glycolysis
Link
Reaction
You are here
Krebs
Cycle
Oxidative
Phosphorylation

3
- The **5C molecule** is then converted to a **4C molecule**. (There are some intermediate compounds formed during this conversion, but you don't need to know about them.)
- **Decarboxylation** and **dehydrogenation** occur, producing **one** molecule of **reduced FAD** and **two** of **reduced NAD**.
- **ATP** is **produced** by the **direct transfer** of a **phosphate** group from an **intermediate** compound **to ADP**. When a phosphate group is directly transferred from one molecule to another it's called **substrate-level phosphorylation**. **Citrate** has now been **converted** into **oxaloacetate**.

Aerobic Respiration

Some **Products** of the **Krebs Cycle** are Used in **Oxidative Phosphorylation**

Some products are **reused**, some are **released** and others are used for the **next stage** of respiration:

Product from one Krebs cycle	Where it goes
1 coenzyme A	Reused in the next link reaction
Oxaloacetate	Regenerated for use in the next Krebs cycle
2 CO_2	Released as a waste product
1 ATP	Used for energy
3 reduced NAD	To oxidative phosphorylation
1 reduced FAD	To oxidative phosphorylation

Oxidative Phosphorylation Produces Lots of **ATP**

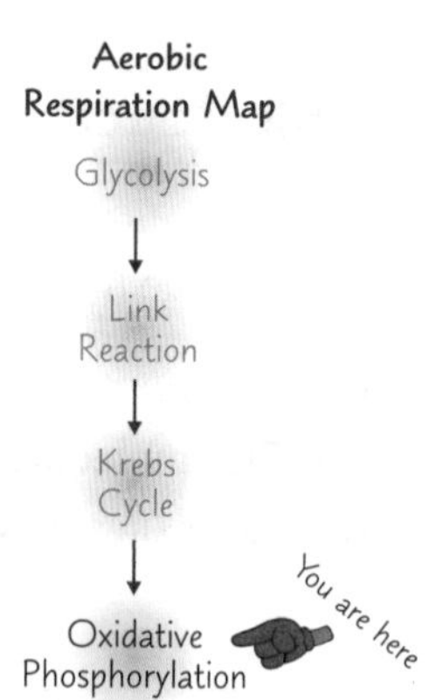

1) Oxidative phosphorylation is the process where the **energy** carried by **electrons**, from **reduced coenzymes** (reduced NAD and reduced FAD), is used to **make ATP**. (The whole point of the previous stages is to make reduced NAD and reduced FAD for the final stage).
2) Oxidative phosphorylation involves the **electron transport chain** and **chemiosmosis** (see below).

Protons are **Pumped Across** the **Inner Mitochondrial Membrane**

So now on to how **oxidative phosphorylation** actually **works**:

The regenerated coenzymes are reused in the Krebs cycle.

1) **Hydrogen atoms** are released from **reduced NAD** and **reduced FAD** as they're **oxidised** to NAD and FAD. The H atoms **split** into **protons (H^+)** and **electrons (e^-)**.
2) The **electrons** move down the **electron transport chain** (made up of **electron carriers**), **losing energy** at each carrier.

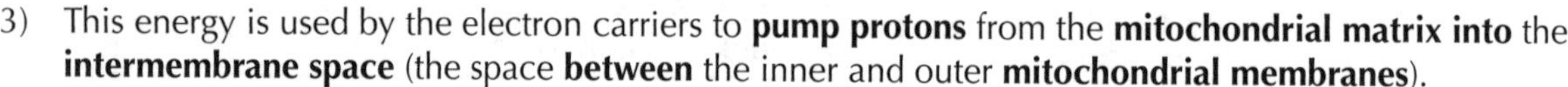

3) This energy is used by the electron carriers to **pump protons** from the **mitochondrial matrix into** the **intermembrane space** (the space **between** the inner and outer **mitochondrial membranes**).
4) The **concentration** of **protons** is now **higher** in the **intermembrane space** than in the mitochondrial matrix — this forms an **electrochemical gradient** (a **concentration gradient** of **ions**).
5) Protons then **move down** the **electrochemical gradient**, back across the inner mitochondrial membrane and into the mitochondrial matrix, via **ATP synthase** (which is embedded in the inner mitochondrial membrane). This **movement** drives the synthesis of **ATP** from **ADP** and **inorganic phosphate** (P_i).
6) This process of ATP production driven by the movement of H^+ ions across a membrane (due to electrons moving down an electron transport chain) is called **chemiosmosis** (which is described by the **chemiosmotic theory**).
7) In the mitochondrial matrix, at the end of the transport chain, the **protons**, **electrons** and **O_2** (from the blood) combine to form **water**. Oxygen is said to be the final **electron acceptor**.

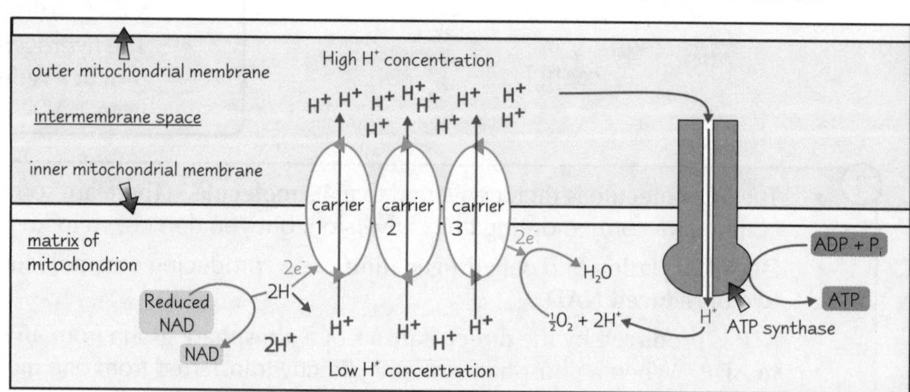

Aerobic Respiration

32 ATP Can be Made from One Glucose Molecule

As you know, **oxidative phosphorylation makes ATP** using energy from the reduced coenzymes — **2.5 ATP** are made from each **reduced NAD** and **1.5 ATP** are made from each **reduced FAD**. The table on the right shows **how much** ATP a cell can make from **one molecule** of **glucose** in **aerobic respiration**. (Remember, one molecule of glucose produces 2 pyruvate, so the link reaction and Krebs cycle happen twice.)

Stage of respiration	Molecules produced	Number of ATP molecules
Glycolysis	2 ATP	2
Glycolysis	2 reduced NAD	2 × 2.5 = 5
Link Reaction (×2)	2 reduced NAD	2 × 2.5 = 5
Krebs cycle (×2)	2 ATP	2
Krebs cycle (×2)	6 reduced NAD	6 × 2.5 = 15
Krebs cycle (×2)	2 reduced FAD	2 × 1.5 = 3
		Total ATP = 32

The number of ATP produced per reduced NAD or reduced FAD was thought to be 3 and 2, but new research has shown that the figures are nearer 2.5 and 1.5.

ATP Production Can be Affected by Mitochondrial Diseases

1) **Mitochondrial diseases** affect the **functioning** of **mitochondria**. They can affect how **proteins** involved in **oxidative phosphorylation** or the **Krebs cycle** function, **reducing ATP production**.
2) This may cause **anaerobic respiration** to **increase**, to try and make up some of the **ATP shortage**.
3) This results in **lots** of **lactate** being produced, which can cause **muscle fatigue** and **weakness**.
4) Some lactate will also **diffuse** into the **bloodstream**, leading to **high lactate concentrations** in the **blood**.

Other Respiratory Substrates Can also be Used in Aerobic Respiration

It's not just glucose that can be used as the **substrate** in aerobic respiration. Some products resulting from the **breakdown** of **other** molecules, such as **fatty acids** from **lipids** and **amino acids** from **proteins**, can be converted into molecules that are **able** to **enter** the **Kreb's cycle** (usually acetyl CoA).

Warm-Up Questions

Q1 Where in the cell does glycolysis occur?

Q2 Is glycolysis an anaerobic or aerobic process?

Q3 How many ATP molecules are used up in glycolysis?

Q4 What are the products of the link reaction?

Q5 Where in the cell does the Krebs cycle occur?

Q6 How many times does decarboxylation happen during one turn of the Krebs cycle?

Q7 What do the electrons lose as they move along the electron transport chain in oxidative phosphorylation?

Exam Questions

Q1 At the end of a 100 m sprint, runners will have built up lactate in their muscle cells.
  a) Name the reduced coenzyme regenerated by lactate production. [1 mark]
  b) What is the advantage for the runner of producing lactate in anaerobic respiration? [2 marks]

Q2 Carbon monoxide inhibits the final electron carrier in the electron transport chain.
  a) Explain how this affects ATP production via the electron transport chain. [2 marks]
  b) Explain how this affects ATP production via the Krebs cycle. [2 marks]

Q3 Describe how a 6-carbon molecule of glucose is converted to pyruvate. [6 marks]

The electron transport chain isn't just a FAD with the examiners...

Oh my gosh, I didn't think it could get any worse... You may be wondering how to learn these pages of crazy chemistry. Basically, you have to put in the time and go over and over it. Don't worry though, it WILL pay off and before you know it, you'll be set for the exams. And once you know this lot you'll be able to do anything, e.g. world domination.

Respiration Experiments

You can use experiments to test how quickly respiration is taking place. Here are a few examples for you.

You can **Investigate Factors Affecting Respiration** in **Single-celled Organisms**

Yeast are single-celled organisms that can be grown in **culture**. They can respire **aerobically** when **plenty** of **oxygen** is **available** and **anaerobically** when **oxygen isn't available**. Both aerobic and anaerobic respiration in yeast produce **CO_2**, so the **rate of CO_2 production** gives an indication of the yeast's **respiration rate**. One way to measure CO_2 production is by using a **gas syringe** to collect the CO_2.

The methods below show you how to investigate the **effects** of **temperature** on **yeast respiration**. You'll need to decide what **temperatures** you're going to **investigate before** you **start** (e.g. 10 °C, 20 °C and 25 °C).

Aerobic Respiration

1) Put a **known volume** and **concentration** of **substrate solution** (e.g. glucose) in a test tube. Add a known volume of **buffer solution** to keep the **pH constant**. (Choose the optimum pH for the yeast you're testing — usually 4-6.)
2) Place the test tube in a **water bath** set to one of the **temperatures** being **investigated**. Leave it there for **10 minutes** to allow the temperature of the **substrate** to **stabilise**.
3) Add a **known mass** of **dried yeast** (e.g. *Saccharomyces cerevisiae*) to the test tube and **stir** for two minutes.
4) After the yeast has dissolved into the solution, put a **bung** with a **tube attached** to a **gas syringe** in the top of the test tube. The **gas syringe** should be set to **zero**.
5) **Start** a **stop watch** as soon as the bung has been put in the test tube.
6) As the yeast **respire**, the **CO_2 formed** will travel up the tube and into the **gas syringe**, which is used to measure the **volume** of **CO_2 released**.
7) At **regular time intervals** (e.g. every minute), record the **volume** of **CO_2** that is **present** in the **gas syringe**. Do this for a set amount of time (e.g. 10 minutes).
8) A **control** experiment should also be set up at each temperature, where **no yeast** is present. **No CO_2** should be formed without the yeast.
9) **Repeat** the experiment three times at **each temperature** you're investigating. Use your data to **calculate** the **mean rate of CO_2 production** at each temperature.

bung with tube
gas syringe (held by stand and clamp)
test tube
water bath
yeast culture and substrate solution

The yeast will only respire aerobically until the oxygen trapped in the tube is all used up. If you wanted to run the experiment for more time or with more yeast or glucose, you could use a conical flask that can trap more oxygen.

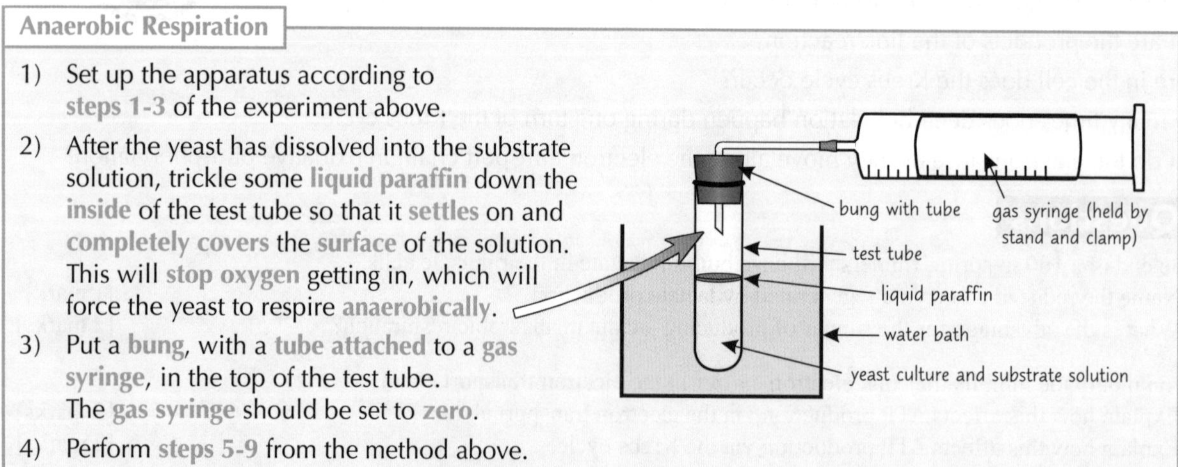

Anaerobic Respiration

1) Set up the apparatus according to **steps 1-3** of the experiment above.
2) After the yeast has dissolved into the substrate solution, trickle some **liquid paraffin** down the **inside** of the test tube so that it **settles** on and **completely covers** the **surface** of the solution. This will **stop oxygen** getting in, which will force the yeast to respire **anaerobically**.
3) Put a **bung**, with a **tube attached** to a **gas syringe**, in the top of the test tube. The **gas syringe** should be set to **zero**.
4) Perform **steps 5-9** from the method above.

You can also easily **adapt** these methods to investigate the **effects** of other **variables**, such as **substrate concentration** and the use of **different respiratory substrates** (e.g. sucrose) on the **respiration rate**.

Just remember that you should **only change one variable at a time** (the independent variable, see page 224). **All** the **other variables** that could affect your results need to be **controlled** (kept the same) or your results won't be **valid**.

Respiration Experiments

The Rate of Oxygen Consumption can be Measured using a Respirometer

Respirometers can be used to indicate the **rate** of **aerobic respiration** by measuring the **amount** of **oxygen consumed** by an organism over a **period** of **time**. The example below shows how a respirometer can be used to measure the respiration rate of **woodlice**. You could also use it to measure the respiration rate of other small organisms or of plant seeds.

1) The apparatus is set up as shown on the right, partially submerged in a **water bath** at **15 °C** to provide the **optimum temperature** for the **woodlice** and therefore, the **optimum temperature** for the **enzymes** involved in their **respiration**.
2) The **control tube** is set up in exactly the **same way** as the woodlouse tube, except that the woodlice are substituted with **glass beads** of the **same mass**.
3) For **ten minutes**, the **tap** is **left open** and the **syringe** is **removed** to allow the **apparatus** to **equilibrate** (accounting for any **expansion** that might cause the **pressure** to **change inside**) and the **respiration rate** of the woodlice to **stabilise** in their new environment.
4) When the ten minutes is up, the **tap** is **closed** and the **syringe** is **attached**.
5) The **syringe** is used to **reset** the **manometer**, so that the **ends** of the **fluid** are at the **same level** on **either side** of the '**U**' and the reading from the **volume scale** on the **syringe** (usually in cm^3) is **recorded**.
6) As respiration occurs, the **volume** of the **air** in the test tube containing woodlice will decrease, due to the **oxygen consumed** during respiration (all the CO_2 produced is **absorbed** by the **potassium hydroxide**).
7) The decrease in the volume of the air will **reduce the pressure** in the test tube, causing the **coloured fluid** in the capillary tube of the manometer to **move towards** it.
8) After **leaving** the apparatus to **run** for a **set** period of time (e.g. 10 minutes), the syringe is used to **reset** the **manometer** and the **reading** on the **syringe's volume scale** is **recorded again**. The **difference** between **this figure** and the **figure taken** at the **start** of the **experiment** is the **oxygen consumption** for this **time period**. You can use this to calculate a **rate of respiration**.
9) To check the **precision** of the results, the experiment is **repeated** and a **mean volume** of O_2 is calculated.

manometer (a capillary tube filled with coloured fluid, with a calibrated scale)
syringe
closed tap
woodlice on gauze
glass beads
water bath
potassium hydroxide solution
potassium hydroxide solution
Test tube
Control tube

Alfred the aphid thought holding his breath in the respirometer would be really funny. The students didn't.

Oxygen consumption can also be calculated by recording the movement of the fluid in the manometer, read from the scale on the manometer itself.

Warm-Up Questions

Q1 What does a respirometer measure?

Exam Question

PRACTICE QUESTIONS

Q1 A student was trying to find the optimum pH for yeast to produce ethanol. She set up three test tubes, each containing a solution of glucose buffered to a different pH. She then dissolved some dried *Saccharomyces cerevisiae* in the solution and trickled some liquid paraffin down the inside of the test tubes. Immediately after, she put a bung in the top of each test tube, with a tube attached to a gas syringe. Every 60 seconds, she recorded how much CO_2 had been released into the gas syringe.

a) Why did the student trickle liquid paraffin down the inside of the test tubes? [1 mark]

b) Why would measuring the rate of CO_2 production help her to find out how quickly ethanol was being produced? [2 marks]

c) Give two variables that should have been controlled in this experiment and describe how each of these variables should have been controlled. [2 marks]

d) What negative control should have been included in this experiment and why? [2 marks]

Respiration experiments — they're a gas...

Examiners love to ask you questions on experiments. Remember how these ones work in case something similar comes up in the exams. When you've got them stuck in your head, do something more interesting like learn to play the tuba.

Energy Transfer in Ecosystems

Some organisms get their energy from the Sun, some get it from other organisms, and it's all very friendly. Yeah right.

Plants Photosynthesise and Produce Biomass

1) An **ecosystem** includes all the **organisms** living in a particular area and all the **non-living** (abiotic) conditions (see p. 182).
2) In all ecosystems, there are **producers** — organisms that make their **own food**, e.g. plants and algae produce their own food through **photosynthesis**.
3) During **photosynthesis** plants use **energy** (from sunlight) and **carbon dioxide** (from the atmosphere in land-based ecosystems, or dissolved in water in aquatic ecosystems) to make **glucose** and other sugars (see pages 114-117).
4) Some of the sugars produced during photosynthesis are used in **respiration**, to release **energy** for **growth**.
5) The **rest** of the glucose is used to make other **biological molecules**, such as **cellulose** (a component of plant cell walls). These biological molecules make up the plant's **biomass** — the mass of living material.
6) **Biomass** can also be thought of as the **chemical energy stored** in the **plant**.
7) Energy is **transferred** through the **living organisms** of an ecosystem when organisms **eat** other organisms, e.g. producers are eaten by organisms called **primary consumers**. Primary consumers are then eaten by **secondary consumers** and secondary consumers are eaten by **tertiary consumers**. This is a **food chain** (see p. 13

Biomass can be Measured as Dry Mass or Using a Calorimeter

Biomass can be measured in terms of the **mass of carbon** that an organism contains or the **dry mass** of its tissue per **unit area**.

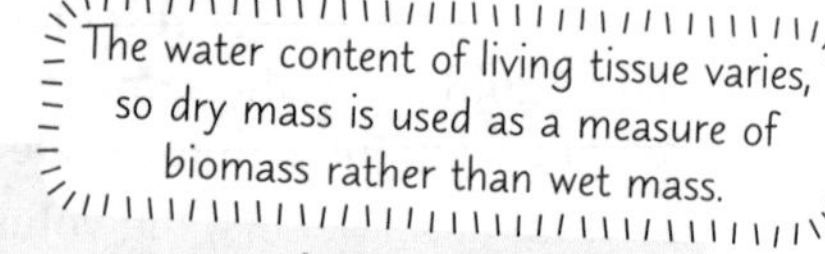

1) **Dry mass** is the **mass** of the organism with the **water removed**.
2) To **measure** the dry mass, a **sample** of the organism is **dried**, often in an **oven** set to a low temperature. The sample is then weighed at **regular** intervals (e.g. every day). Once the **mass** becomes **constant** you know that all the water has been removed.
3) If needed, the result from the sample can be **scaled up** to give the dry mass (biomass) of the **total population** or the **area** being investigated. A **typical unit** for dry mass might be **kg m^{-2}**.
4) The **mass** of **carbon** present is generally taken to be **50%** of the dry mass.

You can **estimate** the amount of **chemical energy** stored in biomass by **burning** the **biomass** in a **calorimeter**. The amount of **heat given off** tells you **how much** energy is in it. Energy is measured in joules (J) or kilojoules (kJ).

1) A sample of dry biomass is **burnt** and the **energy released** is used to **heat** a **known volume** of **water**.
2) The **change in temperature** of the water is used to calculate the **chemical energy** of the **dry biomass**.

GPP and NPP are Chemical Energy Stores

Remember plants convert light energy to chemical energy during photosynthesis.

1) **Gross primary production** (***GPP***) is the **total** amount of **chemical energy** converted from light energy by **plants**, in a given area.
2) Approximately 50% of the gross primary production is **lost to the environment** as **heat** when the plants **respire**. This is called **respiratory loss** (***R***).
3) The **remaining** chemical energy is called the **net primary production** (***NPP***). So ***NPP* = *GPP* – *R***.
4) The **NPP** is the energy available to the plant for **growth** and **reproduction** — the energy is stored in the plant's biomass. It is also the energy available to **organisms** at the **next stage** in the **food chain** (the next trophic level, see page 130). These include **herbivores** (animals that eat the plants) and **decomposers**.
5) Often primary production is expressed as a **rate** — i.e. the total amount of chemical energy (or biomass) in a given area, in a given **time**. Typical units might be **kJ ha^{-1} year^{-1}** (kilojoules per hectare per year) or **kJ m^{-2} yr^{-1}**. When primary production is expressed as a rate, it is called **primary productivity**.

net primary production = gross primary production – respiratory loss

EXAMPLE: The grass in an ecosystem has a gross primary productivity of **20 000 kJ m^{-2} yr^{-1}**. It loses **8 000 kJ m^{-2} yr^{-1}** as heat from **respiration**.

net primary productivity = 20 000 – 8 000
= 12 000 kJ m^{-2} yr^{-1}

Energy Transfer in Ecosystems

You Can Also **Calculate Net Production** for **Consumers**

Gus felt he needed to compensate for the 90% of energy he was not getting from his food.

1) **Consumers** also **store chemical energy** in their **biomass**.
2) Consumers get **energy** by **ingesting** plant material, or animals that have eaten plant material.
3) However, **not all** the chemical energy stored in the consumers' food is **transferred** to the **next** trophic level — around **90%** of the **total available energy** is **lost** in various ways.
4) Firstly, not all of the food is **eaten** (e.g. plant roots, bones) so the energy it contains is **not taken in**. Then, of the parts that **are ingested**:
   - Some are indigestible, so are **egested** as **faeces**. The **chemical energy** stored in these parts is therefore **lost** to the **environment**.
   - Some energy is also **lost** to the **environment** through **respiration** or excretion of **urine**.
5) The energy that's **left** after all this is **stored** in the **consumers' biomass** and is available to the next trophic level. This energy is the consumers' **net production**.
6) The **net production** of **consumers** can be **calculated** using the following **formula**:

Remember, when production is expressed as a rate it is called productivity.

$$N = I - (F + R)$$

N = Net production

I = Chemical energy in ingested food

F = Chemical energy lost in faeces and urine

R = Energy lost through respiration

EXAMPLE: The rabbits in an ecosystem ingest **20 000 kJ m^{-2} yr^{-1}** of energy, but lose **12 000 kJ m^{-2} yr^{-1}** of it in faeces and urine. They lose a further **6000 kJ m^{-2} yr^{-1}** using energy for **respiration**. You can use this to **calculate** the **net productivity** of the rabbits:

net productivity = 20 000 – (12 000 + 6000)
= 20 000 – 18 000
= 2000 kJ m^{-2} yr^{-1}

The net production of consumers can also be called **secondary production** (or **secondary productivity** when it's expressed as a **rate**).

7) You might also be asked to **calculate** how **efficient energy transfer** from one trophic level to another is:

The rabbits receive **20 000 kJ m^{-2} yr^{-1}**, and their **net productivity** is **2000 kJ m^{-2} yr^{-1}**. So the **percentage efficiency** of **energy transfer** is:

(2000 ÷ 20 000) × 100 = 10%

Warm-Up Questions

Q1 What is biomass?

Q2 How is energy transferred through an ecosystem?

Q3 State the formula for net primary production.

Q4 Briefly explain why not all the energy from one trophic level gets transferred to the next trophic level.

Exam Questions

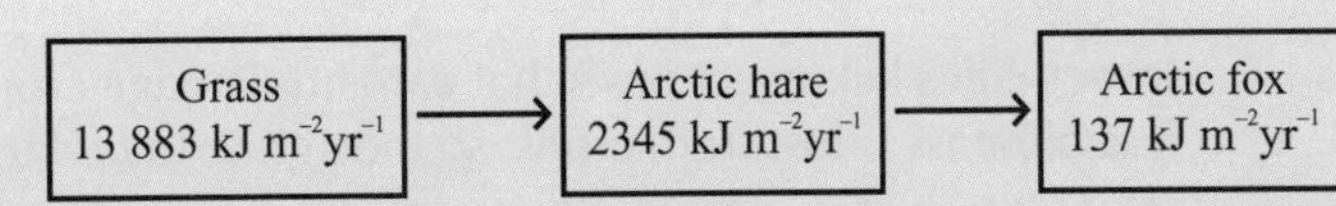

Q1 The diagram above shows the net productivity of different trophic levels in a food chain.
   a) Explain why the net productivity of the Arctic hare is less than the net primary productivity of the grass. [4 marks]
   b) The Arctic hare ingests 18 905 kJ m^{-2} yr^{-1} of food. Calculate the total energy loss of the Arctic hare. [2 marks]

Q2 A farmer grows cabbages in one of his fields.
   a) Suggest how he could estimate the chemical energy store in the dry mass of one of his cabbages. [3 marks]
   b) Using this estimate, the energy of the cabbage field was calculated as 15 600 kJ m^{-2}. Does this represent the gross or net primary production? Give a reason for your answer. [2 marks]

Boy, do I need an energy transfer this morning...

Golly, lots of similar sounding terms on these pages. Plants are primary producers, so you calculate their net primary production. They get energy from the Sun and lose some through respiration. For consumers it's just net production — they eat, then lose energy from respiration and faeces and urine. Simple. I mean, you never saw a plant on the loo...

Farming Practices and Production

Farmers know the theory behind energy transfers and try to use it to maximise production — smart thinking. You don't have to milk the cows, but you do need to know how to increase the efficiency of energy transfer...

Food Webs Show How **Energy** is **Transferred** Between **Organisms**

1) **Food chains** and **food webs** show how energy is **transferred** through an ecosystem.
2) **Food chains** show **simple lines** of energy transfer. Each of the stages in a food chain is called a **trophic level**.
3) **Food webs** show **lots** of **food chains** in an ecosystem and how they **overlap**.
4) **Decomposers** (e.g. fungi) are also part of food webs. Decomposers **break down dead** or **undigested** material, allowing nutrients to be recycled (see page 132).

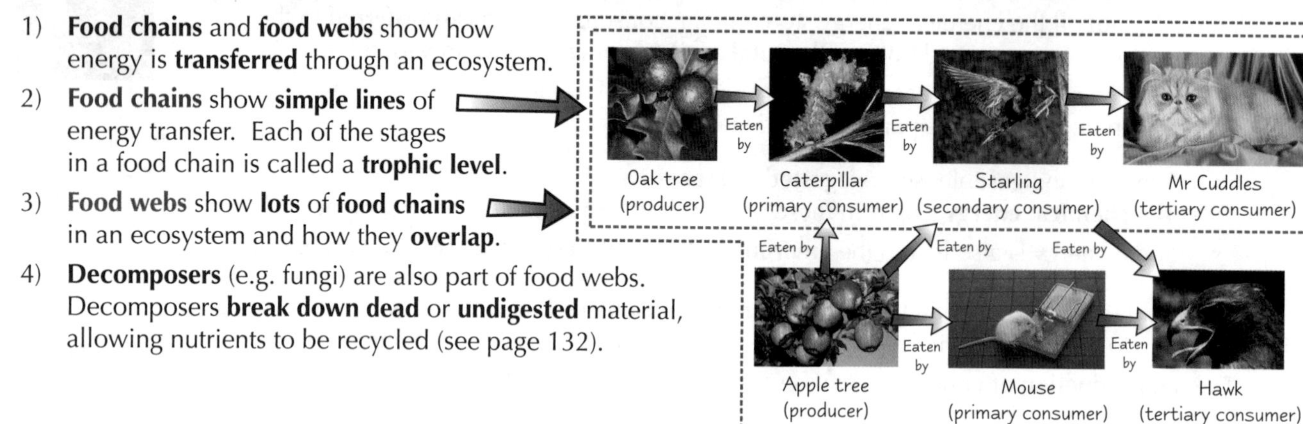

Farming Practices Increase The **Efficiency** of **Energy Transfer**

Most farming practices aim to **increase** the amount of **energy** that is **available** for **human consumption**. There are different ways this can be done. You need to know about **two** of them.

1) The **energy lost** to other **organisms**, e.g. pests, can be **reduced**.
2) The **energy lost** through **respiration** can be **reduced**.

(1) **Simplifying Food Webs Reduces Energy Loss** to Other **Organisms**

Here's an example of a simplified **food web** involving a **crop plant** grown for **human consumption**:

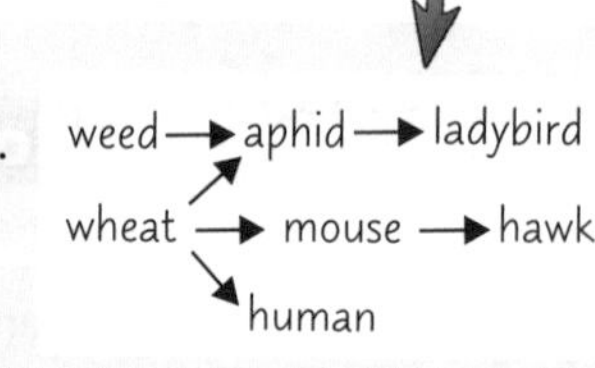

The weed, the mouse and the aphid are **pests** — organisms that **reduce** the amount of **energy** available for crop **growth** and therefore the **net primary production** (NPP). This ultimately **reduces** the amount of **energy** available for **humans**. By **simplifying** the food web, i.e. getting rid of food chains that **don't** involve humans, **energy losses** will be **reduced** and the **NPP** of the crop will **increase**.

Simplifying the food web means **getting rid of pests** — and for that, farmers need **pest control**.

1) Farmers can **reduce pest numbers** using **chemical pesticides**. For example:
   - **Insecticides** kill **insect** pests that **eat** and **damage** crops. Killing insect pests means **less biomass** is **lost** from crops, so they grow to be **larger**, which means **NPP** is **greater**.
   - **Herbicides** kill **weeds** (unwanted plant species). Killing weeds can **remove direct competition** with the crop for energy from the Sun. It can also remove the preferred habitat or food source of the **insect** pests, helping to **further reduce** their numbers and **simplify** the food web.
2) **Biological agents also** reduce the **numbers of pests**, so crops lose **less energy** and **biomass**, **increasing** the efficiency of energy transfer to humans.
   - **Parasites** live in or lay their **eggs** on a **pest insect**. Parasites either **kill** the insect or **reduce** its ability to **function**, e.g. some wasp species lay their eggs inside caterpillars — the eggs hatch and **kill** the caterpillars.
   - **Pathogenic** (disease-causing) **bacteria** and **viruses** are used to kill pests, e.g. the bacterium ***Bacillus thuringiensis*** produces a **toxin** that kills a wide range of **caterpillars**.

Natural predators can also be introduced to the ecosystem to eat the pest species, e.g. ladybirds eat aphids — this is useful but doesn't really simplify the food web.

3) Farmers can use **integrated systems** that combine **both** chemical and biological methods. The **combined effect** of using both can reduce pest numbers **even more** than either method **alone**, meaning **NPP** is **increased** even more.

Farming Practices and Production

2) Reducing **Respiratory Losses** Means **Energy** is **Transferred** More **Efficiently**

1) One way that farmers **increase** the **net production** of their **livestock** is by **controlling** the **conditions** that they live in, so that **more** of their **energy** is used for **growth** and **less** is **lost** through **respiration** (and activities that **increase** the **rate** of respiration). For example:

- Movement increases the rate of respiration, so animals may be kept in pens where their **movement is restricted.**
- The pens are often **indoors** and **kept warm**, so **less energy** is **wasted by** generating body heat.

Increasing production was not an issue that was easy to raise with Herbert.

2) This means that **more biomass** is produced and **more chemical energy** can be stored, **increasing net production** and the **efficiency** of **energy transfer** to humans.

The benefits are that **more food** can be produced in a **shorter** space of time, often at **lower cost.** However, enhancing net production by keeping animals in pens raises **ethical issues.** For example, some people think that the **conditions** intensively reared animals are kept in cause the animals **pain, distress** or restricts their **natural behaviour**, so it **shouldn't be done.**

Warm-Up Questions

Q1 What is a food web?

Q2 What is the role of decomposers in a food web?

Q3 How does simplifying a food web involving a crop increase the NPP of the crop?

Exam Question

Q1 The graph below shows the yearly percentage loss of a crop to three different insect pests before and after a chemical pesticide was used on the crop.

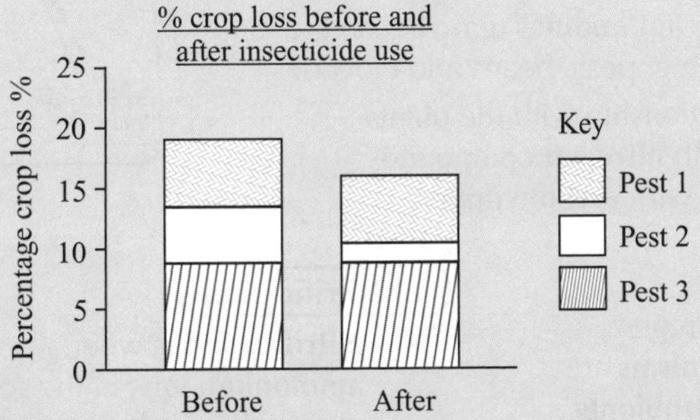

a) How do insect pests reduce the NPP of crops? *[1 mark]*

b) What conclusions can be drawn about the effectiveness of the chemical pesticide from this graph? *[2 marks]*

c) Suggest two ways in which the farmer growing this crop could further reduce the percentage crop loss to insect pests. *[2 marks]*

d) Explain two ways in which livestock farmers can increase the net production of their animals. *[2 marks]*

Farming practices — baa-aa-aa-rmy...

Crikey, so farming's not just about getting up early to feed the chicks then — farmers want to produce as much food as they can, so they try to eliminate energy losses to pests and respiration. Remember, farmers really want to maximise production — the more energy available for crop and livestock growth, the better.

Nutrient Cycles

Organisms don't need to worry about which recycling bin to use. Ecosystems have developed a much better system to make sure necessary elements like nitrogen and phosphorus can be recycled and don't run out.

Fungi and **Bacteria** Have an **Important Role** in **Nutrient Recycling**

1) A **natural ecosystem** is one that hasn't been **changed** by **human activity**. In **natural ecosystems** nutrients are recycled though the food webs, but **human activity** often **disrupts** the cycling of nutrients.
2) **Microorganisms**, such as **bacteria** and **fungi**, are an important part of food webs. Many are **saprobionts** (a type of decomposer) — they feed on the **remains** of **dead plants** and **animals** and on their **waste products** (faeces, urine), breaking them down. This allows important **chemical elements** in the remains to be **recycled**.
3) Saprobionts **secrete enzymes** and **digest** their **food externally**, then **absorb** the **nutrients** they need. This is known as **extracellular digestion**. During this process, **organic molecules** are broken down into **inorganic ions**. Obtaining nutrients from dead organic matter using extracellular digestion is known as **saprobiotic nutrition**.
4) Some fungi form **symbiotic relationships** with the **roots** of **plants**. These relationships are known as **mycorrhizae**.
   - The fungi are made up of **long**, **thin strands** called **hyphae**, which **connect** to the **plant's roots**.
   - The hyphae greatly **increase** the **surface area** of the plant's root system, helping the plant to absorb **ions** from the soil that are usually **scarce** (e.g. **phosphorus**). Hyphae also **increase** the uptake of **water** by the plant.
   - In turn, the fungi obtain **organic compounds**, such as glucose, from the plant.

The **Nitrogen Cycle** shows how **Nitrogen** is **Recycled** in **Ecosystems**

Plants and animals **need nitrogen** to make **proteins** and **nucleic acids** (DNA and RNA). The atmosphere's made up of about 78% nitrogen gas, but plants and animals **can't use it** in that form — they need **bacteria** to **convert** it into **nitrogen-containing compounds** first. The **nitrogen cycle** shows how nitrogen is **converted** into a usable form and then **passed** on between different **living** organisms and the **non-living** environment.

The nitrogen cycle includes **food chains** (nitrogen is passed on when organisms are eaten), and four different processes that involve bacteria — **nitrogen fixation**, **ammonification**, **nitrification** and **denitrification**:

1) Nitrogen fixation

- **Nitrogen fixation** is when nitrogen **gas** in the atmosphere is turned into nitrogen-containing compounds. Biological nitrogen fixation is carried out by **bacteria** such as ***Rhizobium***. They turn nitrogen into **ammonia**, which goes on to form ammonium ions in solution that can then be **used** by plants.
- *Rhizobium* are found inside **root nodules** (growths on the roots) of **leguminous** plants (e.g. peas, beans and clover).
- They form a **mutualistic** relationship with the plants — they provide the plant with **nitrogen compounds** and the plant provides them with **carbohydrates**.

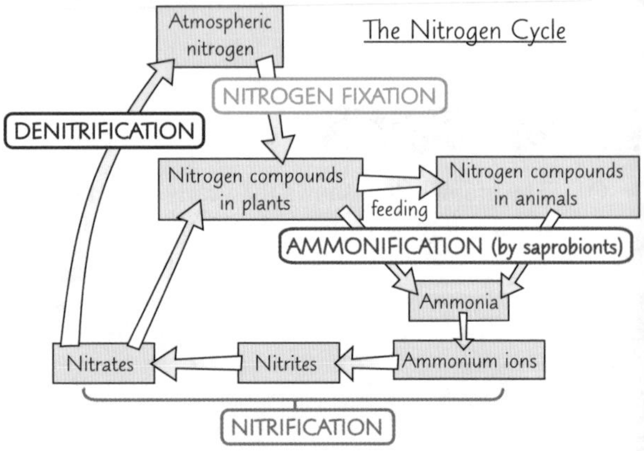

The Nitrogen Cycle

Atmospheric nitrogen
NITROGEN FIXATION
DENITRIFICATION
Nitrogen compounds in plants
feeding
Nitrogen compounds in animals
AMMONIFICATION (by saprobionts)
Ammonia
Ammonium ions
Nitrites
Nitrates
NITRIFICATION

2) Ammonification

- **Ammonification** is when nitrogen compounds from **dead organisms** are turned into **ammonia** by **saprobionts**, which goes on to form **ammonium ions**.
- Animal **waste** (**urine** and **faeces**) also contains nitrogen compounds. These are also turned into ammonia by saprobionts and go on to form ammonium ions.

3) Nitrification

- **Nitrification** is when **ammonium ions** in the soil are **changed** into **nitrogen compounds** that can then be **used** by plants (nitrates).
- First **nitrifying bacteria** called ***Nitrosomonas*** change **ammonium ions** into **nitrites**.
- Then other nitrifying bacteria called ***Nitrobacter*** change **nitrites** into **nitrates**.

Don't worry — you don't need to learn the names of the microorganisms.

4) Denitrification

- **Denitrification** is when nitrates in the soil are **converted** into **nitrogen gas** by **denitrifying bacteria** — they use nitrates in the soil to carry out **respiration** and produce nitrogen gas.
- This happens under **anaerobic conditions** (where there's **no** oxygen), e.g. in **waterlogged** soils.

Other ways that **nitrogen** gets into an **ecosystem** are by **lightning** (which **fixes atmospheric nitrogen**) or by **artificial fertilisers** (they're **produced from atmospheric nitrogen** on an **industrial scale** in the **Haber process**).

Nutrient Cycles

Phosphorus is Passed Through the Food Web in The Phosphorus Cycle

Plants and **animals** need **phosphorus** to make **biological molecules** such as **phospholipids** (which make up cell membranes), **DNA** and **ATP**. Phosphorus is found in **rocks** and **dissolved** in the **oceans** in the form of **phosphate ions** (PO_4^{3-}). Phosphate ions dissolved in water in the soil can be **assimilated** (absorbed and then used to make more complex molecules) by **plants** and other **producers**.

The phosphorus cycle shows how phosphorus is passed through an **ecosystem**.

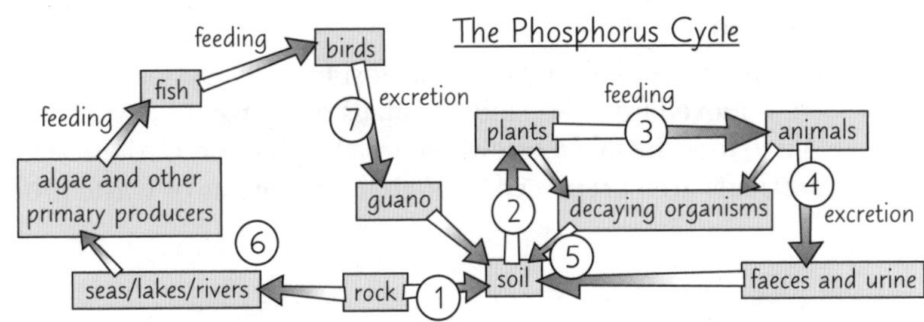

1) Phosphate ions in **rocks** are released into the soil by **weathering**.
2) Phosphate ions are **taken into** the plants through the **roots**. **Mycorrhizae** (see previous page) greatly **increase** the **rate** at which phosphorus can be assimilated.
3) Phosphate ions are **transferred** through the **food chain** as animals eat the plants and are in turn eaten by other animals.
4) Phosphate ions are **lost** from the **animals** in waste **products**.
5) When plants and animals **die**, **saprobionts** are involved in **breaking down** the organic compounds, releasing **phosphate ions** into the soil for **assimilation** by **plants**. These microorganisms also release the phosphate ions from **urine** and **faeces**.
6) **Weathering** of **rocks** also releases phosphate ions into **seas**, **lakes** and **rivers**. This is taken up by **aquatic producers**, such as algae, and passed along the food chain to **birds**.
7) The waste produced by **sea birds** is known as **guano** and contains a **high proportion** of phosphate ions. Guano returns a significant amount of phosphate ions to **soils** (particularly in coastal areas). It is often used as a **natural fertiliser**.

Warm-Up Questions

PRACTICE QUESTIONS

Q1 What are saprobionts?

Q2 How do mycorrhizae benefit plants?

Q3 Why do plants and animals need nitrogen?

Q4 Briefly describe the process of nitrification.

Q5 How do animals obtain phosphate ions?

Q6 How are phosphate ions transferred from the sea to the land?

Exam Question

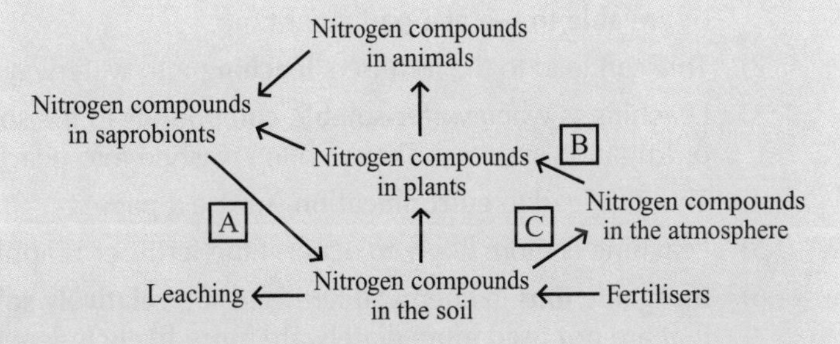

Q1 The diagram on the right shows the nitrogen cycle.

a) Name the processes labelled A, B and C in the diagram. [3 marks]

b) i) Describe the role of saprobionts in process A. [2 marks]

ii) Describe how saprobionts obtain their nutrients. [2 marks]

Nitrogen fixation — cheaper than a shoe fixation...

The nitrogen cycle's not as bad as it seems. Divide up the four processes of nitrogen fixation, ammonification, nitrification and denitrification and learn them separately, then hey presto — you've learnt the whole cycle. Learning the phosphorus cycle ain't that bad either — it's got a few rocks in it as well as all the plants and animals though.

Fertilisers and Eutrophication

Every silver lining has a dark cloud — using fertilisers to replace lost nutrients is all fine and dandy till they don't stay where you put 'em and end up killing all the fish...

Nutrients are **Lost** when **Crops** are **Harvested**

1) Crops **take in** minerals from the soil as they **grow** and use them to build their own tissues.
2) When **crops** are **harvested**, they're **removed** from the field where they're grown rather than being allowed to die and decompose there. This means the **mineral ions** that they contain (e.g. **phosphates** and **nitrates**) are not returned to the **soil** by **decomposers** in the **nitrogen** or **phosphorus cycles**.
3) **Phosphates** and **nitrates** are also lost from the system when **animals** or **animal products** are **removed** from the land. Animals eat **grass** and other plants, **taking in** their **nutrients**. When they are taken elsewhere for slaughter or transferred to a different field, the **nutrients aren't replaced** through their **remains** or **waste products**.

Fertilisers are **Added** to **Soils** to **Replace Lost Nutrients**

Adding fertiliser **replaces** the lost minerals, so **more energy** from the ecosystem can be used for growth, **increasing** the **efficiency** of energy transfer. Fertilisers can be **artificial** or **natural**.

Artificial fertilisers are **inorganic** — they contain **pure chemicals** (e.g. ammonium nitrate) as powders or pellets.

Natural fertilisers are **organic** matter — they include **manure**, **composted vegetables**, **crop residues** (the parts left over after the harvest) and **sewage sludge**.

True, Daisy had a dippy fringe, but she sure could produce a lot of fertiliser.

Using **Fertilisers** Raises **Environmental Issues**

1) Sometimes **more** fertiliser is **applied** than the plants **need** or are **able to use** at a particular time.
2) This can lead to the fertilisers **leaching** into waterways.
3) Leaching is when **water-soluble compounds** in the **soil** are washed away, e.g. by rain or irrigation systems. They're often washed into **nearby ponds** and **rivers**.
4) This can lead to **eutrophication** (see next page).
5) Leaching is more likely to occur if the fertiliser is applied **just before heavy rainfall**.
6) **Inorganic ions** in chemical fertilisers are **relatively soluble**. This means that excess minerals that are **not used immediately** are **more likely** to leach into waterways. In **natural fertilisers**, the **nitrogen** and **phosphorus** are still contained in **organic molecules** that need to be **decomposed** by microorganisms before they can be absorbed by plants. This means that their **release** into the soil for uptake by plants is more **controlled**, and leaching is **less likely**.
7) The **leaching** of **phosphates** is **less likely** than the leaching of **nitrates** because phosphates are **less soluble** in water.
8) Using fertilisers also changes the **balance** of **nutrients** in the soil — **too much** of a particular nutrient can cause crops and other plants to **die**.

Fertilisers and Eutrophication

Eutrophication is Caused by Excess Nutrients

This is the process of **eutrophication**:

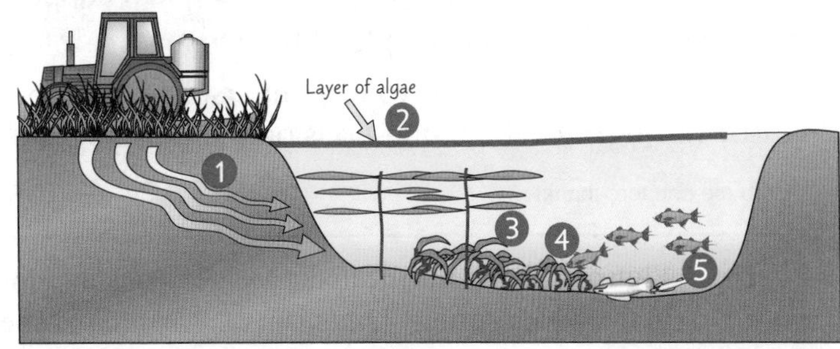

1. **Mineral ions leached** from **fertilised fields** stimulate the **rapid growth** of **algae** in ponds and rivers.

2. Large amounts of algae **block light** from reaching the plants below.

3. Eventually the **plants die** because they're **unable** to **photosynthesise** enough.

4. **Bacteria** feed on the dead plant matter. The **increased** numbers of **bacteria reduce** the **oxygen** concentration in the water by carrying out **aerobic respiration.**

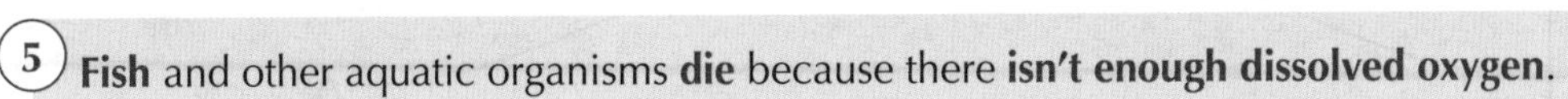

5. **Fish** and other aquatic organisms **die** because there **isn't enough dissolved oxygen.**

Hey, who turned out the lights?

Warm-Up Questions

Q1 Why are nutrients lost when plants are harvested?

Q2 What are artificial fertilisers?

Q3 What is leaching?

Q4 Briefly describe the process of eutrophication.

PRACTICE QUESTIONS

Exam Question

Q1 A study was conducted to investigate the effect, on a nearby river, of adding fertiliser to farmland. The oxygen and algal content of a river that runs past a field where nitrate fertiliser had been applied was measured at the field and up to a distance of 180 m away. A similar control river next to an unfertilised field was also studied. The results are shown in the graphs below.

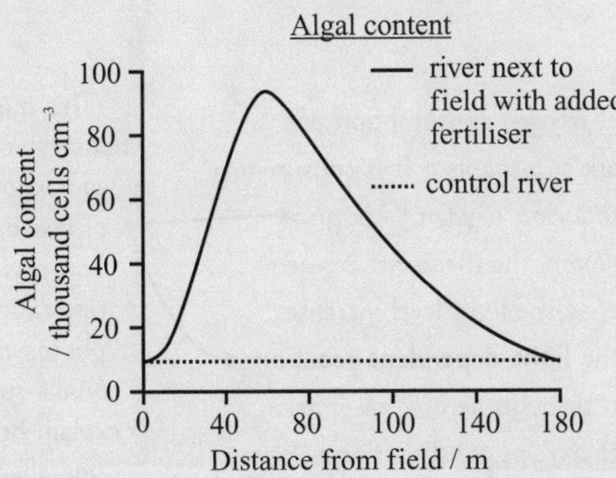

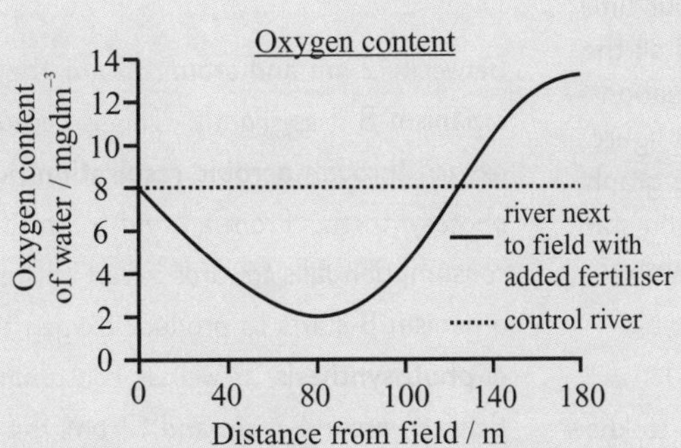

a) Explain the purpose of the control river in the study. [1 mark]

b) Calculate the percentage increase in algal content from 0 to 60 m away from the fertilised field. [1 mark]

c) Describe the relationship between the algal content of the water and the oxygen content of the water in the river next to the fertilised field. [1 mark]

d) Suggest an explanation for the relationship you described in part c). [4 marks]

Help — everything I just learnt is leaching out of my brain...

Fertilisers are important for giving plants all the nutrients they need but, as with a lot of things, a little goes a long way. Too much fertiliser and you can find yourself struggling to breathe. Literally. If you're a fish that is. Nitrogen and phosphorus are good for algae as well as plants and if they get hold of it, by 'eck do their numbers explode...

Extra Exam Practice

Phew, Topic 5 was a tricky one — there're lots of complex processes to remember, that's for sure. I find the best way to check I've understood everything is to just dive right in to some practice exam questions. Ok, maybe re-energise with a quick cuppa and a biscuit* first — I'll wait...

- Have a look at this example of how to answer a tricky exam question.
- Then check how much you've understood from Topic 5 by having a go at the questions on the next page.

*Grab me one too, thanks.

When you're feeling ready, there are synoptic questions on p.233-241 that will test your knowledge of the whole course.

1 **Figure 1** shows the mean net oxygen consumption for two organisms (**A** and **B**), over 24 hours, in the UK in spring. Organism **A** is an animal species and organism **B** is a plant species.

Figure 1

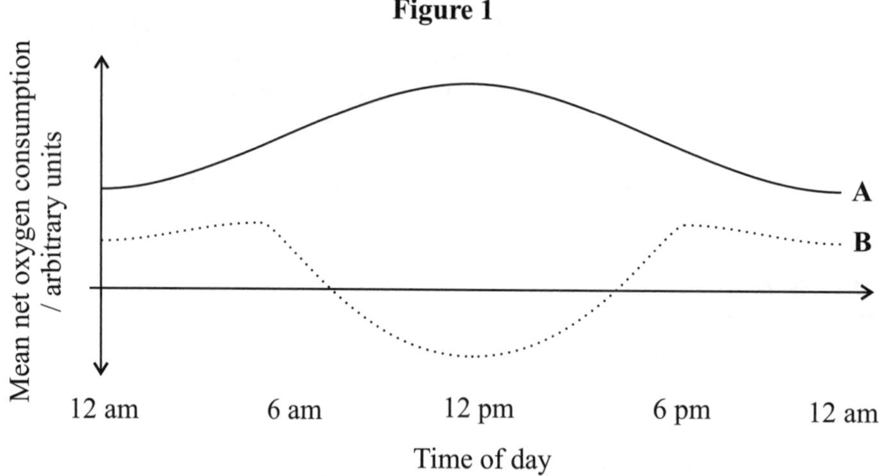

1.1 Explain what the graph shows for organism **B**, between the hours of 12 am and 12 pm.

(6 marks)

1.2 Explain how the rate of acetyl coenzyme A production in organism **A** will change between 12 pm and 6 pm.

(2 marks)

Take your time to read all the information and to figure out the graph before you start your answer.

1.1

Between 12 am and around 5 am, the mean net oxygen consumption of organism B rises slightly. This is because it's dark so organism B is consuming oxygen through **aerobic respiration** but not producing oxygen through photosynthesis. From around 5 am to around 7 am, the mean net oxygen consumption falls towards zero. This is because, as the light level increases, organism B starts to produce oxygen through the **light-dependent reaction** of **photosynthesis**, as well as consuming oxygen through aerobic respiration. Between around 7 am and 12 pm, the mean net oxygen consumption is below zero, meaning that organism B is producing more oxygen through photosynthesis than it is using up through aerobic respiration.

This question requires you to use your knowledge of two different processes — respiration and photosynthesis. Make sure you explain how each one affects the graph in your answer.

1.2

Acetyl coenzyme A is produced during the **link reaction** of **aerobic respiration**. As the mean net oxygen consumption in organism A falls between 12 pm and 6 pm, this suggests that the rate of aerobic respiration is falling and therefore that the rate of the link reaction and acetyl coenzyme A production is falling too.

You need to know the names of lots of different substances as part of your course. If you get asked about a specific substance in a question, take time to think about its role and how it relates to what you're being asked.

Extra Exam Practice

2 Aquaponics is a farming technique in which fish and plants are grown together in a controlled environment. This technique reduces the need for artificial fertilisers, as waste products from the fish can supply the plants with essential nutrients. **Figure 2** shows the basic set-up of an aquaponics system.

Figure 2

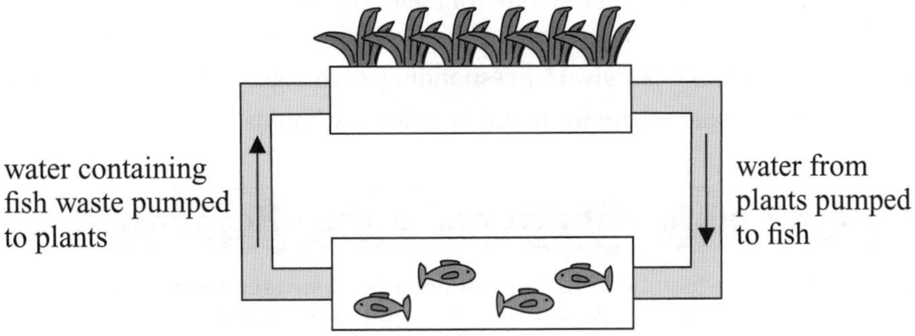

water containing fish waste pumped to plants

water from plants pumped to fish

In an aquaponics system, constant artificial lighting can be used to aid photosynthesis.

2.1 Describe how light energy is used to produce ATP during photosynthesis. Details of electron transfer chains are not required.

(2 marks)

2.2 Explain how constant artificial lighting could increase the production of organic substances in the plants.

(4 marks)

Microorganisms are an essential component of an aquaponics system.

2.3 Explain why saprobionts and nitrifying bacteria are needed to ensure the plants are able to synthesise molecules containing nitrogen and phosphorus.

(5 marks)

3 A simplified food chain is shown in **Figure 3**.

Figure 3

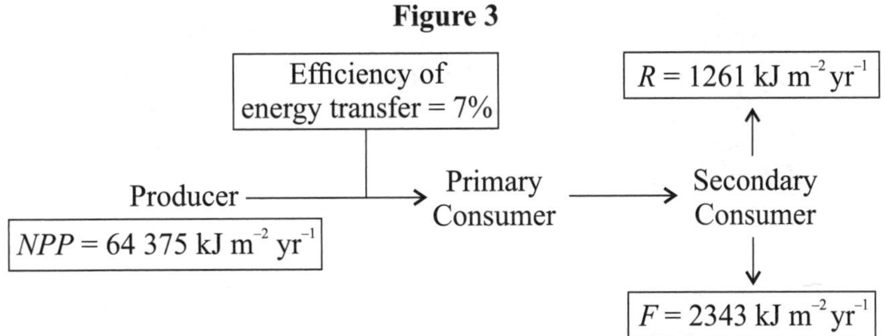

3.1 During summer, the *NPP* of the producer increases. This is partly due to an increased activity rate of rubisco. Explain how an increase in the activity rate of rubisco could lead to a higher *NPP*.

(3 marks)

3.2 The secondary consumer does not ingest 10% of the material available in the primary consumer. Using this information and **Figure 3**, calculate the net productivity of the secondary consumer.

(3 marks)

3.3 The breathing rate of the secondary consumer increases when it is hunting for food. Explain how this helps to maximise the rate of ATP synthesis in the secondary consumer.

(3 marks)

Nervous Communication

Your body has an amazing network of nerve cells which constantly send electrical signals — a bit like a big circuit board.

Responding to their **Environment** Helps **Organisms Survive**

1) **Animals increase** their **chances** of **survival** by **responding** to **changes** in their **external environment**, e.g. by **avoiding harmful environments** such as places that are too hot or too cold.
2) They also **respond** to **changes** in their **internal environment** to make sure that the **conditions** are always **optimal** for their **metabolism** (all the chemical reactions that go on inside them).
3) **Plants** also **increase** their **chances** of **survival** by **responding** to **changes** in their **environment** (see p. 140).
4) Any **change** in the internal or external **environment** is called a **stimulus**.

Receptors Detect **Stimuli** and **Effectors** Produce a **Response**

1) **Receptors detect stimuli** — they can be **cells** or **proteins** on **cell surface membranes**. There are **loads** of **different types** of receptors that detect **different stimuli**.
2) **Effectors** are cells that bring about a **response** to a **stimulus**, to produce an **effect**. Effectors include **muscle cells** and cells found in **glands**, e.g. the **pancreas**.
3) Receptors **communicate** with effectors via the **nervous system** or the **hormonal system**, or sometimes using **both**.

Receptors are specific to one type of stimulus — see p. 142.

The **Nervous System** Sends Information as **Electrical Impulses**

1) The **nervous system** is made up of a **complex network** of cells called **neurones**. There are **three main types**:
   - **Sensory neurones** transmit electrical impulses from **receptors** to the **central nervous system (CNS)** — the **brain** and **spinal cord**.
   - **Motor neurones** transmit electrical impulses from the **CNS** to **effectors**.
   - **Relay neurones** transmit electrical impulses **between** sensory neurones and motor neurones.

Electrical impulses are also called nerve impulses or action potentials (see p. 146).

2) A stimulus is detected by **receptor cells** and an **electrical impulse** is sent along a **sensory neurone**.
3) When an **electrical impulse** reaches the end of a neurone, chemicals called **neurotransmitters** take the information across to the **next neurone**, which then sends an **electrical impulse** (see p. 149).
4) The **CNS** (the **coordinator**) **processes** the information and sends impulses along **motor neurones** to an **effector**.

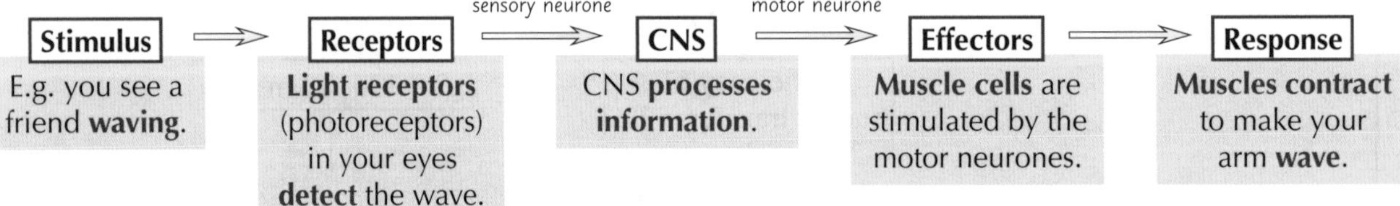

5) The **nervous system** is split into two different systems:

You don't need to learn the structure of the nervous system, but understanding it'll help with the rest of the section.

- The **central nervous system (CNS)** — made up of the **brain** and the **spinal cord**.
- The **peripheral nervous system** — made up of the neurones that connect the CNS to the **rest** of the **body**. It also has two different systems:
  - The **somatic nervous system** controls **conscious** activities, e.g. running and playing video games.
  - The **autonomic nervous system** controls **unconscious** activities, e.g. digestion. It's got two divisions that have **opposite effects** on the body:
    - The **sympathetic** nervous system gets the body **ready for action**. It's the '**flight or fight**' system.
    - The **parasympathetic** nervous system **calms** the body down. It's the '**rest and digest**' system.

Harold thought it was about time his sympathetic nervous system took over.

Nervous Communication

Reflexes are Rapid, Automatic Responses to Stimuli

1) A **reflex** is where the body **responds** to a stimulus **without** making a **conscious decision** to respond.
2) Because you don't have to **spend time deciding** how to respond, information travels **really fast** from **receptors** to **effectors**.
3) So simple reflexes help organisms to **protect** the body because they're **rapid**.
4) The **pathway** of neurones linking receptors to effectors in a reflex is called a **reflex arc**. You need to **learn** a **simple reflex arc** involving three neurones — a **sensory**, a **relay** and a **motor** neurone.

E.g. the hand-withdrawal response to heat

- **Thermoreceptors** in the skin **detect** the **heat** stimulus.
- The **sensory neurone** carries impulses to the **relay neurone**.
- The **relay neurone** connects to the **motor neurone**.
- The **motor neurone** sends **impulses** to the **effector** (your biceps muscle).
- Your **muscle contracts** to withdraw your hand and **stop** it being **damaged**.

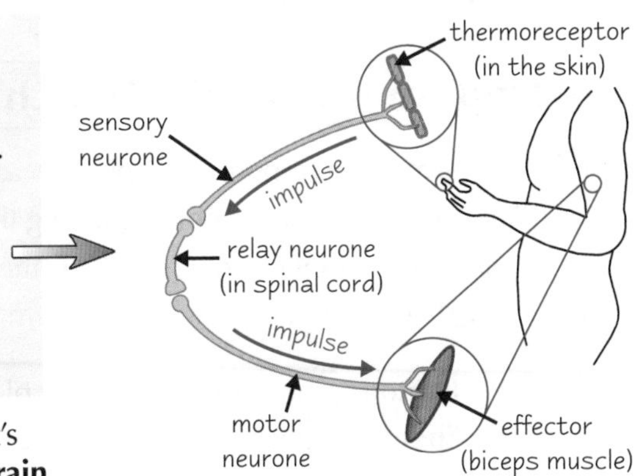

5) If there's a **relay neurone** involved in the simple reflex arc then it's possible to **override** the reflex, e.g. in the example above your **brain** could tell your hand to **withstand** the **heat**.

Nervous System Communication is Localised, Short-lived and Rapid

1) When an **electrical impulse** reaches the end of a neurone, **neurotransmitters** (see page 149) are **secreted directly** onto **target cells** (e.g. muscle cells) — so the nervous response is **localised**.
2) **Neurotransmitters** are **quickly removed** once they've done their job, so the response is **short-lived**.
3) Electrical impulses are **really fast**, so the response is **rapid** — this allows animals to **react quickly** to stimuli.

Warm-Up Questions

Q1 What is a stimulus?

Q2 Name the three main types of neurone.

Q3 What is a reflex?

Q4 How can reflexes help protect the body?

PRACTICE QUESTIONS

Exam Questions

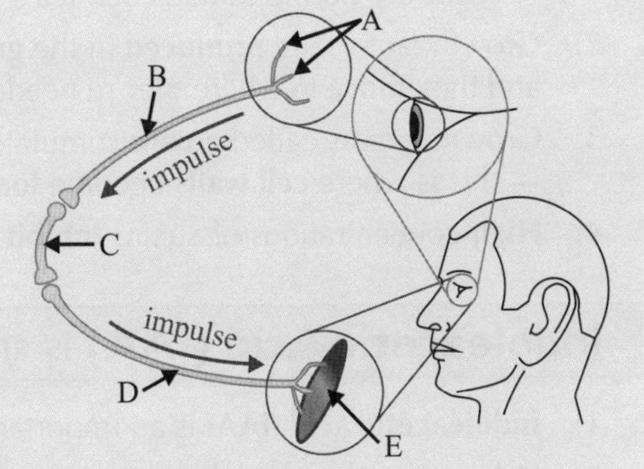

Q1 An animal responds to a stimulus in its environment.
State the role of receptors and effectors in this response. [2 marks]

Q2 The human blink reflex is an involuntary response, which results in the automatic closing of the eyelids (a blink) when an object touches the surface of the eye. A reflex arc for the blink reflex is shown in the diagram above.

a) Using the diagram, describe the reflex arc involved in this response. [4 marks]

b) The knee-jerk is another reflex response. You can test for it by tapping someone just below their patella (knee cap). Suggest why the absence of this response could indicate some damage to a person's CNS. [1 mark]

Q3 Polio is a virus that can cause damage to the CNS. In severe cases, the virus can damage motor neurones. Suggest and explain how this might lead to paralysis. [3 marks]

Responding to questions in an exam helps you to pass...

Actually, this stuff is really quite fascinating once you realise just how much your body can do without you even knowing. Just sit back and let your nerves do the work... Ah, apart from the whole revision thing — your body can't do that without you knowing, unfortunately. Get your head around these pages before you tackle the rest of the section.

Responses in Plants and Animals

Plants and simple animals respond to uncomplicated things like gravity and light. This helps them to survive in their environment...

Plants Need to **Respond** to **Stimuli Too**

Flowering plants, like animals, **increase** their chances of **survival** by **responding** to changes in their **environment**, e.g.

- They sense the direction of **light** and **grow** towards it to **maximise** light absorption for **photosynthesis**.
- They can sense **gravity**, so their roots and shoots **grow** in the **right direction**.
- **Climbing** plants have a sense of **touch**, so they can find things to climb up and **reach** the **sunlight**.

A **Tropism** is a **Plant's Growth Response** to an **External Stimulus**

1) A **tropism** is the **response** of a plant to a **directional stimulus** (a stimulus coming from a particular direction).
2) Plants respond to stimuli by **regulating** their **growth**.
3) A **positive tropism** is growth **towards** the stimulus.
4) A **negative tropism** is growth **away** from the stimulus.

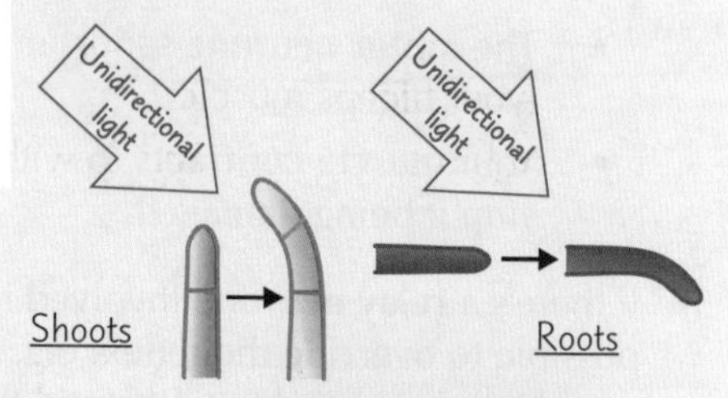

- **Phototropism** is the growth of a plant in response to **light**.
- **Shoots** are **positively phototropic** and grow **towards** light.
- **Roots** are **negatively phototropic** and grow **away** from light.

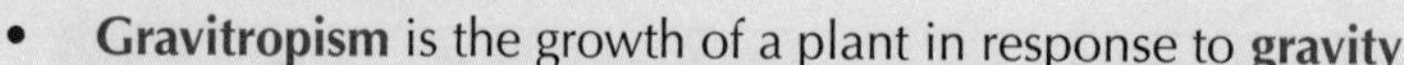

- **Gravitropism** is the growth of a plant in response to **gravity**.
- **Shoots** are **negatively gravitropic** and grow **upwards**.
- **Roots** are **positively gravitropic** and grow **downwards**.

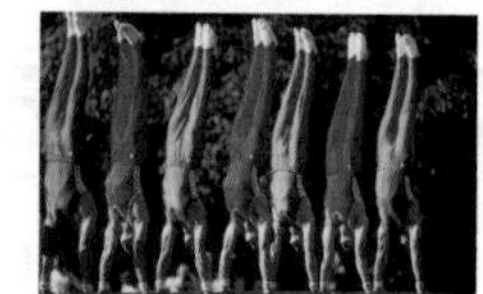

The men's gymnastics team were negatively gravitropic.

Responses are **Brought About** by **Growth Factors**

1) Plants **respond** to directional stimuli using specific **growth factors** — these are hormone-like chemicals that **speed up** or **slow down** plant **growth**.
2) Growth factors are **produced** in the **growing regions** of the plant (e.g. shoot tips, leaves) and they **move** to where they're needed in the **other parts** of the plant.
3) Growth factors called **auxins** stimulate the **growth** of shoots by **cell elongation** — this is where **cell walls** become **loose** and **stretchy**, so the cells get **longer**.
4) **High** concentrations of auxins **inhibit growth** in **roots** though.

Indoleacetic Acid (IAA) is an Important **Auxin**

1) **Indoleacetic acid** (**IAA**) is an important **auxin** that's produced in the **tips** of **shoots** in flowering plants.
2) IAA is **moved** around the plant to **control tropisms** — it moves by **diffusion** and **active transport** over short distances, and via the **phloem** over long distances.
3) This results in **different parts** of the plant having **different concentrations** of IAA. The **uneven distribution** of IAA means there's **uneven growth** of the plant, e.g:

Phototropism — IAA moves to the more **shaded** parts of the **shoots** and **roots**, so there's uneven growth.

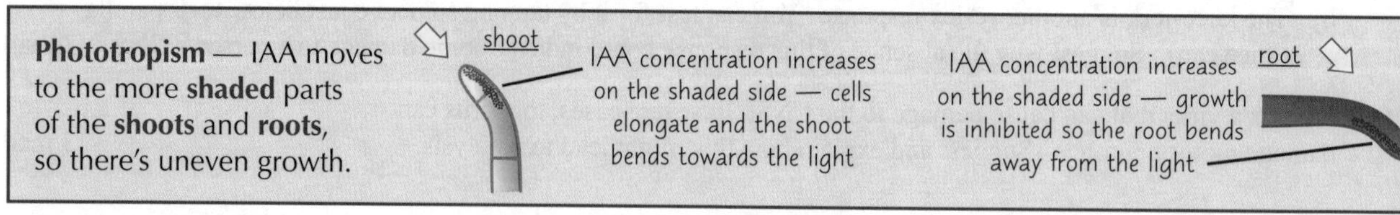

Gravitropism — IAA moves to the **underside** of **shoots** and **roots**, so there's uneven growth.

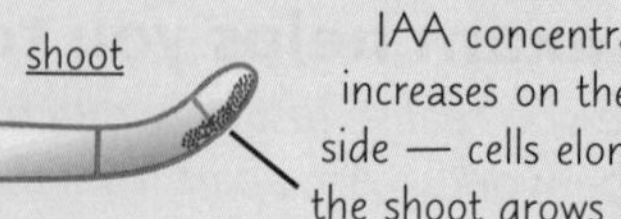

IAA concentration increases on the lower side — growth is inhibited so the root grows downwards

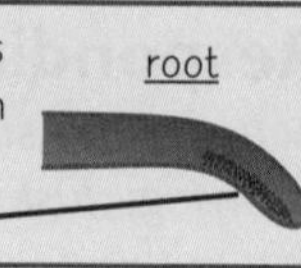

Responses in Plants and Animals

Simple Responses Keep Simple Organisms in a Favourable Environment

Simple mobile organisms, e.g. woodlice and earthworms, have **simple responses** to keep them in a **favourable environment**. Their **response** can either be **tactic** or **kinetic**:

- **Tactic responses (taxes)** — the organisms move towards or away from a **directional stimulus**, e.g. **light**.

  For example, **woodlice** show a **tactic** response to light (**phototaxis**) — they move **away from** a **light source**. This helps them **survive** as it keeps them **concealed** under stones during the day (where they're **safe** from predators) and keeps them in **damp conditions** (which reduces water loss).

- **Kinetic responses (kineses)** — the organisms' movement is affected by a **non-directional** stimulus, e.g. **humidity**.

  For example, **woodlice** show a **kinetic** response to **humidity**. In **high humidity** they move **slowly** and **turn less** often, so that they **stay where they are**. As the air gets **drier**, they move **faster** and **turn more** often, so that they move into a **new area**. This response **increases** the **chance** that a woodlouse will move to an area with **higher humidity**. This **improves** the **survival** chances of the organism — it **reduces** their **water loss** and it helps to keep them **concealed**.

You Can Use Choice Chambers to Investigate Animal Responses

PRACTICAL SKILLS

A **choice chamber** is a container with different compartments, in which you can create different **environmental conditions**. It can be used to investigate how animals, such as woodlice, **respond** to conditions like **light intensity** or **humidity** in the **laboratory**. Here's how you can use a choice chamber:

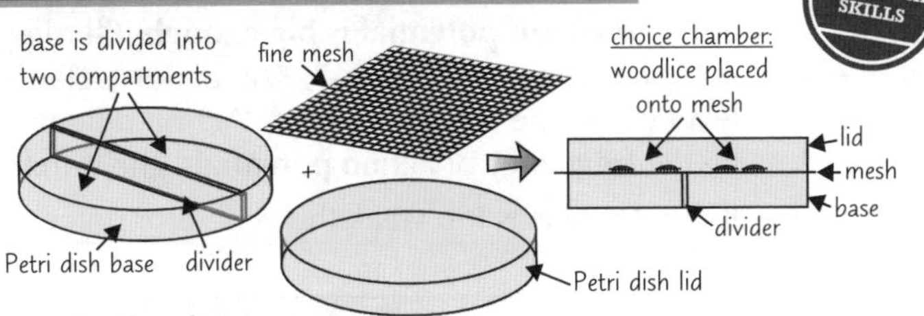

1) Construct a choice chamber using the **equipment** shown in the diagram.
2) To investigate the effect of **light intensity** on woodlouse movement, cover **one half** of the lid (including the sides) with **black paper**. This will make **one side** of the chamber **dark**. Put **damp filter paper** in **both** sides of the base.
3) Place **10 woodlice** on the mesh in the centre of the chamber and cover the chamber with the lid.
4) After **10 minutes**, take off the lid and record the number of woodlice on each side of the chamber.
   Try to minimise the amount of time the lid is off, so that the environmental conditions created aren't disturbed.
5) **Repeat** the experiment after gently moving the woodlice back to the centre. You should find that most woodlice **end up** on the **dark side** of the choice chamber (a **tactic response** to light).
   You can use a small, soft paintbrush to help with moving the woodlice if necessary. For ethical reasons, you should handle the woodlice carefully and return them to their natural habitat as soon as possible. Make sure you wash your hands after handling the woodlice.
6) To investigate **humidity**, place some **damp filter paper** in one side of the base and a **desiccating (drying) agent** in the other side. **Don't cover** the **lid** with paper. Put the **lid on** and leave the chamber for 10 minutes to stabilise before carrying out steps 3)-5) above.
7) You can do a similar experiment using a **maze** instead of a choice chamber.

PRACTICE QUESTIONS

Warm-Up Questions

Q1 What is positive gravitropism?

Q2 Describe the difference between taxes and kineses.

Exam Question

Q1 The table shows the results some students obtained when they investigated the effect of providing plants with auxins.

Week	Height of plant not given auxins / cm	Height of plant provided with auxins / cm
1	1	2
2	2	5
3	4	8
4	6	9
5	9	13

a) Describe and explain what the data shows. [2 marks]

b) Suggest why this data might be useful to a commercial tomato producer. [1 mark]

c) Explain the role of auxins in the control of phototropism in the shoots. [3 marks]

IAA Productions — do you have the growth factor — with Simon Trowel...

The tactic response to revision — when you see your revision notes, you always move away from them. Or if you were a plant, I guess we could say you were negatively revisi-tropic. You've still got to learn this lot for your exams though.

Receptors

Receptors are the front line in animal responses — they detect what's going on and pass on information about it.

Receptors are Specific to One Kind of Stimulus

1) Receptors are **specific** — they only **detect one particular stimulus**, e.g. light, pressure or glucose concentration.
2) There are **many different types** of receptor that each detect a **different type of stimulus**.
3) Some receptors are **cells**, e.g. photoreceptors are receptor cells that connect to the nervous system. Some receptors are **proteins** on **cell surface membranes**, e.g. glucose receptors are proteins found in the cell membranes of some pancreatic cells.
4) Here's a bit more about how receptor cells that communicate information via the **nervous system** work:
   - When a nervous system receptor is in its **resting state** (not being stimulated), there's a **difference in charge** between the **inside** and the **outside** of the cell — this is generated by ion pumps and ion channels (see p. 14 This means that there's a **voltage** across the membrane. Voltage is also known as **potential difference**.
   - The **potential difference** when a cell is at **rest** is called its **resting potential**. When a stimulus is detected, the cell membrane is **excited** and becomes **more permeable**, allowing **more ions** to move **in** and **out** of the cell — **altering** the **potential difference**. The **change** in **potential difference** due to a stimulus is called the **generator potential**.
   - A **bigger stimulus** excites the membrane more, causing a **bigger movement** of ions and a **bigger change** in potential difference — so a **bigger generator potential** is produced.
   - If the **generator potential** is **big enough** it'll trigger an **action potential** — an electrical impulse along a neurone (see pages 146-147). An action potential is only triggered if the generator potential reaches a certai level called the **threshold** level. Action potentials are all one size, so the **strength** of the **stimulus** is measure by the **frequency** of **action potentials** (the number of action potentials triggered during a certain time period)
   - If the stimulus is **too weak** the generator potential **won't reach** the **threshold**, so there's **no action potential**.

Pacinian Corpuscles are Pressure Receptors in Your Skin

1) **Pacinian corpuscles** are **mechanoreceptors** — they detect **mechanical stimuli**, e.g. **pressure** and **vibrations**. They're found in your **skin**.
2) Pacinian corpuscles contain the end of a **sensory neurone**, imaginatively called a **sensory nerve ending**. The sensory nerve ending is **wrapped** in loads of layers of connective tissue called **lamellae**.
3) When a Pacinian corpuscle is **stimulated**, e.g. by a tap on the arm, the lamellae are **deformed** and **press** on the **sensory nerve ending**.
4) This causes the sensory neurone's cell membrane to stretch, deforming the **stretch-mediated sodium ion channels**. The channels **open** and **sodium ions diffuse into** the cell, creating a **generator potential**.
5) If the **generator potential** reaches the **threshold**, it triggers an **action potential**.

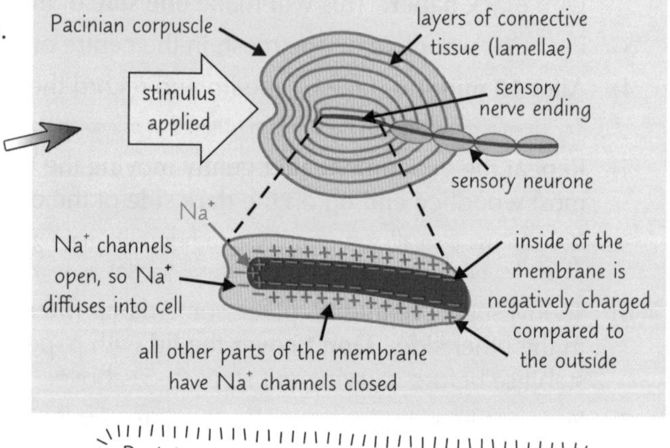

Pacinian corpuscles only respond to mechanical stimuli — this is a good example of how receptors only respond to specific stimuli.

Photoreceptors are Light Receptors in Your Eye

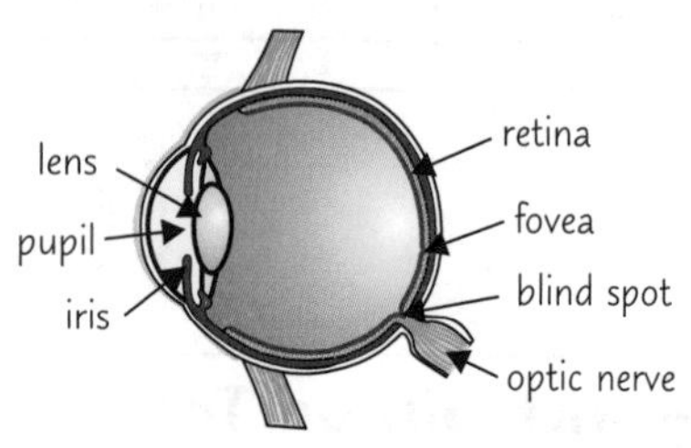

1) **Light** enters the eye through the **pupil**. The **amount** of light that enters is **controlled** by the muscles of the **iris**.
2) Light rays are **focused** by the **lens** onto the **retina**, which lines the inside of the eye. The retina contains **photoreceptor cells** — these **detect light**.
3) The **fovea** is an area of the retina where there are **lots** of **photoreceptors**.
4) **Nerve impulses** from the photoreceptor cells are carried from the **retina** to the **brain** by the **optic nerve**, which is a bundle of **neurones**. Where the optic nerve leaves the eye is called the **blind spot** — there **aren't** any **photoreceptor cells**, so it's **not sensitive** to **light**.

Receptors

Photoreceptors Convert Light into an Electrical Impulse

1) **Light** enters the eye, hits the **photoreceptors** and is **absorbed** by **light-sensitive optical pigments.**
2) Light bleaches the pigments, causing a **chemical change** and altering the **membrane permeability** to **sodium** ions.
3) A **generator potential** is created and if it reaches the threshold, a nerve impulse is sent along a **bipolar neurone.**
4) Bipolar neurones connect **photoreceptors** to the **optic nerve**, which takes impulses to the **brain.**

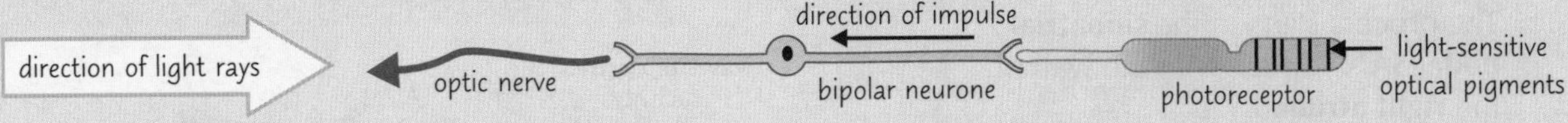

Light passes straight through the optic nerve and bipolar neurone to get to the photoreceptor.

5) The human eye has **two types** of photoreceptor — **rods** and **cones.**
6) Rods are mainly found in the **peripheral** parts of the **retina**, and cones are found **packed together** in the **fovea.**
7) Rods and cones contain **different optical pigments** making them **sensitive** to **different wavelengths** of light.
8) Rods only give information in **black and white** (monochromatic vision), but cones give information in **colour** (trichromatic vision). There are three types of cones, each containing a different **optical pigment** — **red-sensitive, green-sensitive** and **blue-sensitive.** When they're stimulated in **different proportions** you see different colours.

Rods are More Sensitive, but Cones let you See More Detail

Sensitivity

- Rods are **very sensitive to light** (they work well in **dim light**). This is because **many rods** join **one neurone**, so many weak **generator potentials combine** to **reach** the **threshold** and trigger an action potential.
- Cones are **less sensitive** than rods (they work best in **bright light**). This is because **one cone** joins **one neurone**, so it takes more light to reach the threshold and trigger an action potential.

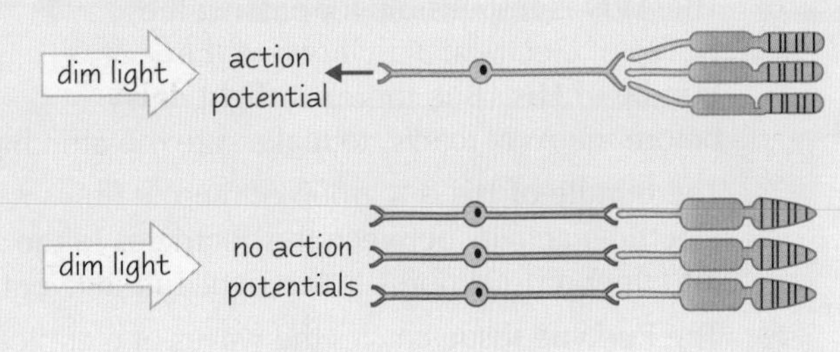

Visual acuity (the ability to tell apart points that are close together)

- Rods give **low visual acuity** because **many rods** join the **same neurone**, which means light from two points close together **can't** be told apart.
- Cones give **high visual acuity** because cones are **close together** and **one cone** joins **one neurone.** When light from two points hits two cones, two action potentials (one from **each cone**) go to the brain — so you can distinguish two points that are close together as **two separate points.**

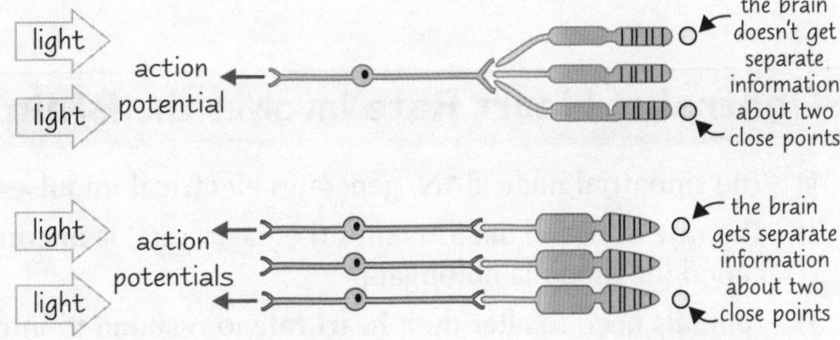

Warm-Up Questions

Q1 Why are receptors described as specific?

Q2 In the human eye, which type of photoreceptor is more sensitive to light?

PRACTICE QUESTIONS

Exam Questions

Q1 Explain how a generator potential is created when a Pacinian corpuscle is stimulated. [3 marks]

Q2 Explain how the human eye can provide high visual acuity. [3 marks]

Pacinian corpuscles love deadlines — they work best under pressure...

Wow, loads of stuff here, so cone-gratulations if you manage to remember it all. Receptors are really important because without them you wouldn't be able to see this book, and without this book revision would be way trickier.

Control of Heart Rate

You don't have to think about making your heart beat — thankfully your heart does it by itself. However, your body has systems to control your heart beat which speed it up or slow it down.

Cardiac Muscle Controls the **Regular Beating** of the Heart

Cardiac (heart) muscle is '**myogenic**' — it can contract and relax without receiving signals from nerves. This pattern of contractions controls the **regular heartbeat**.

1) The process starts in the **sinoatrial node (SAN)**, which is in the wall of the **right atrium**.
2) The SAN is like a pacemaker — it sets the **rhythm** of the heartbeat by sending out regular **waves of electrical activity** to the atrial walls.
3) This causes the right and left **atria** to **contract at the same time**.
4) A band of non-conducting **collagen tissue** prevents the waves of electrical activity from being passed directly from the atria to the ventricles.
5) Instead, these waves of electrical activity are transferred from the SAN to the **atrioventricular node (AVN)**.
6) The AVN is responsible for passing the waves of electrical activity on to the **bundle of His**. But, there's a **slight delay** before the AVN reacts, to make sure the atria have emptied **before** the ventricles contract.

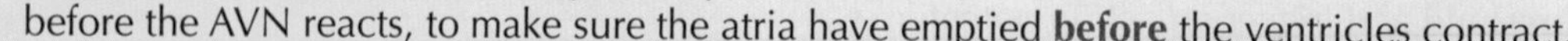

7) The **bundle of His** is a group of muscle fibres responsible for conducting the waves of electrical activity between the ventricles to the apex (bottom) of the heart. The bundle splits into finer muscle fibres in the right and left ventricle walls, called the **Purkyne tissue**.

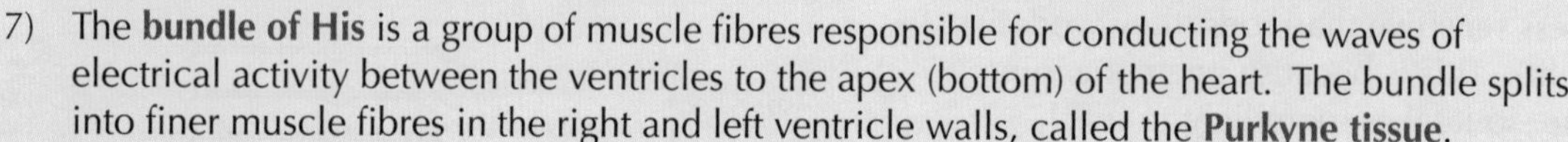

8) The Purkyne tissue carries the waves of electrical activity into the muscular walls of the right and left ventricles, causing them to **contract simultaneously**, from the bottom up.

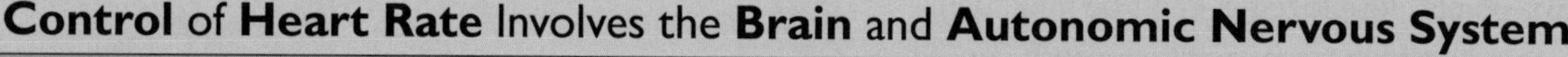

Control of Heart Rate Involves the **Brain** and **Autonomic Nervous System**

1) The **sinoatrial node (SAN)** generates **electrical impulses** that cause the **cardiac muscles** to **contract**.
2) The **rate** at which the SAN fires (i.e. heart rate) is **unconsciously controlled** by a part of the **brain** called the **medulla oblongata**.
3) Animals need to **alter** their **heart rate** to **respond** to **internal stimuli**, e.g. to prevent fainting due to low blood pressure or to make sure the heart rate is high enough to supply the body with enough oxygen.
4) **Stimuli** are **detected** by **pressure receptors** and **chemical receptors**:
   - There are **pressure receptors** called **baroreceptors** in the **aorta** and the **carotid arteries** (major arteries in the neck). They're stimulated by **high** and **low blood pressure**.
   - There are **chemical receptors** called **chemoreceptors** in the **aorta**, the **carotid arteries** and in the **medulla**. They **monitor** the **oxygen** level in the **blood** and also **carbon dioxide** and **pH** (which are indicators of O_2 level).

There's more about the autonomic nervous system on page 138.

5) Electrical impulses from receptors are sent **to the medulla** along **sensory** neurones. The medulla processes the information and sends impulses to the SAN along **sympathetic** or **parasympathetic** neurones (which are part of the **autonomic nervous system**). There's more on this on the next page.

Control of Heart Rate

Stimuli Detected by Receptors Cause Heart Rate to Speed Up or Slow Down

This table shows how the heart responds to different **stimuli**:

Stimulus	Receptor	Neurone and transmitter	Effector	Response
High blood pressure.	**Baroreceptors** detect **high** blood pressure.	Impulses are sent to the medulla, which sends impulses along **parasympathetic** neurones. These secrete **acetylcholine** (a neurotransmitter), which binds to receptors on the SAN.	Cardiac muscles	Heart rate **slows** down to **reduce blood pressure** back to normal.
Low blood pressure.	**Baroreceptors** detect **low** blood pressure.	Impulses are sent to the medulla, which sends impulses along **sympathetic** neurones. These secrete **noradrenaline** (a neurotransmitter), which binds to receptors on the SAN.	Cardiac muscles	Heart rate **speeds up** to **increase blood pressure** back to normal.
High blood O_2, **low** CO_2 or **high** pH levels.	**Chemoreceptors** detect chemical changes in the blood.	Impulses are sent to the medulla, which sends impulses along **parasympathetic** neurones. These secrete **acetylcholine**, which binds to receptors on the SAN.	Cardiac muscles	Heart rate **decreases** to return O_2, CO_2 and pH levels back to normal.
Low blood O_2, **high** CO_2 or **low** pH levels.	**Chemoreceptors** detect chemical changes in the blood.	Impulses are sent to the medulla, which sends impulses along **sympathetic** neurones. These secrete **noradrenaline**, which binds to receptors on the SAN.	Cardiac muscles	Heart rate **increases** to return O_2, CO_2 and pH levels back to normal.

For more about neurotransmitters see page 149-151.

When Ed did that special thing to her beak, Polly's sympathetic neurones went into overdrive.

Warm-Up Questions

Q1 Why is heart muscle described as 'myogenic'?

Q2 What is the function of the bundle of His?

Q3 Why do animals need to alter their heart rate?

Q4 Name the effectors that are involved in increasing or decreasing heart rate.

Exam Questions

PRACTICE QUESTIONS

Q1 The control of heart rate is coordinated by specific parts of the heart. Describe the function of:

a) the sinoatrial node. *[1 mark]*

b) the Purkyne tissue. *[1 mark]*

Q2 Exercise causes an increase in the levels of carbon dioxide in the blood.

a) Explain how increased blood CO_2 leads to an increased heart rate. *[4 marks]*

b) State two other chemical stimuli that cause the heart rate to increase during exercise. *[2 marks]*

Q3 Atrial fibrillation (AF) is a condition that can result in a fast and irregular heartbeat because an abnormally high number of impulses are passed from the atria to the ventricles. Surgical treatment of AF can involve AVN ablation, which involves injuring the AVN so it no longer functions.

a) Suggest how this treatment helps to manage the condition. *[2 marks]*

b) After undergoing AVN ablation, patients also need to have a pacemaker implanted (an electronic device that sends out electrical impulses to control heart rate). Suggest why this is necessary. *[2 marks]*

My heart rate seems to be controlled by the boy next door...

It's also rising rapidly at the sight of so much to learn. You've got to properly learn it though — it's no good just having a rough idea. The SAN, baroreceptors, chemoreceptors — make sure you know what they are and what they do. Try drawing each row of the table above as a flow diagram, showing the route from stimulus to response.

Neurones

Ah, on to the good stuff. Notepad at the ready, motor neurones fired up, OK — lights, camera, action potentials...

Neurone **Cell Membranes** are **Polarised** at **Rest**

1) In a neurone's **resting state** (when it's not being stimulated), the **outside** of the membrane is **positively charged** compared to the **inside**. This is because there are **more positive ions outside** the cell than inside.
2) So the membrane is **polarised** — there's a **difference in charge** (called a **potential difference** or **voltage**) across it.
3) The voltage across the membrane when it's at rest is called the **resting potential** — it's about **–70 mV** (millivolts).
4) The resting potential is created and maintained by **sodium-potassium pumps** and **potassium ion channels** in a neurone's membrane:
   - The **sodium-potassium pumps** move **sodium ions out** of the neurone, but the membrane **isn't permeable** to **sodium ions**, so they **can't diffuse back in**. This creates a **sodium ion electrochemical gradient** (a **concentration gradient** of **ions**) because there are **more** positive sodium ions **outside** the cell than inside.
   - The sodium-potassium pumps also move **potassium ions in** to the neurone, but the membrane **is permeable** to **potassium ions** so they **diffuse back out** through **potassium ion channels**.
   - This makes the **outside** of the cell **positively charged** compared to the inside.

<u>Sodium-potassium pump</u> — These pumps use **active transport** to move **three sodium ions** (Na^+) **out** of the neurone for every **two potassium ions** (K^+) moved in. **ATP** is needed to do this.

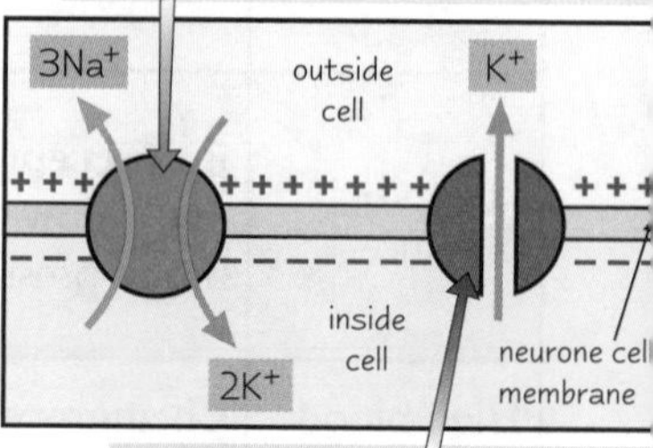

<u>Potassium ion channel</u> — These channels allow **facilitated diffusion** of **potassium ions** (K^+) **out** of the neurone, down their **concentration gradient**.

The sodium-potassium pump, potassium ion channel and sodium ion channel (see below) are all types of transport protein.

Neurone **Cell Membranes** Become **Depolarised** when They're **Stimulated**

A **stimulus** triggers other ion channels, called **sodium ion channels**, to **open**. If the stimulus is big enough, it'll trigger a **rapid change** in **potential difference**. The sequence of events is known as an **action potential**:

<u>Changes in potential difference during an action potential</u>

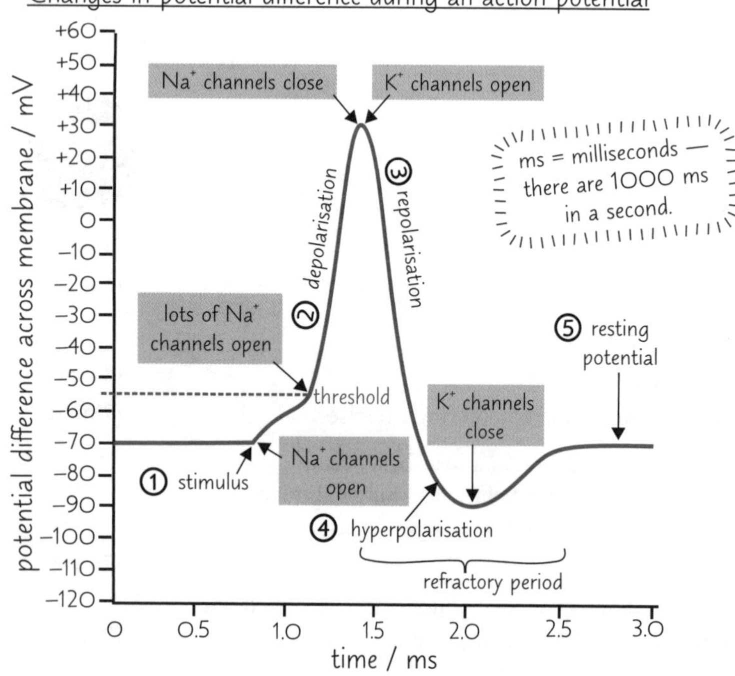

ms = milliseconds — there are 1000 ms in a second.

① **Stimulus** — this **excites** the neurone cell membrane, causing **sodium ion channels** to **open**. The membrane becomes **more permeable** to sodium, so **sodium ions diffuse into** the neurone down the sodium ion electrochemical gradient. This makes the **inside** of the neurone **less negative**.

② **Depolarisation** — if the potential difference reaches the **threshold** (around **–55 mV**), **more sodium ion channels open**. **More sodium ions diffuse rapidly into** the neurone.

③ **Repolarisation** — at a potential difference of around **+30 mV** the **sodium ion channels close** and **potassium ion channels open**. The membrane is **more permeable** to potassium so **potassium ions diffuse out** of the neurone down the potassium ion concentration gradient. This starts to get the membrane **back** to its **resting potential**.

The sodium channels have to close or the membrane will remain depolarised.

Neurones

④ **Hyperpolarisation** — **potassium ion channels** are **slow to close** so there's a slight '**overshoot**' where too many potassium ions diffuse out of the neurone. The potential difference becomes **more negative** than the **resting potential** (i.e. less than –70 mV).

⑤ **Resting potential** — the ion channels are **reset**. The **sodium-potassium pump** returns the membrane to its **resting potential** and maintains it until the membrane's excited by another stimulus.

After an **action potential**, the neurone cell membrane **can't** be **excited** again straight away. This is because the ion channels are **recovering** and they **can't** be made to **open** — **sodium ion channels** are **closed** during repolarisation and **potassium ion channels** are **closed** during hyperpolarisation. This period of recovery is called the **refractory period**.

The **Action Potential** Moves **Along** the **Neurone** as a **Wave** of **Depolarisation**

1) When an **action potential** happens, some of the **sodium ions** that enter the neurone **diffuse sideways**.
2) This causes **sodium ion channels** in the **next region** of the neurone to **open** and **sodium ions diffuse into** that part.
3) This causes a **wave of depolarisation** to travel along the neurone.
4) The **wave** moves **away** from the parts of the membrane in the **refractory period** because these parts **can't fire** an action potential.

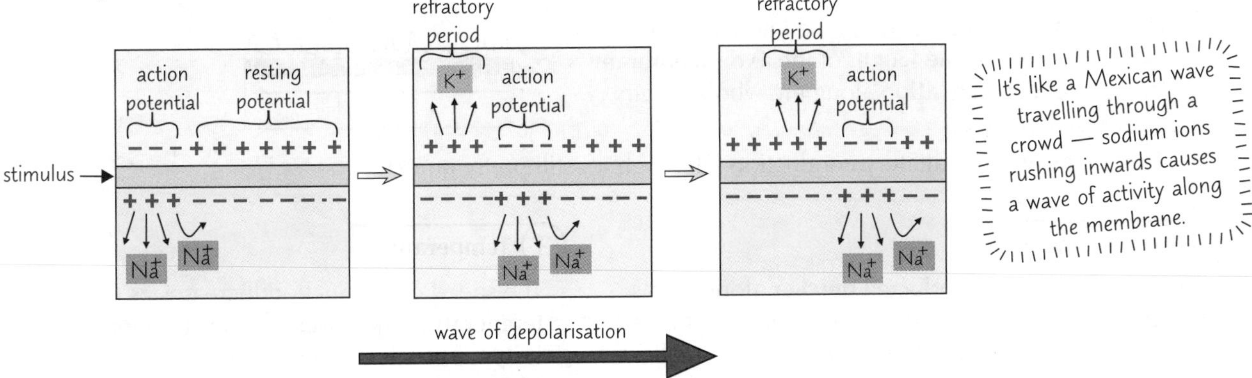

The **Refractory Period** Produces **Discrete Impulses**

1) During the **refractory period**, **ion channels** are **recovering** and **can't** be **opened**.
2) So the refractory period acts as a **time delay** between one action potential and the next. This means that:
   - **action potentials don't overlap**, but pass along as **discrete** (separate) **impulses**.
   - there's a limit to the **frequency** at which the nerve impulses can be transmitted.
   - **action potentials** are **unidirectional** (they only travel in **one direction**).

Action Potentials have an **All-or-Nothing Nature**

1) Once the threshold is reached, an action potential will **always fire** with the **same change in voltage**, no matter how big the stimulus is.
2) If the **threshold isn't reached**, an action potential **won't fire**. This is the **all-or-nothing** nature of action potentials.
3) A **bigger stimulus** won't cause a bigger action potential, but it will cause them to fire **more frequently**.

small stimulus

big stimulus

Neurones

Three Factors Affect the Speed of Conduction of Action Potentials

1) Myelination

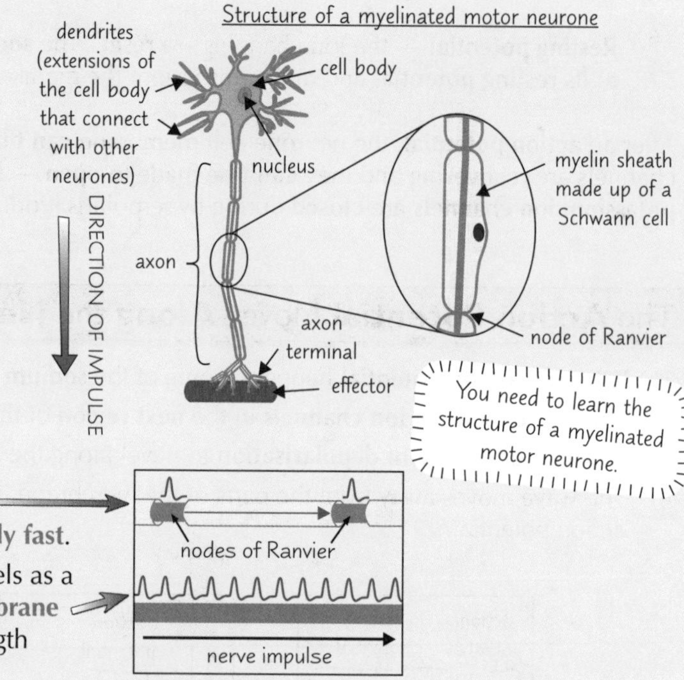

1) Some neurones are **myelinated** — they have a **myelin sheath**.
2) The myelin sheath is an **electrical insulator**.
3) In the peripheral nervous system, the sheath is made of a type of cell called a **Schwann cell**.
4) Between the Schwann cells are tiny patches of **bare membrane** called the **nodes of Ranvier**. **Sodium ion channels** are **concentrated** at the nodes.
5) In a **myelinated** neurone, **depolarisation** only happens at the **nodes of Ranvier** (where sodium ions can get through the membrane).
6) The neurone's **cytoplasm conducts** enough electrical charge to **depolarise** the **next node**, so the impulse '**jumps**' from node to node.
7) This is called **saltatory conduction** and it's **really fast**.
8) In a **non-myelinated** neurone, the impulse travels as a **wave** along the **whole length** of the **axon membrane** (so you get **depolarisation** along the whole length of the membrane).
9) This is **slower** than saltatory conduction (although it's still pretty quick).

2) Axon diameter

Action potentials are conducted **quicker** along axons with **bigger diameters** because there's **less resistance** to the **flow of ions** than in the cytoplasm of a smaller axon. With less resistance, **depolarisation reaches** other parts of the neurone cell membrane **quicker**.

3) Temperature

The speed of conduction increases as the **temperature increases** too, because **ions diffuse faster**. The speed only increases up to around **40 °C** though — after that the **proteins** begin to **denature** and the speed decreases.

Warm-Up Questions

PRACTICE QUESTIONS

Q1 Give one function of the refractory period.

Q2 What is meant by the 'all-or-nothing' nature of action potentials?

Q3 What is the function of Schwann cells on a neurone?

Q4 Give three factors that affect the speed of conduction of action potentials.

Exam Question

Q1 The graph shows an action potential across an axon membrane following the application of a stimulus.

a) Explain what causes the change in potential difference between point A and point B. [2 marks]

b) The same stimulus was applied consistently for over one hour. The next action potential fired at 4.5 ms. Calculate how many action potentials fired in one hour. Give your answer in standard form. [2 marks]

c) The strength of the stimulus was increased by 50%. Give the maximum potential difference across the membrane that would be experienced with this stronger stimulus. [1 mark]

I'm feeling a bit depolarised after all that...

All this stuff about neurones can be a bit tricky to get your head around. Take your time and try scribbling it all down a few times till it starts to make some kind of sense. Neurones work because there's an electrical charge across their membrane, which is set up by ion pumps and ion channels. It's a change in this charge that transmits an action potential.

Synaptic Transmission

When an action potential arrives at the end of a neurone, the information has to be passed on to the next cell — this could be another neurone, a muscle cell or a gland cell.

A **Synapse** is a **Junction** Between a **Neurone** and the **Next Cell**

1) A **synapse** is the junction between a **neurone** and another **neurone**, or between a **neurone** and an **effector cell**, e.g. a muscle or gland cell.
2) The **tiny gap** between the cells at a synapse is called the **synaptic cleft**.
3) The **presynaptic neurone** (the one before the synapse) has a **swelling** called a **synaptic knob**. This contains **synaptic vesicles** filled with **chemicals** called **neurotransmitters**.
4) When an **action potential** reaches the end of a neurone it causes **neurotransmitters** to be **released** into the synaptic cleft. They **diffuse across** to the **postsynaptic membrane** (the one after the synapse) and **bind** to **specific receptors**.
5) When neurotransmitters bind to receptors they might **trigger** an **action potential** (in a neurone), cause **muscle contraction** (in a muscle cell), or cause a **hormone** to be **secreted** (from a gland cell).
6) Because the receptors are **only** on the postsynaptic membranes, synapses make sure impulses are **unidirectional** — the impulse can only travel in **one direction**.
7) Neurotransmitters are **removed** from the **cleft** so the **response** doesn't keep happening, e.g. they're taken back into the **presynaptic neurone** or they're **broken down** by **enzymes** (and the products are taken into the neurone).
8) There are many **different** neurotransmitters, e.g. **acetylcholine** (**ACh**) and **noradrenaline**. Synapses that use acetylcholine are called **cholinergic synapses**. Their structure is exactly the **same** as in the diagram above.

Typical structure of a synapse

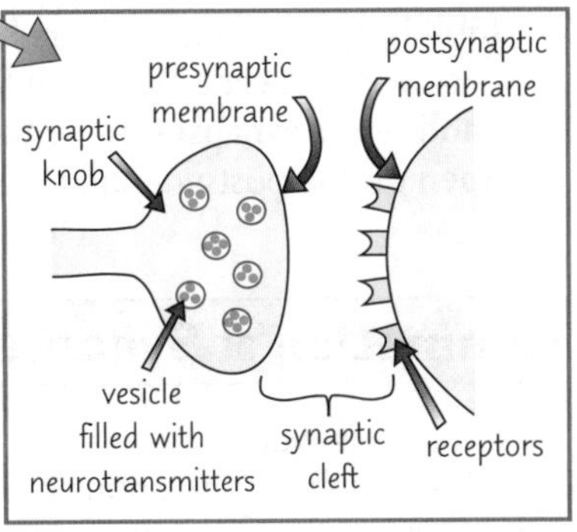

synaptic knob
presynaptic membrane
postsynaptic membrane
vesicle filled with neurotransmitters
synaptic cleft
receptors

ACh Transmits the Nerve Impulse **Across** a **Cholinergic Synapse**

This is how a **nerve impulse** is transmitted across a **cholinergic synapse**:

1) An action potential (see pages 146-147) arrives at the **synaptic knob** of the **presynaptic neurone**.
2) The action potential stimulates **voltage-gated calcium ion channels** in the **presynaptic neurone** to **open**.
3) **Calcium ions diffuse into** the synaptic knob. (They're pumped out afterwards by active transport.)

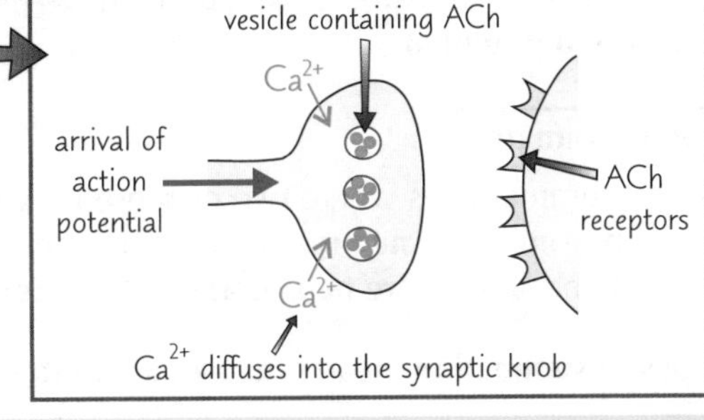

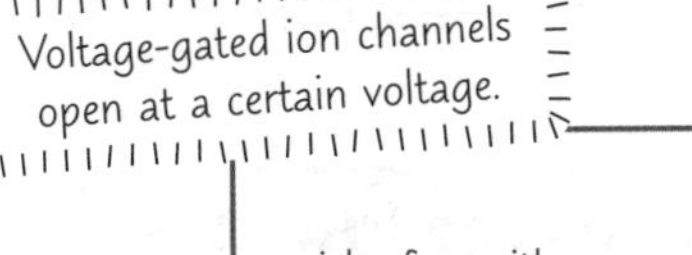

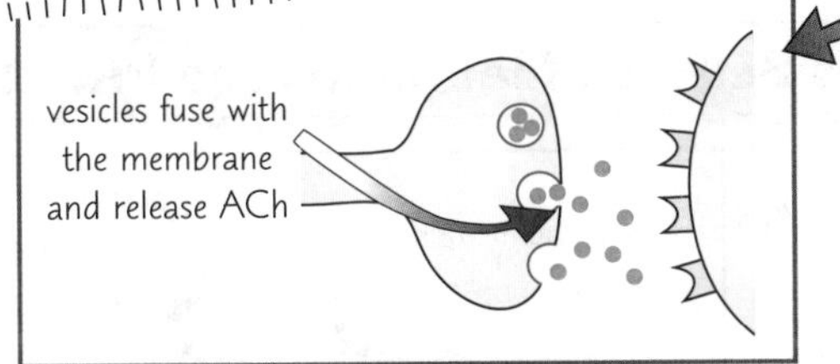

4) The influx of **calcium ions** into the synaptic knob causes the **synaptic vesicles** to **move** to the **presynaptic membrane**. They then **fuse** with the presynaptic membrane.
5) The **vesicles release** the neurotransmitter **acetylcholine** (**ACh**) into the **synaptic cleft** — this is called **exocytosis**.

6) ACh **diffuses** across the **synaptic cleft** and **binds** to specific **cholinergic receptors** on the **postsynaptic membrane**.
7) This causes **sodium ion channels** in the **postsynaptic neurone** to **open**.
8) The **influx** of **sodium ions** into the postsynaptic membrane causes **depolarisation**. An **action potential** on the **postsynaptic membrane** is generated if the **threshold** is reached.
9) ACh is **removed** from the **synaptic cleft** so the **response** doesn't keep happening. It's **broken down** by an **enzyme** called **acetylcholinesterase** (**AChE**) and the products are **re-absorbed** by the **presynaptic neurone** and used to make more ACh.

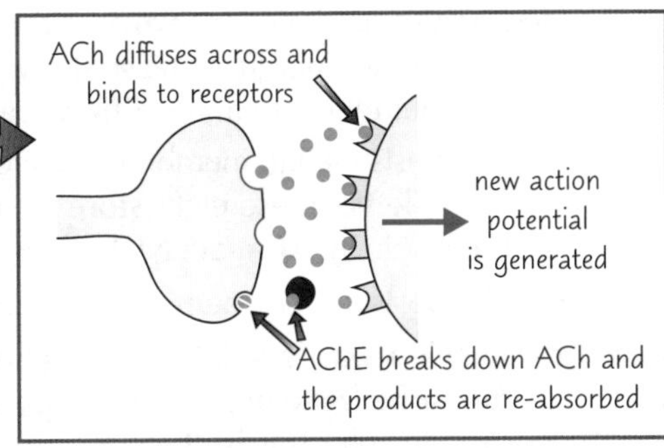

Synaptic Transmission

Neurotransmitters Can be Excitatory, Inhibitory or Both

1) **Excitatory** neurotransmitters **depolarise** the postsynaptic membrane, making it fire an **action potential** if the **threshold** is reached. E.g. **acetylcholine** is an excitatory neurotransmitter at **cholinergic synapses** in the **CNS** — it binds to cholinergic receptors to cause an **action potential** in the postsynaptic membrane — and at **neuromuscular junctions** (see below).
2) **Inhibitory** neurotransmitters **hyperpolarise** the postsynaptic membrane (make the potential difference more negative), **preventing** it from firing an action potential. E.g. **acetylcholine** is an **inhibitory** neurotransmitter at **cholinergic synapses** in the **heart**. When it binds to receptors here, it can cause **potassium ion channels** to **open** on the postsynaptic membrane, **hyperpolarising** it.

Summation at Synapses Finely Tunes the Nervous Response

If a stimulus is **weak**, only a **small amount** of **neurotransmitter** will be released from a neurone into the synaptic cleft. This might not be enough to **excite** the postsynaptic membrane to the **threshold** level and stimulate an action potential. **Summation** is where the effect of neurotransmitter released from many neurones (or one neurone that's stimulated a lot in a short period of time) is **added together**. There are two types of summation:

Spatial summation

1) Sometimes **many** neurones **connect** to **one** neurone.
2) The small amount of **neurotransmitter** released from **each** of these neurones can be enough **altogether** to **reach** the **threshold** in the postsynaptic neurone and **trigger** an **action potential**.
3) If some neurones release an **inhibitory neurotransmitter** then the total effect of all the neurotransmitters might be **no action potential**.

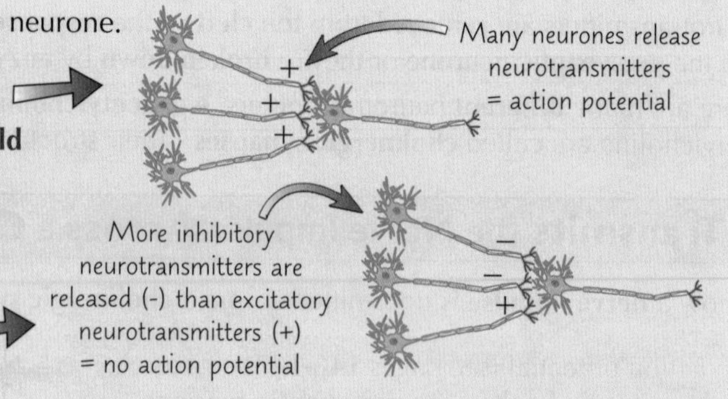

Temporal summation

Temporal summation is where **two or more** nerve impulses arrive in **quick succession** from the **same presynaptic neurone**. This makes an action potential **more likely** because **more neurotransmitter** is released into the **synaptic cleft**.

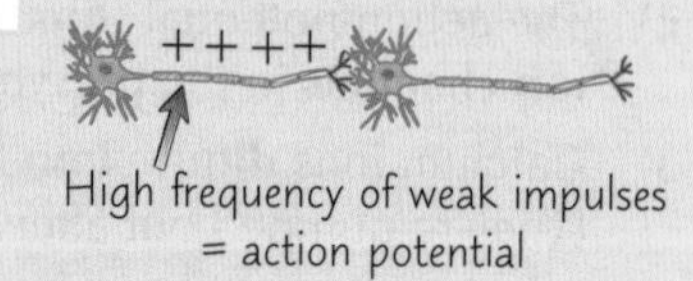

Both types of **summation** mean synapses **accurately process information**, **finely tuning** the response.

Neuromuscular Junctions are Synapses Between Neurones and Muscles

1) A **neuromuscular junction** is a **synapse** between a **motor neurone** and a **muscle cell**.
2) Neuromuscular junctions use the neurotransmitter **acetylcholine** (**ACh**), which binds to cholinergic receptors called **nicotinic cholinergic receptors**.
3) Neuromuscular junctions **work** in basically the **same way** as the **cholinergic synapse** shown on the previous page — but there are a few **differences**:
   - The postsynaptic membrane has lots of **folds** that form **clefts**. These clefts **store** the **enzyme** that breaks down **ACh** (**acetylcholinesterase** — **AChE**).
   - The postsynaptic membrane has **more receptors** than other synapses.
   - ACh is **always excitatory** at a **neuromuscular junction**. So when a **motor neurone** fires an **action potential**, it normally triggers a **response** in a muscle cell. This **isn't** always the case for a synapse between two neurones.

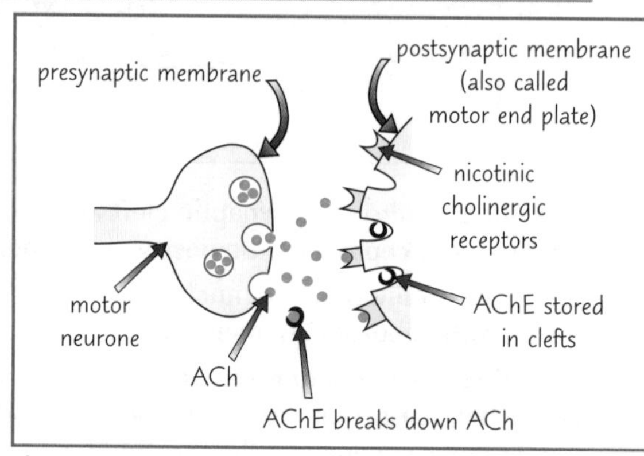

You need to be able to compare transmission across a cholinergic synapse and a neuromuscular junction.

Synaptic Transmission

Drugs Affect the Action of Neurotransmitters at Synapses in Various Ways

Some **drugs affect synaptic transmission**. You might have to **predict** the **effects** that a drug would have at a synapse in your exam. Here are some **examples** of how drugs can affect synaptic transmission:

You don't need to learn the names of the drugs.

1. Some drugs are the **same shape** as neurotransmitters so they **mimic** their action at receptors (these drugs are called **agonists**). This means **more receptors** are **activated**. E.g. **nicotine** mimics **acetylcholine** so binds to nicotinic cholinergic receptors in the brain.

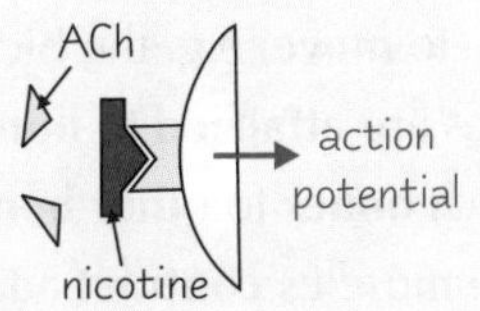

2. Some drugs **block receptors** so they **can't be activated** by neurotransmitters (these drugs are called **antagonists**). This means **fewer receptors** (if any) can be **activated**. E.g. **curare** blocks the effects of acetylcholine by blocking nicotinic cholinergic receptors at neuromuscular junctions, so muscle cells can't be stimulated. This results in the muscle being **paralysed**.

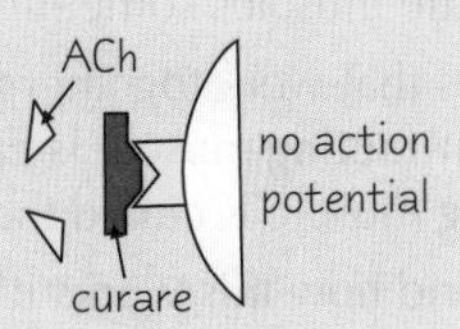

3. Some drugs **inhibit** the **enzyme** that breaks down neurotransmitters (they stop it from working). This means there are **more neurotransmitters** in the synaptic cleft to **bind** to **receptors** and they're there for **longer**. E.g. **nerve gases** stop acetylcholine from being broken down in the synaptic cleft. This can lead to **loss** of **muscle control**.

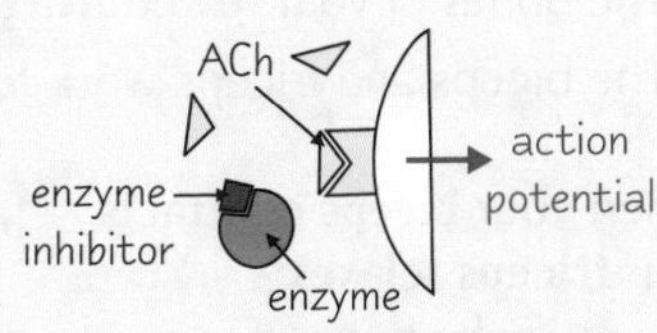

4. Some drugs **stimulate** the release of **neurotransmitter** from the presynaptic neurone so **more receptors** are activated, e.g. **amphetamines**.

5. Some drugs **inhibit** the release of neurotransmitters from the presynaptic neurone so **fewer receptors** are activated, e.g. **alcohol**.

Warm-Up Questions

Q1 How do synapses ensure that nerve impulses are unidirectional?

Q2 Give one way that neurotransmitters are removed from the synaptic cleft.

Q3 Which neurotransmitter do you find at cholinergic synapses?

Q4 Why are calcium ions important in synaptic transmission?

Q5 What do inhibitory neurotransmitters do at synapses?

Q6 What kind of receptors are found at neuromuscular junctions?

Exam Questions

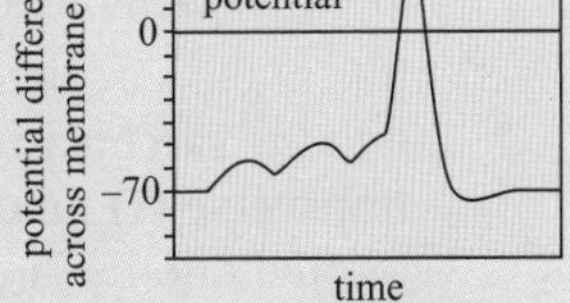

Q1 The graph on the right shows the potential difference across a postsynaptic membrane against time.

a) Suggest why a potential difference of –45 mV is significant for this postsynaptic membrane. [1 mark]

b) The action potential shown on the graph was fired as a result of temporal summation. Use the graph and your own knowledge to explain how this action potential was created. [4 marks]

Q2 Myasthenia gravis is a disease in which the body's immune system gradually destroys receptors at neuromuscular junctions. This leads to weaker muscular responses than normal. Explain why. [3 marks]

Q3 Galantamine is a drug that inhibits the enzyme acetylcholinesterase (AChE). Predict the effect of galantamine at a neuromuscular junction and explain your answer. [3 marks]

Neurotransmitter revision inhibits any excitement...

Some more pretty tough pages here — lovely. And lots more diagrams to have a go at drawing and re-drawing. Don't worry if you're not the world's best artist, just make sure you add labels to your drawings to explain what's happening.

Muscle Contraction

I reckon muscle cells are the spoilt brats of the Biology world. They're so special that everything muscly has to have its own special name — there's none of this "cell membrane" malarkey, oh no, it's "sarcolemma" if you please...

Muscles Act in **Antagonistic Pairs**

1) **Skeletal muscle** (also called striated, striped or voluntary muscle) is the type of muscle you use to **move**, e.g. the biceps and triceps move the lower arm.
2) Skeletal muscles are **attached** to **bones** by **tendons**.
3) **Ligaments attach bones** to **other bones**, to hold them together.
4) Pairs of skeletal muscles **contract** and **relax** to **move bones** at a **joint**. The bones of the skeleton are **incompressible** (rigid) so they act as **levers**, giving the muscles something to **pull** against.
5) Muscles that work together to move a bone are called **antagonistic pairs**. The **contracting** muscle is called the **agonist** and the **relaxing** muscle is called the **antagonist**.

To understand how this works it's best to look at an example:

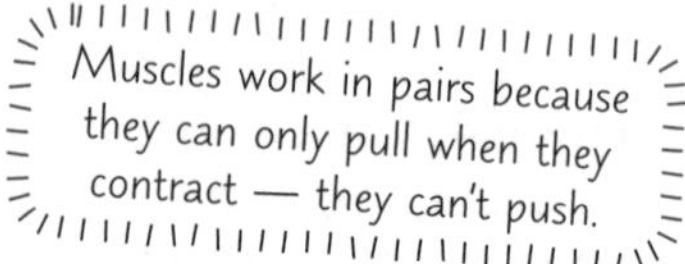

- The bones of your **lower arm** are attached to a **biceps** muscle and a **triceps** muscle by **tendons**.
- The biceps and triceps **work together** to move your arm — as one **contracts**, the other **relaxes**.

When your **biceps contracts** your **triceps relaxes**. This pulls the bone so your **arm bends** (**flexes**) at the elbow. Here, the **biceps** is the **agonist** and the **triceps** is the **antagonist**.

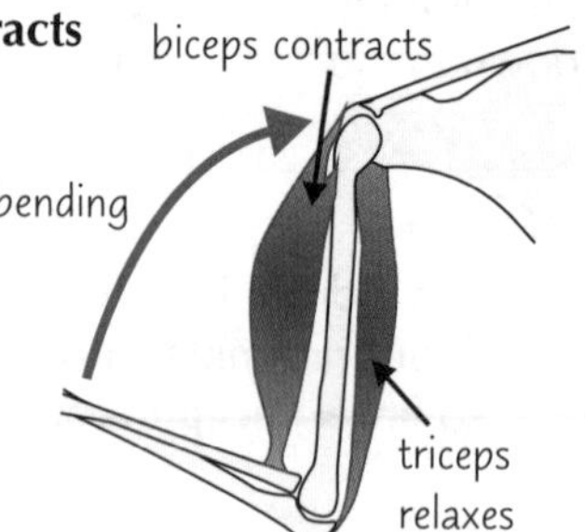

When your **triceps contracts** your **biceps relaxes**. This pulls the bone so your **arm straightens** (**extends**) at the **elbow**. Here, the **triceps** is the **agonist** and the **biceps** is the **antagonist**.

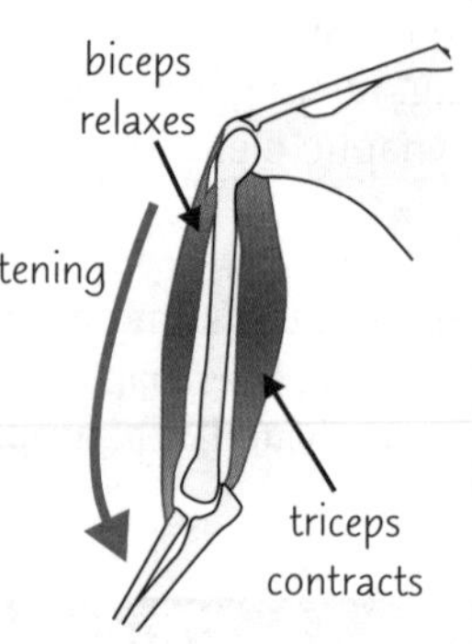

Skeletal Muscle is Made Up of **Long Muscle Fibres**

Muscles act as **effectors** and are **stimulated** to **contract** by neurones.

1) Skeletal muscle is made up of **large bundles** of **long cells**, called **muscle fibres**.
2) The cell membrane of muscle fibre cells is called the **sarcolemma**.
3) Bits of the sarcolemma **fold inwards** across the muscle fibre and stick into the **sarcoplasm** (a muscle cell's cytoplasm). These folds are called **transverse (T) tubules** and they help to **spread electrical impulses** throughout the sarcoplasm so they **reach** all parts of the **muscle fibre**.
4) A network of **internal membranes** called the **sarcoplasmic reticulum** runs through the sarcoplasm. The sarcoplasmic reticulum **stores** and **releases calcium ions** that are needed for muscle contraction (see p. 154).
5) Muscle fibres have lots of **mitochondria** to **provide** the **ATP** that's needed for **muscle contraction**.
6) Muscle fibres are **multinucleate** (contain many nuclei).
7) Muscle fibres have lots of **long, cylindrical organelles** called **myofibrils**. They're made up of proteins and are **highly specialised** for **contraction**.

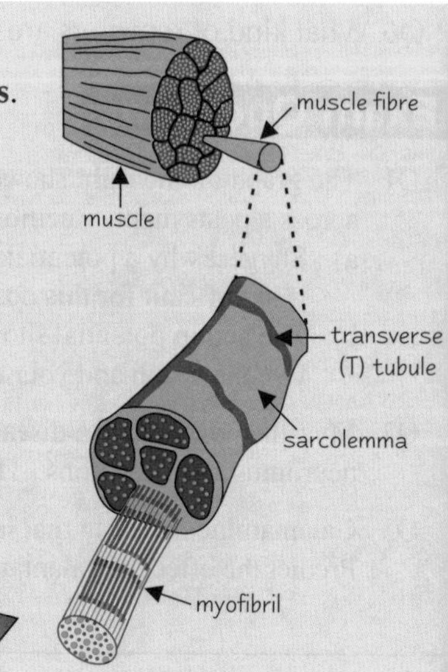

Muscle Contraction

Myofibrils Contain **Thick Myosin** Filaments and **Thin Actin** Filaments

1) Myofibrils contain bundles of **thick** and **thin myofilaments** that **move past each other** to make muscles **contract**.
   - **Thick myofilaments** are made of the protein **myosin**.
   - **Thin myofilaments** are made of the protein **actin**.
2) If you look at a **myofibril** under an electron **microscope**, you'll see a pattern of alternating **dark** and **light bands**:
   - **Dark** bands contain the **thick myosin filaments** and some overlapping thin actin filaments — these are called **A-bands**.
   - **Light** bands contain **thin actin filaments** only — these are called **I-bands**.
3) A myofibril is made up of many short units called **sarcomeres**.
4) The **ends** of each **sarcomere** are marked with a **Z-line**.
5) In the **middle** of each sarcomere is an **M-line**. The M-line is the **middle** of the **myosin** filaments.
6) **Around** the M-line is the **H-zone**. The H-zone **only** contains **myosin** filaments.

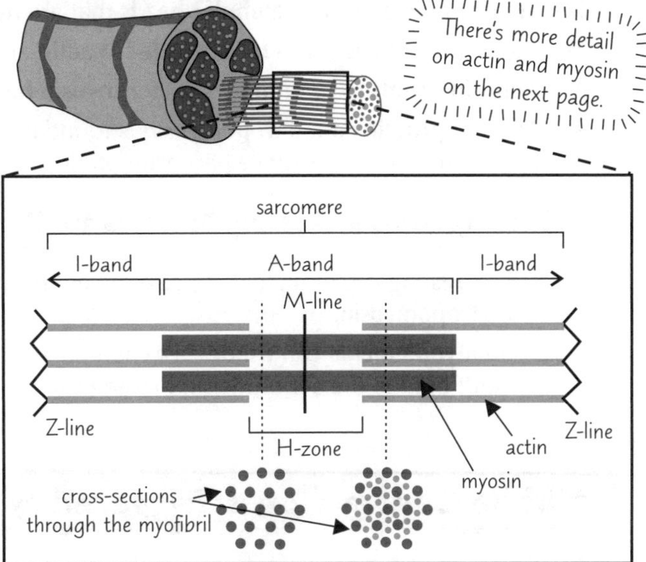

Muscle Contraction is Explained by the **Sliding Filament Theory**

1) **Myosin** and **actin** filaments **slide** over one another to make the **sarcomeres contract** — the myofilaments themselves **don't** contract.
2) The **simultaneous contraction** of lots of **sarcomeres** means the **myofibrils** and **muscle fibres contract**.
3) Sarcomeres return to their **original length** as the muscle **relaxes**.

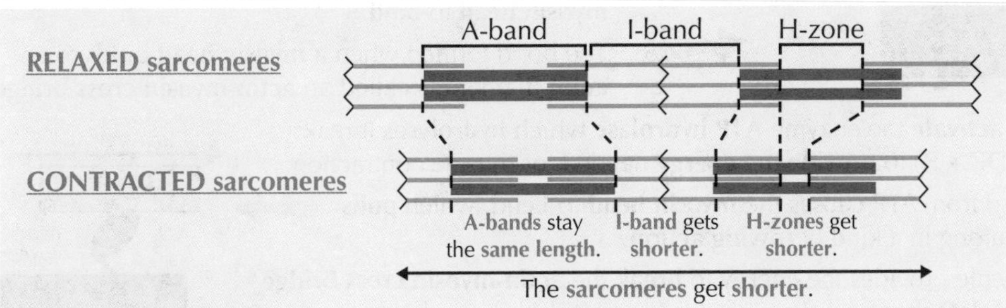

If only the sliding filament theory was as much fun...

Warm-Up Questions

Q1 Describe one example of how muscles act in antagonistic pairs.

Q2 What are transverse (T) tubules?

Q3 Name the two proteins that make up myofibrils.

PRACTICE QUESTIONS

Exam Question

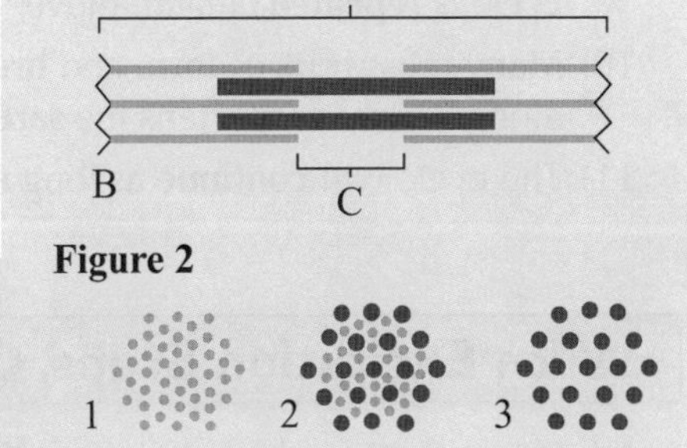

Q1 A muscle myofibril was examined under an electron microscope and a sketch was drawn (Figure 1).

a) What are the correct names for labels A, B and C? [3 marks]

b) Describe how the lengths of the different bands in a myofibril change during muscle contraction. [2 marks]

c) The myofibril was then cut through the M-line (Figure 2). State which of the cross-section drawings you would expect to see and explain why. [3 marks]

Sarcomere — a French mother with a dry sense of humour...

Blimey, there are an awful lot of similar-sounding names to learn on these pages. And then you've got your A-band, I-band, what-band, who-band to memorise too. But once you've learnt them, these are things you'll never forget.

Muscle Contraction

Myofilaments sliding over one another takes a lot of energy — probably why exercise is such hard work...

Myosin Filaments Have Globular Heads and Binding Sites

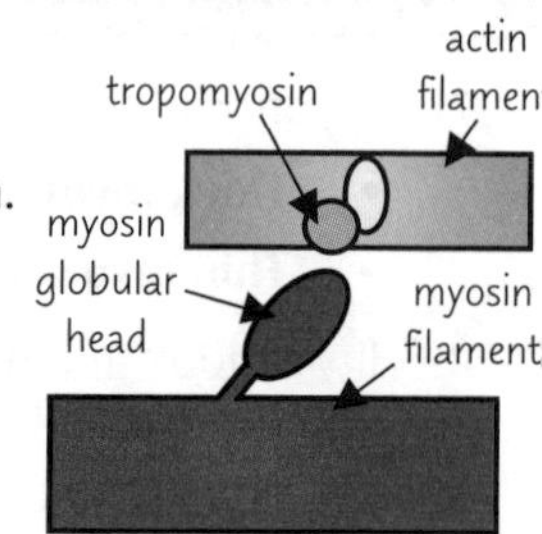

1) **Myosin filaments** have **globular heads** that are **hinged**, so they can move **back** and **forth**.
2) Each myosin head has a **binding site** for **actin** and a **binding site** for **ATP**.
3) **Actin filaments** have **binding sites** for **myosin heads**, called **actin-myosin** binding sites.
4) Another **protein** called **tropomyosin** is found between actin filaments. It **helps** myofilaments **move** past each other.

Binding Sites in Resting Muscles are Blocked by Tropomyosin

1) In a **resting** (unstimulated) muscle the **actin-myosin binding site** is **blocked** by **tropomyosin**.
2) So **myofilaments can't slide** past each other because the **myosin heads can't bind** to the actin-myosin binding site on the actin filaments.

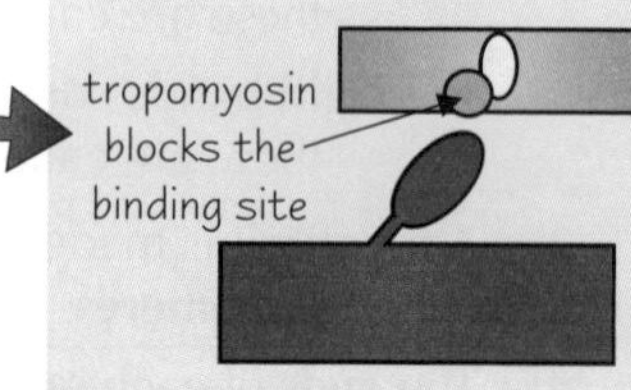

Muscle Contraction is Triggered by an Influx of Calcium Ions

1) When an action potential from a motor neurone **stimulates** a muscle cell, it **depolarises** the **sarcolemma**. Depolarisation **spreads** down the **T-tubules** to the **sarcoplasmic reticulum** (see p. 152).
2) This causes the **sarcoplasmic reticulum** to **release** stored **calcium ions** (Ca^{2+}) into the **sarcoplasm**.

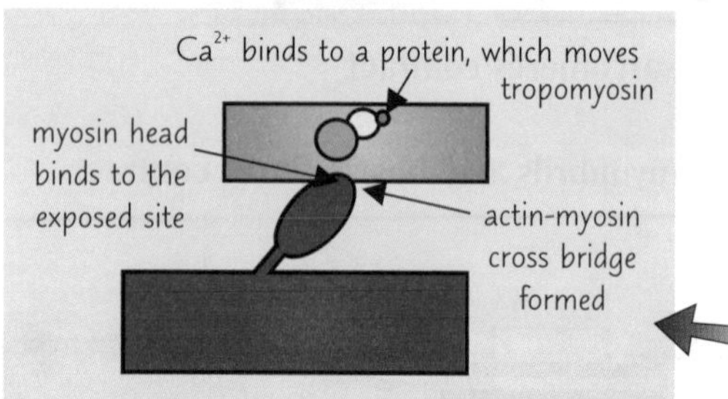

3) Calcium ions **bind** to a protein attached to tropomyosin, causing the protein to **change shape**. This **pulls** the attached **tropomyosin out** of the **actin-myosin binding site** on the actin filament.
4) This **exposes** the **binding site**, which allows the **myosin head** to **bind**.
5) The bond formed when a **myosin head** binds to an **actin filament** is called an **actin-myosin cross bridge**.
6) **Calcium** ions also **activate** the enzyme **ATP hydrolase** which **hydrolyses** (breaks down) **ATP** (into ADP + P_i) to **provide** the **energy** needed for muscle contraction.
7) The **energy** released from ATP causes the **myosin head** to **bend**, which **pulls** the **actin filament** along in a kind of **rowing action**.

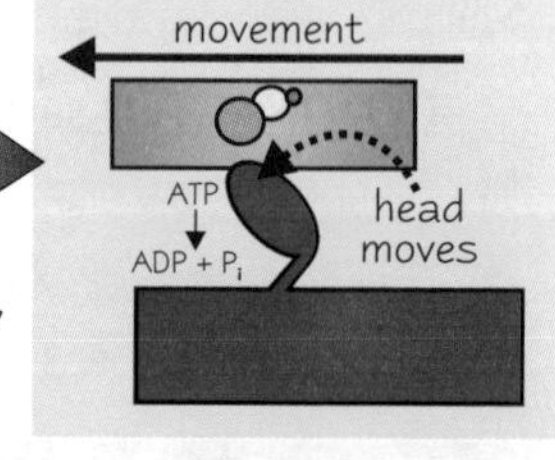

8) Another **ATP** molecule provides the **energy** to **break** the **actin-myosin cross bridge**, so the **myosin head detaches** from the actin filament **after** it's moved.
9) The **myosin head** then **reattaches** to a **different binding site** further along the actin filament. A **new actin-myosin cross bridge** is formed and the **cycle** is **repeated** (attach, move, detach, reattach to new binding site...).
10) **Many** cross bridges **form** and **break** very **rapidly**, pulling the actin filament along — which **shortens** the **sarcomere**, causing the **muscle** to **contract**.
11) The cycle will **continue** as long as **calcium ions** are **present**.

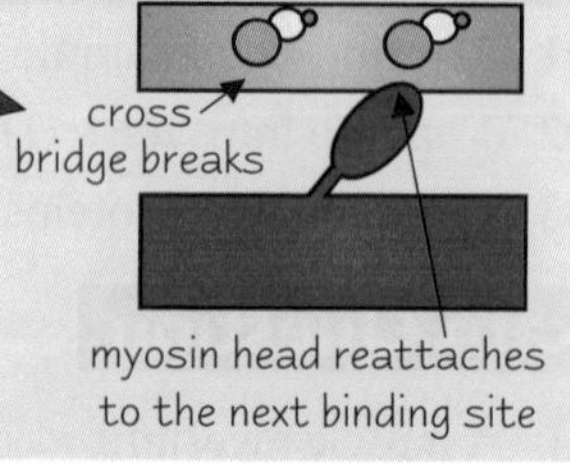

When Excitation Stops, Calcium Ions Leave

1) When the muscle **stops** being **stimulated**, **calcium ions leave** their **binding sites** and are moved by **active transport** back into the **sarcoplasmic reticulum** (this needs **ATP** too).
2) This causes the **tropomyosin** molecules to **move back**, so they **block** the **actin-myosin binding sites** again.
3) Muscles **aren't contracted** because **no myosin heads** are **attached** to **actin** filaments (so there are no actin-myosin cross bridges).
4) The **actin** filaments **slide back** to their **relaxed** position, which **lengthens** the **sarcomere**.

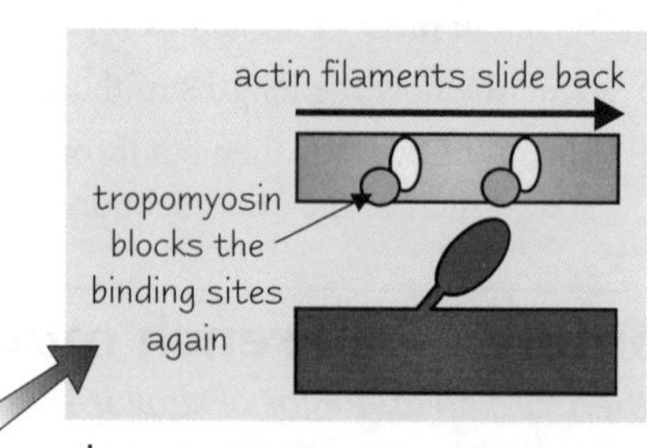

Muscle Contraction

ATP and Phosphocreatine Provide the Energy for Muscle Contraction

So much **energy** is **needed** when muscles contract that **ATP** gets **used up very quickly**.
ATP has to be **continually generated** so exercise can continue — this happens in **three main ways**:

1) Aerobic respiration
   - **Most ATP** is generated via **oxidative phosphorylation** in the cell's **mitochondria**.
   - **Aerobic** respiration only works when there's **oxygen** so it's good for **long periods** of **low-intensity exercise**.

2) Anaerobic respiration

See pages 122-125 for more on aerobic and anaerobic respiration.

   - ATP is made **rapidly** by **glycolysis**.
   - The **end product** of glycolysis is **pyruvate**, which is converted to **lactate** by **lactate fermentation**.
   - Lactate can **quickly build up** in the muscles and cause **muscle fatigue**.
   - Anaerobic respiration is good for **short periods** of **hard exercise**, e.g. a **400 m sprint**.

3) ATP-Phosphocreatine (PCr) System
   - **ATP** is made by **phosphorylating ADP** — adding a phosphate group taken from **PCr**.

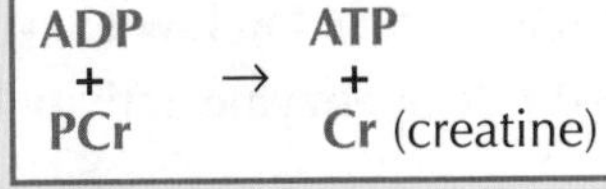

ADP + PCr → ATP + Cr (creatine)

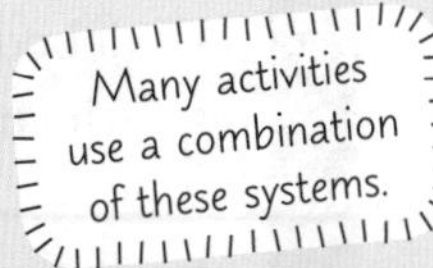

   - **PCr** is **stored** inside cells and the ATP-PCr system **generates ATP** very **quickly**.
   - **PCr runs out** after a few seconds so it's used during **short bursts** of **vigorous exercise**, e.g. a **tennis serve**.
   - The ATP-PCr system is **anaerobic** (it doesn't need oxygen) and it's **alactic** (it doesn't form any lactate).

Some of the creatine (Cr) gets broken down into **creatinine**, which is removed from the body via the **kidneys**. Creatinine levels can be higher in people who **exercise regularly** and those with a **high muscle mass**. High creatinine levels may also indicate **kidney damage**.

Skeletal Muscles are Made of Slow Twitch and Fast Twitch Muscle Fibres

Skeletal muscles are made up of **two types** of **muscle fibres** — **slow twitch** and **fast twitch**.
Different muscles have **different proportions** of slow and fast twitch fibres. The two types have **different properties**:

SLOW TWITCH MUSCLE FIBRES	FAST TWITCH MUSCLE FIBRES
Muscle fibres that contract slowly.	Muscle fibres that contract very quickly.
Muscles you use for posture, e.g. those in the back, have a high proportion of them.	Muscles you use for fast movement, e.g. those in the eyes and legs, have a high proportion of them.
Good for endurance activities, e.g. maintaining posture, long-distance running.	Good for short bursts of speed and power, e.g. eye movement, sprinting.
Can work for a long time without getting tired.	Get tired very quickly.
Energy's released slowly through aerobic respiration. Lots of mitochondria and blood vessels supply the muscles with oxygen.	Energy's released quickly through anaerobic respiration using glycogen (stored glucose). There are few mitochondria or blood vessels.
Reddish in colour because they're rich in myoglobin — a red-coloured protein that stores oxygen.	Whitish in colour because they don't have much myoglobin (so can't store much oxygen).

Warm-Up Questions

PRACTICE QUESTIONS

Q1 Describe one way that ATP can be generated in contracting muscles.

Q2 State three differences between slow and fast twitch skeletal muscle fibres.

Exam Questions

Q1 Rigor mortis is the stiffening of muscles in the body after death. It happens when ATP reserves are exhausted. Explain why a lack of ATP leads to muscles being unable to relax. [3 marks]

Q2 Bepridil is a drug that blocks calcium ion channels.
Describe and explain the effect this drug will have on muscle contraction. [3 marks]

What does muscle contraction cost? 80p...

Sorry, that's my favourite sciencey joke so I had to fit it in somewhere — a small distraction before you revisit this page. It's tough stuff but you know the best way to learn it. That's right, grab yourself a nice felt-tip pen and a pad of paper...

Homeostasis Basics

Ah, there's nothing like learning a nice long word to start you off on a new section — welcome to homeostasis.

Homeostasis is the Maintenance of a Stable Internal Environment

1) **Changes** in your **external environment** can affect your **internal environment** — the blood and tissue fluid that surrounds your cells.
2) **Homeostasis** involves **control systems** that keep your **internal environment** roughly **constant** (within **certain limi**
3) **Keeping** your internal environment **stable** is vital for cells to **function normally** and to **stop** them being **damaged**
4) It's particularly important to **maintain** the right **core body temperature** and **blood pH**. This is because temperatu and pH affect **enzyme activity**, and enzymes **control** the **rate** of **metabolic reactions**:

Temperature

- If body temperature is **too high** (e.g. 40 °C) **enzymes** may become **denatured**. The enzyme's molecules **vibrate too much**, which **breaks** the **hydrogen bonds** that hold them in their **3D shape**. The **shape** of the enzyme's **active site** is **changed** and it **no longer works** as a **catalyst**. This means **metabolic reactions** are **less efficient**.
- If body temperature is **too low enzyme activity** is **reduced**, **slowing** the rate of **metabolic reactions**.
- The **highest rate** of **enzyme activity** happens at their **optimum temperature** (about **37 °C** in humans).

pH

- If blood pH is **too high** or **too low** (highly alkaline or acidic) **enzymes** become **denatured**. The **hydrogen bonds** that hold them in their 3D shape are broken, so the **shape** of the enzyme's **active site** is **changed** and it **no longer works** as a **catalyst**. This means **metabolic reactions** are **less efficient**.
- The **highest rate** of **enzyme activity** happens at their **optimum pH** — usually **around pH 7** (neutral), but some enzymes work best at other pHs, e.g. enzymes found in the stomach work best at a low pH.

5) It's important to **maintain** the right **concentration** of **glucose** in the **blood** because cells need glucose for **energy**. Blood glucose concentration also affects the **water potential** of blood — this is the potential (likelihood) of water molecules to **diffuse** out of or into a solution.

Glucose

- If blood glucose concentration is **too high** the **water potential** of blood is **reduced** to a point where **water** molecules **diffuse out** of cells into the blood by osmosis. This can cause the cells to **shrivel up** and **die**.
- If blood glucose concentration is **too low**, cells are **unable** to carry out **normal activities** because there **isn't enough glucose** for **respiration** to provide **energy**.

Homeostatic Systems Detect a Change and Respond by Negative Feedback

1) Homeostatic systems involve **receptors**, a **communication system** and **effectors** (like the nervous system — see page 138).
2) Receptors detect when a level is **too high** or **too low**, and the information's communicated via the **nervous** system or the **hormonal** system to **effectors**.
3) The effectors respond to **counteract** the change — bringing the level **back** to **normal**.
4) The mechanism that **restores** the level to **normal** is called a **negative feedback** mechanism.
5) Negative feedback **keeps** things around the **normal** level, e.g. body temperature is usually kept **within 0.5 °C** above or below **37 °C**.
6) Negative feedback only works within **certain limits** though — if the change is **too big** then the **effectors** may **not** be able to **counteract** it, e.g. a huge drop in body temperature caused by prolonged exposure to cold weather may be too large to counteract.

Control of **body temperature** by **negative feedback**:

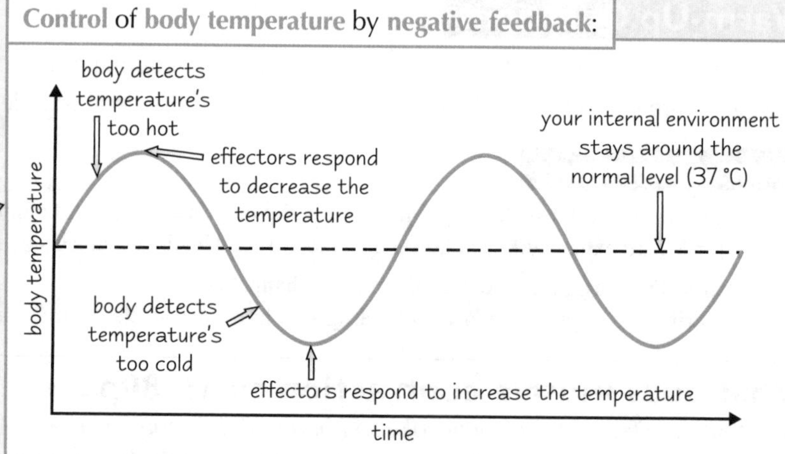

Homeostasis Basics

Multiple Negative Feedback Mechanisms Give More Control

1) Homeostasis involves **multiple negative feedback mechanisms** for each thing being controlled. This is because having more than one mechanism gives **more control** over changes in your internal environment than just having one negative feedback mechanism.
2) Having multiple negative feedback mechanisms means you can **actively increase** or **decrease a level** so it returns to **normal**, e.g. you have feedback mechanisms to reduce your body temperature and you also have mechanisms to increase it.
3) If you only had **one negative feedback mechanism**, all you could do would be **turn it on** or **turn it off**. You'd only be able to actively change a level in **one direction** so it returns to normal, e.g. it's a bit like trying to slow down a car with only an accelerator — all you can do is take your foot off the accelerator (you'd have more control with a brake too).
4) Only **one** negative feedback mechanism means a **slower response** and **less control**.

There was plenty of negative feedback when Carl wore his new vest-pants combo out for dinner.

Positive Feedback Mechanisms Amplify a Change from the Normal Level

1) Some changes trigger a **positive feedback** mechanism, which **amplifies** the change.
2) The effectors respond to **further increase** the level **away** from the **normal** level.
3) Positive feedback is useful to **rapidly activate** something, e.g. a **blood clot** after an injury.

- **Platelets** become **activated** and release a **chemical** — this triggers **more platelets** to be activated, and so on.
- Platelets **very quickly** form a **blood clot** at the injury site.
- The process **ends with negative feedback**, when the body detects the **blood clot** has been **formed**.

4) Positive feedback can also happen when a **homeostatic system breaks down**, e.g. if you're too cold for too long:

Hypothermia involves **positive feedback**:

- **Hypothermia** is **low body temperature** (below 35 °C).
- It happens when **heat's lost** from the body **quicker** than it can be **produced**.
- As body temperature **falls** the **brain doesn't work** properly and **shivering stops** — this makes body temperature **fall even more**.
- **Positive feedback** takes body temperature **further away** from the normal level, and it continues to decrease unless action is taken.

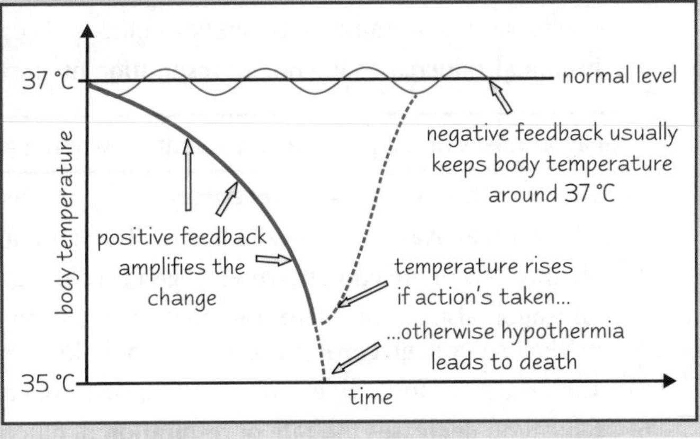

5) Positive feedback **isn't** involved in **homeostasis** because it **doesn't** keep your internal environment **stable**.

PRACTICE QUESTIONS

Warm-Up Questions

Q1 What is homeostasis and why is it necessary?

Q2 Why is it important to control blood pH?

Q3 Why is it important to control blood glucose concentration?

Exam Questions

Statement A: "Hyperthermia happens when the brain can't work properly and body temperature continues to increase."
Statement B: "When body temperature is low, mechanisms return the temperature to normal."

Q1 Look at statements A and B in the box.

a) Which statement is describing a positive feedback mechanism? Give a reason for your answer. [1 mark]

b) Describe and explain what effect a very high body temperature has on metabolic reactions. [2 marks]

Q2 Describe the importance of multiple negative feedback mechanisms in homeostasis. [2 marks]

Homeostasis works like a teacher — everything always gets corrected...

The key to understanding homeostasis is to get your head around negative feedback. Basically, if one thing goes up, the body responds to bring it down — and vice versa. When you're ready, turn over the page for some exciting examples.

Control of Blood Glucose Concentration

These pages are all about how homeostasis sorts out your blood glucose level so you can keep revising.

Eating and **Exercise** Change the **Concentration** of **Glucose** in your **Blood**

1) **All cells** need a constant **energy supply** to work — so **blood glucose concentration** must be carefully **controlled**.
2) The **concentration** of **glucose** in the blood is **normally** around **90 mg per 100 cm³** of blood. It's **monitored** by cells in the **pancreas**.
3) Blood glucose concentration **rises** after **eating food** containing **carbohydrate**. Blood glucose concentration **falls** after **exercise**, as **more glucose** is used in **respiration** to **release energy**.

Insulin and **Glucagon Control** Blood Glucose Concentration

The hormonal system **controls** blood glucose concentration using **two hormones** called **insulin** and **glucagon**. Like all hormones, insulin and glucagon **travel** in the **blood** to their **target cells** (**effectors**). They're both **secreted** by clusters of cells in the **pancreas** called the **islets of Langerhans**:

- **Beta (β) cells** secrete **insulin** into the blood.
- **Alpha (α) cells** secrete **glucagon** into the blooc

Insulin and glucagon act on **effectors**, which respond to **restore** the blood glucose concentration to the **normal level**

Insulin lowers blood glucose concentration when it's **too high**

1) Insulin binds to **specific receptors** on the cell membranes of **liver cells** and **muscle cells**.
2) It **increases** the **permeability** of muscle-cell membranes to glucose, so the cells **take up more glucose**. This involves **increasing** the number of **channel proteins** in the cell membranes (see next page for more).
3) Insulin also **activates enzymes** in liver and muscle cells that convert **glucose** into **glycogen**.
4) The cells are able to **store glycogen** in their cytoplasm, as an **energy source**.
5) The process of **forming glycogen** from glucose is called **glycogenesis**.
6) Insulin also **increases** the **rate** of **respiration** of glucose, especially in muscle cells.

Liver cells are also called hepatocytes.

GLYCOGEN
glycogenesis
glycogenolysis
activated by insulin
GLUCOSE
activated by glucagc
gluconeogenesis
GLYCEROL
AMINO ACIDS

Glucagon raises blood glucose concentration when it's **too low**

1) Glucagon binds to **specific receptors** on the cell membranes of **liver cells**.
2) Glucagon **activates enzymes** in liver cells that **break down glycogen** into **glucose**.
3) The process of **breaking down glycogen** is called **glycogenolysis**.
4) Glucagon also activates **enzymes** that are involved in the formation of glucose from **glycerol** (a component of lipids) and **amino acids**.
5) The process of **forming glucose** from **non-carbohydrates** is called **gluconeogenesis**.
6) Glucagon **decreases** the **rate** of **respiration** of glucose in cells.

Because they **travel in the blood** to their target cells, the responses produced by **hormones** are **slower** than those produced by **nervous impulses** (which are very quick — see page 139). Hormones are not broken down as quickly as neurotransmitters though, so their effects tend to **last for longer**.

Negative Feedback Mechanisms Keep Blood Glucose Concentration **Normal**

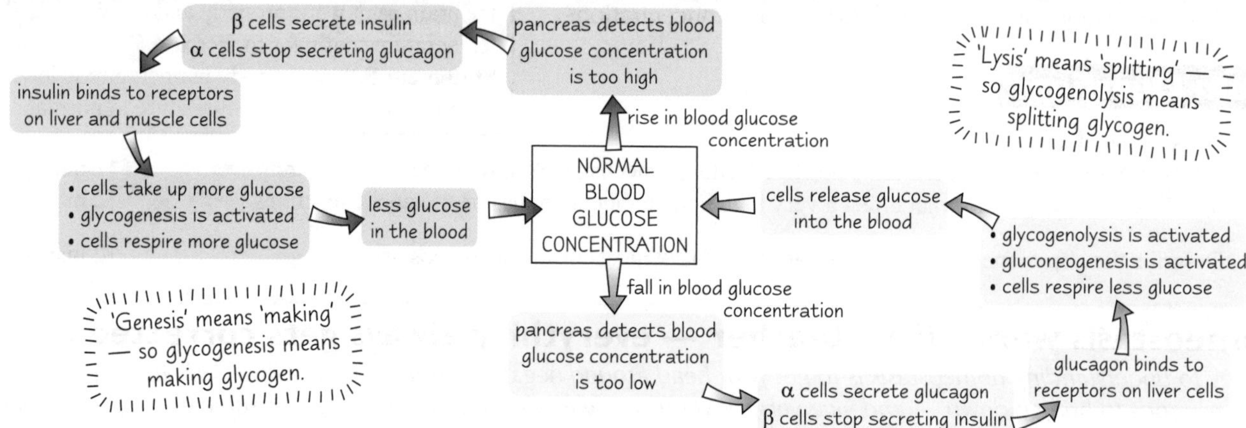

Control of Blood Glucose Concentration

Insulin Makes **Glucose Transporters** Available for **Facilitated Diffusion**

1) **Skeletal** and **cardiac muscle cells** contain a **channel protein** called GLUT4. GLUT4 is a **glucose** transporter.
2) When **insulin** levels are **low**, GLUT4 is **stored** in **vesicles** in the **cytoplasm** of cells.
3) When insulin **binds** to **receptors** on the cell-surface membrane, it triggers the **movement** of GLUT4 **to the membrane**.
4) Glucose can then be transported **into the cell** through the GLUT4 protein, by **facilitated diffusion**.

Like **Glucagon**, **Adrenaline** Also **Increases Blood Glucose Concentration**

1) **Adrenaline** is a **hormone** that's secreted from your **adrenal glands** (found just above your kidneys).
2) It's secreted when there's a **low concentration** of **glucose** in your blood, when you're **stressed** and when you're **exercising**.
3) Adrenaline binds to **receptors** in the cell membrane of **liver cells**:
   - It **activates glycogenolysis** (the breakdown of glycogen to glucose).
   - It **inhibits glycogenesis** (the synthesis of glycogen from glucose).

GLYCOGEN
inhibited by adrenaline
activated by adrenaline
glycogenesis
glycogenolysis
GLUCOSE

4) It also **activates glucagon secretion** and **inhibits insulin secretion**, which increases glucose concentration.
5) Adrenaline gets the **body ready** for **action** by making **more glucose** available for **muscles** to respire.

Adrenaline and **Glucagon** Act via a **Second Messenger**

Both **adrenaline** and **glucagon** can activate glycogenolysis **inside** a cell even though they bind to **receptors** on the **outside** of the cell. Here's **how** they do it:

- The **receptors** for adrenaline and glucagon have specific **tertiary structures** that make them complementary in shape to their respective hormones. Adrenaline and glucagon **bind** to their receptors and **activate** an **enzyme** called **adenylate cyclase** (also known as adenylyl cyclase).
- Activated adenylate cyclase converts **ATP** into a **chemical signal** called a '**second messenger**'.
- The second messenger is called **cyclic AMP** (**cAMP**).
- cAMP **activates** an enzyme called **protein kinase A**. Protein kinase A activates a **cascade** (a chain of reactions) that breaks down glycogen into glucose (**glycogenolysis**).

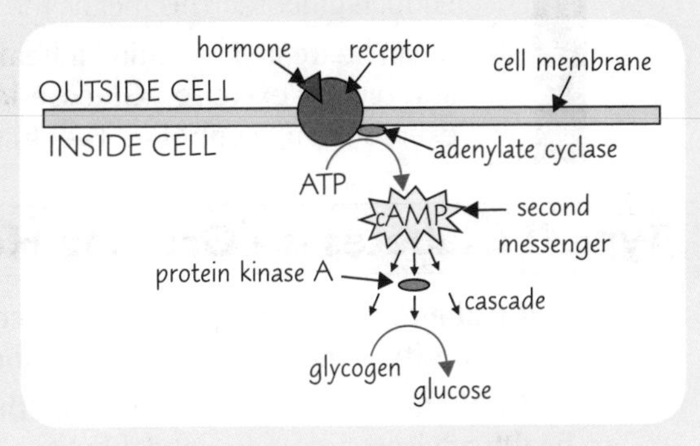

Warm-Up Questions

Q1 Why does your blood glucose concentration fall after exercise?

Q2 What's the process of breaking down glycogen into glucose called?

Q3 Give two effects of glucagon on liver cells.

PRACTICE QUESTIONS

Exam Questions

Q1 The pancreas secretes hormones that control blood glucose concentration.

a) What type of feedback mechanism is involved in the control of blood glucose concentration? Give a reason for your answer. [1 mark]

b) Describe the role of insulin in this feedback mechanism. [3 marks]

Q2 Glucagon and adrenaline trigger glycogenolysis when they bind to receptors on cell membranes. Explain how glycogenolysis is triggered inside the cell when these hormones bind to receptors on the cell surface. [3 marks]

My alpha cells detect low glucose — urgent tea and biscuit break needed...

Aaaaargh there are so many stupidly complex names to learn and they all look and sound exactly the same to me. You can't even get away with sneakily misspelling them all in your exam — like writing 'glycusogen' or 'gluconesisolysis'.

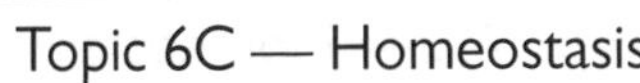

Control of Blood Glucose Concentration

Homeostasis doesn't always work. One example of this is diabetes...

Diabetes Occurs when Blood Glucose Concentration is Not Controlled

Diabetes mellitus is a condition where **blood glucose** concentration **can't** be **controlled** properly. There are **two typ**

Type I

1) In **Type I** diabetes, the immune system attacks the **β cells** in the islets of Langerhans so they **can't produce** any **insulin.** No one knows exactly what **causes** the immune system to do this. Scientists have found that some people have a **genetic predisposition** to developing Type I diabetes. They also think that the disease may be triggered by a **viral infection.**
2) After **eating**, the blood glucose level **rises** and **stays high** — this is called **hyperglycaemia** and can result in **death** if left untreated. The kidneys **can't reabsorb** all this glucose, so some of it's **excreted** in the urine.
3) Type 1 diabetes is treated with **insulin therapy**. Most people with Type I diabetes need **regular insulin injections** throughout the day, but some people use an **insulin pump** to deliver insulin continuously instead. Insulin therapy has to be **carefully controlled** because too much insulin can produce a **dangerous drop** in blood glucose levels — this is called **hypoglycaemia.**
4) **Eating regularly** and **controlling simple carbohydrate intake** (intake of sugars) helps to **avoid** a **sudden rise** in glucose.

Type II

1) **Type II** diabetes is usually acquired **later** in **life** than Type I. It is often linked with **obesity** and is **more likely** in people with a **family history** of the condition. Other risk factors include **lack of exercise**, **age** and **poor diet**.
2) It occurs when the β cells **don't produce enough insulin** or when the body's **cells don't respond** properly to **insulin.** Cells don't respond properly because the insulin **receptors** on their membranes **don't work** properly, so the cells **don't** take up enough glucose. This means the **blood glucose concentration** is **higher** than normal.
3) It can be treated by **eating a healthy**, **balanced diet**, **losing weight** (if necessary) and **regular exercise. Glucose-lowering medication** can be taken if diet and exercise can't control it. Eventually, **insulin injections** may be needed.

Type II Diabetes is a Growing Health Problem

1) **Type II diabetes** is becoming **increasingly common** in the UK. This has been linked to increasing levels of **obesity**, a move towards more **unhealthy diets** and **low levels of physical activity**.
2) Type II diabetes can cause **additional health problems**, including visual impairment and kidney failure, so **health advisors** are understandably keen to **educate people** about the **risks** and **reduce** the **incidence** of the disease. Some people also think the **food industry** has a role to play in tackling the problem.
3) You need to understand the various **responses** to the increase in Type II diabetes and be able to **evaluate** them.

To **reduce** the **risk** of developing Type II diabetes, **health advisors** recommend that people:

- eat a **diet** that's **low** in **fat**, **sugar** and **salt**, with plenty of **whole grains**, **fruit** and **vegetables**,
- take **regular exercise**,
- **lose weight** if necessary.

Campaigns like the NHS's '**Change4Life**', aim to **educate people** on how to have a **healthier diet** and **lifestyle**, and so reduce their risk of developing conditions like Type II diabetes.

Health advisors have also challenged the food industry to **reduce** the **advertising** of **junk food** (particularly to children), to **improve** the **nutritional value** of their products, and to use **clearer labelling** on products — allowing consumers to make **healthier choices** about what to buy.

In **response** to criticism, some **food companies** have attempted to make their products more **healthy**, e.g.

- using **sugar alternatives** to sweeten food/drinks,
- **reducing** the **sugar**, **fat** and **salt** content of products.

Some people believe that diet varieties are not as good for health as they are claimed to be, e.g. there is some evidence to suggest that artificial sweeteners are linked to weight gain.

However, there is **pressure** on companies to **increase profits**. They say that the industry will only respond fully in the **long term**, as **public perception** about healthy eating **changes**.

Control of Blood Glucose Concentration

Colorimetry is Used to Determine the Concentration of a Glucose Solution

Normally, the **concentration** of **glucose** in **urine** is very **low** — between **0** and **0.8 mM**. **Higher** concentrations than this may indicate **diabetes** (although a blood test would be needed to confirm it). You need to be able to **determine** the **concentration** of **glucose** in a **'urine' sample**, using **colorimetry**. Here's how:

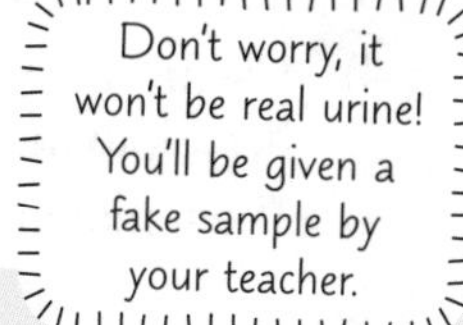

1) **Quantitative Benedict's reagent** is **different** to normal Benedict's reagent. When heated with glucose, the **initial blue colour** is **lost**, but a **brick-red precipitate** is **not** produced.
2) You can use a **colorimeter** to measure the **light absorbance** of the solution **after** the quantitative Benedict's test has been carried out.
3) The **higher** the **concentration** of **glucose**, the **more blue colour** will be **lost** (i.e. the paler the solution will become), **decreasing** the **absorbance** of the solution.

Don't worry, it won't be real urine! You'll be given a fake sample by your teacher.

This is How You Do it:

Initially you need to make up several glucose solutions of **different, known concentrations**. You can do this using a **serial dilution** technique:

This is how you'd make **five serial dilutions** with a **dilution factor of 2**, starting with an initial glucose concentration of **4 mM**...

1) Line up five **test tubes** in a rack.
2) Add **10 cm^3** of the initial **4 mM glucose solution** to the first test tube and **5 cm^3 of distilled water** to the other four test tubes.
3) Then, using a pipette, draw **5 cm^3** of the solution from the **first** test tube, add it to the distilled water in the **second** test tube and **mix** the solution **thoroughly**. You now have **10 cm^3** of solution that's **half as concentrated** as the solution in the first test tube (it's **2 mM**).
4) Repeat this process **three more times** to create solutions of **1 mM**, **0.5 mM** and **0.25 mM**.

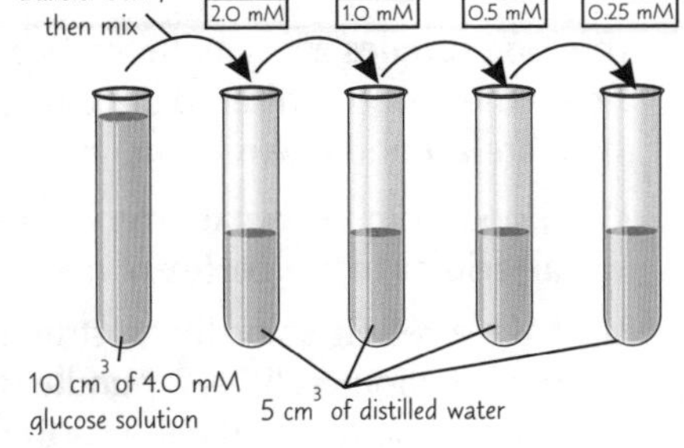

Once you've got your glucose solutions, you need to make a **calibration curve**. Here's how:

1) Do a **quantitative Benedict's test** on each solution (plus a **negative control** of **pure water**). Use the **same amount** of Benedict's solution in each case.
   To do the quantitative Benedict's test, you add quantitative Benedict's reagent to a sample and heat it in a water bath that's been brought to the boil.
2) Use a **colorimeter** (with a **red filter**) to measure the **absorbance** of the Benedict's solution **remaining** in each tube.
3) Use the results to make the **calibration curve**, showing absorbance against glucose concentration.

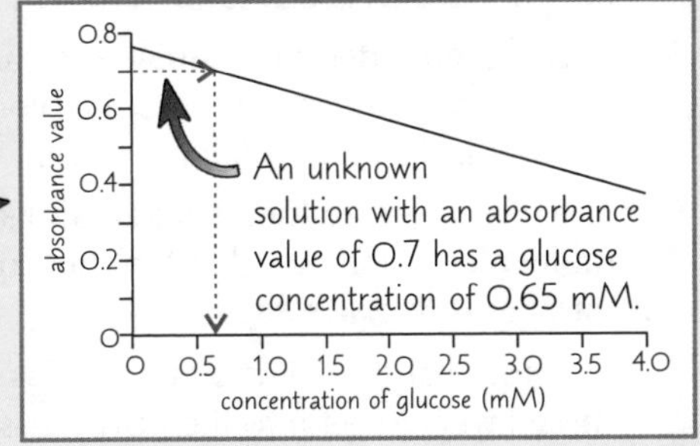

Then you can test the **unknown solution**, i.e. the **'urine' sample**, in the same way as the known concentrations, and use the calibration curve to find the concentration of glucose in the sample.

Warm-Up Questions

Q1 Briefly describe how you would produce a calibration curve to find the concentration of glucose in an unknown solution.

Exam Question

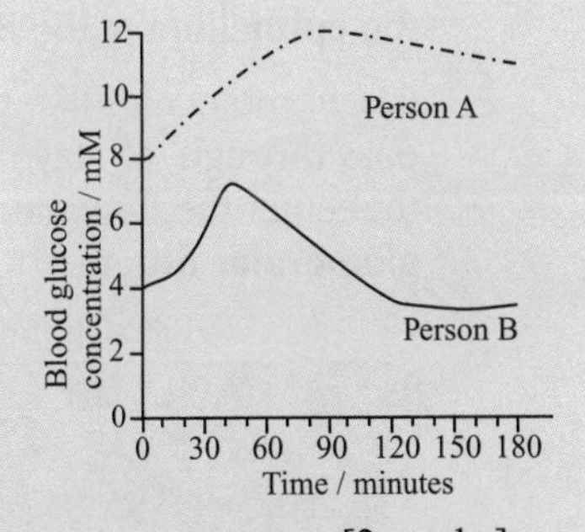

Q1 A glucose tolerance test is a medical test that can indicate the presence of diabetes. After fasting for 12 hours, a drink containing glucose is consumed. The graph shows how the blood glucose concentration of two people changed after having the drink. Person A has Type II diabetes. Person B does not have diabetes.

a) Give two pieces of evidence from the graph that suggest person A has diabetes. [2 marks]

b) Person A produces enough insulin but can't control their blood glucose concentration. Explain why. [2 marks]

Benedict's reagent makes you happy — it causes a loss of blues...

Evaluating can be tricky — you've got to give evidence to support your statements and look at both sides. You can't just say things like 'if only the food industry wasn't producing all that delicious junk food and making us all obese'.

The Kidneys

The kidneys make your urine by filtering waste products out of your blood and reabsorbing the useful stuff.

The **Kidneys Excrete Waste** and **Regulate Blood Water Potential**

1) One of the main functions of the kidneys is to **excrete waste products**, such as **urea**.
2) The kidneys also **regulate** the **water potential** of the **blood** — see pages 164-165.
3) As the **blood** passes through **capillaries** in the **cortex** (outer layer) of the **kidneys**, substances are **filtered out** of the blood and into **long tubules** that surround the capillaries. This process is called **ultrafiltration**.
4) Useful substances, such as **glucose** and the right amount of **water**, are then **reabsorbed** back into the **blood**. This process is called **selective reabsorption**.
5) The remaining **unwanted** substances pass along to the **bladder** and are excreted as **urine**.

Blood is **Filtered** at the **Start** of the **Nephrons**

The **long tubules** along with the bundle of **capillaries** where the blood is **filtered** are called **nephrons** — there are **around one million** nephrons in each kidney.

1) Blood from the **renal artery** enters smaller **arterioles** in the **cortex** of the kidney.
2) Each arteriole splits into a structure called a **glomerulus** (plural, glomeruli) — a **bundle** of **capillaries** looped inside a hollow ball called a **Bowman's capsule**.
3) This is where **ultrafiltration** takes place.
4) The **arteriole** that takes blood **into** each glomerulus is called the **afferent** arteriole, and the arteriole that takes the filtered blood **away** from the glomerulus is called the **efferent** arteriole.
5) The **efferent** arteriole is **smaller** in **diameter** than the afferent arteriole, so the blood in the glomerulus is under **high pressure**.
6) The high pressure **forces liquid** and **small molecules** in the blood **out** of the **capillary** and **into** the **Bowman's capsule**.
7) The liquid and small molecules pass through **three** layers to get into the Bowman's capsule and **enter** the nephron **tubules** — the **capillary wall**, a membrane (called the **basement membrane**) and the **epithelium** of the Bowman's capsule.
8) Larger molecules like **proteins** and **blood cells can't pass through**, so **stay** in the blood. The substances that enter the Bowman's capsule are known as the **glomerular filtrate**.

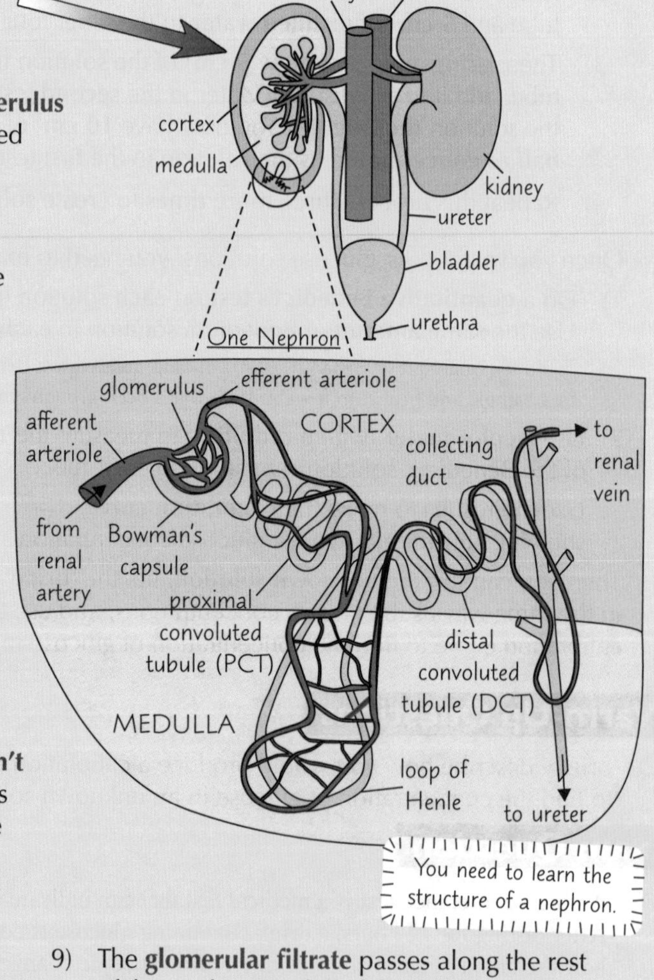

You need to learn the structure of a nephron.

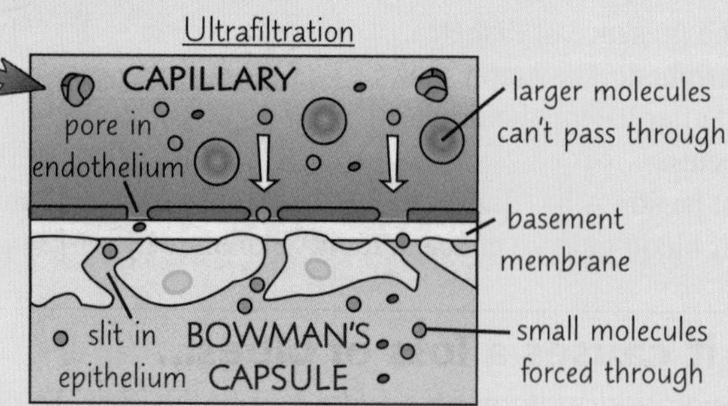

9) The **glomerular filtrate** passes along the rest of the nephron and **useful substances** are **reabsorbed** along the way — see next page.
10) Finally, the filtrate flows through the **collecting duct** and passes out of the kidney along the **ureter**.

The Kidneys

Useful Substances are Reabsorbed Along the Nephron Tubules

1) **Selective reabsorption** takes place as the glomerular filtrate flows along the **proximal convoluted tubule** (**PCT**), through the **loop of Henle**, and along the **distal convoluted tubule** (**DCT**).
2) **Useful substances** leave the tubules of the nephrons and **enter** the capillary network that's **wrapped** around them (see diagram on previous page).
3) The **epithelium** of the wall of the PCT has **microvilli** to provide a **large surface area** for the **reabsorption** of useful materials from the **glomerular filtrate** (in the tubules) into the **blood** (in the capillaries).
4) Useful solutes, like **glucose**, are reabsorbed along the PCT by **active transport** and **facilitated diffusion**.
5) **Water** enters the blood by **osmosis** because the **water potential** of the blood is **lower** than that of the filtrate. Water is reabsorbed from the **PCT**, **loop of Henle**, **DCT** and the **collecting duct** (see next page).
6) The filtrate that remains is **urine**, which passes along the **ureter** to the **bladder**.

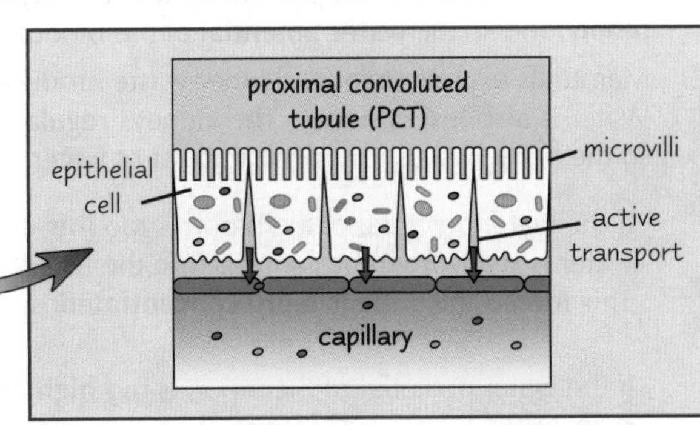

Remember: water potential describes the tendency of water to move from one area to another. Water will move from an area of higher water potential to an area of lower water potential — it moves down the water potential gradient.

Urine is usually **made up of**:
- **Water** and **dissolved salts**.
- **Urea**.
- Other substances such as **hormones** and **excess vitamins**.

Urine **doesn't** usually contain:
- **Proteins** and **blood cells** — they're **too big** to be **filtered out** of the blood.
- **Glucose** because it's **actively reabsorbed** back into the blood (see above).

The volume of water in urine varies depending on how much you've drunk (see p. 165).

Warm-Up Questions

PRACTICE QUESTIONS

Q1 What is selective reabsorption?

Q2 Which blood vessel supplies the kidney with blood?

Q3 What are the bundles of capillaries found in the cortex of the kidneys called?

Q4 By which two processes is glucose reabsorbed from the PCT?

Exam Question

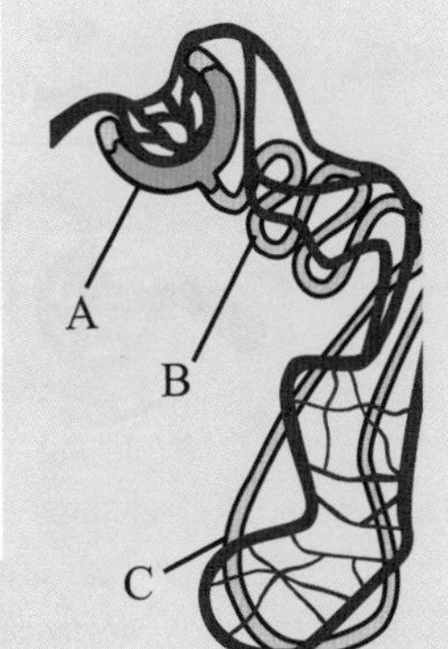

Q1 The diagram on the right shows part of a nephron.

a) Explain how glomerular filtrate is formed at point A. [2 marks]

b) Would you expect the concentration of glucose to be lower at point B or point C on the diagram? Explain your answer. [1 mark]

c) The rate at which the kidneys filter blood is called the glomerular filtration rate (GFR). GFR is normally around 6300 cm^3 $hour^{-1}$. Tests revealed 0 mg of glucose in a person's urine. The same person's blood glucose concentration was 0.9 mg cm^{-3}. Assuming a normal GFR, calculate the rate at which glucose is reabsorbed back into the blood. Give your answer in mg min^{-1}. [1 mark]

Mmm — it's steak and excretion organ pie for dinner...

Excretion is a pretty horrible sounding word I know, but it's gotta be done. Mind you, I've never been able to eat kidney ever since I learnt all about this urine production business. Shame really — I used to love kidney sarnies for lunch. Make sure you can describe how the glomerular filtrate is formed and how glucose and water are reabsorbed.

Controlling Blood Water Potential

The kidneys control the water potential of the blood — osmoregulation, if you're being posh.

The **Kidneys** Regulate the **Water Potential** of the **Blood**

1) Water is **essential** to keep the body **functioning**, so the **amount** of water in the **blood** (and so the **water potential** of the blood) needs to be kept **constant**.
2) Mammals excrete **urea** (and other waste products) in **solution**, which means **water** is **lost** during excretion. Water is also lost in **sweat**. The kidneys **regulate** the **water potential** of the blood (and urine), so the body has just the **right amount** of water — this is called **osmoregulation**.

If the water potential of the blood is too **low** (the body is **dehydrated**), **more** water is **reabsorbed** by osmosis **into** the blood from the tubules of the nephrons. This means the urine is **more concentrated**, so **less** water is **lost** during excretion.

If the water potential of the blood is too **high** (the body is too **hydrated**), **less** water is **reabsorbed** by osmosis **into** the blood from the tubules of the nephrons. This means the urine is **more dilute**, so **more** water is **lost** during excretion (see next page).

Brad liked his urine to be dilute.

3) Water is **reabsorbed** into the blood along **almost all** of the **nephron** (see previous page), but **regulation of water potential** mainly takes place in the **loop of Henle**, **DCT** and **collecting duct**. The **volume** of water reabsorbed by the DCT and collecting duct is controlled by **hormones** (see next page).

The **Loop of Henle** Maintains a **Sodium Ion Gradient**

The **loop of Henle** is located in the **medulla** (inner layer) of the kidneys. It's made up of two '**limbs**' — the **descending** limb and the **ascending** limb. The limbs control the movement of **sodium ions** so that **water** can be **reabsorbed** by the blood.

(1) Near the **top** of the **ascending** limb, Na^+ ions are **pumped out** into the **medulla** using **active transport**. The ascending limb is **impermeable** to **water**, so the water **stays inside** the tubule. This creates a **low water potential** in the **medulla**, because there's a **high concentration** of ions.

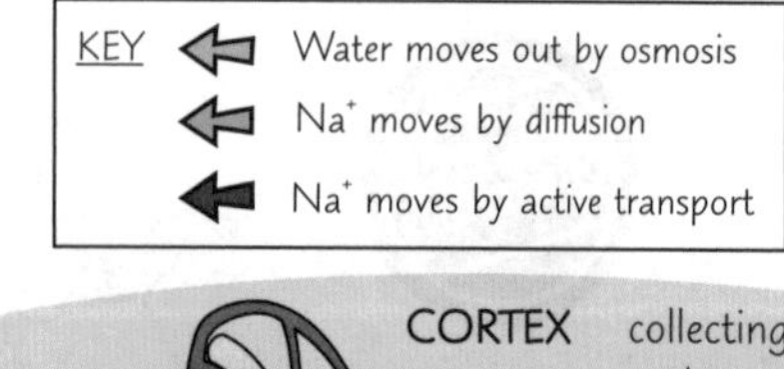

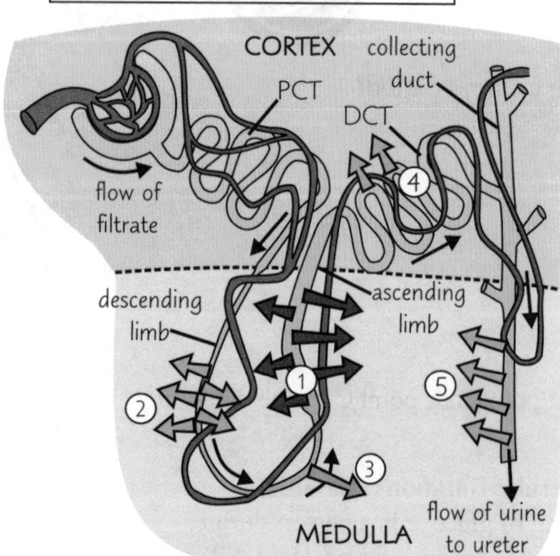

(2) Because there's a **lower** water potential in the **medulla** than in the descending limb, **water** moves **out** of the **descending limb** (which is permeable to water) **into** the **medulla** by **osmosis**. This makes the **filtrate more concentrated** (the ions can't diffuse out — the descending limb isn't permeable to them). The water in the medulla is **reabsorbed** into the **blood** through the **capillary network**.

(3) Near the **bottom** of the **ascending** limb Na^+ ions **diffuse out** into the **medulla**, further **lowering** the **water potential** in the medulla. The ascending limb is **impermeable** to **water**, so it **stays in the tubule**.

(4) Water moves out of the **distal convoluted tubule** (DCT) by osmosis and is reabsorbed into the blood.

(5) The first three stages massively **increase** the **ion concentration** in the **medulla**, which **lowers** the **water potential**. This causes **water** to **move out** of the **collecting duct** by **osmosis**. As before, the water in the medulla is **reabsorbed** into the **blood** through the **capillary network**.

The **volume** of water **reabsorbed** into the capillaries is **controlled** by **changing the permeability** of the **DCT** and the **collecting duct** (see next page).

Controlling Blood Water Potential

Water Reabsorption is Controlled by Hormones

1) The water potential of the blood is **monitored** by cells called **osmoreceptors** in a part of the **brain** called the **hypothalamus**.
2) When the water potential of the blood **decreases**, water will move **out** of the osmoreceptor cells by **osmosis**. This causes the cells to **decrease** in **volume**. This sends a signal to other cells in the **hypothalamus**, which send a signal to the **posterior pituitary gland**. This causes the posterior pituitary to release a **hormone** called **antidiuretic hormone** (ADH) into the blood.
3) ADH makes the walls of the DCT and collecting duct **more permeable** to **water**.
4) This means **more water** is **reabsorbed** from these tubules **into** the medulla and into the blood by osmosis. A **small** amount of **concentrated urine** is produced, which means **less water** is **lost** from the body.

It's called antidiuretic hormone because diuresis is when lots of dilute urine is produced, so anti means a small amount of concentrated urine is produced.

Here's how ADH changes the **water content** of the **blood** when it's too **low** or too **high**:

1) **Blood ADH** Level **Rises** When You're **Dehydrated**

Dehydration is what happens when you **lose water**, e.g. by sweating during exercise, so the **water content** of the blood needs to be **increased**:

1) The **water content** of the blood **drops**, so its **water potential drops**.
2) This is detected by **osmoreceptors** in the **hypothalamus**.
3) The **posterior pituitary gland** is stimulated to release **more ADH** into the blood.
4) **More ADH** means that the DCT and collecting duct become **more permeable**, so **more water** is **reabsorbed** into the blood by osmosis.
5) A **small amount** of **highly concentrated** urine is produced and **less water** is **lost**.

Dehydrated? Me? As if...

2) **Blood ADH** Level **Falls** When You're **Hydrated**

If you're **hydrated**, you've taken in **lots of water**, so the **water content** of the blood needs to be **reduced**:

1) The **water content** of the blood **rises**, so its **water potential rises**.
2) This is detected by the **osmoreceptors** in the **hypothalamus**.
3) The **posterior pituitary gland** releases **less ADH** into the blood.
4) **Less ADH** means that the DCT and collecting duct become **less permeable**, so **less water** is **reabsorbed** into the blood by osmosis.
5) A **large amount** of **dilute** urine is produced and **more water** is **lost**.

Warm-Up Questions

Q1 Describe what happens along the descending limb of the loop of Henle.

Q2 Which cells monitor the water content of the blood?

PRACTICE QUESTIONS

Exam Questions

Q1 The level of ADH in the blood rises during strenuous exercise.
   a) Explain the cause of the increase in ADH. [4 marks]
   b) Explain the effect that the increased ADH levels have on kidney function. [2 marks]

Q2 Gerbils have longer loops of Henle than mice.
   Suggest and explain how this helps gerbils to produce less urine than mice. [4 marks]

If you don't understand what ADH does, ur-ine trouble...

There are two main things to learn here — how a sodium ion gradient lets the kidneys reabsorb so much water into the blood and how the water content of the blood is regulated by osmoreceptors in the hypothalamus. Now I need a wee.

Extra Exam Practice

Before you tick Topic 6 off your list, have a go at these questions mixing different parts of the section.

- Have a look at this example of how to answer a tricky exam question.
- Then check how much you've understood from Topic 6 by having a go at the questions on the next page.

1 The autonomic nervous system controls heart rate in response to internal stimuli.

1.1 Pain can increase heart rate. Pain can be felt following stimulation of receptors called nociceptors. Stimulation of a nociceptor increases the permeability of its membrane to ions. Suggest an explanation for how stimulation of nociceptors leads to the brain rapidly sensing pain.

(3 marks)

Changes in heart rate affect the volume of blood that is pumped around the body in a given time. **Figure 1** shows the volume of blood in the left ventricle of a person over time.

Figure 1

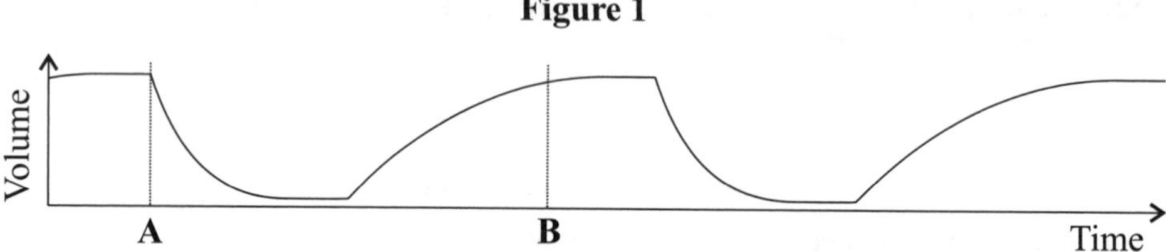

1.2 With reference to the electrical activity of the heart, explain the change in the volume of blood in the ventricle between points **A** and **B** in **Figure 1**.

(4 marks)

1.3 The action of acetylcholine in the heart reduces the rate at which cells in the SAN depolarise. Atropine is a drug that blocks cholinergic receptors on the SAN. Suggest how atropine causes an increase in heart rate.

(3 marks)

1.1

The increased permeability of the membrane to ions will alter the **potential difference** of the membrane. This will lead to a **generator potential** big enough to trigger an **action potential**. Action potentials will travel along sensory neurones to the brain, which will interpret the signals as feelings of pain.

You're not expected to know anything specific about nociceptors — you just have to apply what you know about what happens when a receptor is stimulated.

1.2

At **point A**, a wave of electrical activity is passing through the **Purkyne tissue**. This causes the ventricle to contract and push blood out of the heart, until the volume drops to the **lowest point**. The **volume** then **plateaus** until the **SAN** sends out another wave of electrical activity that stimulates the atrium to contract, pushing blood back into the ventricle and increasing the volume of blood it contains, until it reaches **point B**.

Don't get carried away writing more than you need to — you've been asked to describe the changes between points A and B so only mention this section of the graph.

You'll need to mention these specific structures in the heart to pick up marks.

1.3

Atropine will prevent acetylcholine from affecting the SAN, as it blocks the **cholinergic receptors** that **acetylcholine acts upon**. Therefore, neurotransmitters from the **sympathetic branch** of the autonomic nervous system will have **more of an effect** on the rate at which the SAN depolarises, so **heart rate** will **increase**.

Always double-check you're answering the question — here you've got to explain why heart rate might increase (not just why it might be prevented from decreasing).

From your knowledge of cholinergic synapses you should know that acetylcholine works at cholinergic receptors.

Extra Exam Practice

2 It is estimated that over 4 million people in the UK have diabetes.

Diabetic retinopathy is an eye condition associated with diabetes. It can cause blurred vision by damaging the retina. The extent of vision impairment partly depends on which areas of the retina are most damaged, as different areas of the retina have different proportions of cones and rods.

2.1 With reference to the connections cones and rods make to the optic nerve, explain why vision may be more blurred if diabetic retinopathy affects areas of the retina with a greater proportion of cones than rods.

(3 marks)

SGLT2 inhibitors are drugs that are used to treat type 2 diabetes. They block glucose transport proteins in the proximal convoluted tubules of the kidneys.

2.2 Explain how SGLT2 inhibitors will affect the glucose concentration in the urine.

(2 marks)

Colorimetry and a glucose calibration curve were used to determine the glucose concentration in the urine of a patient taking SGLT2 inhibitors. **Figure 2** shows the glucose calibration curve used.

Figure 2

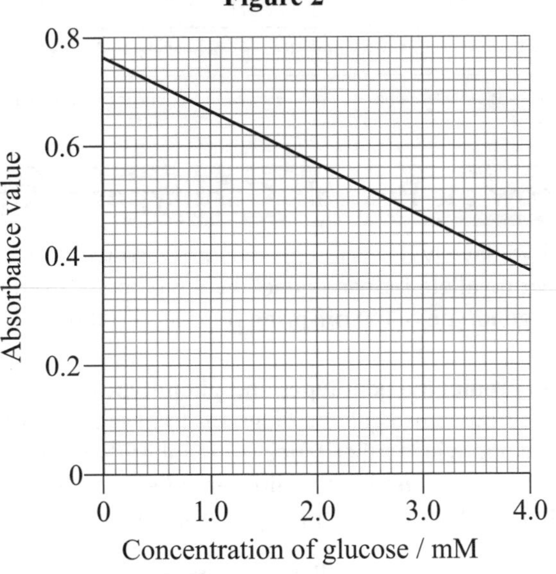

2.3 The absorbance value of the urine was 0.5. The colorimeter had an uncertainty of ±0.05. Using **Figure 2** and the uncertainty, determine the range in which the glucose concentration of this sample may lie.

(2 marks)

3 A scientist investigated muscle contraction during exercise.

3.1 Explain how the phosphocreatine stored in a cell can be used during muscle contraction.

(2 marks)

3.2 Explain how the movement of calcium ions lengthens the sarcomere following muscle contraction.

(3 marks)

3.3 Repeated muscle contraction during exercise can lead to a reduction in blood water potential. Diabetes insipidus is a condition in which a person secretes lower levels of ADH than normal. Would a person with diabetes insipidus be more or less at risk of dehydration during exercise than a person without the condition? Explain your answer.

(4 marks)

Inheritance

Nope, this isn't about who gets Mum's best china — we're talking genetic inheritance here...

You **Need to Know** These **Genetic Terms**

'Codes for' means 'contains the instructions for'.

TERM	DESCRIPTION
Gene	A sequence of bases on a DNA molecule that codes for a protein (polypeptide), which results in a characteristic, e.g. a gene for eye colour.
Allele	A different version of a gene. There can be many different alleles of a single gene, but most plants and animals, including humans, only carry two alleles of each gene, one from each parent. The order of bases in each allele is slightly different — they code for different versions of the same characteristic. They're represented using letters, e.g. the allele for brown eyes (B) and the allele for blue eyes (b).
Genotype	The genetic constitution of an organism — the alleles an organism has, e.g. BB, Bb or bb for eye colour.
Phenotype	The expression of the genetic constitution and its interaction with the environment — an organism's characteristics, e.g. brown eyes.
Dominant	An allele whose characteristic appears in the phenotype even when there's only one copy. Dominant alleles are shown by a capital letter. E.g. the allele for brown eyes (B) is dominant — if a person's genotype is Bb or BB, they'll have brown eyes.
Recessive	An allele whose characteristic only appears in the phenotype if two copies are present. Recessive alleles are shown by a lower case letter. E.g. the allele for blue eyes (b) is recessive — if a person's genotype is bb, they'll have blue eyes.
Codominant	Alleles that are both expressed in the phenotype — neither one is recessive, e.g. the alleles for haemoglobin.
Locus	The fixed position of a gene on a chromosome. Alleles of a gene are found at the same locus on each chromosome in a pair.
Homozygote	An organism that carries two copies of the same allele, e.g. BB or bb.
Heterozygote	An organism that carries two different alleles, e.g. Bb.
Carrier	A person carrying an allele which is not expressed in the phenotype but that can be passed on to offspring.

Genetic Diagrams Show the **Possible Genotypes** of **Offspring**

Humans are **diploid** organisms (we have two sets of chromosomes) so we have **two alleles** for **each gene**. **Gametes** (sex cells) contain only **one allele** for each gene. When gametes from two parents fuse together, the alleles they contain form the **genotype** of the **offspring** produced. At each **locus**, the genotype can be **homozygous** or **heterozygous**.

Genetic diagrams can be used to **predict** the **genotypes** and **phenotypes** of the offspring produced if two parents are **crossed** (bred). You need to know how to use genetic diagrams to interpret or predict the results of various crosses, including **monohybrid crosses**. **Monohybrid inheritance** is the inheritance of a **characteristic** controlled by a **single gene**. **Monohybrid crosses** show the **likelihood** of the **different alleles** of that gene (and so different versions of the characteristic) being **inherited** by offspring of certain parents. This genetic diagram shows how **wing length** is inherited in fruit flies:

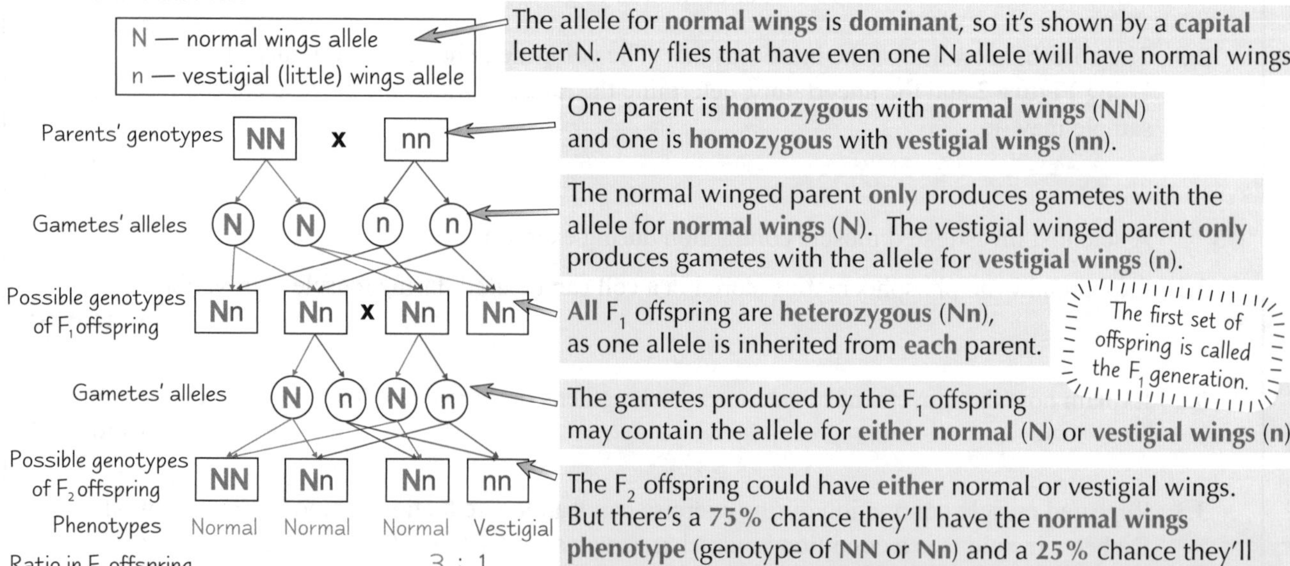

Inheritance

A **Punnett square** is just another way of showing a **genetic diagram** — they're also used to predict the **genotypes** and **phenotypes** of offspring. The Punnett squares below show the same crosses from the previous page:

1) First work out the alleles the **gametes** would have.

Parents' genotypes: NN nn

Gametes' alleles: N N n n

2) Next **cross the parents' gametes** to show the possible genotypes of the F_1 **generation** — **all heterozygous**, Nn.

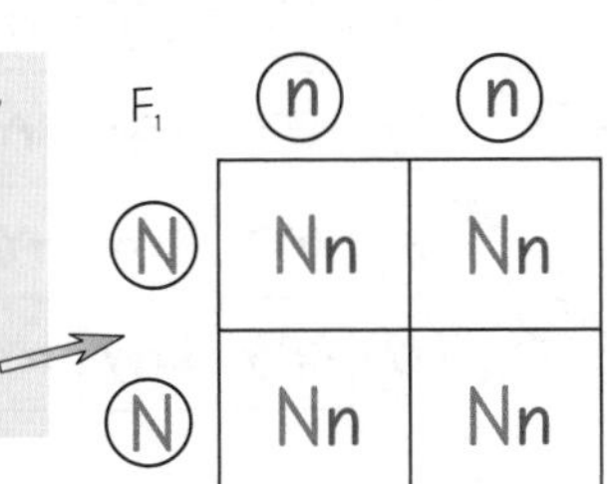

3) Then **cross the gametes' alleles** of the F_1 **generation** to show the possible **genotypes** of the F_2 **generation**. The Punnett square shows a **75%** chance that offspring will have **normal wings** and a **25%** chance that they'll have **vestigial wings**, i.e. a **3:1 ratio**.

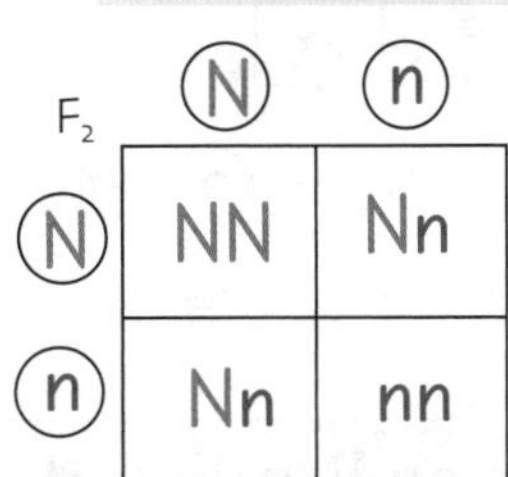

- 1 in 4 chance of offspring having the genotype NN (normal wings)
- 2 in 4 chance of offspring having the genotype Nn (normal wings)
- 1 in 4 chance of offspring having the genotype nn (vestigial wings)
- So, phenotype ratio normal:vestigial = 3:1

Some **Genes** Have **Codominant Alleles**

Occasionally, alleles show **codominance** — **both alleles** are expressed in the **phenotype**, **neither one** is recessive. One example in humans is the allele for **sickle-cell anaemia**:

1) People who are **homozygous** for **normal haemoglobin** (H^NH^N) don't have the disease.
2) People who are **homozygous** for **sickle haemoglobin** (H^SH^S) have **sickle-cell anaemia** — all their **blood cells** are **sickle-shaped** (crescent-shaped).
3) People who are **heterozygous** (H^NH^S) have an **in-between** phenotype, called the **sickle-cell trait** — they have **some** normal haemoglobin and some sickle haemoglobin. The two alleles are **codominant** because they're **both** expressed in the **phenotype**.
4) The **genetic diagram** on the right shows the possible offspring from **crossing** two parents with **sickle-cell trait** (**heterozygous**).

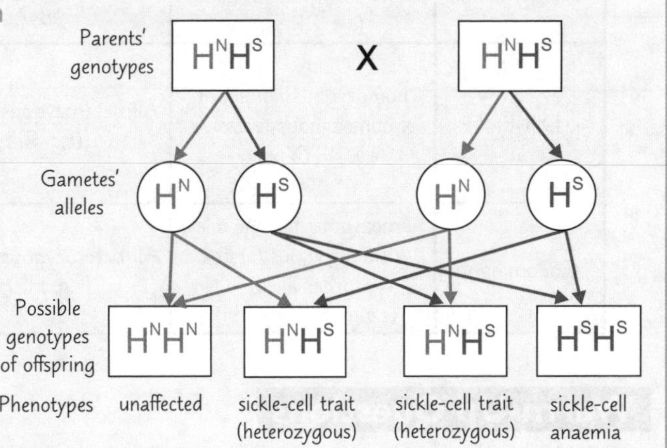

Some **Genes** Have **Multiple Alleles**

Inheritance is **more complicated** when there are **more than two** alleles of the same gene (**multiple alleles**).

Example In the **ABO blood group system** in humans there are **three alleles** for blood type:

I^O is the allele for blood group **O**. I^A is the allele for blood group **A**. I^B is the allele for blood group **B**.

Allele I^O is **recessive**. Alleles I^A and I^B are **codominant** — people with genotype I^AI^B will have blood group **AB**.

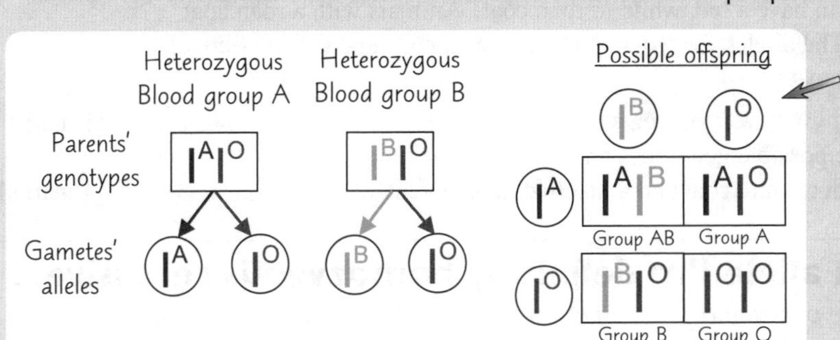

The genetic diagram shows a cross between a **heterozygous** person with blood group **A** and a **heterozygous** person with blood group **B**. Any offspring could have one of **four** different blood groups — **A**, **B**, **O** or **AB**.

Recessive blood groups are normally really rare, but it just so happens that loads of people in Britain are descended from people who were $I^O I^O$, so O's really common.

Inheritance

Genetic Diagrams can Show how More Than One Characteristic is Inherited

You can use genetic diagrams to work out the chances of offspring inheriting certain **combinations** of characteristics. For example, you can use a **dihybrid cross** to look at how **two different genes** are inherited at the same time. The diagram below is a **dihybrid cross** showing how seed texture **and** colour are inherited in **pea plants**.

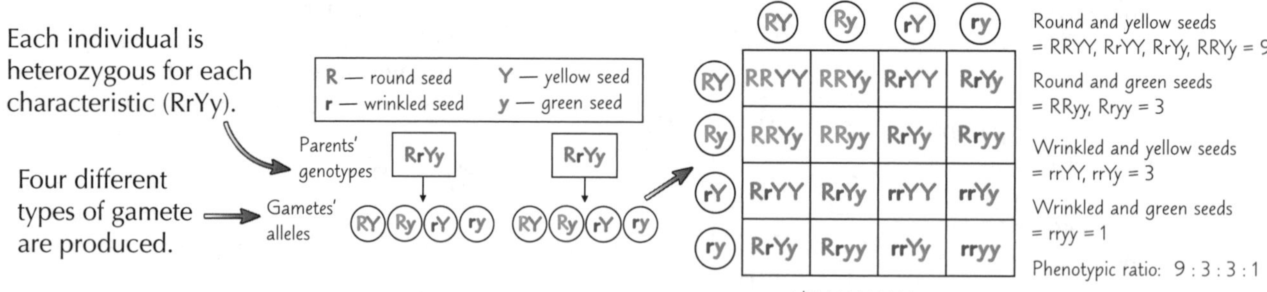

You can also do dihybrid crosses with codominant alleles. They work in the same way as this one but the phenotypic ratios produced are different.

Phenotypic Ratios can be Predicted

The **phenotypic ratio** is the **ratio** of **different phenotypes** in offspring.
Genetic diagrams allow you to **predict** the phenotypic ratios in $\mathbf{F_1}$ **and** $\mathbf{F_2}$ **offspring**.
Here's a handy summary table of ratios for the following crosses:

Type of Cross	Parents	Phenotypic Ratio in F_1	Phenotypic Ratio in F_2
Monohybrid	homozygous dominant x homozygous recessive (e.g. RR x rr)	All heterozygous offspring (e.g. Rr)	3 : 1 dominant : recessive
Dihybrid	homozygous dominant x homozygous recessive (e.g. RRYY x rryy)	All heterozygous offspring (e.g. RrYy)	9 : 3 : 3 : 1 dominant both : dominant 1st recessive 2nd : recessive 1st dominant 2nd : recessive both
Codominant	homozygous for one allele x homozygous for the other allele (e.g. H^NH^N x H^SH^S)	All heterozygous offspring (e.g. H^NH^S)	1 : 2 : 1 homozygous for one allele : heterozygous for the other allele

Sometimes you **won't** get the **expected** (predicted) phenotypic ratio — it'll be quite different. This can be because of **sex linkage**, **autosomal linkage** or **epistasis** — all of which are covered on pages 171-173.

Warm-Up Questions

Q1 What is meant by the term genotype?
Q2 What is meant by the term phenotype?
Q3 What does a dihybrid cross show you?

PRACTICE QUESTIONS

Exam Questions

Q1 In pea plants, seed texture (round or wrinkled) is passed from parent to offspring by monohybrid inheritance. The allele for round seeds is represented by R and the allele for wrinkled seeds is represented by r.

Draw a genetic diagram to show the possible genotypes of F_1 offspring produced by crossing a homozygous round-seed pea plant with a homozygous wrinkled seed pea plant. [3 marks]

Q2 Individuals of a particular breed of cow can have a red, white or roan coat. Animals with a roan coat have patches of both red and white hair. The alleles for red and white coats are C^R and C^W respectively. Heterozygotes for these alleles have roan coats.

a) Explain why heterozygotes for C^R and C^W have roan coats. [1 mark]

b) Draw a genetic diagram to predict the possible genotypes and phenotypes of the F_1 offspring produced by a parent with a white coat and a heterozygous parent. [4 marks]

If there's a dominant revision allele I'm definitely homozygous recessive...

OK, so there are a lot of fancy words on these pages and yes, you do need to know them all. Sorry. But don't despair — once you've learnt what the words mean and know how genetic diagrams work it'll all just fall into place.

Linkage and Epistasis

Right, this stuff is fairly hard, so if you don't get it first time don't panic just work through it again until you do...

Some **Characteristics** are **Sex-linked**

1) The genetic information for **gender** is carried on two **sex chromosomes**.
2) In mammals, **females** have **two X** chromosomes (XX) and **males** have **one X** and **one Y** chromosome (XY).
3) A **characteristic** is said to be **sex-linked** when the allele that codes for it is located on a **sex chromosome**.
4) The **Y chromosome** is **smaller** than the X chromosome and carries **fewer genes**. So most genes on the sex chromosomes are **only carried** on the X chromosome (called **X-linked** genes).
5) As **males** only have **one X chromosome**, they often only have **one allele** for sex-linked genes. So because they **only** have one copy, they **express** the **characteristic** of this allele even if it's **recessive**. This makes males **more likely** than females to show **recessive phenotypes** for genes that are sex-linked.
6) Genetic disorders caused by **faulty alleles** on sex chromosomes include **colour blindness** and **haemophilia**. The faulty alleles for both of these disorders are carried on the X chromosome — they're called **X-linked disorders**.

Example

1) **Colour blindness** is a **sex-linked disorder** caused by a faulty allele carried on the **X** chromosome.
2) As it's sex-linked **both** the chromosome and the allele are **represented** in the **genetic diagram**, e.g. X^n, where **X** represents the **X chromosome** and **n** the **faulty allele** for **colour vision**.
3) The **Y chromosome** doesn't have an allele for colour vision so is **just** represented by **Y**.
4) **Females** would need **two copies** of the **recessive allele** to be colour blind, while **males** only need **one copy**. This means colour blindness is **much rarer** in **women** than **men**.

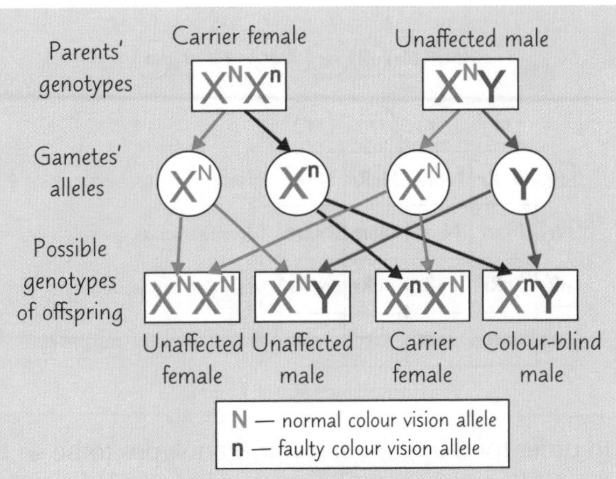

7) In the example above, there's a **3 : 1 ratio** of offspring **without** colour blindness : offspring **with** colour-blindness. But when a **female carrier** and a **male without colour-blindness** have children (as in this example), only their **male offspring** are at risk of being colour-blind. So you can also say that there's a predicted **2 : 1 : 1 ratio** — of **female** offspring **without** colour-blindness : **male** offspring **without** colour-blindness : **male** offspring **with** colour-blindness.
8) This ratio will **change** if a female carrier (X^NX^n) and a male **with** colour-blindness (X^nY) have children. The predicted ratio will then be **1 : 1** — of offspring **with** colour-blindness : offspring **without** colour-blindness. The ratio will be the **same** for offspring of **each gender**. You only end up with this predicted ratio for a monohybrid F_2 cross with a **sex-linked characteristic**.

Some **Autosomal** Genes are **Linked**

1) **Autosome** is the fancy name for any chromosome that **isn't** a sex chromosome. **Autosomal genes** are the genes located on the autosomes.
2) Genes on the **same autosome** are said to be **linked** — because they're on the same autosome they'll stay together during the **independent segregation of chromosomes** in meiosis I, and their alleles will be **passed on to the offspring together**. The only reason this won't happen is if **crossing over** splits them up first.
3) The **closer together** two genes are on the autosome, the **more closely** they are said to be **linked**. This is because **crossing over** is **less likely** to split them up.
4) If two genes are autosomally linked, you **won't get** the phenotypic ratio you expect in the offspring of a cross.
5) For example, in a **dihybrid cross** between two heterozygous parents you'd expect a **9 : 3 : 3 : 1 ratio** in the offspring (see previous page). Instead, the phenotypic ratio is more likely to be that expected for a **monohybrid cross** between two heterozygous parents (**3 : 1**) because the two autosomally-linked alleles are **inherited together**. This means that a **higher proportion** of the **offspring** will have their **parents'** (heterozygous) **genotype** and **phenotype**.
6) This allows you to use the **predicted phenotypic ratio** to **identify** autosomal linkage.

Crossing over is when two homologous (paired) chromosomes 'swap bits'. It happens in meiosis I before independent segregation. You'll have learnt about this in Year 1 of your course.

An autosome
Genes A, B and C are all linked.
Genes A and B are more closely linked than genes A and C.
A
B
C

Linkage and Epistasis

Genetic Cross Results Can Show Autosomal Linkage

In the exam you might get some **genetic cross results** that show **linkage** and have to explain them.

Example

A scientist was investigating **linkage** between the **genes** for **eye colour** and **wing length** in **fruit flies**. The gene for **normal wings** (**N**) is dominant to the gene for **vestigial wings** (**n**) and the gene for **red eyes** (**R**) is dominant to the gene for **purple eyes** (**r**). The **first cross** the scientist carried out was between flies **homozygous dominant** for both normal wings and red eyes (**NNRR**) and flies **homozygous recessive** for both vestigial wings and purple eyes (**nnrr**). The resulting offspring were all **heterozygous** for normal wings and red eyes (**NnRr**).

The **second cross** the scientist carried out was between these offspring (**NnRr**) and the flies homozygous recessive for vestigial wings and purple eyes (**nnrr**).

This is known as a back cross (crossing the offspring with one of the parents).

He expected a **1 : 1 : 1 : 1** ratio as shown below:

Parents' alleles: NnRr × nnrr

Gametes' alleles: (NR) (Nr) (nR) (nr) — (nr) (nr) (nr) (nr)

	nr	nr	nr	nr	
NR	NnRr	NnRr	NnRr	NnRr	Normal wings, red eyes = 4
Nr	Nnrr	Nnrr	Nnrr	Nnrr	Normal wings, purple eyes = 4
nR	nnRr	nnRr	nnRr	nnRr	Vestigial wings, red eyes = 4
nr	nnrr	nnrr	nnrr	nnrr	Vestigial wings, purple eyes = 4

Phenotypic ratio = 1 : 1 : 1 : 1

However, the results he got for this cross show a **8 : 1 : 1 : 8** ratio as in the **table**:

	Number of offspring
Normal wings, red eyes (NnRr)	1216
Normal wings, purple eyes (Nnrr)	152
Vestigial wings, red eyes (nnRr)	148
Vestigial wings, purple eyes (nnrr)	1184

Phenotypic ratio = 8 : 1 : 1 : 8

In order for the **NnRr** and **nnrr** genotypes to be so **common** in the offspring, the **NR** alleles and the **nr** alleles in the **NnRr parent** must have been **linked**. This means that the **NnRr** parent produced mostly **NR** and **nr gametes**. Some **Nr** and **nR gametes** were still made due to **crossing over**, but there were **fewer Nnrr** and **nnRr** offspring overall. As a result, a higher proportion of the offspring have their parents' phenotypes.

An Epistatic Gene Masks the Expression of Another Gene

1) **Many different genes** can control the **same** characteristic — they **interact** to form the phenotype.
2) This can be because the **allele** of one gene **masks** (blocks) **the expression** of the alleles of other genes — this is called **epistasis**.

Example 1 In humans a **widow's peak** (see picture) is controlled by one gene and **baldness** by others. If you have the **alleles** that code for baldness, it **doesn't matter** whether you have the allele for a widow's peak or not, as you have **no hair**. The baldness genes are **epistatic** to the widow's peak gene, as the baldness genes **mask** the expression of the widow's peak gene.

Example 2 **Flower pigment** in a plant is controlled by two genes. **Gene 1** codes for a **yellow pigment** (Y is the dominant yellow allele) and **gene 2** codes for an enzyme that **turns** the yellow pigment **orange** (R is the dominant orange allele). If you **don't have** the **Y** allele it **won't matter** if you have the R allele or not as the flower **will be colourless**. Gene 1 is **epistatic** to gene 2 as it can **mask** the expression of gene 2.

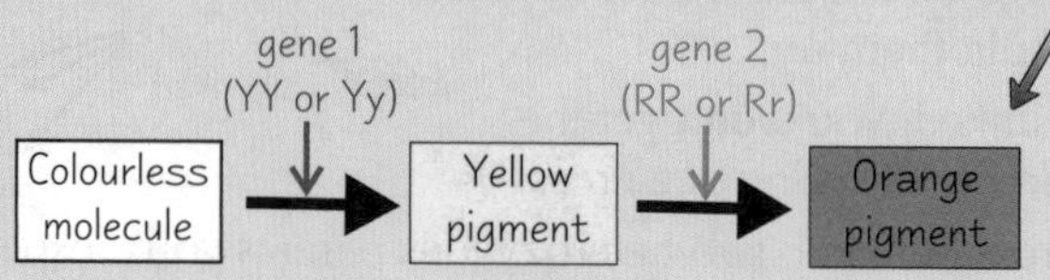

3) **Crosses** involving epistatic genes **don't result** in the **expected phenotypic ratios** given above, e.g. if you cross **two heterozygous orange** flowered plants (YyRr) from the above example you wouldn't get the expected **9 : 3 : 3 : 1** phenotypic ratio for a **normal dihybrid cross**.

Linkage and Epistasis

You can Predict the Phenotypic Ratios for Some Epistatic Genes

Just as you can **predict** the phenotypic ratios for a **normal dihybrid cross** (see page 170), you can predict the phenotypic ratios for dihybrid crosses involving some **epistatic genes** too:

A dihybrid cross involving a recessive epistatic allele — 9 : 3 : 4

Having **two copies** of the **recessive** epistatic allele **masks (blocks)** the expression of the **other gene**. If you cross a **homozygous recessive** parent with a **homozygous dominant** parent you will get a **9 : 3 : 4** phenotypic ratio of **dominant both : dominant epistatic recessive other : recessive epistatic** in the F_2 **generation.**

E.g. the **flower example** from the **previous page** is an example of a **recessive epistatic allele**. If a plant is **homozygous recessive** for the **epistatic gene** (yy) then it will be **colourless**, **masking** the expression of the orange gene. So if you cross homozygous parents, you should get a **9 : 3 : 4** ratio of **orange : yellow : white** in the F_2 **generation**. You can check the **phenotypic ratio** is right **using a genetic diagram**:

F_1 cross
YYRR × yyrr = all YyRr

F_2 cross (YyRr × YyRr)

	YR	Yr	yR	yr
YR	YYRR	YYRr	YyRR	YyRr
Yr	YYRr	YYrr	YyRr	Yyrr
yR	YyRR	YyRr	yyRR	yyRr
yr	YyRr	Yyrr	yyRr	yyrr

Orange = YYRR, YYRr, YyRR, YyRr = 9
Yellow = Yyrr, YYrr = 3
White = yyRR, yyRr, yyrr = 4
Phenotypic ratio: 9 : 3 : 4

A dihybrid cross involving a dominant epistatic allele — 12 : 3 : 1

Having **at least one** copy of the **dominant epistatic** allele **masks (blocks)** the expression of the other gene. Crossing a **homozygous recessive** parent with a **homozygous dominant** parent will produce a **12 : 3 : 1** phenotypic ratio of **dominant epistatic : recessive epistatic dominant other : recessive both** in the F_2 generation.

E.g. **squash colour** is controlled by two genes — the **colour epistatic gene (W/w)** and the **yellow gene (Y/y)**. The **no-colour, white** allele (**W**) is **dominant** over the **coloured** allele (**w**), so **WW** or **Ww** will be **white** and **ww** will be **coloured**. The yellow gene has the **dominant yellow** allele (**Y**) and the **recessive green** allele (**y**). So if the plant has **at least one W**, then the squash **will be white**, **masking** the expression of the yellow gene. So if you cross **wwyy** with **WWYY**, you'll get a **12 : 3 : 1** ratio of **white : yellow : green** in the F_2 generation. Here's a **genetic diagram** to prove it:

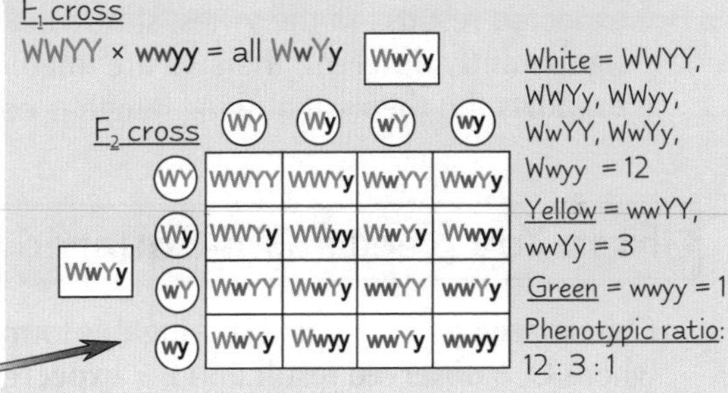

Warm-Up Questions

PRACTICE QUESTIONS

Q1 What is a sex-linked characteristic?

Q2 Two genes are autosomally-linked. What does this mean?

Q3 What is an epistatic gene?

Q4 A dihybrid cross produces the phenotypic ratio 9 : 3 : 4 in the F_2 generation. What does this indicate about the genes involved?

Exam Questions

Homozygous curly hair (hhss) crossed with a homozygous bald (HHSS)

Phenotypes of the F_2 offspring produced		
Bald	Straight hair	Curly hair
36	9	3

Q1 Haemophilia A is a sex-linked genetic disorder caused by a recessive allele carried on the X chromosome (X^h).

a) Draw a genetic diagram for a female carrier and a male with haemophilia A to predict the possible genotypes of their offspring. [3 marks]

b) Explain why haemophilia is more common in males than females. [3 marks]

Q2 Hair type in organism A is controlled by two genes: hair (H bald, h hair) and type (S straight, s curly). The F_2 offspring of a cross are shown in the table above.
Use your knowledge of epistasis to explain these results. [3 marks]

Biology students — 9 : 1 phenotypic ratio normal : geek...

I don't know about you but I think I need a lie-down after these pages. Epistasis is a bit of a tricky topic, but you just need to understand what it is and learn the phenotypic ratios for the two types of epistasis — dominant and recessive.

The Chi-Squared Test

Just when you thought it was safe to turn the page... I stick in some maths. Surprise!

The **Chi-Squared Test** Can Be Used to **Check** the **Results** of **Genetic Crosses**

1) The **chi-squared** (χ^2) **test** is a **statistical test** that's used to see if the **results** of an experiment **support** a **theory**.
2) First, the theory is used to **predict** a **result** — this is called the **expected result**.
   Then, the experiment is carried out and the **actual result** is recorded — this is called the **observed result**.
3) To see if the results support the theory you have to make a **hypothesis** called the **null hypothesis**.
4) The null hypothesis is always that there's **no significant difference** between the observed and expected results (your experimental result will usually be a bit different from what you expect, but you need to know if the difference is just **due to chance**, or because your **theory is wrong**).
5) The χ^2 **test** is then carried out to compare the 'goodness of fit' of the observed and expected results (i.e. to compare how well the observed results match the expected results). The **outcome** either **supports** or **rejects** the **null hypothesis**.
6) You can use the χ^2 test in **genetics** to test theories about the **inheritance** of **characteristics**. For example:

Theory: **Wing length** in fruit flies is controlled by a **single gene** with **two alleles (monohybrid inheritance)**. The **dominant** allele (N) gives **normal** wings, and the **recessive** allele (n) gives **vestigial** wings.

Expected results: With monohybrid inheritance, if you cross a **homozygous dominant** parent with a **homozygous recessive** parent, you'd expect a **3 : 1 phenotypic ratio** of **normal : vestigial** wings in the F_2 generation (see p. 168).

Observed results: The **experiment** (of crossing a homozygous dominant parent with a homozygous recessive parent) is **carried out** on fruit flies and the **number of F_2 offspring** with normal and vestigial wings is **counted**.

Null hypothesis: There's **no significant difference** between the observed and expected results. (If the χ^2 test shows the observed and expected results are **not significantly different**, then we are **unable to reject** the null hypothesis — the data **supports** the **theory** that wing length is controlled by **monohybrid inheritance**.)

In this kind of statistical test, you can never prove that the null hypothesis is true — you can only 'fail to reject it'. This just means that the evidence doesn't give you a reason to think the null hypothesis is wrong.

First, You Need a **Chi-Squared Value...**

Chi-squared χ^2 is calculated using this formula:
where **O** = **observed** result and **E** = **expected** result.
Σ just means '**the sum of...**'.

$$\chi^2 = \sum \frac{(O-E)^2}{E}$$

Although you won't be expected to calculate a chi-squared value in the written exams, you do need to **understand how the test works**, so that you can **interpret the results**. Here's an example for testing the **wing length** of **fruit flies** as explained above:

Homozygous dominant (NN) flies are crossed with homozygous recessive (nn) flies.
160 offspring are produced in the F_2 generation.

(1) First, the **number of offspring** (out of a total of 160) **expected** for each phenotype is worked out.
E for normal wings: 160 (total) ÷ 4 (ratio total) × 3 (predicted ratio for normal wings) = 120.
E for vestigial wings: 160 ÷ 4 × 1 = 40.

Phenotype	Ratio	Expected Result (E)	Observed Result (O)
Normal wings	3	120	
Vestigial wings	1	40	

(2) Then the **actual number** of offspring **observed** with each phenotype (out of the 160 offspring) is **recorded**, e.g. 111 with normal wings.

Phenotype	Ratio	Expected Result (E)	Observed Result (O)
Normal wings	3	120	111
Vestigial wings	1	40	49

(3) The results are used to work out χ^2:

Phenotype	Ratio	Expected Result (E)	Observed Result (O)	O – E	$(O-E)^2$	$\frac{(O-E)^2}{E}$
Normal wings	3	120	111	–9	81	0.675
Vestigial wings	1	40	49	9	81	2.025
					$\sum\frac{(O-E)^2}{E}=$	2.7

The total for this column (2.7) = χ^2

The Chi-Squared Test

...Then **Compare** it to the **Critical Value**

1) To find out if there is a **significant difference** between your observed and expected results you need to **compare** the **χ^2 value** to a **critical value**.
2) The critical value is the value of χ^2 that corresponds to a 0.05 (**5%**) level of **probability** that the **difference** between the observed and expected results is **due to chance**.
3) If your χ^2 value is **larger** than or equal to the critical value then there **is a significant difference** between the observed and expected results (something **other than chance** is causing the difference) — and the **null hypothesis** can be **rejected**.
4) If your χ^2 value is **smaller** than the critical value then there **is no significant difference** between the observed and expected results — the null hypothesis **can't be rejected**. E.g. for the example on the previous page the χ^2 value is **2.7**, which is **smaller** than the critical value of **3.84** (see table below) — there's **no significant difference** between the observed and expected results. We've failed to reject the null hypothesis, so the **theory** that wing length in fruit flies is controlled by **monohybrid inheritance** is **supported**.
5) In the exam you might be **given** the **critical value** or asked to **find it** from a **table**:

Using a χ^2 table:

A **χ^2 table** shows a range of **probabilities** that correspond to different **critical values** for different **degrees of freedom** (explained below). Biologists normally use a **probability** level (P value) of **0.05** (5%), so you only need to look in that column.

There's more on P values on page 227.

- First, the **degrees of freedom** for the experiment are worked out — this is the **number of classes** (number of phenotypes) **minus one**. E.g. 2 – 1 = 1.
- Next, the **critical value** corresponding to a **probability** of **0.05** at **one degree of freedom** is found in the table — here it's **3.84**.
- Then just **compare** your χ^2 value of **2.7** to this critical value, as explained above.

degrees of freedom	no. of classes	Critical values					
1	2	0.46	1.64	2.71	3.84	6.64	10.83
2	3	1.39	3.22	4.61	5.99	9.21	13.82
3	4	2.37	4.64	6.25	7.82	11.34	16.27
4	5	3.36	5.99	7.78	9.49	13.28	18.47
probability that result is due to chance only		0.50 (50%)	0.20 (20%)	0.10 (10%)	0.05 (5%)	0.01 (1%)	0.001 (0.1%)

Warm-Up Questions

Q1 What is a χ^2 test used for?

Q2 What can the results of the χ^2 test tell you?

Q3 How do you tell if the difference between your observed and expected results is due to chance?

Exam Question

Q1 A scientist is investigating petal colour in a flower. It's thought to be controlled by two separate genes (dihybrid inheritance), the colour gene — B = blue, b = purple, and the spots gene — W = white, w = yellow. A cross involving a homozygous dominant parent and a homozygous recessive parent should give a 9 : 3 : 3 : 1 ratio in the F_2 generation. The scientist observes the number of offspring showing each of four phenotypes in 240 F_2 offspring. Her results are shown in the table, along with the chi-squared value the scientist calculated for the experiment.

a) State the null hypothesis for this experiment. [1 mark]

b) The critical value for this experiment is 7.82. Based on the information in the table, is this likely to be a case of dihybrid inheritance or not? Explain your answer. [2 marks]

Phenotype	Ratio	Expected Result (E)	Observed Result (O)	$\frac{(O - E^2)}{E}$
Blue with white spots	9	135	131	0.12
Purple with white spots	3	45	52	1.09
Blue with yellow spots	3	45	48	0.20
Purple with yellow spots	1	15	9	2.4
			Chi-squared =	3.81

The expected result of revising these pages — boredom...

...the observed result — boredom. Remember, the null hypothesis (that there's no difference between the observed and expected results) can only be rejected if the value for chi-squared is higher than or equal to the critical value.

The Hardy-Weinberg Principle

Sometimes you need to look at the genetics of a whole population, rather than a cross between just two individuals. And that's where those spiffing fellows Hardy and Weinberg come in...

Members of a **Population Share** a **Gene Pool**

1) A **species** is defined as a group of **similar organisms** that can **reproduce** to give **fertile offspring**.
2) A **population** is a group of organisms of the **same species** living in a **particular area** at a **particular time** — so they have the potential to **interbreed**.
3) Species can exist as **one** or **more populations**, e.g. there are populations of the American black bear (*Ursus americanus*) in parts of America and in parts of Canada.
4) The **gene pool** is the complete range of **alleles** present in a **population**.
5) How **often** an **allele occurs** in a population is called the **allele frequency**. It's usually given as a **percentage** of the total population, e.g. 35%, or a **number**, e.g. 0.35.

Yogi wanted everyone to kno what population he was in.

The **Hardy-Weinberg Principle** Predicts **Allele Frequencies Won't Change**

1) The **Hardy-Weinberg principle** is a mathematical model. It predicts that the **frequencies** of **alleles** in a population **won't change** from **one generation** to the **next**.
2) But this prediction is **only true** under **certain conditions** — it has to be a **large population** where there's **no immigration**, **emigration**, **mutations** or **natural selection**. There also needs to be **random mating** — all possible genotypes can breed with all others.
3) The **Hardy-Weinberg equations** (see below) can be used to **calculate the frequency** of particular **alleles**, **genotypes** and **phenotypes** within populations.
4) The equations can also be used to test whether or not the Hardy-Weinberg principle **applies** to **particular alleles** in **particular populations**, i.e. to test whether **selection** or any **other factors** are **influencing** allele frequencies. — if frequencies **do change** between generations in a large population then there is an influence of some kind.

The **Hardy-Weinberg Equations** Can be Used to **Predict Allele Frequency...**

When a gene has two alleles, you can **figure out** the frequency of one of the alleles of the gene if you **know the frequency of the other allele**, using this equation:

$$p + q = 1$$

Where: **p** = the **frequency** of one allele, usually the **dominant** one
q = the **frequency** of the other allele, usually the **recessive** one

The total frequency of all possible alleles for a characteristic in a certain population is 1.0. So the frequencies of the individual alleles (e.g. the dominant one and the recessive one) must add up to 1.0.

E.g. a species of plant has either **red** or **white** flowers. Allele **R** (red) is **dominant** and allele **r** (white) is **recessive**. If the frequency of **R** is **0.4**, then the frequency of **r** is: 1 – 0.4 = **0.6**.

... **Predict Genotype** and **Phenotype Frequency...**

You can **figure out** the frequency of one genotype if you **know the frequencies of the others**, using this equation:

$$p^2 + 2pq + q^2 = 1$$

Where: p^2 = the **frequency** of the **homozygous dominant genotype**
$2pq$ = the **frequency** of the **heterozygous genotype**
q^2 = the **frequency** of the **homozygous recessive genotype**

The total frequency of all possible genotypes for one characteristic in a certain population is 1.0. So the frequencies of the individual genotypes must add up to 1.0.

p^2 is the homozygous dominant genotype frequency if p is the dominant allele.

E.g. if there are **two alleles** for **flower colour** (R and r), there are **three possible genotypes** — **RR**, **Rr** and **rr**. If the frequency of genotype **RR** (p^2) is **0.34** and the frequency of genotype **Rr** (**2pq**) is **0.27**, the frequency of genotype **rr** (q^2) must be: 1 – 0.34 – 0.27 = **0.39**.

Genotype frequencies can then be used to work out **phenotype frequencies**.

E.g. the frequency of **red flowers** is equal to the genotype frequencies of **RR** and **Rr** added together (0.34 + 0.27 = **0.61**) and the frequency of **white flowers** is equal to the genotype frequency of **rr** (**0.39**).

The Hardy-Weinberg Principle

...Predict the Percentage of a Population that has a Certain Genotype...

EXAMPLE

The frequency of cystic fibrosis (genotype ff) in the UK is currently approximately 1 birth in every 2500. From this information you can estimate the percentage of people in the UK that are cystic fibrosis carriers (Ff). To do this you need to find the frequency of heterozygous genotype Ff, i.e. 2pq, using both equations:

$$p + q = 1 \qquad p^2 + 2pq + q^2 = 1$$

First calculate q:
Frequency of cystic fibrosis (homozygous recessive, ff) is 1 in 2500
$ff = q^2 = 1 \div 2500 = 0.0004$
So, $q = \sqrt{0.0004} = 0.02$

Next calculate p:
using $p + q = 1$, $p = 1 - q$
$p = 1 - 0.02 = 0.98$

Then calculate 2pq:
$2pq = 2 \times 0.98 \times 0.02 = 0.039$

The frequency of genotype Ff is 0.039, so the percentage of the UK population that are carriers is 3.9%.

...and Show if External Factors are Affecting Allele Frequency

EXAMPLE

If the **frequency** of **cystic fibrosis** is measured **50 years later** it might be found to be **1 birth in 3500**. From this information you can estimate the **frequency** of the **recessive allele** (f) in the population, i.e. **q**.

To calculate q:
Frequency of cystic fibrosis (homozygous recessive, ff) is 1 in 3500
$ff = q^2 = 1 \div 3500 = 0.00029$
So, $q = \sqrt{0.00029} = 0.017$

The frequency of the recessive allele is now **0.017**, compared to **0.02** currently (see above).
As the frequency of the allele has **changed** between generations the **Hardy-Weinberg principle doesn't apply** so there must have been some **factors** affecting **allele frequency**, e.g. **immigration**, **emigration**, **mutations** or **natural selection**.

Warm-Up Questions

Q1 What is a population?

Q2 What is a gene pool?

Q3 What conditions are needed for the Hardy-Weinberg principle to apply?

Q4 Which term usually represents the frequency of the homozygous recessive genotype in the Hardy-Weinberg equations?

Q5 Which term represents the frequency of the heterozygous genotype in the Hardy-Weinberg equations?

Exam Questions

Q1 Cleft chins are controlled by a single gene with two alleles. The allele coding for a cleft chin (T) is dominant over the allele coding for a non-cleft chin (t). In a particular population the frequency of the homozygous dominant genotype for cleft chin is 0.14.

a) What is the frequency of the recessive allele in the population? [2 marks]

b) What is the frequency of the homozygous recessive genotype in the population? [1 mark]

c) What percentage of the population have a cleft chin? [1 mark]

Q2 In Erminette chickens, feather colour is controlled by a single gene with two codominant alleles — F^B (black feathers) and F^W (white feathers). In a population of Erminette chickens, 43% of birds have the F^B allele. Calculate the frequency of the heterozygous genotype. [2 marks]

This stuff's surely not that bad — Hardly worth Weining about...

Two equations that you absolutely have to know — so learn 'em. And whilst you're at it make sure that you learn what each of the terms means as well. You'll feel like a right wally if you know that $p^2 + 2pq + q^2 = 1$ but haven't got a clue what p^2, $2pq$ and q^2 stand for. It's the kind of stuff that falls out of your head really easily so learn it, learn it, learn it.

Variation and Selection

You might remember a lot of this stuff from Topic 4. Well you need to learn it all again now but with a bit of extra detail for Topic 7. Great. At least there's some extra new stuff to get your teeth stuck into...

Variation Can be **Caused** by **Genes**, the **Environment**, or **Both**

1) **Variation** is the **differences** that exist between individuals.
2) Variation **within a species** means that **individuals** in a population can show a wide range of **different phenotypes**.
3) Although individuals of the **same species** have the **same genes**, they have **different alleles** (versions of genes) — this causes **genetic variation** within a species.
4) The main **source** of this genetic variation is **mutation**, e.g. when changes in the DNA base sequence lead to the production of **new alleles** — see page 194. But genetic variation is also introduced during **meiosis** (through the **crossing over** of chromatids and the **independent segregation** of chromosomes) and because of the **random fertilisation** of gametes during sexual reproduction.
5) Variation within a species can also be caused by differences in the **environment**, e.g. climate, food, lifestyle.
6) Most variation within a species is caused by a **combination** of **genetic** and **environmental** factors. But only **genetic variation** results in **evolution**.

Variation within a species is also called intraspecific variation.

Evolution is a **Change** in **Allele Frequencies Over Time**

The **frequency** of an **allele** in a population **changes** over time — this is **evolution**. **Natural selection** is **one method** by which evolution occurs. Here's a reminder of how it works:

Evolution also occurs by genetic drift. See page 181.

1) **Individuals** of the same species **vary** because they have **different alleles**.
2) **Predation, disease** and **competition** (**selection pressures**) create a **struggle for survival**.
3) Because individuals vary, some are **better adapted** to the selection pressures than others.
4) This means there are **differential levels** of **survival** and **reproductive success** in a population. Individuals with a **phenotype** that **increases** their **chance of survival** are **more likely** to **survive**, **reproduce** and **pass on** their genes (including the **beneficial alleles** that determine their phenotype), than individuals with a different phenotype.
5) This means that a **greater proportion** of the next generation **inherit** the **beneficial alleles**.
6) They, in turn, are **more likely** to **survive**, **reproduce** and **pass on** their genes.
7) So the **frequency** of the **beneficial alleles** in the gene pool **increases** from generation to generation.

A selection pressure is anything that affects an organism's chance of survival and reproduction.

Different Types of **Natural Selection** Lead to **Different Frequency Patterns**

Stabilising selection and **directional selection** are **types** of **natural selection** that affect **allele frequency** in different ways. You'll have covered these in Topic 4, but now there's an extra one to learn about — **disruptive selection**.

STABILISING SELECTION is where individuals with alleles for characteristics towards the **middle** of the range are more likely to **survive** and **reproduce**. It occurs when the environment **isn't changing**, and it **reduces the range** of possible **phenotypes**.

Example: In any **mammal population** there's a **range of fur length**. In a **stable climate**, having fur at the **extremes** of this range **reduces** the **chances** of **surviving** as it's harder to maintain the **right body temperature**. Animals with alleles for **average fur length** are the **most** likely to **survive**, **reproduce** and **pass on** their alleles. So these alleles **increase** in **frequency**. The **proportion** of the **population** with **average fur length increases** and the **range** of fur lengths **decreases**.

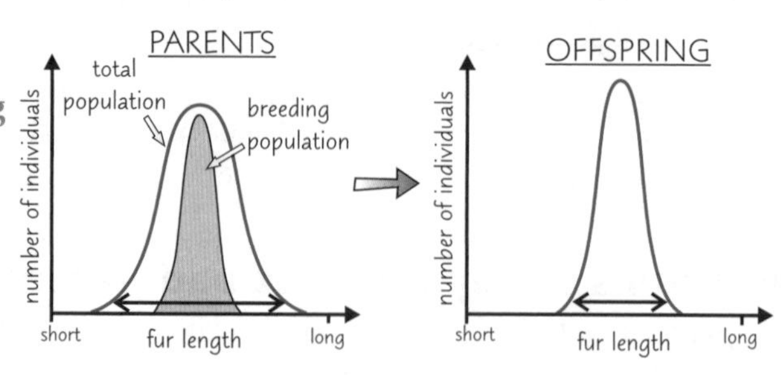

Variation and Selection

DIRECTIONAL SELECTION is where individuals with alleles for a single **extreme phenotype** are more likely to **survive** and **reproduce**. This could be in response to an **environmental change**.

Example: **Cheetahs** are the **fastest** animals on land. It's likely that this characteristic was developed through **directional selection**, as individuals that have **alleles** for increased **speed** are **more likely** to **catch prey** than slower individuals. So they're **more likely** to **survive**, **reproduce** and **pass on** their alleles. Over time the **frequency** of alleles for **high speed increases** and the population becomes **faster**.

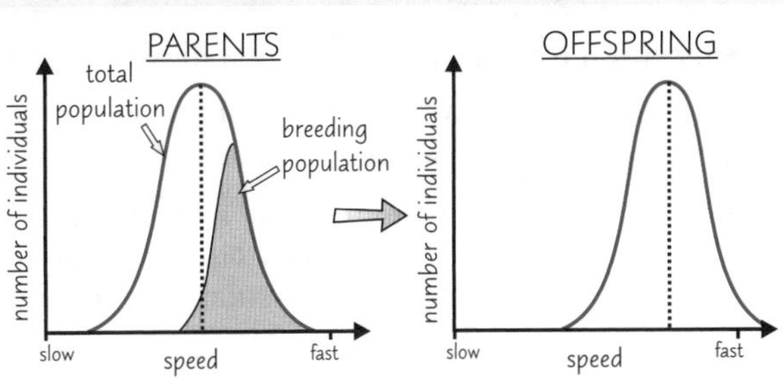

DISRUPTIVE SELECTION is where individuals with alleles for **extreme phenotypes** at **either end of the range** are more likely to **survive** and **reproduce**. It's the **opposite** of **stabilising selection** because characteristics towards the **middle** of the range are **lost**. It occurs when the environment favours **more than one phenotype**.

Example: In **bird populations** there's a **range** of **beak sizes**. Birds with **large beaks** are specialised to eat **large seeds** and birds with **small beaks** are specialised to eat **small seeds**. In an environment where the majority of seeds are large or small and very few (if any) are medium-sized, birds with **medium-sized beaks** may have a **reduced chance of survival**. This is because they are **unable to eat either** large or small seeds effectively. Birds with **large or small** beaks are most likely to **survive**, **reproduce** and **pass on** their alleles. So the alleles for a **large beak** and a **small beak increase in frequency**, but the alleles for a **medium-sized beak decrease in frequency**. Over time the proportion of the population that have **either small or large beaks increases**.

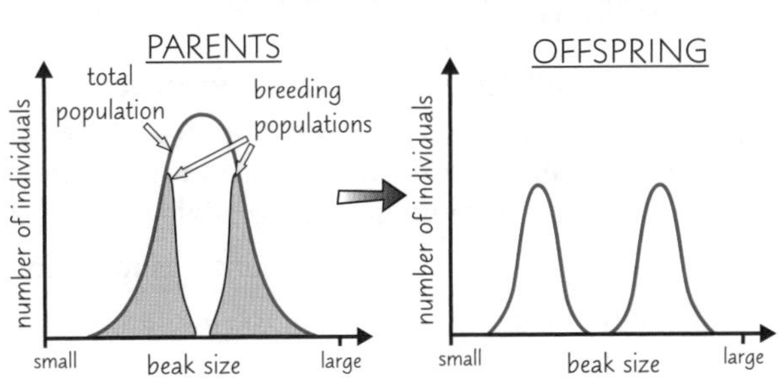

The mating shown here is assortative (non-random) — the birds with small beaks are more likely to mate with other birds with small beaks than they are with large-beaked birds (and vice versa). That's why you end up with two breeding populations.

Warm-Up Questions

Q1 Give two possible sources of genetic variation.

Q2 In terms of alleles, what is evolution?

Q3 What is directional selection?

Q4 What is disruptive selection?

Exam Question

Q1 The table on the right shows the results of an investigation into hair length of golden hamsters in a climate where the temperature is decreasing. Hair length is controlled by a single gene with two alleles. H represents the allele for short hair, which is dominant over the allele for long hair, represented by h.

Average Temp / °C	Frequency of h allele
22	0.11
21	0.13
19	0.19
18	0.20
16	0.23

a) Describe the relationship between the frequency of the recessive long hair allele and temperature. Suggest an explanation for this relationship. [4 marks]

b) What type of selection is responsible for this change in allele frequency? [1 mark]

Directional selection — when all the nutty ones are left in a box of chocs...

Ah. more stuff about alleles... it's actually a pretty nice word isn't it? Allele... just rolls off the tongue... Anyway, back to evolution and all that. A key thing to take on board here is that evolution is all about a change in allele (ooh, there it is again) frequency in a population — and natural selection is one way that this can happen.

Speciation and Genetic Drift

Ever wondered how there are so many different species on planet Earth? Well read on and learn, my friend...

Speciation is the **Development** of a **New Species**

1) **Speciation** is the development of a **new species** from an existing species.
2) Speciation occurs when populations of the same species become **reproductively isolated** — **changes in allele frequency** cause **changes in phenotype**, which mean they can **no longer interbreed** to produce **fertile offspring**.
3) This can happen when a **physical barrier**, e.g. a flood or an earthquake, **divides** a population of a species, causing some individuals to become **separated** from the main population. This is known as **geographical isolation** and leads to **allopatric speciation**.
4) Alternatively, speciation can also occur when a population becomes reproductively isolated **without** any **physical separation**. This is known as **sympatric speciation**.

Allopatric Speciation Requires **Geographical Isolation**

1) Populations that are geographically separated will experience slightly **different conditions**. For example, there might be a **different climate** on each side of the physical barrier.
2) This means the populations will experience **different selection pressures** and so **different changes** in allele frequencies could occur:

- Different **alleles** will be **more advantageous** in the different populations. For example, if geographical separation places one population in a **colder climate** than before, **longer fur length** will be **beneficial**. **Directional selection** will then act on the **alleles** for fur length in this population, increasing the frequency of the allele for **longer fur length**.
- Allele frequencies will also change as **mutations** (see p. 194) occur **independently** in each population.
- **Genetic drift** may also affect the allele frequencies in one or both populations (see next page).

3) The changes in allele frequency will lead to **differences** accumulating in the **gene pools** of the separated populations, causing changes in **phenotype frequencies**.
4) Eventually, individuals from the different populations will have changed so much that they won't be able to breed with one another to produce **fertile** offspring — they'll have become **reproductively isolated**.
5) The two groups will have become **separate species**, as shown in the diagram below.

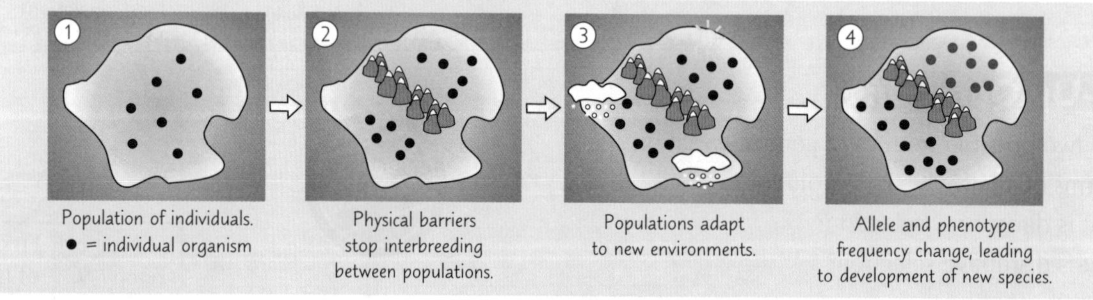

1 Population of individuals. ● = individual organism
2 Physical barriers stop interbreeding between populations.
3 Populations adapt to new environments.
4 Allele and phenotype frequency change, leading to development of new species.

Sympatric Speciation **Doesn't** Require **Geographical Isolation**

A population **doesn't** have to become **geographically isolated** to become **reproductively isolated**. Random mutations could occur **within a population**, **preventing** members of that population breeding with other members of the species.

There are some more examples of how organisms can become reproductively isolated on the next page.

Example

1) Most eukaryotic organisms are **diploid** — they have **two sets of homologous** (matched) **chromosomes** in their cells. Sometimes, **mutations** can occur that **increase** the number of **chromosomes**. This is known as **polyploidy**.
2) Individuals with different numbers of chromosomes **can't reproduce** sexually to give **fertile offspring** — so if a polyploid organism emerges in a diploid population, the polyploid organism will be **reproductively isolated** from the diploid organisms.
3) If the polyploid organism then reproduces **asexually**, a **new species** could develop.
4) Polyploidy can only lead to speciation if it **doesn't prove fatal** to the organism and more polyploid organisms can be produced. It's **more common** in **plants** than animals.

Speciation and Genetic Drift

Reproductive Isolation Occurs in Many Ways

Reproductive isolation occurs because **changes** in alleles and phenotypes in **some individuals prevent** them from **breeding successfully** with individuals **without** these changes. These changes include:

1) **Seasonal** — individuals from the same population develop different **flowering** or **mating** seasons, or become **sexually active** at different times of the year.
2) **Mechanical** — changes in **genitalia** prevent successful mating.
3) **Behavioural** — a group of individuals develop **courtship rituals** that **aren't attractive** to the main population.

Genetic Drift Can Lead to Speciation

1) Different **selection pressures** can change the **allele frequencies** in two geographically isolated species (see previous page). This is evolution by **natural selection**.
2) But evolution can also occur by **genetic drift**. This is when **chance**, rather than **environmental factors**, dictates which individuals **survive**, **breed** and **pass on** their **alleles**:

- Individuals within a population show **variation** in their **genotypes** (e.g. A and B).
- By **chance**, the **allele** for **one genotype** (B) is **passed on** to the offspring **more often** than others.
- So the number of individuals with the allele **increases**.
- **Changes** in allele frequency in two isolated populations could eventually lead to **reproductive isolation** and **speciation**.

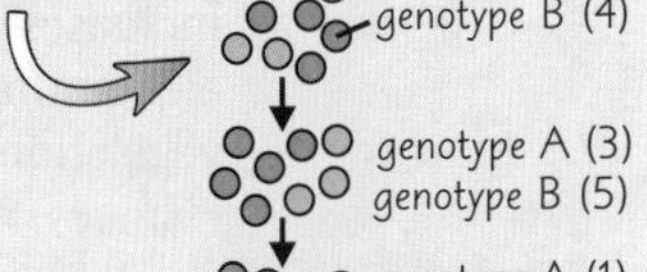

3) Natural selection and genetic drift work **alongside each other** to drive evolution, but one process can drive evolution **more** than the other depending on the **population size**.
4) **Evolution by genetic drift** usually has a **greater effect** in **smaller populations** where **chance** has a **greater influence**. In larger populations, any **chance variations** in allele frequency tend to **even out** across the whole population.

Evolutionary Change Has Resulted in a Great Diversity of Organisms

1) The diversity of life on Earth today is the result of **speciation** and **evolutionary change over millions of years**.
2) To start with there was **one population** of organisms. The population was **divided** and the new populations **evolved** into **separate species**. The new species were then **divided again** and the new populations **evolved** into more separate species.
3) This process has been **repeated** over a long period of time to create millions of new species.

1 species

2 species

5 species etc.

Warm-Up Questions

PRACTICE QUESTIONS

Q1 What is speciation?

Q2 Describe the process of genetic drift.

Exam Question

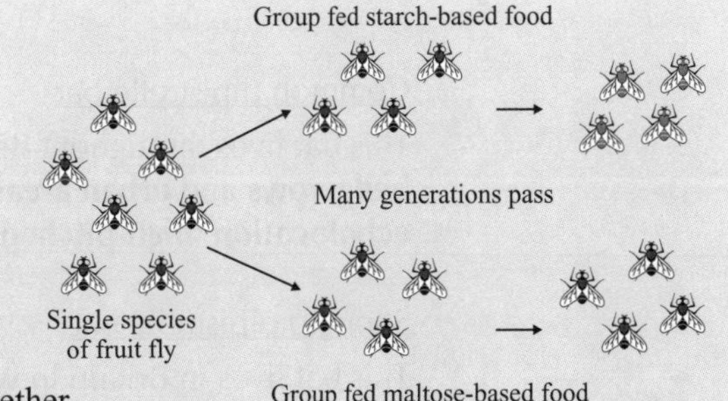

Q1 The diagram shows an experiment conducted with fruit flies. One population was split in two and each population was fed a different food. After many generations the two populations were placed together and it was observed that they were unable to breed together.

a) What evidence shows that speciation occurred? [1 mark]

b) Explain why the experiment resulted in speciation. [3 marks]

If they were ever separated, Al and Patrick would be heartbroken...

So, there are two types of speciation that you need to learn about here. To remember which one allopatric is, I imagine someone on an island shouting 'allo (hello) to their friend Patrick on a separate island. Just thought it might help...

Ecosystems

Ecosystems are amazing — that's why there are all those documentaries about them on TV. You don't have to watch a TV documentary to learn about them though, cos everything you need to know is in this topic.

You Need to **Learn Some Definitions** to get you **Started**

Habitat — The **place** where an organism **lives**, e.g. a rocky shore or a field.

Population — **All** the organisms of **one species** in a **habitat**.

Community — Populations of **different species** in a habitat make up a **community**.

Ecosystem — A **community**, plus all the **non-living** (abiotic) **conditions** in the area in which it lives. Ecosystems can be **small**, e.g. a pond, or **large**, e.g. an entire ocean.

Abiotic conditions — The **non-living** features of the ecosystem, e.g. **temperature** and **availability of water**.

Biotic conditions — The **living** features of the ecosystem, e.g. the presence of **predators** or **food**.

Niche — The **role** of a species within its habitat, e.g. what it eats, where and when it feeds.

Adaptation — A **feature** that members of a species have that **increases** their chance of **survival** and **reproduction**, e.g. **giraffes** have **long necks** to help them reach vegetation that's high up. This increases their chances of survival when food is **scarce**.

Being a member of the undead made it hard for Mumra to know whether he was a living or a non-living feature of the ecosystem.

Every Species Occupies a **Different Niche**

1) The **niche** a species occupies within its habitat includes:

- Its **biotic** interactions — e.g. the organisms it **eats**, and those it's **eaten by**.
- Its **abiotic** interactions — e.g. the **oxygen** an organism breathes in, and the **carbon dioxide** it breathes out.

2) Every species has its own **unique niche** — a niche can only be occupied by **one species**.

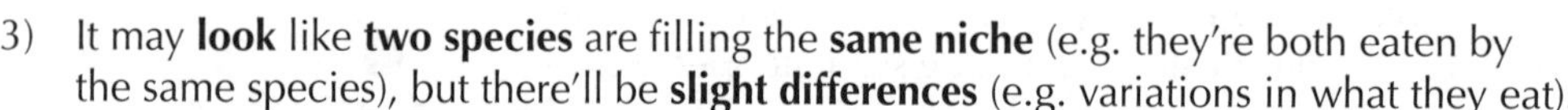

3) It may **look** like **two species** are filling the **same niche** (e.g. they're both eaten by the same species), but there'll be **slight differences** (e.g. variations in what they eat).
4) If two species **try** to occupy the **same niche**, they will **compete** with each other. One species will be **more successful** than the other, until **only one** of the species is **left**.
5) Here are a couple of examples of niches:

Don't get confused between habitat (where a species lives) and niche (what it does in its habitat).

<u>Common pipistrelle bat</u>

This bat lives throughout Britain on **farmland**, **open woodland**, **hedgerows** and **urban areas**. It feeds by **flying** and catching **insects** using **echolocation** (**high-pitched sounds**) at a **frequency** of around **45 kHz**.

<u>Soprano pipistrelle bat</u>

This bat lives in Britain in **woodland** areas, close to **lakes** or **rivers**. It feeds by **flying** and catching **insects** using **echolocation**, at a **frequency** of **55 kHz**.

It may **look like** both species are filling the **same niche** (e.g. they both eat insects), but there are **slight differences** (e.g. they use **different frequencies** for their echolocation).

Ecosystems

Organisms are Adapted to Biotic and Abiotic Conditions

1) As you know, **adaptations** are features that **increase** an organism's chance of **survival** and **reproduction**.
2) They can be **physiological** (processes **inside** their body), **behavioural** (the way an organism **acts**) or **anatomical** (**structural features** of their body).
3) Organisms with better adaptations are **more likely** to **survive**, **reproduce** and **pass on** the alleles for their adaptations, so the adaptations become **more common** in the population. This is called **natural selection**.
4) Every species is adapted to **use** an **ecosystem** in a way that **no other** species can — it has it's own **unique niche** (see previous page). For example, only giant anteaters can **break into** ant nests and **reach** the ants. They have **claws** to rip open the nest, and a **long**, **sticky tongue** which can move **rapidly** in and out of its mouth to **pick up** the ants.
5) Organisms are **adapted** to both the **abiotic conditions** (e.g. how much **water** is available) and the **biotic conditions** (e.g. what **predators** there are) in their ecosystem.

Here are a few ways that **different organisms** are **adapted** to the **abiotic** or the **biotic** conditions in their ecosystems:

Adaptations to abiotic conditions

- **Otters** have **webbed paws** — this means they can both **walk** on land and **swim** effectively. This increases their chance of survival because they can **live** and **hunt** both on land and in water.
- **Seals** have a **thick layer** of **blubber** (fat) — this helps to keep them **warm** in the **coldest seas**. This increases their chance of survival because they can **live** in places where food is plentiful.
- **Hedgehogs hibernate** — they **lower their rate of metabolism** (all the chemical reactions taking place in their body) over **winter**. This increases their chance of survival because they can **conserve energy** during the **coldest** months.

Adaptations to biotic conditions

- **Sea otters** use **rocks** to **smash open** shellfish and clams. This increases their chance of survival because it gives them **access** to **another source** of food.
- **Male frogs** produce **mating calls** to **attract females** — this makes sure they **attract** a **mate** of the **same species**. This increases their chance of reproduction by making **successful mating** more likely.
- Some **bacteria** produce **antibiotics** — these **kill other species** of bacteria in the **same area**. This increases their chance of survival because there's **less competition** for **resources**.

Warm-Up Questions

PRACTICE QUESTIONS

Q1 What is the name given to all the organisms of one species in a habitat?

Q2 Define a community.

Q3 Give the term for the non-living features of an ecosystem.

Q4 What happens when two species try to occupy the same niche in an ecosystem?

Exam Question

Q1 Common pipistrelle bats have light, flexible wings, which means they can fly fast and are manoeuvrable. They hunt insects at night using echolocation and live on farmland, in open woodland, hedgerows and urban areas. They make unique mating calls to find mates, hibernate through the winter, and roost in cracks in trees and buildings during the day.

Explain how the common pipistrelle bat is adapted to the biotic conditions in its ecosystem. [3 marks]

Unique quiche niche — say it ten times really fast...

All this population and ecosystem stuff is pretty wordy I'm afraid, but I'll tell you what, you'll be missing it when you get back to the really sciencey stuff later. You just need to learn and relearn all the key words here, then when they ask you to interpret some bat-related babble in the exam, you'll know exactly what they're talking about. Niche work.

Variation in Population Size

Uh-oh, anyone who loves cute little bunnies look away now — these pages are about how the population sizes of organisms fluctuate and the reasons why. One of the reasons, I'm sad to say, is because the little rabbits get eaten.

Population Size Varies Because of Abiotic Factors...

Remember — abiotic factors are the non-living features of the ecosystem.

1) **Population size** is the **total number** of organisms of **one species** in a **habitat**.
2) The **maximum stable population size** of a species that an ecosystem can **support** is called the **carrying capacity**. Carrying capacity varies as a result of both **abiotic** and **biotic factors** (see below).
3) **Abiotic** factors include the amount of **light**, **water** or **space** available, the **temperature** of the surroundings or the **chemical composition** of the surroundings. When abiotic conditions are **ideal** for a species, organisms can **grow fast** and **reproduce successfully**.

E.g. when the temperature of a mammal's surroundings is the ideal temperature for **metabolic reactions** to take place, they don't have to **use up** as much energy **maintaining** their **body temperature**. This means more energy can be used for **growth** and **reproduction**, so their population size will **increase**.

4) When abiotic conditions **aren't ideal** for a species, organisms **can't** grow as **fast** or reproduce as **successfully**.

E.g. when the temperature of a mammal's surroundings is significantly **lower** or **higher** than their **optimum** body temperature, they have to **use** a lot of **energy** to maintain the right **body temperature**. This means less energy will be available for **growth** and **reproduction**, so their population size will **decrease**.

...and Because of Biotic Factors

Biotic factors are the living features of the ecosystem.

1 Interspecific Competition — Competition Between Different Species

1) Interspecific competition is when organisms of **different species compete** with each other for the **same resources**, e.g. **red** and **grey** squirrels compete for the same **food sources** and **habitats** in the **UK**.
2) Interspecific competition between two species can mean that the **resources available** to **both** populations are **reduced**, e.g. if they share the **same** source of food, there will be **less** available to both of them. This means both populations will be **limited** by a lower amount of food. They'll have less **energy** for **growth** and **reproduction**, so the population sizes will be **lower** for both species. E.g. in areas where both **red** and **grey** squirrels live, both populations are **smaller** than they would be if there was **only one** species there.
3) If **two** species are competing but one is **better adapted** to its surroundings than the other, the less well adapted species is likely to be **out-competed** — it **won't** be able to **exist** alongside the better adapted species. E.g. since the introduction of the **grey squirrel** to the UK, the native **red squirrel** has **disappeared** from large areas. The grey squirrel has a better chance of **survival** because it's **larger** and can store **more fat** over winter. It can also eat a **wider range** of **food** than the red squirrel.

Never mind what the doctors said, Nutkin knew his weight problem would increase his chance of survival.

2 Intraspecific Competition — Competition Within a Species

Intraspecific competition is when organisms of the **same species compete** with each other for the **same resources**.

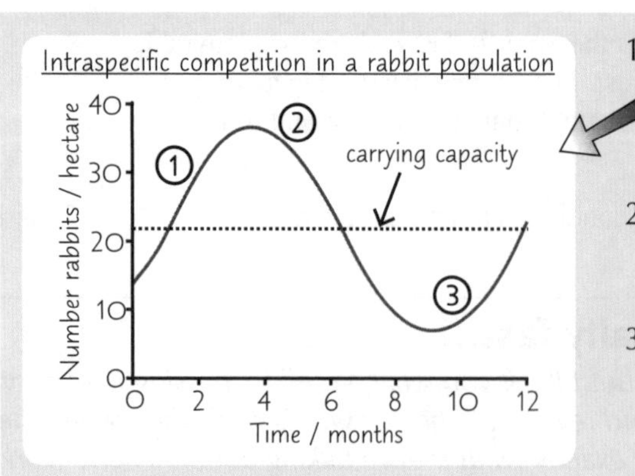

1) The **population** of a species (e.g. rabbits) **increases** when resources are **plentiful**. As the population increases, there'll be **more** organisms competing for the **same amount** of space and **food**.
2) Eventually, resources such as food and space become **limiting** — there **isn't enough** for all the organisms. The population then begins to **decline**.
3) A **smaller** population then means that there's **less competition** for space and food, which is **better** for **growth** and **reproduction** — so the population starts to **grow** again.

Variation in Population Size

3 Predation — Predator and Prey Population Sizes are Linked

Predation is where an organism (the predator) kills and eats another organism (the prey), e.g. lions kill and eat (**predate** on) buffalo. The **population sizes** of predators and prey are **interlinked** — as the population of one **changes**, it **causes** the other population to **change**:

1) As the **prey** population **increases**, there's **more food** for predators, so the **predator** population **grows**. E.g. in the graph on the right the **lynx** population **grows** after the **snowshoe hare** population has **increased** because there's **more food** available.
2) As the **predator** population **increases**, **more prey** is **eaten** so the **prey** population then begins to **fall**. E.g. **greater numbers** of lynx eat lots of snowshoe hares, so their population **falls**.
3) This means there's **less food** for the **predators**, so their population **decreases**, and so on. E.g. **reduced** snowshoe hare numbers means there's **less food** for the lynx, so their population **falls**.

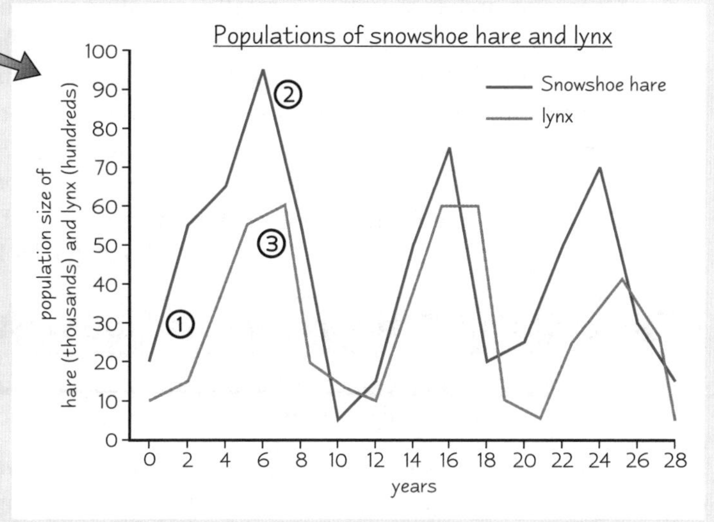

Predator-prey relationships are usually more **complicated** than this though because there are **other factors** involved, like availability of **food** for the **prey**. E.g. it's thought that the population of snowshoe hare initially begins to **decline** because there's **too many** of them for the amount of **food available**. This is then **accelerated** by **predation** from the lynx.

Warm-Up Questions

Q1 What is the carrying capacity of an ecosystem?

Q2 Give one example of how an abiotic factor can affect population size.

Q3 What is interspecific competition?

Q4 What will be the effect of interspecific competition on the population size of a species?

Q5 What does it mean when a species is out-competed?

Q6 Define intraspecific competition.

Exam Question

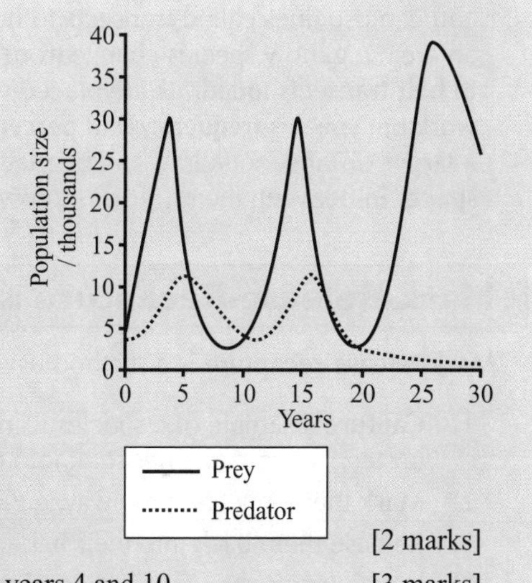

Q1 The graph on the right shows the population size of a prey species and a predator species over a period of 30 years.

a) Calculate the rate at which the prey population increased over the first 4 years. [2 marks]

b) Explain the changes in both the predator and prey populations between years 4 and 10. [3 marks]

c) During what time period was prey population size likely to have been most heavily influenced by intraspecific competition? Give a reason for your answer. [2 marks]

Predator-prey relationships — they don't usually last very long...

You'd think they could have come up with names a little more different than inter- and intraspecific competition. I always remember it as int-er means diff-er-ent species. The factors that affect population size are divided up nicely for you here — just like predators like to nicely divide up their prey into bitesize chunks.

Investigating Populations

Don't just take my word about all this population stuff — you can go to a field and find out for yourself...

You need to take a **Random Sample** from the **Area You're Investigating**

Most of the time it'd be too **time-consuming** to measure the **number of individuals** in a species (population size) and the **distribution** of that species (i.e. where it's found) in the **entire area** you're investigating. Instead you take **samples**:

1) **Choose** an **area** to **sample** — a **small** area **within** the area being investigated.
2) Samples should be **random** to **avoid bias**, e.g. if you were investigating a field you could pick random sample sites by dividing the field into a **grid** and using a **random number generator** to select **coordinates**.
3) Use an **appropriate technique** to take a sample of the population (see below).
4) **Repeat** the process, taking as many samples as possible. This will **reduce** the **likelihood** that your results are down to **chance**.
5) The **number of individuals** for the **whole area** can then be **estimated** by taking the **mean** of the data collected in each sample and **multiplying** it by the size of the whole area.

Quadrats and **Transects** are used to **Investigate Non-Motile Organisms**

Non-motile organisms are ones that **don't move** about — like **plants**. Quadrats and transects can also be used to investigate **slow-moving** organisms, which include things like **limpets**.

1) A quadrat is a square frame, usually divided into a grid of 100 smaller squares by strings attached across the frame.
2) Quadrats are placed on the ground at different points within the area you're investigating.
3) The species frequency (how often a species is found) or the number of individuals of each species is recorded in each quadrat.
4) The percentage cover of a species can also be measured by counting how much of the quadrat is covered by the species — you count a square if it's more than half-covered. Percentage cover is a quick way to investigate populations and you don't have to count all the individual organisms.

the area of this quadrat is $0.25\ m^2$
0.5 m
0.5 m

Measuring % cover
Species A 42 squares = 42%
Species B 12 squares = 12%
Species C 47 squares = 47%

You can use **lines** called **transects** to help find out how plants are **distributed across** an area, e.g. how species change from a hedge towards the middle of a field. In **belt transects**, **quadrats** are placed next to each other **along** the transect to work out **species frequency** and **percentage cover** along the transect. To cover a **larger** distance, quadrats can be placed at **intervals** along the line (i.e. with **spaces** in between them). This is known as an **interrupted** belt transect.

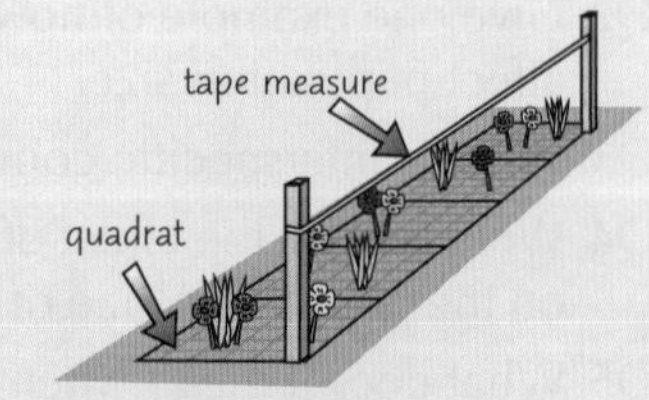

Mark-Release-Recapture is Used to **Investigate More Motile** Species

Mark-release-recapture is a method used to measure the **abundance** of more **motile** species. Here's how it's done:

1) **Capture** a sample of a species using an **appropriate technique**, e.g. you could use pitfall traps (a steep sided container sunk into the ground) to capture ground insects, and **count** them.
2) **Mark** them in a harmless way, e.g. by putting a spot of **paint** on them, or by **removing** a tuft of **fur**.
3) **Release** them back into their habitat.
4) Wait a week, then take a **second sample** from the **same population**.
5) **Count** how many of the second sample **are marked**. You can then use this **equation** to **estimate** the **total** population size.

$$\text{Total population size} = \frac{\text{Number caught in 1st sample} \times \text{Number caught in 2nd sample}}{\text{Number marked in 2nd sample}}$$

When using this method, you have to make a few **assumptions**:

1) The marked sample has had enough **time** and **opportunity** to **mix** back in with the population.
2) The marking hasn't affected the individuals' **chances of survival** (e.g. by making them more visible to predators), and the marking itself is **still visible** (e.g. it hasn't rubbed off).
3) There are **no changes** in **population size** due to **births**, **deaths** and **migration** during the period of the study.

Investigating Populations

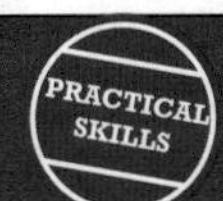

You can **Investigate Environmental Factors** and **Species Distribution**

The **distribution** of species often changes within a particular area. E.g. you might find more **shade-loving** plants at the **edge** of a field where they're **sheltered** by a tree, than in the **centre** where they're **exposed** to full sunlight. You need to be able to **investigate** the **effect** of an **environmental factor**, such as shade, on the distribution of a species. Here's an **example** of an investigation you could do to investigate the effect of **soil pH** on **marram grass** in a **coastal ecosystem**:

1) Place a **tape measure** in a straight line from the shore, heading inland. This will be your **transect**.
2) Take a 1 m^2 **quadrat** divided into 100 squares (10 by 10).
3) Starting from the shore, place the quadrat **next to** the tape measure. It doesn't matter where you position the quadrat relative to the tape measure, but you should make sure that you do it the same way each time.
4) **Count** the **squares** containing **marram grass** and record the result in a table as **percentage cover** (as shown below). If you have time, take two repeat quadrat samples next to your initial quadrat and take a mean of your results. Alternatively, you could take a mean of the data from your whole class.
5) At each **sample point**, you should also measure the **pH** and record the results in the table.
6) **Repeat** the observations every 10 m along the transect.

<u>To measure pH:</u>
If you have one, you can use a digital **pH probe** to take pH readings of sand or soil in the field. If you don't have one, you can test the sand/soil back at school. Take a **sample** for testing. When you get back to school, **sieve** it to remove any debris, like twigs and leaves, and place it in a test tube. Add some **barium sulfate**, distilled water and **pH indicator**. **Shake** thoroughly and then leave to it **settle**. Check the **colour** against a **pH chart** and record the result.

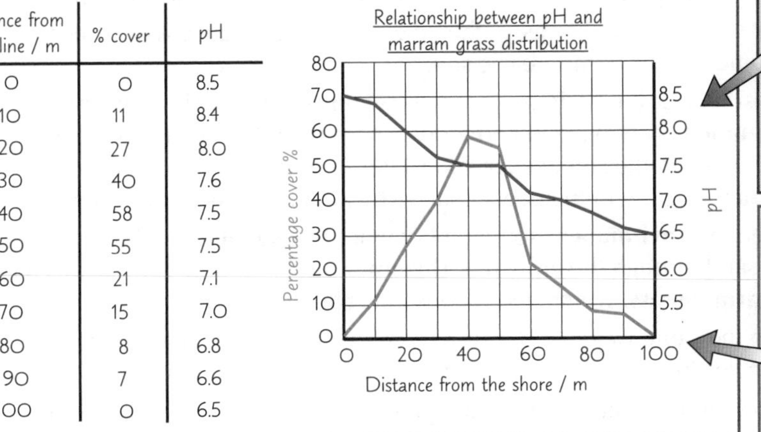

Distance from shoreline / m	% cover	pH
0	0	8.5
10	11	8.4
20	27	8.0
30	40	7.6
40	58	7.5
50	55	7.5
60	21	7.1
70	15	7.0
80	8	6.8
90	7	6.6
100	0	6.5

pH decreases as you move **inland**. This is because near the shore the sand/soil contains lots of **shell** fragments which are made of **calcium carbonate**, an **alkaline** compound. Further inland, the **rotting vegetation** adds **organic matter** to the soil, which is more **acidic**.

At first, as **pH decreases** from 8.5 to 7.5, the percentage cover of **marram grass increases**. After pH 7.5, **marram grass** percentage cover **decreases** as **pH** continues to **decrease**. You can't say pH **caused** these trends in marram grass cover though — there could be other factors affecting it, including **soil moisture content, salinity,** and **competition** from other species.

<u>Safety issues:</u>
You need to think about what risks you'll be exposed to during fieldwork, so you can plan ways to reduce the chance of them happening. For example, you need to:

- use tide timetables, so you know what the local tide times are when you're working on a beach. Low tide is the best time to work.
- wear suitable clothing and footwear for the weather and terrain, e.g. a sun hat if it's hot and sturdy shoes to stop you slipping.
- wash your hands before eating, especially after handling soil.

<u>Ethical issues:</u>
All fieldwork affects the environment where it's carried out, e.g. lots of people walking around may cause soil erosion and marram grass can be killed by people trampling all over it. Investigations should be planned to have the smallest impact possible, e.g. people should restrict where they walk to the area being studied and try to avoid treading on the plants themselves.

Warm-Up Questions

PRACTICE QUESTIONS

Q1 Give the formula for calculating population size from the mark-release-recapture method.

Exam Question

An accurate result is one that's close to the true answer (see p. 224).

Q1 A student is investigating the population size of clover plants in a field.
a) Describe how she could estimate the population size of the clover plants using random samples. [4 marks]
b) Explain how incorrect identification of plant species could reduce the accuracy of the results. [1 mark]

What did the quadrat say to the policeman — I've been framed...

If you want to know what it's really like doing these investigations, then read these pages outside in the pouring rain. Doing it while you're tucked up in a nice warm, dry exam hall won't seem so bad after that, take my word for it.

Succession

Repeat after me: successful succession involves several simple successive stages.

Succession is the **Process of Ecosystem Change**

Ecosystems are **dynamic** — they are constantly **changing**. **Succession** is the process by which an **ecosystem** (see p. 182) **changes** over **time**. The **biotic conditions** (e.g. **plant** and **animal communities**) change as the **abiotic conditions** (e.g. **water** availability) change. There are **two** types of succession:

Remember — biotic = living things abiotic = non-living.

1) **Primary succession** — this happens on land that's been **newly formed** or **exposed**, e.g. where a **volcano** has erupted to form a **new rock surface**, or where **sea level** has **dropped** exposing a new area of land. There's **no soil** or **organic material** to start with, e.g. just bare rock.
2) **Secondary succession** — this happens on land that's been **cleared** of all the **plants**, but where the **soil remains**, e.g. after a **forest fire** or where a forest has been **cut down by humans**.

Succession Occurs in a **Series** of **Stages**

1) **Primary succession** starts when species **colonise** a new land surface. **Seeds** and **spores** are blown in by the **wind** and begin to **grow**. The **first species** to colonise the area are called **pioneer species**.
    - The **abiotic conditions** are **hostile** (**harsh**), e.g. there's no soil to **retain water**. Only pioneer species **grow** because they're **specially adapted** to cope with the harsh conditions, e.g. **marram grass** can grow on sand dunes near the sea because it has **deep roots** to get water and can **tolerate** the salty environment.
    - The pioneer species **change** the **abiotic conditions** — they **die** and **microorganisms decompose** the dead **organic material** (**humus**). This forms a **basic soil**.
    - This makes conditions **less hostile**, e.g. the basic soil helps to **retain water**, which means **new organisms** with **different adaptations** can move in and grow. These then die and are decomposed, adding **more** organic material, making the soil **deeper** and **richer in minerals**. This means **larger plants** like **shrubs** can start to grow in the deeper soil, which retains **even more** water.
    - Some new species may **change** the **environment** so that it becomes **less suitable** for the previous species. E.g. **sand sedge stabilises** the sand through the growth of **rhizomes** (underground stems). This makes the conditions **less suitable** for **marram grass**, which needs constant **reburial** by **sand** in order to grow healthily.
2) **Secondary succession** happens in the **same way**, but because there's already a **soil layer** succession starts at a **later stage** — the pioneer species in secondary succession are **larger plants**, e.g. shrubs.
3) At each stage, **different** plants and animals that are **better adapted** for the improved conditions move in, **out-compete** the plants and animals that are already there, and become the **dominant species** in the ecosystem.
4) As succession goes on, the ecosystem becomes **more complex**. New species move in **alongside** existing species which means that **biodiversity** (the variety of living organisms) **increases**.
5) The **final stage** is called the **climax community** — the ecosystem is supporting the **largest** and **most complex** community of plants and animals it can. It **won't change** much more — it's in a **steady state**.

This example shows primary succession on bare rock, but succession also happens on sand dunes, salt marshes and even on lakes.

Example of primary succession — bare rock to woodland

1) **Pioneer species colonise** the rocks. E.g. **lichens** grow **on** and **break down** rocks, **releasing minerals**.
2) The lichens **die** and are **decomposed** helping to form a **thin soil**, which thickens as more **organic material** is formed. This means other species such as **mosses** can **grow**.
3) **Larger plants** that need **more water** can move in as the soil **deepens**, e.g. **grasses** and **small flowering plants**. The soil **continues to deepen** as the larger plants die and are decomposed.
4) **Shrubs**, **ferns** and **small trees** begin to grow, **out-competing** the grasses and smaller plants to become the **dominant** species. **Diversity increases**.
5) Finally, the soil is **deep** and **rich** enough in **nutrients** to support **large trees**. These become the dominant species, and the **climax community** is formed.

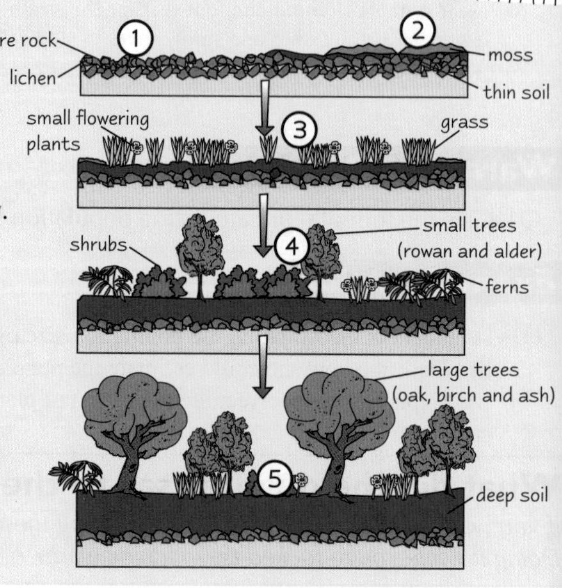

Succession

Different Ecosystems have Different Climax Communities

Which species make up the climax community depends on what the **climate** is like in an ecosystem. The climax community for a **particular** climate is called its **climatic climax**. For example:

- In a **temperate climate** there's **plenty** of **available water**, **mild temperatures** and not much **change** between the seasons. The climatic climax will contain **large trees** because they **can grow** in these conditions once **deep soils** have developed.
- In a **polar climate** there's **not much available water**, temperatures are **low** and there are **massive changes** between the seasons. Large trees **won't ever** be able to grow in these conditions, so the climatic climax contains only **herbs** or **shrubs**, but it's still the **climax community**.

Conservation Often Involves Managing Succession

Human activities can **prevent succession**, stopping a climax community from **developing**. When succession is stopped **artificially** like this the climax community is called a **plagioclimax**. For example:

A **regularly mown** grassy field **won't develop** shrubs and trees (**woody plants**), even if the climate of the ecosystem could support them. The **growing points** of the woody plants are **cut off** by the lawnmower, so larger plants **can't establish** themselves. The **longer** the interval between mowing, the **further** succession can progress and the more **diversity increases**. But with **more frequent** mowing, succession can't progress and diversity will be **lower** — only the grasses can **survive** being mowed.

A mighty weapon with which to tame the forces of nature.

Conservation (the **protection** and **management** of ecosystems) sometimes involves preventing succession in order to **preserve** an ecosystem in its **current** stage of succession. For example, there are large areas of **moorland** in **Scotland** that provide **habitats** for many species of plants and animals. If the moorland was left to **natural processes**, succession would lead to a **climax community** of **spruce forest**. This would mean the **loss** of the moorland habitat and could lead to the loss of some of the plants and animals that **currently** live there. Preventing succession keeps the moorland ecosystem **intact**. There are a couple of ways to **manage succession** to **conserve** the moorland ecosystem:

1) **Animals** are allowed to **graze** on the land. This is similar to **mowing** — the animals eat the **growing points** of the shrubs and trees, which **stops** them from establishing themselves and helps to keep vegetation **low**.
2) **Managed fires** are lit. After the fires, **secondary succession** will occur on the moorland — the species that grow back **first** (**pioneer species**) are the species that are being **conserved**, e.g. heather. Larger species will take **longer** to grow back and will be **removed again** the next time the moor is burnt.

Warm-Up Questions

Q1 What is the difference between primary and secondary succession?

Q2 What is the name given to species that are the first to colonise an area during succession?

Q3 What is meant by a climax community?

Exam Question

Q1 Succession occurs on sand dunes.
You can often see the different stages of succession as you move further inland from the shoreline.

a) Name the type of succession that is taking place when the first grasses start to appear on the dune. Give a reason for your answer. [2 marks]

b) Explain how the growth of grasses can lead to the colonisation of the dune by larger plants like shrubs. [2 marks]

Revision succession — bare brain to a woodland of knowledge...

When answering questions on succession, examiners are pretty keen on you using the right terminology — that means saying "pioneer species" instead of "the first plants to grow there". If you can manage that, then you'll be just fine.

Conservation

Who'd have thought conservation could be such a tricky business — cos I'm feeling nice, I'll try and explain why...

There Can be **Conflict** Between **Human Needs** and **Conservation**

1) Conservation is the **protection** and **management** of **species** and **habitats** (**ecosystems**) in a **sustainable** way. Sustainable means that enough resources are taken to meet the **needs** of people **today**, without **reducing the ability** of people in the **future** to meet their own needs.
2) **Not everyone** agrees with every conservation measure though — there's often **conflict** between **human needs** and **conservation**. **Careful management** is needed to find a **balance** between the two and maintain the **sustainability** of **natural resources**. Here's an example:

- The **Maasai Mara** is a national reserve in **Kenya**. It's a large area of **grassland** (**savannah**) with lots of wildlife.
- The Maasai people traditionally **earn a living** by raising **livestock**, such as **cattle**. This can bring them into **conflict** with conservationists — e.g. **overgrazing** by **livestock** can destroy grassland for **wildlife**.
- Conservation trusts are working with the Maasai to help them **make money** from their land through **conservation** and **ecotourism projects** rather than farming, and to **farm** in a **sustainable way**. So the **economic needs** of the Maasai are met, while still allowing the area and its wildlife to be conserved.

3) There are many **different methods** of conservation. Some focus on conserving a particular **species**, whilst others protect the **habitat** for all the species that live there. Here are some **examples** of conservation techniques:

- Plants can be conserved using **seedbanks**, which are **stores** of lots of **seeds** from lots of **different plant species**. If the plants become **extinct** in the wild, the stored seeds can be used to **grow new plants**.
- **Fishing quotas** are **limits** to the **amount** of certain fish species that fishermen are **allowed** to **catch**. Fishing quotas help to **conserve** fish species by **reducing** the numbers that are **caught** and **killed**.
- **Protected areas** such as **national parks** and **nature reserves** protect habitats (and so protect the **species** in them) by **restricting urban development**, **industrial development** and **farming**.
- **Endangered species** can be **bred** in **captivity** (e.g. a zoo) to **increase** their numbers, then returned to the **wild**.

You May Have to **Evaluate Evidence** and **Data** About **Conservation Issues**

You need to be able to **evaluate** any **evidence** or **data** about **conservation** projects and research that the examiners throw at you — so here's an example I made earlier:

In recent years, **native British bluebells** have become **less common** in woodland areas. It's thought that this is due to the presence of **non-native Spanish bluebells**, which compete with the native species for a **similar niche**. An experiment was carried out to see if **removing** the invasive Spanish species would help to **conserve** the native species. Each year for 15 years the **percentage cover** of native species was estimated in a **50 m by 50 m** area of **woodland** using random sampling and **250, 1 m² quadrats**. After five years, **all** the Spanish bluebells were **removed**. A **similar-sized control woodland** in which the Spanish bluebells remained **untouched** was also studied. The results are shown on the right. You might be asked to:

1) **Describe the data:**
   - For the first **five years**, the **percentage cover** of **native bluebells fell** from **50%** to around **25%**. After the Spanish species was **removed**, it **increased** from around **25%** to around **45%** in **ten years**.
   - The **control experiment** shows a fairly **steady drop** in native bluebell percentage cover from **60%** to **20%** over the 15 years.

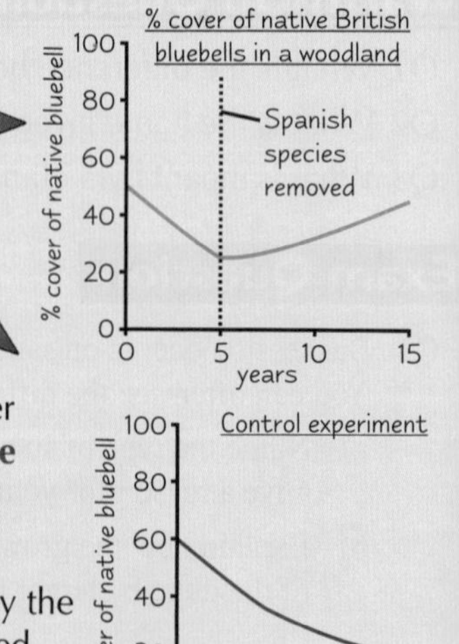

2) **Draw conclusions:**
   The removal of Spanish bluebells **resulted** in an **increase** in the percentage cover of **native bluebells** over a **ten year period**. This suggests that the **recent decrease** in native British bluebells is due to **competition** with the Spanish bluebells.
3) **Evaluate the method:**
   - The effects of some **other variables** (e.g. **changing weather**) were **removed** by the **control experiment**, where the percentage cover of native bluebells continued to fall throughout the 15-year study. This increases the **validity** of the results.
   - The **study area** and **sample size** were quite **large**, giving **more accurate** data.
   - **Random sampling** removed bias — the data's **more likely** to be an **accurate estimate** of the **whole area**.

Conservation

You Need to be Able to Consider Conflicting Evidence

1) The **evidence** from **one study** alone **wouldn't usually be enough** to conclude that there's a **link** between decreasing percentage cover of native bluebells, and the presence of Spanish bluebells.
2) **Similar studies** would be carried out to **investigate** the link. If these studies came to the **same conclusion**, the conclusion would become **increasingly accepted**.
3) Sometimes studies come up with **conflicting evidence** though — evidence that leads to a **different conclusion** than other studies. For example:

Another study was carried out to **investigate** the effect on native bluebells of **removing** Spanish bluebells. It was **similar** to the study above except a **20 m by 20 m** area was sampled using a random sample of **20 quadrats**, and **no control** woodland was used. You might be asked to:

1) **Describe the data:**
   In the first five years, the **percentage cover** of **native bluebells fell** from **50%** to around **25%**. After the Spanish species was **removed**, it **kept decreasing** to around **15%** after the **full 15** years.
2) **Draw conclusions:**
   The **removal** of the Spanish bluebells had **no effect** on the **decreasing** percentage cover of native bluebells — which **conflicts** with the study on the previous page.
3) **Evaluate the method:**
   - There **wasn't** a **control** woodland, so the **continuing decrease** in native bluebell cover after the removal of the Spanish bluebells could be due to **another factor**, e.g. cold weather in years 5-10.
   - The **study area** and **sample size** were quite small, giving a **less accurate** total percentage cover.

Warm-Up Questions

PRACTICE QUESTIONS

Q1 What is conservation?

Q2 Briefly describe why conflict can occur over conservation issues.

Q3 Suggest one conservation technique that could be used to protect plant species.

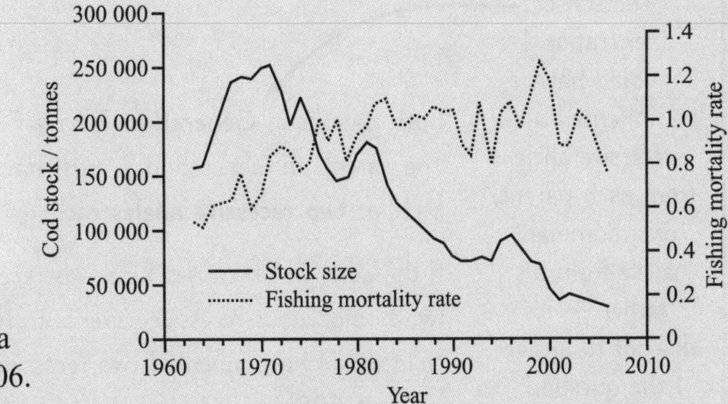

Exam Questions

Q1 The graph shows the stock of spawning cod in the North Sea and the rate of mortality caused by fishing from 1963 to 2006.

a) Suggest a conclusion that could be drawn from the graph. [2 marks]

b) How might this data be used to make informed decisions about the conservation of cod stocks? [1 mark]

c) Suggest why there might be conflict between conservationists and the North Sea fishing industry. [2 marks]

Q2 Read the following passage and then answer the questions that follow.

Wood, or timber, is an important resource in the UK. It is used in the building industry, to make furniture and as a fuel. Woodlands are also important habitats for many native species. Some deciduous woodland in the UK is managed through a technique called coppicing with standards. When a woodland is managed in this way, just over half of the trees in the woodland are coppiced. This means that the trees are cut down to the stump, and allowed to regrow from shoots which grow from the base of the stump. The rest of the trees are not cut down and are left to grow and mature as normal. These trees are called standards. It's recommended that no more than 40% of the canopy is made up of standard trees.

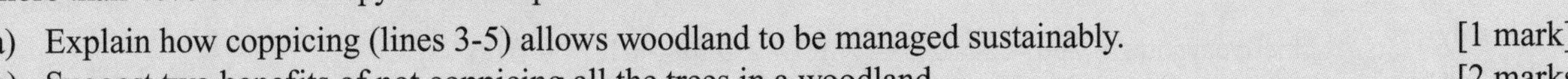

a) Explain how coppicing (lines 3-5) allows woodland to be managed sustainably. [1 mark]

b) Suggest two benefits of not coppicing all the trees in a woodland. [2 marks]

c) Suggest why it is necessary to restrict how much of the canopy is made up of standards (lines 6-7). [1 mark]

I'm considering conflict after these pages, I tell you...

Ah hah ha, aaaah ha ha ha... oh, I think I need to stop my evil laugh now. Evaluating evidence and data's an important nut to crack — you might have to do it in your exams for conservation, or for another topic altogether.

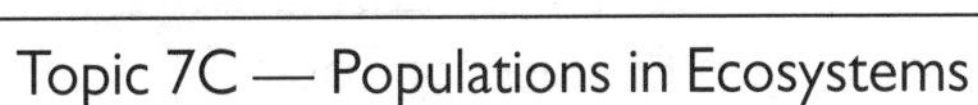

Extra Exam Practice

It's almost time to crawl back to my favourite niche — the sofa.
Topic 7 is nearly all wrapped up, but there's one more thing you need to do. That's right...
Give these questions mixing parts of Topic 7 a go...

- Have a look at this example of how to answer a tricky exam question.
- Then check how much you've understood from Topic 7 by having a go at the questions on the next page.

Pssst... There are some synoptic questions on p.233-241 for you to have a go at when you feel confident with all the topics.

1 Epidermolysis bullosa simplex (EBS) is a skin disorder caused by a dominant allele. **Figure 1** shows the incidence of EBS in a family, over three generations. Horizontal lines between two individuals link mating parents and vertical lines link parents and offspring.

Figure 1

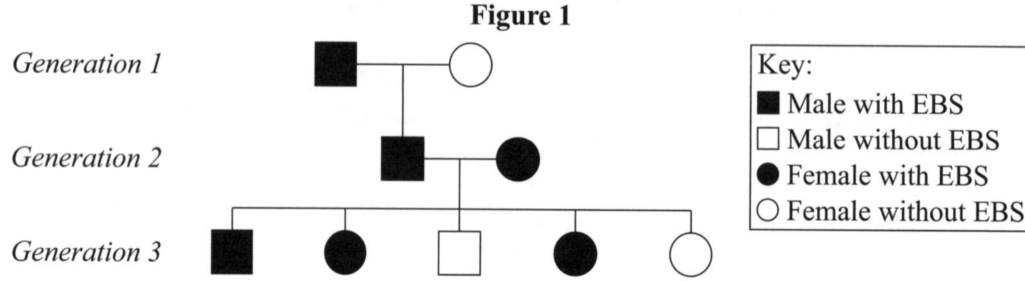

1.1 Explain how **Figure 1** shows that the allele which causes EBS is dominant.

(2 marks)

1.2 It is estimated that 1 in every 17 000 children that are born in the UK, will inherit EBS. Use the Hardy-Weinberg equations to calculate the percentage of people in the UK that are heterozygous for EBS. Show your working.

(3 marks)

The son of Generation 1 could have inherited a recessive allele from each parent, or a dominant allele from his father — it's difficult to answer the question until you look at the offspring of Generation 2.

1.1

Both parents in **Generation 2** have EBS but two of their children don't. The parents in Generation 2 must have, either, at least **one dominant allele** each, or **two recessive alleles** each.

If the allele which causes EBS was recessive then it would be **impossible** for two homozygous recessive parents to have any children without EBS, as each child would have inherited two recessive EBS alleles. As there are **two children without EBS**, it can be concluded that the allele which causes EBS is dominant.

You'd pick up 1 mark for identifying that both parents in Generation 2 must carry a dominant allele and 1 mark for explaining how you know this based on the phenotypes of Generation 3.

1.2

Use the information from the question to work out where to start.

EBS is **coded for by a dominant allele** so children with the disease could have the homozygous dominant genotype or the heterozygous genotype.

So, $p^2 + 2pq = 1 \div 17\,000 = 5.88... \times 10^{-5}$

$\mathbf{p^2 + 2pq + q^2 = 1}$, so $q^2 = 1 - \mathbf{5.88... \times 10^{-5}} = 9.99... \times 10^{-1}$

$q = \sqrt{9.99... \times 10^{-1}} = 9.99... \times 10^{-1}$

$\mathbf{p + q = 1}$, so $p = 1 - 9.99... \times 10^{-1} = 2.94... \times 10^{-5}$

Heterozygotes $= 2pq = 2 \times 2.94... \times 10^{-5} \times 9.99... \times 10^{-1} = 5.88... \times 10^{-5}$

Percentage $= 5.88... \times 10^{-5} \times \mathbf{100} = 0.00588...\% = \mathbf{0.006\%}$

You'd get 3 marks for the correct answer, but if your final answer was wrong you could still pick up 1 mark for correctly identifying $p^2 + 2pq$ as children with EBS, and 1 mark for correctly identifying $2pq$ as heterozygotes.

You should know the Hardy-Weinberg equations and what they mean — see pages 176-177.

Don't round until the end. You can use the 'Ans' button on your calculator to carry the result of one step over to the next.

You've been asked for a percentage so convert from a decimal by multiplying by 100.

Extra Exam Practice

2 **Figure 2** shows the biomass of three different types of plants during succession of a woodland.

Figure 2

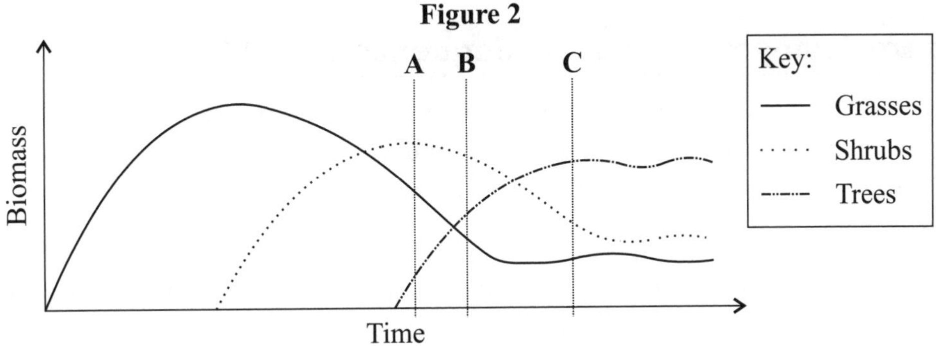

2.1 The grasses shown in **Figure 2** require soil in order to grow.
Explain why this means that **Figure 2** does not show primary succession.

(2 marks)

2.2 Explain how removing trees at point **A** may work as a method of habitat conservation in the woodland.

(2 marks)

2.3 The carrying capacity of several shrub species in the woodland is lower at point **C** than it is at point **B**. Using **Figure 2**, explain why interspecific competition may be responsible for this.

(2 marks)

Wildfires are rapidly spreading fires that can occur naturally in wild areas such as forests. These fires have shaped the evolution of many tree species.

2.4 Trees growing in areas with frequent wildfires often have thicker bark than trees growing in areas with very infrequent wildfires. The thicker bark helps to protect the trees when there is a wildfire. Explain the type of natural selection that may have acted on populations of trees frequently exposed to wildfires.

(3 marks)

A team of scientists wanted to assess the effect of a wildfire on a woodland squirrel population. They used a mark-release-recapture method to sample the squirrels shortly after the wildfire. They also obtained data from similar study that had been done several months before the wildfire. The data they collected is shown in **Figure 3**.

Figure 3

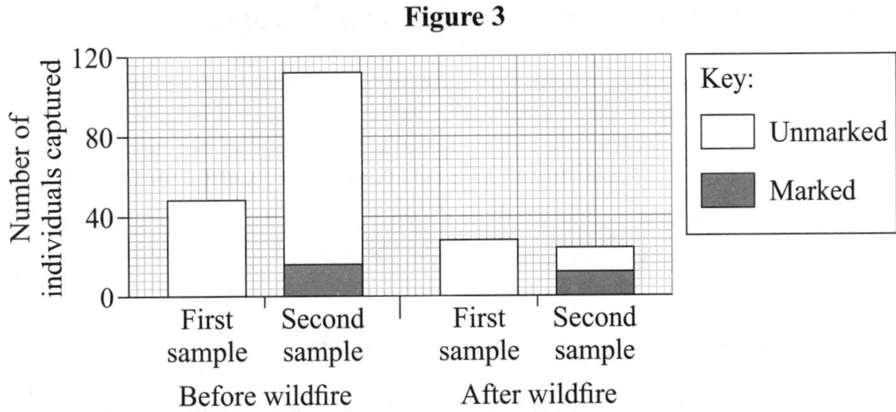

2.5 Before the wildfire, the squirrel population was calculated as 336.
Use **Figure 3** to calculate the percentage change in the population size of the squirrels following the wildfire. Show your working.

(3 marks)

Mutations

Unfortunately, mutations don't usually give you special powers like in superhero movies — in fact, they can be quite harmful. You've already covered mutations in Topic 4, but now you need to know about them in more detail.

Mutations are **Changes** to the **Base Sequence** of **DNA**

1) Any change to the **base (nucleotide) sequence** of DNA is called a **mutation**.
2) Mutations can be caused by **errors** during **DNA replication**.
3) The rate of mutation can be increased by **mutagenic agents** (see next page).
4) The **types** of mutations that can occur include:

- **Substitution** — one or more bases are swapped for another, e.g. ATGCCT becomes ATTCCT.
- **Deletion** — one or more bases are removed, e.g. ATGCCT becomes ATCCT.
- **Addition** — one or more bases are added, e.g. ATGCCT becomes ATGACCT.
- **Duplication** — one or more bases are repeated, e.g. ATGCCT becomes ATGCCCCT.
- **Inversion** — a sequence of bases is reversed, e.g. ATGCCT becomes ACCGTT.
- **Translocation** — a sequence of bases is moved from one location in the genome to another. This could be movement within the same chromosome or movement to a different chromosome.

5) The **order** of **DNA bases** in a gene determines the **sequence of amino acids** in a particular **polypeptide**. If a mutation occurs in a gene, the **sequence** of **amino acids** in the **polypeptide** that it **codes for** could be **changed**.

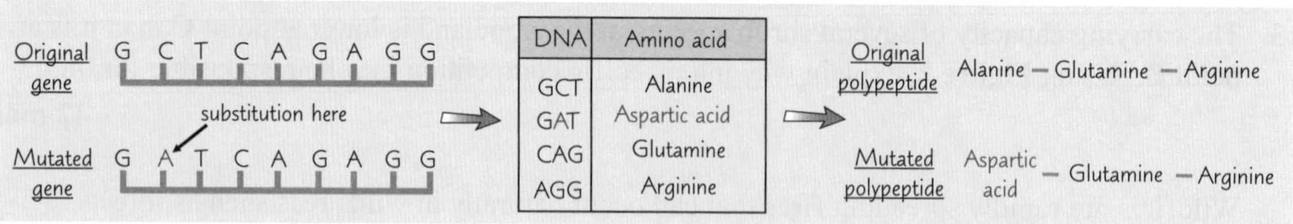

6) Polypeptides make up proteins. A change in the amino acid sequence of a polypeptide may **change** the final **3D shape** of the **protein**, which could mean that it **doesn't work** properly. E.g. a mutation in a polypeptide that makes up an **enzyme** may **change the shape** of the enzyme's **active site**. This may **stop substrates** from being able to **bind to the active site**, leaving the enzyme **unable to catalyse** the reaction.
7) Some mutations can cause **genetic disorders** — inherited disorders caused by **abnormal genes** or **chromosomes**, e.g. cystic fibrosis. Some mutations can **increase** the **likelihood** of developing certain **cancers**, e.g. mutations of the gene **BRCA1** can increase the chances of developing **breast cancer**.
8) If a **gamete** (sex cell) containing a mutation for a genetic disorder or a type of cancer is **fertilised**, the mutation will be present in the new **fetus** formed — these are called **hereditary mutations** because they are passed on to the offspring.

Not all hereditary mutations are harmful — beneficial hereditary mutations drive evolution (see page 178).

Not All Mutations Affect the **Order** of **Amino Acids...**

The **degenerate nature** of the genetic code means that some amino acids are coded for by **more than one DNA triplet** (e.g. tyrosine can be coded for by TAT or TAC in DNA). This means **not all** types of mutation will **always** result in a change to the **amino acid sequence** of the **polypeptide**. For example, some **substitutions** will still **code** for the **same amino acid**:

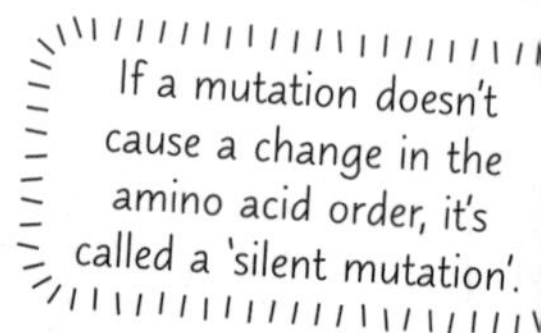

DNA	Amino acid
TAT	Tyrosine
TAC	Tyrosine
AGT	Serine
CTT	Leucine
CTG	Leucine
TTG	Leucine
GTT	Valine
GTC	Valine

Sometimes, **inversion** mutations **don't** cause a **change** in the amino acid **sequence** either:

Mutations

...but Some Types of Mutation Do

1) **Additions**, **duplications** and **deletions** within a gene will almost always **change** the **amino acid sequence** of a polypeptide.
2) That's because these mutations all **change** the **number** of **bases** in the DNA code.
3) This causes a **shift** (called a **frameshift**) in the **base triplets** that **follow**, so that the **triplet code** is read in a **different way**.
4) Here's how a **deletion** can cause a **frameshift** and change the amino acid order:

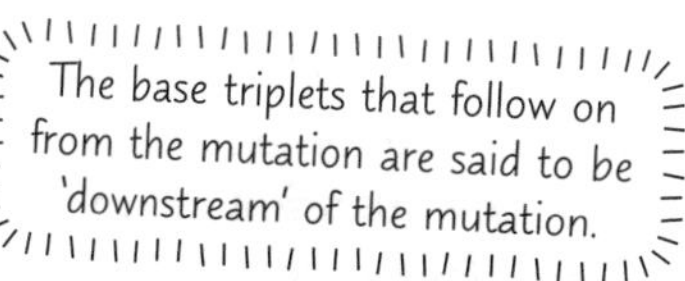

DNA	Amino acid
TAT	Tyrosine
TAC	Tyrosine
AGT	Serine
CTT	Leucine
GTC	Valine

Mutagenic Agents Increase the Rate of Mutation

Mutations occur **spontaneously**, e.g. when DNA is **misread** during **replication**. But some things can **increase** the **rate** of **mutations** — these are called **mutagenic agents**. **Ultraviolet radiation**, **ionising radiation**, some **chemicals** and some **viruses** are examples of mutagenic agents. They can increase the rate of mutations by:

1) **Acting as a base** — chemicals called **base analogs** can **substitute** for a base during DNA replication, **changing** the **base sequence** in the new DNA. E.g. **5-bromouracil** is a base analog that can substitute for **thymine**. It can pair with **guanine** (**instead** of **adenine**), causing a **substitution mutation** in the new DNA.
2) **Altering bases** — some chemicals can **delete** or **alter bases**. E.g. **alkylating agents** can add an alkyl group to **guanine**, which **changes** the **structure** so that it pairs with **thymine** (**instead** of **cytosine**).
3) **Changing** the **structure of DNA** — some types of **radiation** can change the structure of DNA, which causes **problems** during DNA replication. E.g. **UV radiation** can cause adjacent **thymine** bases to **pair up** together.

It may have been the sunniest summer on record, but nobody expected the extra UV radiation to have such disturbing effects on the pumpkin patch.

Warm-Up Questions

Q1 What is a substitution mutation?

Q2 What is the difference between a duplication and an addition mutation?

Q3 What is an inversion mutation?

Q4 What are mutagenic agents?

Q5 List three common mutagenic agents.

Exam Question

Before exposure: A G T T A T C A G G C T

After exposure: A G G T A T G A G G C C

DNA	Amino acids	DNA	Amino acids
AGT	Serine	GAG	Glutamic acid
AGG	Arginine	GCT	Alanine
TAT	Tyrosine	GCC	Alanine
CAG	Glutamine		

Q1 The order of bases in a gene before and after exposure to a mutagenic agent is shown above.

a) Underline any mutation(s) that have occurred. [1 mark]

b) Use the table to explain the changes that the mutations would cause to the sequence of amino acids. [4 marks]

Just hope your brain doesn't have a deletion mutation during the exam...

Right, there's plenty to learn on these pages and some of it's a bit complicated, so you know the drill. Don't read it all through at once — take the sections one by one and get all the facts straight. There could be nothing more fun...

Cancer

Cancer is a disease that affects animals and people of all ages. There are lots of different types of cancer, but they all involve uncontrolled cell growth and all have potentially devastating effects. Here's more on how cancer can occur...

Mutations in Genes Can Cause Uncontrolled Cell Growth

1) Mutations that occur in individual cells **after** fertilisation (e.g. in adulthood) are called **acquired mutations**.
2) If these mutations occur in the **genes** that **control** the rate of **cell division** (by mitosis), it can cause **uncontrolled cell division**.
3) If a cell divides uncontrollably the result is a **tumour** — a mass of abnormal cells. Tumours that **invade** and **destroy surrounding tissue** are called **cancers** (see below).
4) There are **two types** of **gene** that control cell division — **tumour suppressor genes** and **proto-oncogenes**. Mutations in these genes can cause cancer:

Tumour suppressor genes can be **inactivated** if a **mutation** occurs in the DNA sequence.

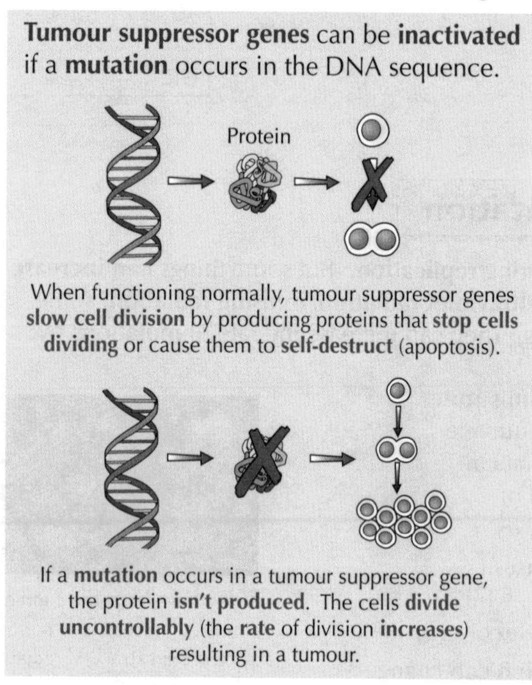

When functioning normally, tumour suppressor genes **slow cell division** by producing proteins that **stop cells dividing** or cause them to **self-destruct** (apoptosis).

If a **mutation** occurs in a tumour suppressor gene, the protein **isn't produced**. The cells **divide uncontrollably** (the **rate** of division **increases**) resulting in a tumour.

The **effect** of a **proto-oncogene** can be **increased** if a **mutation** occurs in the DNA sequence. A mutated proto-oncogene is called an **oncogene**.

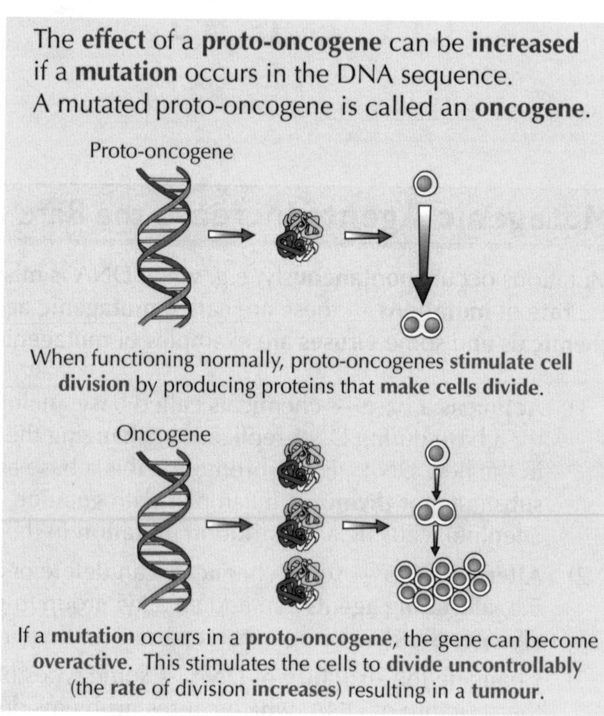

When functioning normally, proto-oncogenes **stimulate cell division** by producing proteins that **make cells divide.**

If a **mutation** occurs in a **proto-oncogene**, the gene can become **overactive**. This stimulates the cells to **divide uncontrollably** (the **rate** of division **increases**) resulting in a **tumour.**

Tumours can be Benign or Malignant (Cancerous)

Tumours can develop for **years** without any obvious symptoms and can be quite **large** by the time they're discovered. **Not** all tumours are **cancerous** — there are **two** different types:

1) **Malignant tumours** are **cancers**. They usually grow **rapidly** and **invade** and **destroy** surrounding tissues. Cells can break off the tumours and **spread** to other parts of the body in the **bloodstream** or **lymphatic system**.
2) **Benign tumours** are not cancerous. They usually grow **slower** than malignant tumours and are often covered in **fibrous tissue** that stops cells invading other tissues. Benign tumours are often **harmless**, but they can cause **blockages** and put **pressure** on organs. Some benign tumours can become **malignant**.

Tumour Cells Look and Function Differently to Normal Cells

Tumour cells **can differ** from normal cells in many **different ways**:

1) They have an **irregular shape**.
2) The **nucleus** is **larger** and **darker** than in normal cells. Sometimes the cells have more than one nucleus.
3) They don't produce all the proteins needed to function correctly.
4) They have **different antigens** on their **surface**.
5) They don't respond to **growth regulating processes**.
6) They divide (by mitosis) **more frequently** than normal cells.

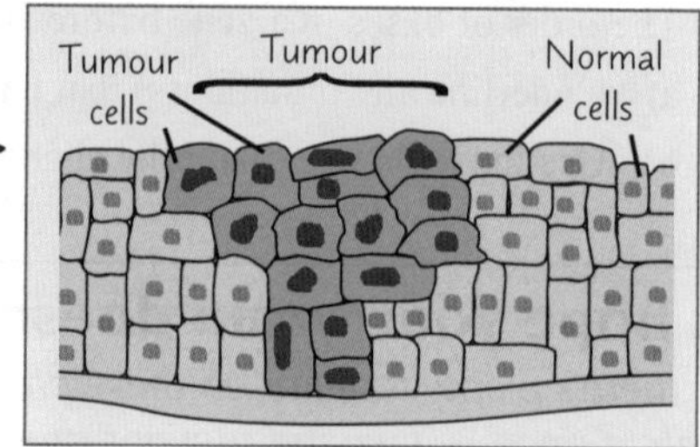

Cancer

Abnormal Methylation of Cancer-Related Genes Can Cause Tumour Growth

1) **Methylation** means **adding** a **methyl** (–CH_3) **group** onto something.
2) **Methylation** of **DNA** is an important method of **regulating gene expression** — it can control **whether** or not a gene is **transcribed** (copied into mRNA) and **translated** (turned into a protein).
3) When methylation is happening **normally**, it plays a **key role** in many processes in the body. It's only when it happens **too much** (**hypermethylation**) or **too little** (**hypomethylation**) that it becomes a **problem**.

For loads more on methylation, see page 207.

4) The **growth** of **tumours** can be **caused** by **abnormal methylation** of certain **cancer-related genes**:

   1) When **tumour suppressor genes** (see previous page) are **hypermethylated**, the genes are **not transcribed** — so the **proteins** they produce to slow cell division **aren't made**. This means that cells are able to **divide uncontrollably** by mitosis and **tumours** can develop.
   2) **Hypomethylation** of **proto-oncogenes** causes them to act as **oncogenes** — **increasing** the **production** of the **proteins** that encourage **cell division**. This stimulates cells to **divide uncontrollably**, which causes the **formation** of **tumours**.

Increased Oestrogen May Contribute to Some Breast Cancers

1) **Increased exposure** to **oestrogen** over an extended period of time is thought to **increase** a woman's **risk** of developing **breast cancer**. (Increased exposure may be the result of starting menstruation earlier than usual or the menopause later than usual. It could also be the result of taking oestrogen-containing drugs, such as HRT.)
2) The **exact** reasons behind this aren't fully understood, but there are a few theories as to how oestrogen can contribute to the **development** of some **breast cancers**:

   1) **Oestrogen** can **stimulate** certain **breast cells** to **divide** and **replicate**. The fact that **more cell divisions** are taking place naturally **increases** the chance of **mutations** occurring, and so **increases** the chance of cells **becoming cancerous**.
   2) This ability to **stimulate division** could also mean that if cells do become **cancerous**, their rapid **replication** could be **further assisted** by **oestrogen**, helping **tumours** to **form quickly**.
   3) Other research suggests that oestrogen is actually able to **introduce mutations directly** into the **DNA** of certain **breast cells**, again **increasing** the chance of these cells becoming **cancerous**.

Warm-Up Questions

PRACTICE QUESTIONS

Q1 What is a tumour suppressor gene?

Q2 What is the difference between a proto-oncogene and an oncogene?

Q3 What is hypermethylation?

Exam Question

Q1 A woman has been diagnosed with cancer. Her doctor has told her that she has a malignant tumour in her left breast.

a) Describe two differences between benign and malignant tumours. [2 marks]

b) Describe how tumours can arise from mutations in DNA. [5 marks]

c) Increased exposure to oestrogen has been linked to some breast cancers. How might oestrogen contribute to causing breast cancer? [4 marks]

Remember, only malignant tumours are cancerous...

You need to understand the difference between benign and malignant. You should never say benign cancers — there's no such thing. Only malignant tumours are cancerous. Make sure that you also know all about the roles that oncogenes and tumour suppressor genes play in causing cancer, as well as the roles of DNA methylation and oestrogen.

Interpreting Data on Cancer

Okay... these pages are a bit daunting. Nevertheless, they're important. Some of the stuff is pretty hard to get your head around, so you'll have to concentrate. After that, take a break and relax. Maybe cut your toenails.

Genetic and **Environmental** Factors Affect the **Risk** of **Cancer**

There's **no single cause** for cancer but scientists have identified lots of different '**risk factors**' — things that **increase** a person's **chance** of getting cancer. Risk factors can be either **genetic** or **environmental**:

1) **Genetic factors** — some cancers are linked with **specific inherited alleles**. If you **inherit** that allele you're **more likely** to get that type of cancer (but it **doesn't mean** you'll **definitely** get that type of cancer).
2) **Environmental factors** — exposure to **radiation**, **lifestyle choices** such as **smoking**, increased **alcohol consumption**, and a **high-fat diet** have all been **linked** to an increased chance of developing some cancers.

It's Difficult to **Interpret** the **Relative Contributions** of **Genes** and **Environment**

1) Data on variation can be very tricky to **interpret** because some characteristics can be affected by **many different genes** (they're polygenic) and **many environmental factors**.
2) It's difficult to know **which factors** (genes or environment) are having the **greatest effect**.
3) This makes it **hard** to **draw conclusions** about the **causes of variation**.

Example: The Effects of Genetic and Environmental Factors on Breast Cancer

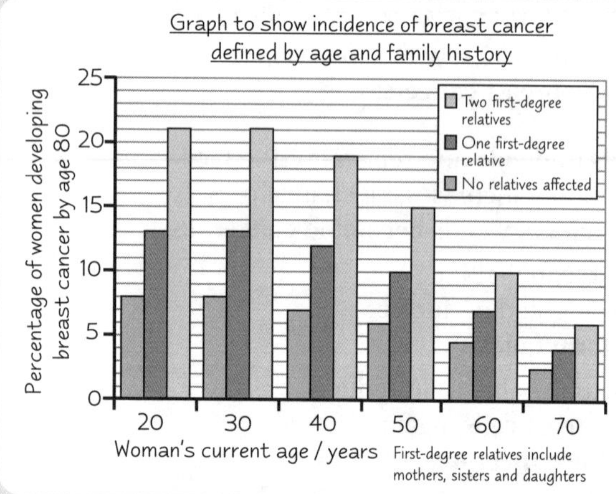

This graph shows how the **incidence** of **breast cancer** is affected by both **age** and **family history**. There's a **positive correlation** between **incidence** of breast cancer in women and the number of their **first-degree relatives** who have also had breast cancer. The effect of family history **decreases** with **age**, but the incidence of breast cancer is always **higher** in women with a **close family history** of the disease. A woman is **more likely** to develop breast cancer if **members of her family** have had breast cancer, which suggests a **genetic link**.

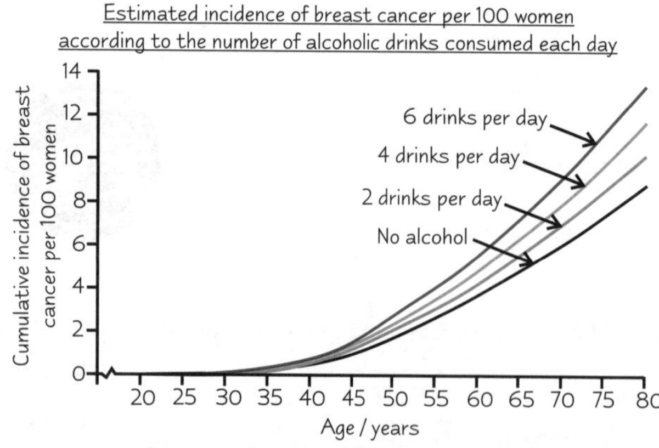

This graph shows that the **incidence** of **breast cancer** is linked to both **age** and **alcohol consumption**. The graph shows that the **incidence** of breast cancer in women **increases** with **age** — i.e. there's a **positive correlation** between incidence of breast cancer and age. There's also a **positive correlation** between the **number** of **alcoholic drinks** consumed **each day** and **incidence** of breast cancer. Alcohol consumption is an **environmental factor.**

1) If you only saw **one** of these graphs you may think **only genetics and age**, or **only alcohol consumption and age**, affect your **risk** of developing **breast cancer**.
2) When you look at **both sets of data** you can see that **all** these things affect the risk.
3) It's **difficult** to tell **which factor** (genes or alcohol) has the **largest effect**.
4) Also, there are **other environmental factors** that are thought to be involved in increasing the risk of developing breast cancer (e.g. **diet**, **exercise**, etc.) that aren't considered here.

There's more on correlations and cause on page 229.

Interpreting Data on Cancer

Knowing the Mutation is Useful for the Prevention and Treatment of Cancer

1) **Cancer** is caused by **mutations** in **proto-oncogenes** and **tumour suppressor genes** (see page 196).
2) **Understanding** the **role** that these genes play in **causing cancer**, and knowing exactly **how** they **work**, can be really **helpful** for coming up with ways to **prevent**, **treat** and **cure** cancer.
3) Here are a few examples:

Prevention

1) If a **specific** cancer-causing **mutation** is **known**, then it is possible to **screen for** (look for) the mutation in a person's DNA (see page 218). E.g. it's possible to screen for the mutated allele of the **BRCA1 tumour suppressor gene**, which greatly increases a woman's risk of developing **breast cancer** in her lifetime.
2) Knowing about this increased risk means that **preventative steps** can be taken to reduce it. E.g. a woman with the BRCA1 mutation may choose to have a **mastectomy** (removal of one or both breasts) to significantly **reduce the risk** of breast cancer **developing**. Women with this mutation may also be **screened** for **signs of breast cancer more often** than the rest of the population, as **early diagnosis increases** the chances of **recovery**.
3) Knowing about specific mutations also means that **more sensitive tests** can be **developed**, which can lead to **earlier** and **more accurate diagnosis**. For example, there's a **mutation** in the **RAS proto-oncogene** in around **half** of all **bowel cancers**. Bowel cancer can be **detected early** by looking for RAS mutations in the DNA of **bowel cells**.

Treatment and Cure

1) The **treatment** for **cancer** can be **different** for different mutations, so knowing how **specific mutations** actually **cause cancer** can be very useful for **developing drugs** to **effectively target** them. For example, **breast cancer** caused by a mutation of the **HER2 proto-oncogene** can be treated with a drug called **Herceptin®**. This drug binds **specifically** to the altered HER2 protein receptor and **suppresses cell division** and **tumour growth**. Breast cancer caused by other mutations is **not** treated with this drug as it doesn't work.
2) Some cancer-causing mutations require **more aggressive treatment** than others, so understanding how the mutation that causes them works can help **produce** the **best treatment plan**. E.g. if a mutation is known to cause an **aggressive** (**fast-growing**) cancer, it may be treated with **higher doses** of **radiotherapy** or by **removing larger areas** of the **tumour** and **surrounding tissue** during surgery.
3) **Gene therapy** (where **faulty alleles** in a person's cells are **replaced** by working versions of those alleles — see page 217) may also be able to treat cancer caused by some mutations. For example, if you know that the cancer is being caused by **inactivated tumour suppressor genes**, it's hoped that gene therapy could be used in the future to provide **working versions** of the genes.

Warm-Up Questions

Q1 Give three environmental factors that have been linked to an increased risk of developing cancers.

Q2 How can understanding a specific mutation in a cancer-related gene help treat cancer?

Exam Question

Q1 Possessing a faulty allele of the BRCA1 tumour suppressor gene significantly increases the chance of a woman developing breast cancer in her lifetime. A woman may have her DNA screened for this faulty allele if she has a close family history of breast cancer.

Explain why the ability to screen DNA for the faulty allele may help to prevent a woman with this mutation dying from breast cancer. [4 marks]

Relative contributions — a tenner on your birthday...

In the exam, you might have to evaluate evidence showing correlations between genetic and environmental factors and cancer. Just remember that there are usually several factors at work and that correlation doesn't always mean cause — see page 229 for more. Take a proper look at these examples to help get yourself into the right way of thinking.

Stem Cells

Stem cells — they're the daddy of all cells, the big cheese, the top dog, and the head honcho. And here's why...

Totipotent Stem Cells are Able to **Mature** into **Any Type** of **Body Cell**

1) **Multicellular organisms** are made up from many **different cell types** that are **specialised** for their function, e.g. liver cells, muscle cells, white blood cells.
2) **All** these specialised cell types originally came from **stem cells**.
3) Stem cells are **unspecialised** cells that can develop into **other types** of cell.
4) Stem cells divide to become **new** cells, which then become **specialised**.
5) All multicellular organisms have some form of stem cell.
6) Stem cells are found in the **embryo** (where they become all the **specialised cells** needed to form a **fetus**) and in **some adult tissues** (where they become **specialised** cells that need to be **replaced**, e.g. stem cells in the intestines constantly replace intestinal epithelial cells).
7) Stem cells that can mature (develop) into **any type** of **body cell** in an organism, (including the cells that make up the placenta in mammals) are called **totipotent cells**.
8) **Totipotent** stem cells are **only** present in mammals in the **first few cell divisions** of an **embryo**.
9) After this point the **embryonic stem cells** become **pluripotent**. They can still specialise into **any** cell in the body, but **lose** the **ability** to become the cells that make up the **placenta**.
10) The stem cells present in **adult mammals** are either:

- **Multipotent stem cells** — These are able to differentiate into a **few different types of cell**. For example, both **red** and **white blood cells** can be formed from multipotent stem cells found in **bone marrow**.
- **Unipotent stem cells** — These can only differentiate into **one type of cell**. For example, there's a type of unipotent stem cell that can only divide to produce **epidermal skin cells**, which make up the **outer layer** of your **skin**.

Stem Cells Become **Specialised** Because **Different Genes** are **Expressed**

Stem cells become **specialised** because during their development, they only **transcribe** and **translate part** of their **DNA**:

Remember: transcription is when DNA is copied into mRNA. Translation is when proteins are produced using the code in mRNA.

1) **Stem cells** all contain the **same genes** — but during **development**, **not all** of them are **transcribed** and **translated** (expressed).
2) Under the **right conditions**, some **genes** are **expressed** and others are switched off.
3) **mRNA** is only **transcribed** from **specific genes**.
4) The mRNA from these genes is then **translated** into **proteins**.
5) These proteins **modify** the cell — they determine the cell **structure** and **control cell processes** (including the expression of **more genes**, which produces more proteins).
6) **Changes** to the cell produced by these proteins cause the cell to become **specialised**. These changes are **difficult** to **reverse**, so once a cell has specialised it **stays** specialised.

All of the girls expressed different jeans.

Example: Red Blood Cells

1) **Red blood cells** are produced from a type of **stem cell** in the **bone marrow**. They contain lots of **haemoglobin** and have **no nucleus** (to make room for more haemoglobin).
2) The stem cell produces a new cell in which the genes for **haemoglobin production** are **expressed**. Other genes, such as those involved in **removing the nucleus**, are **expressed** too. Many other genes are not expressed (switched off), resulting in a specialised red blood cell.

Stem Cells

Cardiomyocytes Can be Made from **Unipotent Stem Cells**

1) **Cardiomyocytes** are **heart muscle cells** that **make up** a lot of the **tissue** in our hearts. In **mature** mammals, it's thought that they **can't divide** to **replicate** themselves.
2) This meant that for ages, everyone thought that we **weren't able** to **regenerate** our own **heart cells** at all. This is a major problem if the heart becomes **damaged**, e.g. by a **heart attack**, or the cells became **worn out** through age. **Recent research** however, has suggested that our hearts **do** have some **regenerative capability**.
3) Some scientists now think that **old** or **damaged** cardiomyocytes can be **replaced** by **new** cardiomyocytes **derived** from a small supply of **unipotent stem cells** in the heart.
4) Some researchers think that this process could be **constantly occurring**, but haven't yet agreed on **how quickly** it happens.
    - Some believe that it's a **really slow** process and that it's possible that **some** cardiomyocytes are **never replaced** throughout a person's entire **lifetime**.
    - Others think that it's occurring **more quickly**, so that **every** cardiomyocyte in the heart is **replaced several times** in a lifetime.

Stem Cells Can be Used to **Treat Human Disorders**

Since stem cells can divide into a **range** of **specialised cell types**, doctors and scientists think they could be used to **replace** cells **damaged** by illness or injury.

Some **Stem Cell Therapies Already Exist**

1) Some stem cell therapies **already exist** for some diseases affecting the **blood** and **immune system**.
2) **Bone marrow** contains **stem cells** that can become specialised to form **any type** of **blood cell**. **Bone marrow transplants** can be used to replace the **faulty** bone marrow in patients that produce **abnormal blood cells**. The stem cells in the transplanted bone marrow **divide** and **specialise** to produce healthy blood cells.
3) This technique has been used successfully to treat **leukaemia** (a **cancer** of the blood or bone marrow) and **lymphoma** (a cancer of the **lymphatic system**).
4) It has also been used to treat some **genetic disorders**, such as **sickle-cell anaemia** and **severe combined immunodeficiency** (**SCID**):

Example

Severe combined immunodeficiency (**SCID**) is a genetic disorder that affects the immune system. People with SCID have a **poorly functioning immune system** as their **white blood cells** (made in the bone marrow from stem cells) are **defective**. This means they **can't defend** the body against infections by identifying and destroying microorganisms. So SCID sufferers are **extremely susceptible** to **infections**. Treatment with a **bone marrow transplant** replaces the faulty bone marrow with donor bone marrow that contains **stem cells without** the **faulty genes** that cause SCID. These then **differentiate** to produce **functional** white blood cells. These cells can identify and destroy invading pathogens, so the **immune system functions properly**.

Stem Cells Could be Used to **Treat Other Diseases**

Scientists are **researching** the use of stem cells as **treatment** for lots of conditions, including:

- **Spinal cord injuries** — stem cells could be used to replace damaged **nerve tissue**.
- **Heart disease** and **damage caused by heart attacks** — stem cells could be used to replace damaged **heart tissue**.
- **Bladder conditions** — stem cells could be used to grow **whole bladders**, which are then **implanted** in patients to replace diseased ones.
- **Respiratory diseases** — **donated windpipes** can be stripped down to their simple collagen structure and then covered with **tissue** generated by stem cells. This can then be **transplanted** into patients.
- **Organ transplants** — organs could be **grown** from stem cells to provide new organs for people on **donor waiting lists**.

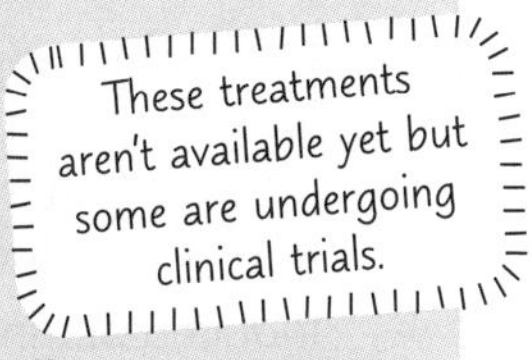

Stem Cells

There are **Huge Benefits** to Using **Stem Cells** in **Medicine**

People who make **decisions** about the **use** of stem cells to treat human disorders have to consider the **potential benefits** of stem cell therapies:

- They could **save** many **lives** — e.g. many people waiting for organ transplants **die** before a **donor organ** becomes available. Stem cells could be used to **grow organs** for those people awaiting transplants.
- They could **improve** the **quality of life** for many people — e.g. stem cells could be used to replace damaged cells in the eyes of people who are **blind**.

Human **Stem Cells** Can Come from **Adult Tissue** or **Embryos**

To **use stem cells** scientists have to get them from somewhere.
There are **three** main potential **sources** of human stem cells:

1 Adult Stem Cells

1) These are obtained from the **body tissues** of an **adult**. For example, adult stem cells are found in **bone marrow**.
2) They can be obtained in a relatively **simple operation** — with very **little risk** involved, but quite **a lot** of **discomfort**.
3) Adult stem cells **aren't** as **flexible** as embryonic stem cells — they can only specialise into a **limited** range of cells, not all body cell types (they're **multipotent**).

2 Embryonic Stem Cells

1) These are obtained from **embryos** at an **early stage of development**.
2) Embryos are created in a **laboratory** using ***in vitro*** **fertilisation** (IVF) — **egg cells** are **fertilised** by sperm **outside the womb**.
3) Once the embryos are approximately **4 to 5 days old**, **stem cells** are **removed** from them and the rest of the embryo is **destroyed**.
4) Embryonic stem cells can divide an **unlimited number** of times and develop into **all types** of body cells (they're **pluripotent**).

3 Induced Pluripotent Stem Cells (iPS Cells)

1) iPS cells are created by scientists in the **lab**. The process involves **'reprogramming' specialised adult body cells** so that they **become pluripotent**.
2) The adult cells are made to **express** a series of **transcription factors** that are normally associated with pluripotent stem cells. The transcription factors cause the adult body cells to **express genes** that are associated with pluripotency.
3) One of the ways that these transcription factors can be **introduced** to the **adult cells** is by **infecting** them with a **specially-modified virus**. The virus has the **genes coding** for the transcription factors within its DNA. When the virus **infects** the **adult cell**, these **genes** are passed **into the adult cell's DNA**, meaning that the cell is **able** to **produce** the **transcription factors**.
4) Induced pluripotent stem cells could become really useful in **research** and **medicine** in the future — see next page. At the moment though, **more research** into how **similar** they actually are to **true pluripotent embryonic stem cells** is needed before they can be properly utilised.

Transcription factors are proteins that control whether or not genes are transcribed — see page 204 for more.

Stem Cells

There are Ethical Issues Surrounding Embryonic Stem Cell Use

1) Obtaining stem cells from **embryos** created by IVF raises **ethical issues** because the procedure results in the **destruction** of an embryo that **could** become a **fetus** if placed in a **womb**.
2) Some people believe that at the moment of **fertilisation** an **individual** is formed that has the **right** to **life** — so they believe that it's **wrong** to **destroy** embryos.
3) Some people have **fewer objections** to stem cells being **obtained** from **egg cells** that **haven't** been fertilised by sperm, but have been **artificially activated** to start **dividing**. This is because the cells **couldn't survive** past a few days and **wouldn't** produce a fetus if placed in a womb.
4) Some people think that **scientists** should **only use** adult stem cells because their production **doesn't** destroy an embryo. But adult stem cells **can't** develop into all the specialised cell types that embryonic stem cells can.
5) This is where **induced pluripotent stem cells** could prove really useful. They have the potential to be **as flexible** as **embryonic stem cells**, but, as they're **obtained** from **adult tissue**, there **aren't** the same **ethical issues** surrounding their use. Good news all round.
6) It's also possible that iPS cells could be made from a **patient's own cells**. These iPS cells, which would be **genetically identical** to the patient's cells, could then be used to **grow** some **new tissue** or **an organ** that the patient's body **wouldn't reject** (rejection of transplants occurs quite often and is caused by the patient's immune system recognising the tissue as **foreign** and **attacking it**).
7) The decision makers in **society** have to take into account **everyone's views** when making decisions about **important scientific work** like stem cell research and its use to treat human disorders.

You might be asked to evaluate the use of stem cells in treating human disorders in the exams — so make sure you know all the pros and cons on pages 201 to 203.

Warm-Up Questions

Q1 At what stage of development can totipotent stem cells be found in mammals?

Q2 Describe the difference between pluripotent and multipotent stem cells.

Q3 How do stem cells become specialised?

Q4 Name two conditions that stem cells could potentially be used to treat.

Q5 Describe one difference between embryonic and adult stem cells.

Tina, Joe and Bex knew their cells were specialised — specialised to look good.

PRACTICE QUESTIONS

Exam Question

Q1 Scientists are currently exploring the potential for the use of stem cells in medicine.

a) Explain one way in which stem cell therapy is currently being used. [4 marks]

b) Explain why some people object to the use of embryonic stem cells in treating human disorders. [2 marks]

It may be possible to use induced pluripotent stem cells (iPS cells) instead of embryonic stem cells to treat human disorders.

c) Describe how induced pluripotent stem cells can be produced. [4 marks]

It's OK — you can grow yourself a new brain especially for this revision...

Stem cells are pretty amazing when you think about it — some can differentiate into absolutely any cell type needed to form an organism. I guess that makes them the cellular equivalent to those giant penknives that have a tool for everything. Totipotent stem cells are the most flexible, followed by pluripotent, multipotent, then unipotent stem cells. Some stem cells are already being used in medicine, but their full potential isn't currently being met. You need to be able to evaluate the use of stem cells in medicine, taking all the benefits and drawbacks on these pages into account.

Regulation of Transcription and Translation

These pages cover some of the ways that transcription and translation are regulated. It's really all incredibly clever.

Transcription Factors Control the Transcription of Target Genes

You should remember from Topic 4 that **transcription** is when a **gene** is **copied** from DNA into **messenger RNA** (**mRNA**). The enzyme responsible for synthesising mRNA from DNA is called **RNA polymerase**.

1) **All** the **cells** in an organism carry the **same genes** (DNA) but the **structure** and **function** of different cells **varies**.
2) This is because **not all** the **genes** in a cell are **expressed** (transcribed and used to make a protein).
3) Because **different genes** are expressed, **different proteins** are made and these proteins modify the cell — they determine the **cell structure** and control **cell processes** (including the expression of more genes, which produce more proteins).
4) The **transcription** of genes is **controlled** by protein molecules called **transcription factors**:

1) In eukaryotes, transcription factors **move** from the **cytoplasm** to the **nucleus**.
2) In the nucleus they **bind** to **specific DNA sites** near the start of their **target genes** — the genes they **control** the expression of.
3) They control expression by controlling the **rate** of transcription.
4) Some transcription factors, called **activators**, **stimulate** or **increase** the **rate of transcription** — e.g. they help **RNA polymerase bind** to the start of the target gene and **activate** transcription.
5) Other transcription factors, called **repressors**, **inhibit** or **decrease** the **rate of transcription** — e.g. they **bind** to the start of the target gene, **preventing RNA polymerase** from **binding**, **stopping** transcription.

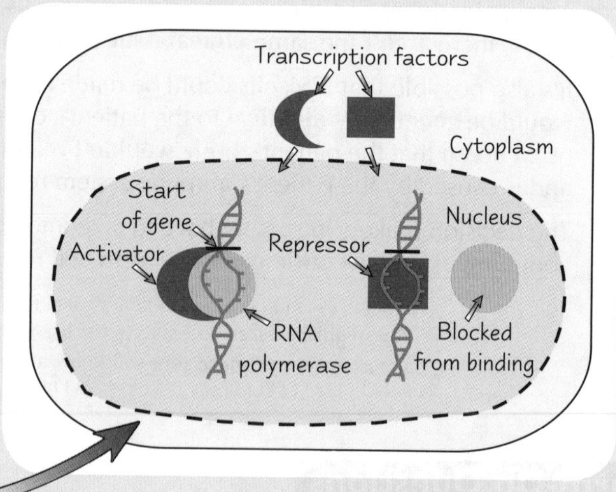

Oestrogen Can Initiate the Transcription of Target Genes

The **expression** of **genes** can also be **affected** by **other molecules** in the cell, e.g. **oestrogen**:

1) Oestrogen is a steroid **hormone** that can affect transcription by **binding** to a **transcription factor** called an **oestrogen receptor**, forming an **oestrogen-oestrogen receptor complex**.
2) The complex moves from the **cytoplasm** into the **nucleus** where it **binds** to **specific DNA sites** near the **start** of the **target gene**.
3) The complex can act as an **activator** of transcription, e.g. **helping** RNA polymerase bind to the start of the target gene.

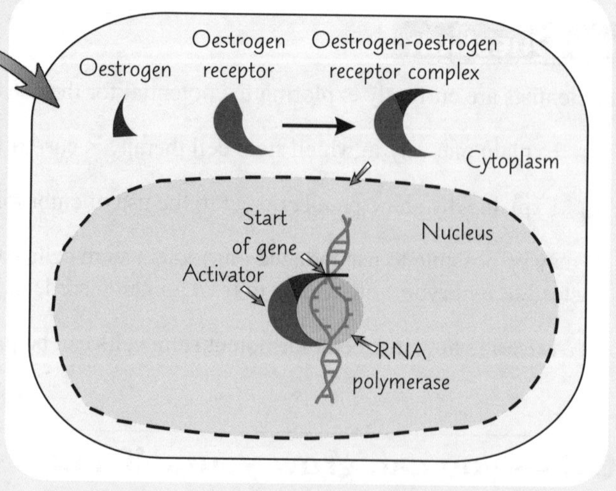

In some cells, the oestrogen-oestrogen receptor complex can act as a repressor of transcription instead of an activator. It depends on the type of cell and the target gene.

Regulation of Transcription and Translation

RNA Interference (RNAi) Can Inhibit the Translation of mRNA

1) In **eukaryotes**, **gene expression** is also affected by **RNA interference** (**RNAi**).
2) RNAi is where small, double-stranded **RNA** molecules **stop mRNA** from target genes being **translated** into **proteins**. A **similar process** to RNAi can also occur in prokaryotes.
3) The molecules involved in RNAi are called **siRNA** (small interfering RNA) and **miRNA** (microRNA).
4) Here's how RNAi works:

RNAi molecules are small lengths of non-coding RNA (they don't code for proteins).

siRNA (and miRNA in plants)

1) Once mRNA has been transcribed, it leaves the nucleus for the **cytoplasm**.
2) In the cytoplasm, double-stranded **siRNA** associates with several **proteins** and unwinds. A single strand then **binds** to the **target mRNA**. The **base sequence** of the **siRNA** is **complementary** to the base sequence in sections of the **target mRNA**.
3) The **proteins** associated with the **siRNA cut** the mRNA into fragments — so it can **no longer** be **translated**. The fragments then move into a **processing body**, which contains 'tools' to **degrade** them.
4) A **similar process** happens with **miRNA** in **plants**.

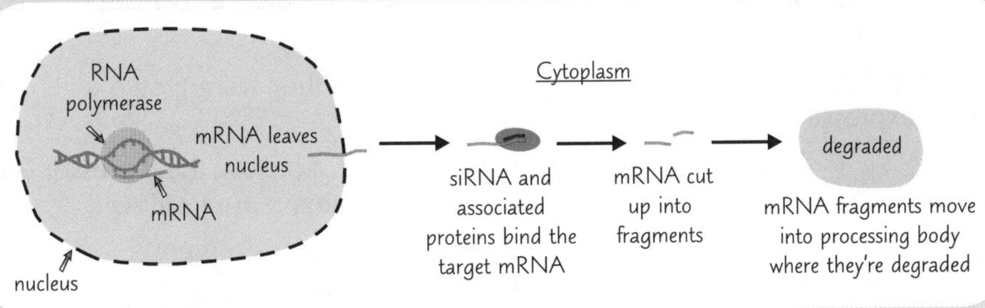

miRNA in mammals

1) In **mammals**, the **miRNA isn't** usually **fully complementary** to the target mRNA. This makes it **less specific** than siRNA and so it may target **more than one** mRNA molecule.
2) Like siRNA, it associates with proteins and **binds** to **target mRNA** in the **cytoplasm**.
3) Instead of the proteins associated with miRNA cutting mRNA into fragments, the miRNA-protein complex physically **blocks** the **translation** of the **target mRNA**.
4) The mRNA is then moved into a **processing body**, where it can either be **stored** or **degraded**. When it's stored, it can be **returned** and **translated** at **another time**.

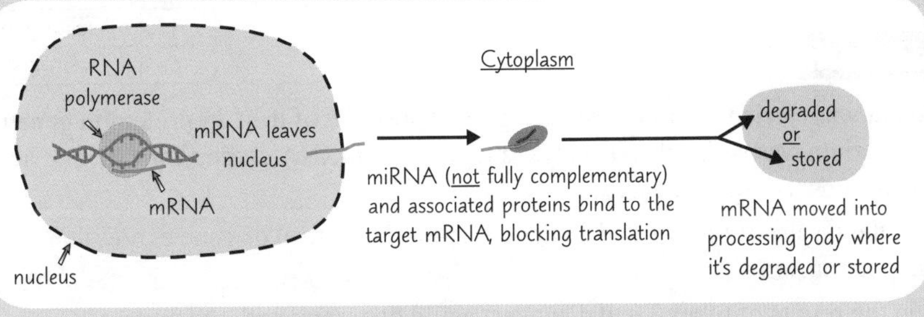

You Need to be Able to Interpret Experimental Data on Gene Expression

You could get a question in the exam where you have to **interpret data** about **gene expression**. It could be on anything you've learnt about on these two pages (e.g. **transcription factors**, **oestrogen** or **RNAi**) or it could be on **epigenetic control** of gene expression (see pages 207-208).

On the next page there's an example of a **gene expression system** in bacteria and an experiment that **investigates** how it works. You **don't** need to **learn** the information, just **understand** what the results of the experiment tell you about how the expression of the gene is **controlled**.

Regulation of Transcription and Translation

The *lac* repressor:

1) ***E. coli*** is a **bacterium** that respires **glucose**, but it can use **lactose** if glucose **isn't available**.
2) If lactose is present, *E. coli* makes an **enzyme** (**β-galactosidase**) to **digest** it. But if there's **no** lactose, it doesn't **waste energy** making an enzyme it **doesn't need**. The enzyme's **gene** is **only expressed** when lactose is **present**.
3) The production of the enzyme is **controlled** by a **transcription factor** — the ***lac* repressor**.
4) When there's **no** lactose, the *lac* repressor **binds** to the **DNA** at the start of the gene, **stopping transcription**.
5) When lactose **is present** it **binds** to the *lac* repressor, **stopping** it binding to the DNA, so the gene **is transcribed**.

Experiment:

1) Different ***E. coli*** mutants were isolated and grown in **different media**, e.g. with lactose or glucose.
2) The mutants have **mutations** (**changes** in their **DNA bases**, see page 194) that mean they **act differently** from normal *E. coli*, e.g. they **produce β-galactosidase** when grown with glucose.
3) To **detect** whether active (working) β-galactosidase was produced, a **chemical** that turns **yellow** in the presence of active β-galactosidase was **added** to the medium.
4) The production of **mRNA** that **codes for β-galactosidase** was also measured. The results are shown in the **table**.

Medium	Mutant	mRNA	Colour
Glucose	Normal	No	No yellow
Lactose	Normal	Yes	Yellow
Glucose	Mutant 1	Yes	Yellow
Lactose	Mutant 1	Yes	Yellow
Glucose	Mutant 2	No	No yellow
Lactose	Mutant 2	Yes	No yellow

5) In **mutant 1**, mRNA and active β-galactosidase **were produced** even when they were grown with **only glucose** — the gene is **always** being expressed.
6) This suggests that mutant 1 has a **faulty *lac* repressor**, e.g. in the **absence** of lactose the repressor **isn't able** to bind DNA, so transcription **can** occur and mRNA and active β-galactosidase **are produced**.
7) In **mutant 2**, **mRNA** is produced but **active β-galactosidase isn't** when **lactose** is present — the **gene** is being **transcribed** but it **isn't** producing **active** β-galactosidase.
8) This suggests mutant 2 is producing **faulty β-galactosidase**, e.g. because a **mutation** has affected its active site.

Warm-Up Questions

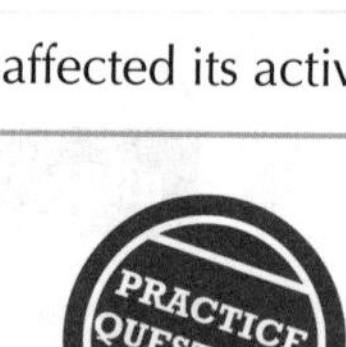

Q1 What is a transcription factor?
Q2 Explain how repressors stop transcription from happening.
Q3 What is RNAi?
Q4 How does RNAi involving siRNA work?
Q5 How does RNAi involving miRNA work?

Exam Question

Tube	Medium	Bacteria	Full length mRNA	Protein
1	+ Oestrogen	Normal	Yes	Active
2	– Oestrogen	Normal	No	No
3	+ Oestrogen	Mutant	No	No
4	– Oestrogen	Mutant	No	No

Q1 An experiment was carried out to investigate gene expression of the Chi protein in genetically engineered bacteria. A mutant bacterium was isolated and analysed to look for mRNA coding for Chi, and active Chi protein production. The results are shown in the table above:

a) What do the results of tubes 1 and 2 suggest about the control of gene expression? Explain your answer. [2 marks]

b) What do the results of tubes 3 and 4 suggest could be wrong with the mutant? Explain your answer. [3 marks]

c) If an siRNA complementary to the Chi gene was added to tube 1, what would you expect the results to be? Explain your answer. [3 marks]

Transcription Factor — not quite as eXciting as that other factor programme...

If it was a competition, oestrogen would totally win — it's very jazzy and awfully controlling. Flexible too — sometimes it helps to activate and other times it helps to repress. Although I'm not sure it can hold a note or wiggle in time to music. Make sure that you understand everything on these pages. Transcription factors are pretty important molecules.

Epigenetic Control of Gene Expression

Epigenetic changes are another way of controlling gene expression. If you thought transcription factors and the like were clever, then you're in for a real treat with this lot. Prepare to have your mind well and truly blown...

Epigenetic Control Can Determine Whether or Not a Gene is Expressed

1) In **eukaryotes**, **epigenetic control** can determine whether a **gene** is **switched on** or **off** — i.e. whether the gene is **expressed** (transcribed and translated) or not.
2) It works through the **attachment** or **removal** of **chemical groups** (known as **epigenetic marks**) to or from **DNA** or **histone proteins** (see below).
3) These epigenetic marks **don't alter** the **base sequence** of **DNA**.
4) Instead, they **alter** how **easy** it is for the **enzymes** and other proteins needed for **transcription** to **interact** with and **transcribe** the **DNA**.
5) Epigenetic changes to gene expression play a **role** in lots of **normal cellular processes** and can **also occur** in **response** to **changes** in the **environment** — e.g. pollution and availability of food.

Epigenetic Changes Can be Inherited by Offspring

1) Organisms **inherit** their **DNA base sequence** from their **parents**.
2) Most **epigenetic marks** on the DNA are **removed** between generations, but **some escape** the **removal process** and are **passed on** to **offspring**.
3) This means that the expression of some genes in the **offspring** can be **affected** by **environmental changes** that affected their **parents or grandparents**.
4) For example, epigenetic changes in some **plants** in **response** to **drought** have been shown to be **passed on** to later generations.

This epigenetic change, caused by environmental exposure to too many cheesy CGP jokes, has been passed on to three generations so far.

Increased Methylation of DNA Switches a Gene Off

One method of **epigenetic control** is **methylation** of **DNA**:

1) This is when a **methyl group** (an example of an **epigenetic mark**) is **attached** to the **DNA** coding for a **gene**.
2) The group always attaches at a **CpG site**, which is where a **cytosine** and **guanine** base are **next to** each other in the DNA (linked by a **phosphodiester bond**).
3) **Increased** methylation **changes** the **DNA structure** so that the **transcriptional** machinery (enzymes, proteins etc.) **can't interact** with the gene — so the gene is **not expressed** (i.e. it's **switched off**).

A methyl group is a $-CH_3$ group.

Decreased Acetylation of Histones Can also Switch Genes Off

Histones are **proteins** that DNA **wraps around** to form **chromatin**, which makes up **chromosomes**. Chromatin can be **highly** condensed or **less** condensed. How **condensed** it is affects the **accessibility** of the **DNA** and whether or not it can be **transcribed**.

1) Histones can be **epigenetically modified** by the **addition** or **removal** of **acetyl groups** (which are another example of an **epigenetic mark**).
2) When histones are **acetylated**, the chromatin is **less condensed**. This means that the transcriptional machinery **can access** the DNA, allowing genes to be **transcribed**.
3) When **acetyl groups** are **removed** from the histones, the chromatin becomes **highly condensed** and genes in the DNA **can't** be **transcribed** because the transcriptional machinery **can't** physically **access** them.
4) **Histone deacetylase** (HDAC) enzymes are responsible for **removing** the **acetyl groups**.

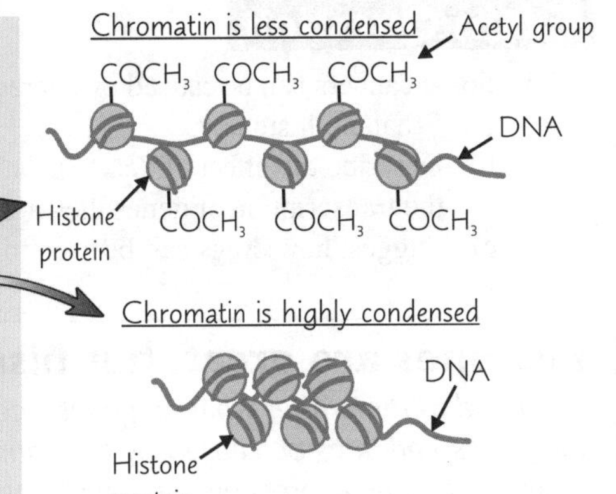

Epigenetic Control of Gene Expression

Epigenetics Can Lead to the Development of Disease

You've already seen on page 197 how **epigenetics** can play a role in the **development** of **disease**, with the fact that **abnormal methylation** of **tumour suppressor genes** and **oncogenes** can cause **cancer**. However, the role of epigenetics in disease **doesn't** stop there. It can play a role in the development of many other diseases, including Fragile X syndrome, Angelman's syndrome and Prader-Willi syndrome.

Example: Fragile-X syndrome

1) **Fragile-X syndrome** is a genetic disorder that can cause symptoms such as **learning** and **behavioural difficulties**, as well as **characteristic physical features**.
2) It's caused by a heritable **duplication mutation** (see page 194) in a **gene** on the **X chromosome**, called **FMR1**. The mutation results in the **short DNA sequence CGG** being **repeated** many **more** times than usual.
3) These repeats mean that there are lots more **CpG sites** (see previous page) in the gene than usual. More CpG sites result in **increased methylation** of the gene, which **switches it off**.
4) Because the gene is **switched off**, the **protein** that it codes for **isn't produced**. It's the **lack** of this **protein** that causes the **symptoms** of the disease.

Drugs May be Able to Treat Diseases Caused by Epigenetic Changes

1) **Epigenetic changes** are **reversible**, which makes them **good targets** for new **drugs** to combat **diseases** they cause.
2) These drugs are designed to **counteract** the epigenetic changes that **cause** the **diseases**.
3) For example, **increased methylation** is an epigenetic change that can lead to a **gene** being **switched off**. Drugs that **stop DNA methylation** can sometimes be used to treat diseases caused in this way. For example, the drug **azacitidine** is used in **chemotherapy** for types of **cancer** that are caused by **increased methylation** of **tumour suppressor genes**.
4) **Decreased acetylation** of **histones** can also lead to **genes** being **switched off**. **HDAC inhibitor drugs**, e.g. **romidepsin**, can be used to **treat diseases** that are caused in this way — including some types of **cancer**. These drugs work by **inhibiting** the **activity** of **histone deacetylase** (**HDAC**) **enzymes**, which are responsible for **removing** the **acetyl groups** from the **histones**. Without the activity of HDAC enzymes, the genes **remain acetylated** and the proteins they code for **can be transcribed**.
5) The problem with developing drugs to counteract epigenetic changes is that these changes take place normally in a lot of cells, so it's important to make sure the drugs are as **specific** as possible. E.g. drugs used in **cancer therapies** can be designed to only **target dividing cells** to avoid damaging normal body cells.

Warm-Up Questions

Q1 What is epigenetic control?

Q2 What are epigenetic marks?

Q3 How can methylation of DNA affect gene expression?

PRACTICE QUESTIONS

Exam Question

Q1 Some cancers can be caused by decreased acetylation of histones associated with genes related to cell division.

a) What are histones? [1 mark]

b) Describe the effect of decreased acetylation of histones on the transcription of genes they are associated with. [3 marks]

c) Suggest how drugs can be used to treat cancers caused in this way. [3 marks]

Histones are great, but hisrhythm is way off...

You need to remember what epigenetic control is all about. It's a method for determining whether a gene is transcribed that can sometimes be caused by environmental changes and can be inherited by your offspring. Sometimes, epigenetic changes to gene expression can cause nasty diseases but, in some cases, drugs can be created to cancel them out.

Evaluating Data on Phenotypes

We're finally at the end of the section... and what a whopper it was — but before you rush off to the next one, there's a little bit to learn about the relative influences of genetics and the environment on phenotype...

You Might Have to **Evaluate Data** About **Influences** on **Phenotypes**

The **phenotype** (characteristics) of an organism is the result of the organism's **genotype** and the **interaction** of its genotype with the **environment** (see page 168). It's not always clear **how much** a phenotype is influenced by genes and how much it's influenced by the environment. Let's take a look at these examples:

Example 1 — Overeating

1) **Overeating** was thought to be caused only by environmental factors, like an **increased availability of food** in developed countries.
2) It was later discovered that food consumption **increases** brain **dopamine** levels in animals.
3) Once enough dopamine was released, people would **stop** eating.
4) Researchers discovered that people with one particular **allele** had **30% fewer** dopamine receptors.
5) They found that people with this particular allele were **more likely** to overeat — they wouldn't stop eating when dopamine levels increased.
6) Based on this evidence, scientists now think that overeating has **both genetic** and **environmental** causes.

Example 2 — Antioxidants

1) Many foods in our diet contain **antioxidants** — compounds that are thought to play a role in **preventing chronic diseases**.
2) Foods such as **berries** contain **high levels** of antioxidants.
3) Scientists thought that the berries produced by different **species** of plant contained **different levels** of antioxidants because of **genetic factors**.
4) But experiments that were carried out to see if **environmental** conditions affected antioxidant levels found that environmental conditions caused a great deal of **variation**.
5) Scientists now believe that antioxidant levels in berries are due to **both genetic** and **environmental** factors.

In the exam, you might have to **evaluate data** on the relative influences of genes and the environment on phenotype. This data may come from twin studies...

Twin Studies Can Help to **Determine** Influences on **Phenotype**

Studies of **identical twins** are extremely **useful** when trying to **determine** what's due to **environmental factors** and what's due to **genetic factors**.

These twins are **genetically identical**, so any **differences** in **phenotype** must be entirely due to **environmental factors**. If a characteristic is **very similar** in **identical twins**, **genetics** probably plays a **more important** role. But if a characteristic is **different between** the **twins**, the **environment** must have a **larger influence**.

Twin studies can be used to determine whether a shared bad taste in hats is genetic or just down to buy one get one free offers.

Data that comes from twin studies involving a large sample size (i.e. lots of pairs of twins) is better for drawing valid conclusions than data based on a small sample size. That's because a large sample size is more representative of the population.

Warm-Up Questions

Q1 Give an example of a characteristic that varies due to both genetic and environmental factors.

Exam Question

Q1 Twin studies have found that stuttering (a speech disorder) of both twins is more common in identical twins than in non-identical twins.

What do these findings suggest about the influence of genetic and environmental factors on stuttering? [1 mark]

I just don't think there's anything funny about this page...

Like I say, it's been a pretty heavy section. Evaluate means look at both sides of argument and give an overall judgement about something. The important thing is to look at the data properly — don't just skim over it and leap to a conclusion.

Genome Projects and Making DNA Fragments

Gene technologies are seriously amazing. From sequencing the entire human genome to chopping out bits of DNA to insert into other organisms, you never know what scientists will get up to next.

Sequencing Projects Have Read Entire Genomes

1) A genome is the **entire set** of DNA, including all the genes in an organism.
2) **Improvements** in **technology** have allowed us to **sequence** the **genomes** of a variety of organisms, from bacteria to humans.
3) Gene sequencing **methods** only work on fragments of DNA. So if you want to sequence the **entire genome** of an organism, you need to chop it up into **smaller pieces** first. The smaller pieces are **sequenced** and then **put back in order** to give the sequence of the whole genome.
4) The **Human Genome Project**, which was completed in 2003, mapped the **entire sequence** of the **human genome** for the first time.

Sequencing the Genome of Simple Organisms Helps Identify their Proteins

1) The **proteome** of an organism is all the **proteins** that are made by it.
2) You might remember from Topic 4 that while some parts of the genome code for **specific proteins**, some parts don't code for anything at all (the DNA is **non-coding**).
3) **Simple** organisms, such as **bacteria**, don't have much non-coding DNA.
4) This means it is relatively **easy** to **determine** their **proteome** from the DNA sequence of their **genome**.
5) This can be useful in **medical research** and **development**. For example, identifying the **protein antigens** on the surface of **disease-causing bacteria** and **viruses** can help in the development of **vaccines** to prevent the disease.

Example: ***N. meningitidis*** **group B bacteria** cause **meningitis B**. **Sequencing** the **genome** of these bacteria helped researchers identify **antigens** for use in developing a **vaccine** against the disease.

Remember: vaccines contain antigens that cause your body to produce memory cells. If you're later infected by a pathogen with the same antigens, your memory cells will quickly recognise it and divide to produce antibodies against it — so you don't get sick.

It's Harder to Translate the Genome of Complex Organisms

1) More **complex organisms** contain **large sections** of **non-coding** DNA.
2) They also contain complex **regulatory genes**, which determine when the genes that code for particular proteins should be **switched on** and **off**.
3) This makes it **more difficult** to **translate** their **genome** into their **proteome**, because it's hard to find the bits that code for proteins among the non-coding and regulatory DNA.
4) However, work is being done on the **human proteome**. The codes for more than 30 000 human proteins have been identified so far.

Yes, Sofia was quite sure she didn't need any more sequin-cing.

Sequencing Methods are Continuously Updated

1) In the **past**, many sequencing methods were **labour-intensive**, **expensive** and could only be done on a **small scale**.
2) Now these techniques are often **automated**, more **cost-effective** and can be done on a **large scale**.
3) For example, **pyrosequencing** is a **recently developed** technique that can sequence around **400 million bases** in a ten hour period (which is **super fast** compared to older techniques).
4) With **newer**, **faster** techniques such as pyrosequencing available, scientists can now sequence **whole genomes** much more quickly.

Genome Projects and Making DNA Fragments

Recombinant DNA Technology Involves Transferring Fragments of DNA

1) **Recombinant DNA technology** involves **transferring** a **fragment** of **DNA** from **one organism** to **another**.
2) Because the genetic code is **universal** (the same DNA base triplets code for the same amino acids in **all living things**), and because **transcription** and **translation mechanisms** are pretty similar too, the transferred DNA can be used to produce a **protein** in the cells of the **recipient organism**. The recipient and donor organisms don't even have to be from the same species. This can be pretty useful — see page 215. Organisms that contain **transferred DNA** are known as **transgenic organisms**.

DNA Fragments Can Be Made in Different Ways

In order to **transfer** a **gene** from one organism to another, you first need to get a **DNA fragment** containing the gene you're **interested** in (the **target gene**). There are **three ways** that DNA fragments can be produced:

1 Using Reverse Transcriptase

1) Most **cells** only contain **two copies** of each gene, making it **difficult** to obtain a DNA fragment containing the target gene. But they can contain **many mRNA** molecules which are complementary to the gene, so mRNA is often **easier** to obtain.
2) The mRNA molecules can be used as **templates** to **make lots of DNA**. The **enzyme**, **reverse transcriptase**, **makes DNA** from an RNA template. The DNA produced is called **complementary DNA** (**cDNA**).
3) For example, **pancreatic cells** produce the protein **insulin**. They have loads of mRNA molecules complementary to the **insulin gene**, but only **two copies** of the gene **itself**. So reverse transcriptase could be used to **make cDNA** from the **insulin mRNA**.
4) To do this, **mRNA** is first isolated from cells. Then it's **mixed** with **free DNA nucleotides** and **reverse transcriptase**. The reverse transcriptase uses the mRNA as a **template** to synthesise a **new strand** of cDNA.

You should remember from Topic 4 that DNA is copied into mRNA during transcription.

DNA sequence
A T G C A A
U A C G U U
complementary mRNA sequence

2 Using Restriction Endonuclease Enzymes

1) Some sections of DNA have **palindromic** sequences of **nucleotides**. These sequences consist of **antiparallel base pairs** (base pairs that read the **same** in **opposite directions**).

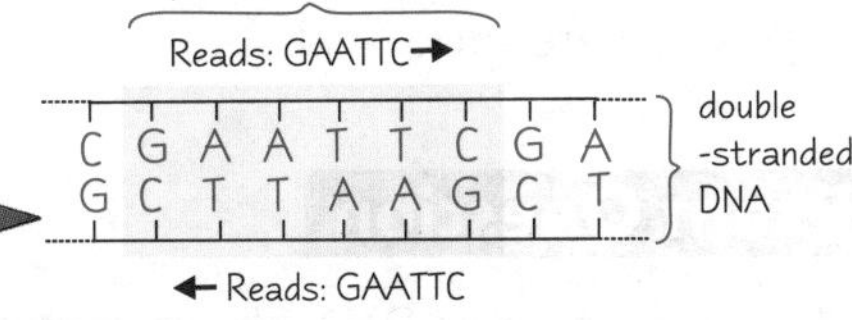

2) **Restriction endonucleases** are enzymes that **recognise specific** palindromic sequences (known as **recognition sequences**) and **cut** (**digest**) the DNA at these places.
3) Different restriction endonucleases cut at **different specific** recognition sequences, because the **shape** of the recognition sequence is **complementary** to the enzyme's **active site**. E.g. the restriction endonuclease *Eco*RI cuts at GAATTC, but *Hind*III cuts at AAGCTT.
4) If recognition sequences are present at **either side** of the DNA fragment you want, you can use restriction endonucleases to **separate** it from the rest of the DNA.
5) The DNA sample is **incubated** with the specific restriction endonuclease, which **cuts** the DNA fragment out via a **hydrolysis reaction**.
6) Sometimes the cut leaves **sticky ends** — **small tails** of **unpaired bases** at **each end** of the fragment. Sticky ends can be used to **bind** (**anneal**) the DNA fragment to another piece of DNA that has sticky ends with **complementary sequences** (there's more about this on p. 213).

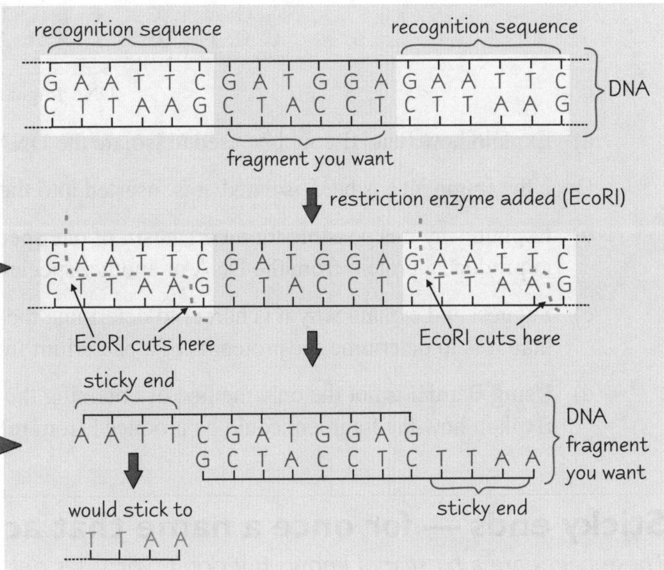

Genome Projects and Making DNA Fragments

3 Using a 'Gene Machine'

1) More **recently**, technology has been developed so that fragments of DNA can be **synthesised** from **scratch**, without the need for a **pre-existing DNA template**.
2) Instead, a **database** contains the necessary information to produce the **DNA fragment**.
3) This means that the DNA sequence does not have to **exist naturally** — **any sequence** can be made.
4) Here's how it's done:

- The **sequence** that is required is **designed** (if one doesn't already exist).
- The first **nucleotide** in the sequence is fixed to some sort of support, e.g. a bead.
- Nucleotides are added **step by step** in the **correct order**, in a cycle of processes that includes adding **protecting groups**. Protecting groups make sure that the nucleotides are **joined** at the **right points**, to prevent **unwanted branching**.
- Short sections of DNA called **oligonucleotides**, roughly 20 nucleotides long, are produced. Once these are complete, they are broken off from the support and all the protecting groups are removed. The oligonucleotides can then be joined together to make **longer DNA fragments**.

A jean machine would be perfect for Rob. He loves his denim.

Warm-Up Questions

Q1 Why can it be useful to determine the proteome of a simple organism?

Q2 What can make it difficult to determine the proteome of more complex organisms?

Q3 Briefly outline how genetic sequencing methods have changed over time.

Q4 What is recombinant DNA technology?

Q5 Give three ways a DNA fragment can be produced.

Q6 What is reverse transcriptase?

Q7 What are sticky ends?

PRACTICE QUESTIONS

Exam Question

Q1 A fragment of DNA (shown below) needs to be isolated from some bacterial DNA. The restriction endonuclease BamHI recognises the sequence GGATCC and cuts between G and G.

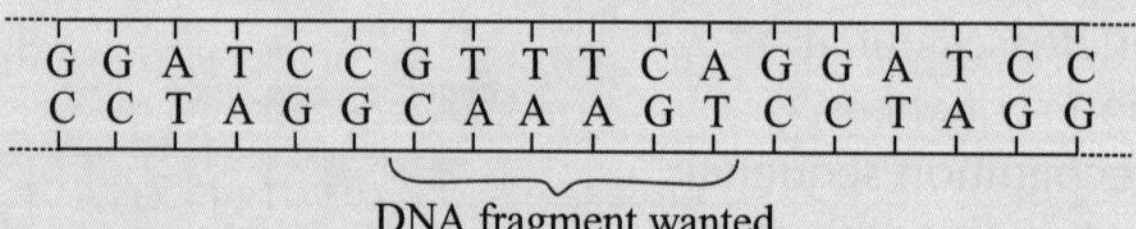

a) Explain how BamHI could be used to isolate the DNA fragment. [2 marks]

Once the fragment has been isolated, it is inserted into the DNA of a plant.

b) Explain why it is possible for an organism of one species to produce a protein from the DNA of another species. [2 marks]

c) Suggest and explain why it is harder to determine the proteome of a plant from its genome than it is to determine the proteome of a bacterium from its genome. [3 marks]

d) Using BamHI is not the only method of obtaining the DNA fragment. Explain how the fragment could be produced from mRNA. [3 marks]

Sticky ends — for once a name that actually makes sense...

These pages are a bit scary I know. But don't worry, it's not as difficult as photosynthesis — you just need to keep going over the steps of the different techniques until they make sense. I know I've said it before, but drawing out the diagrams will help — then you'll know reverse transcriptase and restriction endonucleases like a pro.

Amplifying DNA Fragments

Once you've got your teeny tiny fragment of DNA, you need to amplify it so you've got lots and lots to play with...

In Vivo Amplification Involves Transforming Host Cells

Once you've **isolated** your **DNA fragment** (using one of the techniques on pages 211-212) you need to **amplify** it (make lots of copies of it) so you have a **sufficient quantity** to work with. One way of doing this is to use ***in vivo* cloning** — this is where **copies** of the DNA fragment are made **inside** a **living organism**.

Step 1 — The DNA Fragment is Inserted into a Vector

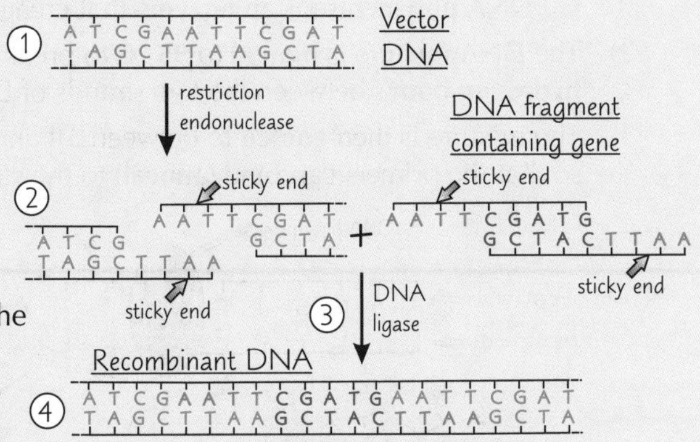

1) The DNA fragment is inserted into vector DNA — a **vector** is something that's used to **transfer DNA** into a **cell**. They can be **plasmids (small, circular molecules** of DNA in **bacteria)** or **bacteriophages** (**viruses** that **infect** bacteria).
2) The vector DNA is **cut open** using the **same** restriction endonuclease that was used to **isolate** the DNA fragment containing the target gene (see p. 211). So the **sticky ends** of the vector are **complementary** to the sticky ends of the DNA fragment containing the gene.
3) The vector DNA and DNA fragment are **mixed together** with **DNA ligase** (another enzyme). DNA ligase **joins** the sticky ends of the DNA fragment to the sticky ends of the vector DNA. This process is called **ligation**.
4) The new combination of bases in the DNA (vector DNA + DNA fragment) is called **recombinant DNA**.

Step 2 — The Vector Transfers the DNA Fragment into Host Cells

1) The **vector** with the **recombinant DNA** is used to **transfer** the gene into **cells** (called **host** cells).
2) If a **plasmid vector** is used, **host cells** have to be **persuaded** to **take in** the plasmid vector and its DNA.
   E.g. host bacterial cells are placed into ice-cold calcium chloride solution to make their cell walls more permeable. The plasmids are added and the mixture is heat-shocked (heated to around 42 °C for 1-2 minutes), which encourages the cells to take in the plasmids.
3) With a **bacteriophage** vector, the bacteriophage will **infect** the host bacterium by **injecting** its **DNA** into it. The phage DNA (with the target gene in it) then **integrates** into the bacterial DNA.
4) **Host cells** that **take up** the vectors containing the gene of interest are said to be **transformed**.

Step 3 — Identifying Transformed Host Cells

Only **around 5%** of host cells will **take up** the vector and its DNA, so it's important to be able to identify which cells have been transformed. **Marker genes** can be used to **identify** the **transformed** cells:

1) **Marker genes** can be inserted into vectors at the **same time** as the gene to be cloned. This means any **transformed host cells** will contain the gene to be cloned **and** the marker gene.
2) Host cells are **grown** on **agar plates**. Each cell **divides** and **replicates** its DNA, creating a **colony** of **cloned cells**. Transformed cells will produce colonies where **all the cells** contain the cloned gene and the marker gene.
3) The marker gene can code for **antibiotic resistance** — host cells are grown on agar plates **containing** the specific **antibiotic**, so **only** transformed cells that have the **marker gene** will **survive** and **grow**. Or it can code for **fluorescence** — when the agar plate is placed under a **UV light only** transformed cells will **fluoresce**.
4) **Identified** transformed cells are allowed to **grow more**, producing **lots** and **lots** of **copies** of the **cloned gene**.

To Produce Proteins You Need Promoter and Terminator Regions

1) If you want the **transformed host cells** to **produce** the **protein** coded for by the **DNA fragment**, you need to make sure that the **vector** contains **specific promoter** and **terminator regions**.
2) **Promoter regions** are **DNA sequences** that tell the enzyme **RNA polymerase** when to **start** producing **mRNA**. **Terminator regions** tell it when to **stop**. Without the **right** promoter region, the DNA fragment **won't** be transcribed by the host cell and a protein **won't** be made.
3) Promoter and terminator regions may be present in the **vector DNA** or they may have to be **added in** along with the **fragment**.

Amplifying DNA Fragments

In Vitro Amplification Uses the **Polymerase Chain Reaction (PCR)**

DNA fragments can also be amplified using ***in vitro*** cloning — this is where copies of the DNA fragments are made **outside** of a living organism using the **polymerase chain reaction** (PCR). PCR can be used to make **millions of copies** of a fragment of DNA in just a few hours. PCR has **several stages** and is **repeated** over and over to make lots of copies.

1) A reaction mixture is set up that contains the **DNA sample**, **free nucleotides**, **primers** and **DNA polymerase.**
   - **Primers** are short pieces of DNA that are **complementary** to the bases at the **start** of the fragment you want.
   - **DNA polymerase** is an **enzyme** that creates new DNA strands.
2) The DNA mixture is **heated** to **95 °C** to break the **hydrogen bonds** between the two strands of DNA.
3) The mixture is then **cooled** to between **50** and **65 °C** so that the primers can **bind (anneal)** to the strands.

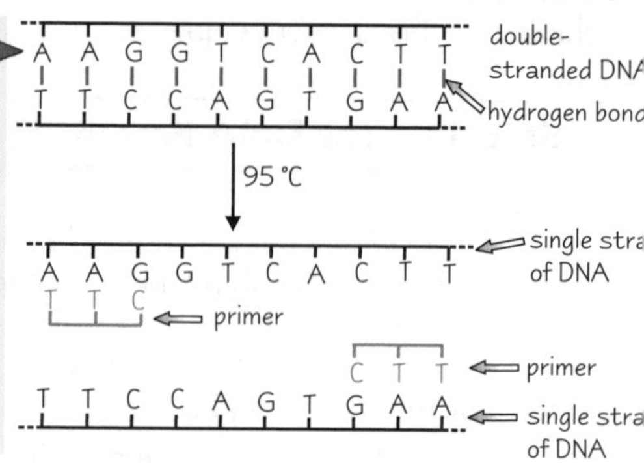

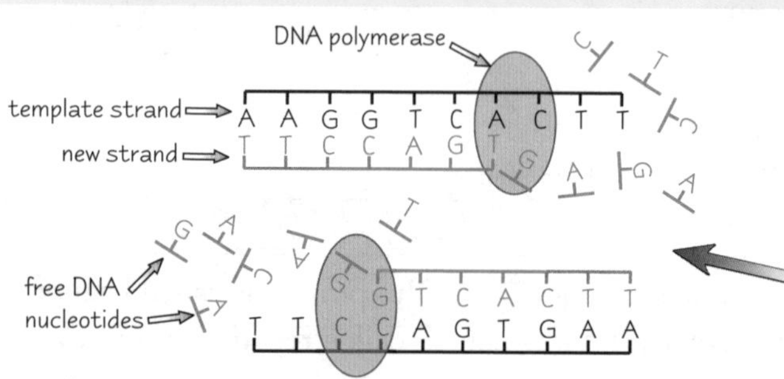

4) The reaction mixture is heated to **72 °C**, so **DNA polymerase** can **work.**
5) The DNA polymerase **lines up** free DNA nucleotides **alongside** each **template strand.** Specific **base pairing** means **new complementary strands** are formed.

6) **Two new copies** of the fragment of DNA are formed and **one cycle** of PCR is **complete.**
7) The cycle starts again, with the mixture being heated to 95 °C and this time **all four strands** (two original and two new) are used as **templates.**
8) Each PCR cycle **doubles** the amount of DNA, e.g. **1st cycle** = 2 × 2 = **4 DNA fragments**, **2nd cycle** = 4 × 2 = **8 DNA fragments**, **3rd cycle** = 8 × 2 = **16 DNA fragments**, and so on.

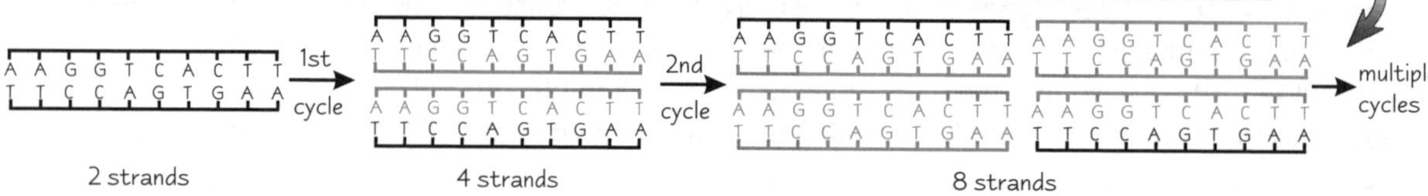

Warm-Up Questions

Q1 In *in vitro* amplification what are vectors used to do?
Q2 What is recombinant DNA?
Q3 What does PCR stand for?
Q4 What is a primer?

PRACTICE QUESTIONS

Exam Question

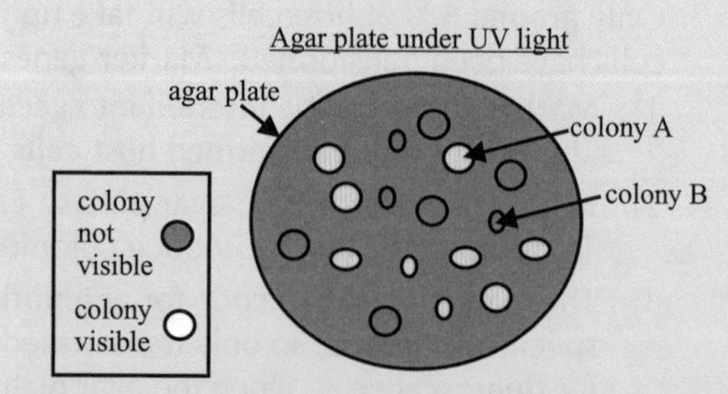

Q1 A scientist amplified a gene by transferring a plasmid containing the target gene and a fluorescent marker gene into some bacterial cells. The cells were grown on an agar plate. The plate was then placed under UV light (see above).

a) Explain why the scientist thinks colony A contains transformed host cells, but colony B doesn't. [2 marks]

b) Explain how the scientist might have inserted the target gene into the plasmid. [3 marks]

If only you could amplify fragments of knowledge — or cake...

Okay, your eyes might have gone funny from seeing so many nucleotides on these pages. But once you've recovered, it's really important to go over these pages as many times as you need to, 'cause examiners love throwing in a few questions on restriction enzymes or PCR. Bless 'em — examiners get excited about the strangest things.

Using Recombinant DNA Technology

Now that you know how to make a DNA fragment and amplify it, it's probably a good time to tell you why you might want to. Don't worry — it's not evil stuff, but I promise to do my evil laugh. Mwah ha hah.

Transformed Organisms Are Made Using **Recombinant DNA Technology**

1) **Microorganisms**, **plants** and **animals** can all be **transformed** using **recombinant DNA technology**. This is called **genetic engineering**.

Transformed organisms are also known as genetically engineered or genetically modified (GM) organisms.

2) **Transformed microorganisms** can be made using the same technology as ***in vivo*** cloning (see page 213). For example, **foreign DNA** can be **inserted** into **microorganisms** to produce **lots** of **useful protein**, e.g. insulin:

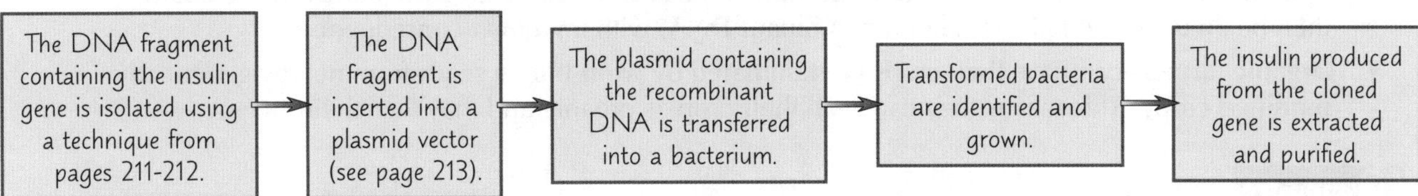

3) **Transformed plants** can also be produced. A gene that codes for a **desirable protein** is inserted into a **plasmid**. The plasmid is added to a **bacterium** and the bacterium is used as a **vector** to get the gene into the **plant cells**. If the right **promoter region** has been added along with the gene (see p. 213), the transformed cells will be able to **produce** the **desired protein**.
4) **Transformed animals** can be produced too — a gene that codes for a **desirable protein** can be inserted into an early animal **embryo** or into the **egg cells** of a female. If the gene is inserted into a **very early embryo**, **all** the **body cells** of the resulting **transformed animal** will end up containing the gene. Inserting it into the **egg cells** means that when the female **reproduces**, **all** the **cells** of her **offspring** will contain the gene.
5) **Promoter regions** that are **only activated** in **specific cell types** can be used to **control** exactly which of an animal's **body cells** the **protein** is produced in. If the protein is only produced in **certain cells**, it can be **harvested** more easily. Producing the protein in the **wrong cells** could also **damage** the organism.

Recombinant DNA Technology Can be Used to **Benefit Humans**

Transformed organisms (microorganisms, plants and animals) can be used in a variety of ways. You need to be able to **interpret information** about how they are used. Here are some examples:

1 Agriculture

- **Agricultural crops** can be **transformed** so that they give **higher yields** or are **more nutritious**. This means these plants can be used to reduce the risk of **famine** and **malnutrition**. Crops can also be transformed to have **pest resistance**, so that **fewer pesticides** are needed. This **reduces costs** and reduces any **environmental problems** associated with using pesticides.
- For example, ***Golden Rice*** is a variety of **transformed rice**. It contains **one gene** from a **maize plant** and **one gene** from a **soil bacterium**, which together enable the rice to produce **beta-carotene**. The beta-carotene is used by our bodies to produce **vitamin A**. *Golden Rice* is being developed to **reduce vitamin A deficiency** in areas where there's a **shortage** of **dietary vitamin A**, e.g. **south Asia**, **Africa**. Vitamin A deficiency is a big problem in these areas, e.g. up to **500 000 children per year** worldwide go **blind** due to vitamin A deficiency.

2 Industry

- **Industrial processes** often use **biological catalysts** (**enzymes**). These enzymes can be produced from **transformed organisms**, so they can be produced in **large quantities** for **less money**, **reducing costs**.
- For example, **chymosin** (or **rennin**) is an enzyme used in **cheese-making**. It used to be made from **rennet** (a substance produced in the **stomach** of **cows**), but it can now be produced by **transformed organisms**. This means it can be made in **large quantities**, relatively **cheaply** and **without killing** any **cows**, making some cheese suitable for **vegetarians**.

3 Medicine

- Many **drugs** and **vaccines** are produced by transformed organisms, using recombinant DNA technology. They can be made **quickly**, **cheaply** and in **large quantities** using this method.
- For example, **insulin** is used to treat **Type 1 diabetes** and used to come from **animals** (cow, horse or pig pancreases). This insulin **wasn't** human insulin though, so it **didn't work quite as well**. Human insulin is now made from **transformed microorganisms**, using a **cloned human insulin gene** (see above).

Using Recombinant DNA Technology

There are **Concerns** About the Use of **Recombinant DNA Technology...**

There are **ethical**, **financial** and **social issues** associated with the **use** of **recombinant DNA technology**:

1 Agriculture

- **Farmers** might plant only **one type** of transformed crop (this is called **monoculture**). This could make the **whole crop vulnerable** to the **same disease** because the plants are **genetically identical**. Environmentalists are also concerned about monocultures **reducing biodiversity**, as this could damage the environment.
- Some people are concerned about the possibility of '**superweeds**' — weeds that are **resistant** to **herbicides**. These could occur if transformed crops **interbreed** with **wild plants**. There could then be an **uncontrolled spread** of **recombinant DNA**, with **unknown consequences**.
- **Organic farmers** can have their crops **contaminated** by **wind-blown seeds** from nearby **genetically modified** crops. This means they can't **sell** their crop as organic and may **lose** their **income**.

2 Industry

- **Anti-globalisation activists oppose globalisation** (e.g. the **growth** of **large multinational companies** at the **expense** of **smaller ones**). A **few**, **large** biotechnology companies **control** some forms of genetic engineering. As the **use** of this technology **increases**, these companies get **bigger** and **more powerful**. This may **force** smaller companies **out of business**, e.g. by making it **harder** for them to **compete**.
- **Without proper labelling**, some people think they **won't** have a **choice** about whether to consume food made using genetically engineered organisms.
- Some **consumer markets**, such as the EU, won't **import GM foods** and products. This can cause an **economic loss** to **producers** who have traditionally sold to those markets.

3 Medicine

- Companies who **own** genetic engineering technologies may **limit** the **use** of technologies that could be **saving lives**.
- Some people worry this technology could be used **unethically**, e.g. to make **designer babies** (babies that have characteristics **chosen** by their parents). This is currently **illegal** though.

Recombinant DNA technology also creates **ownership issues**. Here are some examples:

- There is some debate about who **owns genetic material** from humans once it has been removed from the body — the **donor** or the **researcher**. Some people argue that the **individual** holds the right to their **own genetic information**, however others argue that **value** is **created** by the **researcher** who uses it to develop a **medicine** or in **diagnosis**.
- A small number of **large corporations** own **patents** to particular **seeds**. They can charge **high prices**, sometimes including a 'technology fee', and can require farmers to **repurchase seeds** each year. If non-GM crops are **contaminated** by GM crops, farmers can be **sued** for breaching the patent law.

...But **Humanitarians** Think it will **Benefit People**

Recombinant DNA technology has **many** potential **humanitarian benefits**:

1) **Agricultural crops** could be produced that help **reduce** the risk of **famine** and **malnutrition**, e.g. **drought-resistant** crops for **drought-prone** areas.
2) **Transformed crops** could be used to produce **useful pharmaceutical products** (e.g. **vaccines**) which could make drugs **available** to **more people**, e.g. in areas where **refrigeration** (usually needed for **storing** vaccines) **isn't available**.
3) **Medicines** could be produced more **cheaply**, so more people can **afford** them.
4) Recombinant DNA technology has the potential to be used in **gene therapy** to **treat human diseases** (see next page).

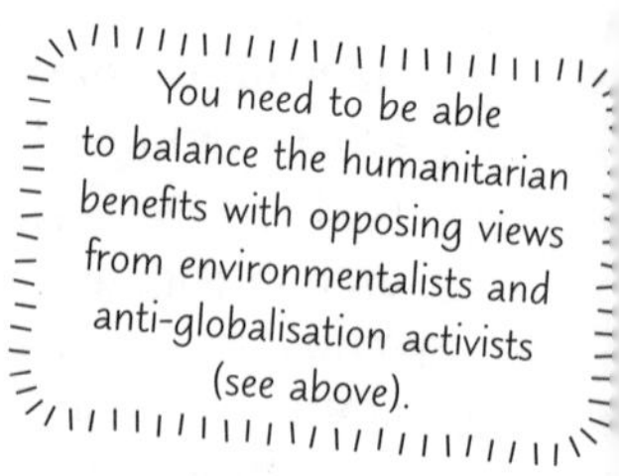

Using Recombinant DNA Technology

Gene Therapy Could be Used to **Treat** or **Cure Genetic Disorders** and **Cancer**

Recombinant DNA technology could also be used to **treat human diseases**. This is known as **gene therapy**.

Gene therapy isn't being used widely yet but there is a form of somatic gene therapy available and other treatments are undergoing clinical trials.

How it works:

1) Gene therapy involves **altering** the **defective genes** (mutated alleles) inside cells to treat **genetic disorders** and **cancer**.
2) How you do this depends on whether the disorder is caused by a mutated **dominant allele** or two mutated **recessive alleles** (see page 168):
   - If it's caused by two mutated **recessive** alleles you can **add** a working **dominant allele** to make up for them (you '**supplement**' the faulty ones).
   - If it's caused by a mutated **dominant** allele you can '**silence**' the **dominant allele** (e.g. by sticking a bit of DNA in the middle of the allele so it doesn't work any more).

   Both of these processes involve **inserting** a **DNA fragment** into the person's **original DNA**.

How you get the 'new' allele (DNA) inside the cell:

1) The allele is **inserted into cells** using **vectors** (see page 213) just like in **recombinant DNA technology**.
2) Different **vectors** can be used, e.g. altered **viruses**, **plasmids** or **liposomes** (spheres made of lipid).

There are two types of gene therapy:

1) **Somatic therapy** — this involves **altering** the **alleles** in **body cells**, particularly the cells that are **most affected** by the disorder. For example, **cystic fibrosis** (CF) is a genetic disorder that's very **damaging** to the **respiratory system**, so somatic therapy for CF **targets** the epithelial cells lining the lungs. Somatic therapy doesn't affect the individual's **sex cells** (sperm or eggs) though, so any **offspring** could still **inherit** the disease.
2) **Germ line therapy** — this involves **altering** the **alleles** in the **sex cells**. This means that **every cell** of **any offspring** produced from these cells will be **affected** by the gene therapy and they **won't suffer from the disease**. Germ line therapy in humans is currently **illegal** though.

There are also many **ethical issues** associated with gene therapy. For example, some people are worried that the technology could be used in ways **other** than for **medical treatment**, such as for treating the **cosmetic effects** of **aging**. Other people worry that there's the potential to do **more harm** than good by using the technology (e.g. risk of overexpression of genes — gene produces too much of the missing protein).

Warm-Up Questions

Q1 What are transformed organisms?

Q2 Give one financial issue associated with the use of recombinant DNA technology in industry.

Q3 What is gene therapy?

Exam Question

Q1 A large agricultural company isolated a gene from bacteria that may increase the drought resistance of wheat plants.

a) Briefly explain how this gene could be used to make a transformed wheat plant. [3 marks]

b) Suggest how the transformed wheat plants might be beneficial to humans. [2 marks]

c) Suggest why anti-globalisation activists may be against the use of this gene. [1 mark]

Neapolitan — recombinant ice cream...

Ahhh, sitting in the Sun, licking an ice cream, exams all over. That's where you'll be in a few months' time. After revising all this stuff that is. As recombinant DNA technology advances, more questions will pop up about its implications. So it's a good idea to know all sides of the argument — you need to know them for the exam anyway.

Gene Probes and Medical Diagnosis

Being able to manipulate DNA is also really useful for diagnosing medical problems...

You can **Look** for **Alleles** Using **DNA Probes** and **Hybridisation**

1) DNA probes can be used to **locate specific alleles** of **genes** (e.g. on **chromosomes**) or to see if a person's DNA contains a **mutated allele** that causes a **genetic disorder**.
2) **DNA probes** are **short strands** of **DNA**. They have a **specific base sequence** that's **complementary** to the base sequence of part of a **target allele** (the allele you're looking for, e.g. an allele that causes a genetic disorder).
3) This means a DNA probe will **bind** (**hybridise**) to the **target allele** if it's **present** in a **sample** of DNA.
4) A DNA probe also has a **label attached**, so that it can be **detected**. The two most common types of label are a **radioactive** label (detected using **X-ray film**) or a **fluorescent** label (detected using **UV light**).
5) Here's how it's done:

- A **sample** of **DNA** is **digested** into fragments using **restriction enzymes** (see p. 211) and separated using **electrophoresis** (see p. 220).
- The separated DNA fragments are then transferred to a **nylon membrane** and **incubated** with the **fluorescently labelled DNA probe**.
- If the allele is **present**, the **DNA probe** will **hybridise** (**bind**) to it.
- The **membrane** is then **exposed** to **UV light** and if the gene is present there will be a **fluorescent band**. E.g. **sample 3** has a visible band, so this patient has the **allele**.

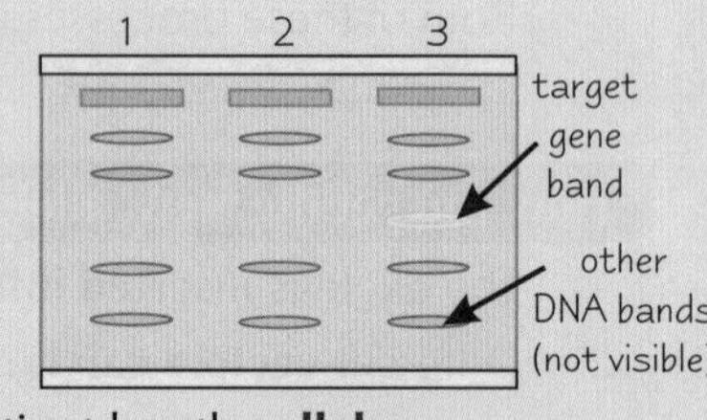

6) Alternatively, the probe can be used as part of a **DNA microarray**, which can screen **lots** of **genes** at the **same tim**

- A **DNA microarray** is a **glass slide** with **microscopic spots** of **different** DNA probes **attached** to it in **rows**.
- A sample of **fluorescently labelled human DNA** is washed over the array.
- If the labelled human DNA **contains** any **DNA sequences** that **match** any of the **probes**, it will **stick** to the array.
- The array is **washed**, to remove any labelled DNA that **hasn't** stuck to it.
- The array is then **visualised** under **UV light** — any **labelled DNA attached** to a probe will **show up** (fluoresce).
- Any spot that fluoresces means that the person's DNA **contains** that specific **allele**. E.g. if the probe is for a mutated allele that causes a **genetic disorder**, this person has the allele.

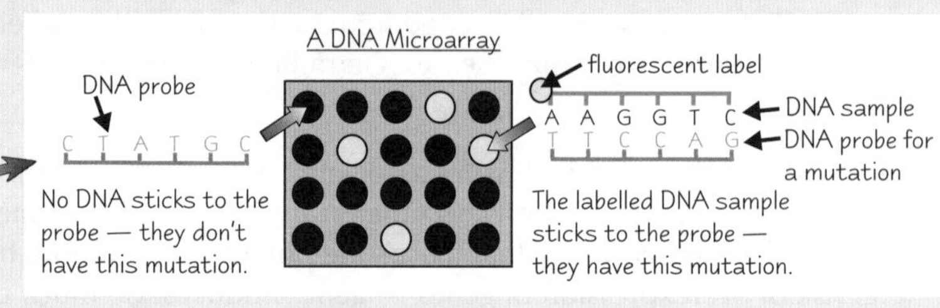

7) To **produce** a DNA probe, you first need to **sequence** the **allele** that you want to screen for (see page 210). You then use **PCR** (see p. 214) to produce **multiple complementary copies** of **part** of the allele — these are the **probe**

Screening Using **DNA Probes** Has **Lots of Uses**

You need to be able to evaluate information about screening for inherited conditions and people's responses to drugs.

For example, screening can be used to...

1) ...help identify **inherited conditions**. E.g. **Huntington's disease** is an inherited condition that affects the nervous system and does not usually start to display symptoms until a person is aged between 30 and 50. People with a family history of the disease may choose to be **screened** for the **mutated allele** to find out if they have inherited i The NHS offers to screen **all newborn babies** for the inherited disorder **cystic fibrosis** (which can cause breathin and digestive difficulties) so that **treatment** for the condition can begin as **soon as possible**.
2) ...help determine how a patient will respond to specific drugs (see next page). E.g. **breast cancer** can be caused by a **mutation** in the **HER2 proto-oncogene** and treated with the drug **Herceptin**® (see page 199). Herceptin® is **only effective** against this type of breast cancer because it **targets** a specific receptor. Screening for this particular mutation helps determine whether Herceptin® will be a useful treatment or not.
3) ...help identify **health risks**. E.g. inheriting particular mutated alleles **increases** your **risk** of developing certain **cancers** (although it doesn't make it certain that you'll develop cancer). If a person knows they have these alleles, it might help them to make **choices** that could **reduce the risk** of the disease developing (see next page). However, some people are concerned that **genetic screening** may lead to **discrimination** by **insurance companies** and **employers** if people are known to have a **high risk** of developing a condition.

Gene Probes and Medical Diagnosis

The **Results** of **Screening** can be used for **Genetic Counselling**...

1) **Genetic counselling** is **advising patients** and their **relatives** about the **risks** of **genetic disorders**.
2) It involves **advising** people about **screening** (e.g. looking for mutated alleles if there's a **history** of **cancer**) and **explaining** the **results** of a screening. Screening can help to **identify** if someone is the **carrier** of a mutated allele, the **type** of **mutated allele** they're carrying (indicating the type of genetic disorder or cancer) and the **most effective treatment**.
3) If the results of a screening are **positive** (an individual **has** the mutation) then genetic counselling is used to advise the patient on the **options** of **prevention** or **treatment** available. Here are two examples:

EXAMPLE 1: A **woman** with a family history of **breast cancer** may have **genetic counselling** to help her **decide** whether or not to be **screened** for **known mutations** that can lead to breast cancer, e.g. a mutation in the BRCA1 **tumour suppressor gene** (see p. 199). If she is screened and the result is **positive**, genetic counsellors might explain that a woman with the mutated BRCA1 gene has a **50 to 85%** chance of developing **breast cancer** in her lifetime. Counselling could also help the woman to **decide** if, for example, she wants to take **surgical steps** to **reduce the risk** of breast cancer developing (by having a **mastectomy**).

EXAMPLE 2: **Sickle-cell anaemia** is a **recessive** genetic disorder caused by a **mutation** in the **haemoglobin gene**. A couple who are **both carriers** of the **sickle-cell allele** may **like** to have **kids**. They may undergo genetic counselling to help them **understand** their **chances** of having a child with sickle-cell anaemia (**one in four**). Genetic counselling also provides **unbiased advice** on the possibility of having **IVF** and **screening** their **embryos** for the allele, so embryos **without the mutation** are **implanted** in the womb. It could also provide information on the **help** and **drugs** available if they have a child with sickle-cell anaemia.

A carrier is a person carrying an allele that is not expressed in their phenotype but that can be passed on to offspring — see page 168.

...and in **Personalised Medicine**

1) Your **genes** determine how your body **responds** to certain **drugs**. **Different people** respond to the **same drug** in **different ways** — which makes certain drugs **more effective** for **some people** than others. This is where **personalised medicines** come in.
2) Personalised medicines are medicines that are **tailored** to an **individual's DNA**. The theory is that if doctors have your **genetic information**, they can use it to **predict** how you will respond to different drugs and only prescribe the ones that will be **most effective** for you.

Warm-Up Questions

Q1 What are DNA probes?

Q2 Give three situations where screening for mutated genes may be useful.

Q3 What is personalised medicine?

Exam Questions

Q1 a) Briefly describe how a DNA probe for a clinically important allele can be produced. [2 marks]

b) Describe how you could screen a person for this allele and many other alleles at the same time. [4 marks]

Q2 A hospital patient has colon cancer. A drug called Cetuximab is used to treat colon cancer caused by a mutation in the KRAS proto-oncogene. The patient is screened and tests negative for the KRAS oncogene.

a) Why is it unlikely that the patient will be treated with Cetuximab? [1 mark]

b) Suggest why the patient will undergo genetic counselling. [2 marks]

DNA probes — don't worry, the DNA doesn't feel a thing...

All of the techniques you've learnt earlier in this section (making and amplifying DNA fragments, PCR) come together nicely in this medical diagnosis stuff — it's good to know that what you've learnt has a point to it.

Genetic Fingerprinting

We've been able to identify people from their fingerprints for over 100 years, but now we can use their DNA instead

Genomes Contain Non-Coding **Variable Number Tandem Repeats**

1) **Not all** of an organism's **genome** (all the genetic material in an organism) **codes** for **proteins**.
2) Some of the genome consists of **variable number tandem repeats** (**VNTRs**) — base sequences that **don't** code for proteins and **repeat** next to each other over and over (sometimes thousands of times), e.g. CATGCATGCATGCATG is a repeat of the non-coding base sequence CATG.
3) The **number of times** these sequences are **repeated differs** from person to person, so the **length** of these sequences in nucleotides differs too. E.g. a **four** nucleotide sequence might be repeated **12 times** in one person = **48 nucleotides** (12 × 4), but repeated **16 times** in another person = **64 nucleotides** (16 × 4).
4) The repeated sequences occur in **lots of places** in the **genome**. The **number** of times a **sequence is repeated** (and so the number of nucleotides) at **different places** in their genome can be **compared** between **individuals** — this is called **genetic fingerprinting**.
5) The **probability** of **two individuals** having the **same** genetic fingerprint is **very low** because the **chance** of **two individuals** having the **same number** of VNTRs at **each place** they're found in DNA is **very low**.

Electrophoresis Separates DNA Fragments to Make a **Genetic Fingerprint**

So **genetic fingerprints** can be **compared** between **different individuals**. Now you need to know how one is **made**:

1) A **sample** of **DNA** is obtained, e.g. from a person's **blood**, **saliva**, etc.
2) **PCR** (see page 214) is used to make **many copies** of the **areas** of DNA that contain the VNTRs — **primers** are used that bind to **either side** of these **repeats** and so the **whole** repeat is amplified.
3) You end up with **DNA fragments** where the **length** (in nucleotides) corresponds to the **number of repeats** the person has at each specific position, e.g. one person may have 80 nucleotides, another person 120.
4) A **fluorescent tag** is added to all the DNA fragments so they can be viewed under **UV light**.
5) The DNA fragments undergo **electrophoresis**:
    - The DNA mixture is placed into a **well** in a slab of **gel** and covered in a **buffer solution** that **conducts electricity**.
    - An **electrical current** is passed through the gel — DNA fragments are **negatively charged**, so they **move towards** the **positive electrode** at the far end of the gel.
    - **Small** DNA fragments move **faster** and **travel further** through the gel, so the DNA fragments **separate** according to **size**.
6) The DNA fragments are viewed as **bands** under **UV light** — this is the **genetic fingerprint**.
7) Two genetic fingerprints can be **compared** — e.g. if both fingerprints have a band at the **same location** on the **gel** it means they have the **same number** of **nucleotides** and so the same number of **VNTRs** at that place — it's a **match**.

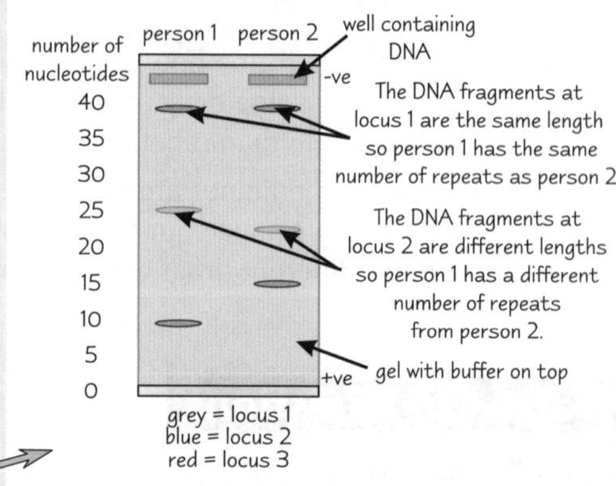

Genetic Fingerprinting is Used to **Determine Relationships** and **Variability**

Genetic fingerprinting has **many uses**, which include:

- **Determining genetic relationships** — We **inherit** VNTR base sequences from our **parents**. Roughly **half** of the sequences come from **each parent**. This means the **more bands** on a genetic fingerprint that match, the more **closely related** (**genetically similar**) two people are. E.g. **paternity tests** are used to determine the **biological father** of a child by comparing genetic fingerprints. If lots of bands on the fingerprint **match**, then that person is **most probably** the child's father. The **higher** the **number** of places in the genome compared, the more **accurate** the test result.
- **Determining genetic variability within a population** — The **greater** the **number of bands** that **don't** match on a genetic fingerprint, the more **genetically different** people are. This means you can **compare** the **number of repeats** at **several places** in the genome for a population to find out how **genetically varied** that population is. E.g. the **more** the **number of repeats** varies at **several places**, the **greater** the **genetic variability** within a population.

Genetic Fingerprinting

Genetic Fingerprinting can be Used in Forensic Science...

Forensic scientists use genetic fingerprinting to **compare** samples of **DNA** collected from **crime scenes** (e.g. DNA from **blood**, **semen**, **skin cells**, **saliva**, **hair**, etc.) to samples of DNA from **possible suspects**, which could **link them** to crime scenes.

1) The **DNA** is **isolated** from all the collected samples (from the crime scene and from the suspects).
2) Each sample is **replicated** using **PCR** (see p. 214).
3) The **PCR products** are run on an **electrophoresis gel** and the genetic fingerprints produced are **compared** to see if any match.
4) If the samples match, it **links** a **person** to the **crime scene**. E.g. this gel shows that the genetic fingerprint from **suspect C matches** that from the crime scene, **linking** them to the crime scene. All five bands match, so suspect C has the **same number** of repeats (nucleotides) at **five** different places.

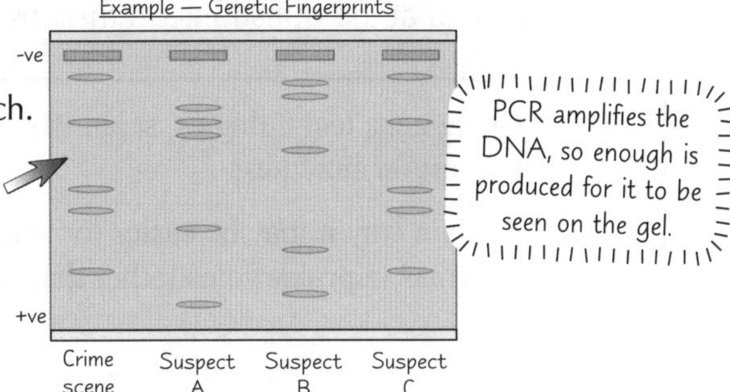

PCR amplifies the DNA, so enough is produced for it to be seen on the gel.

...Medical Diagnosis...

- In medical diagnosis, a genetic fingerprint can refer to a **unique pattern** of **several alleles**.
- It can be used to **diagnose genetic disorders** and **cancer**. It's useful when the **specific** mutation **isn't** known or where **several mutations** could have caused the disorder, because it identifies a **broader, altered** genetic pattern.

  EXAMPLE 1: **Preimplantation genetic haplotyping (PGH) screens embryos** created by **IVF** for genetic disorders **before** they're **implanted** into the uterus. The **faulty regions** of the **parents' DNA** are used to produce **genetic fingerprints**, which are **compared** to the genetic fingerprint of the **embryo**. If the fingerprints **match**, the embryo has **inherited** the **disorder**. It can be used to screen for **cystic fibrosis**, **Huntington's disease**, etc.

  EXAMPLE 2: Genetic fingerprinting can be used to **diagnose sarcomas** (types of **tumour**). Conventional methods of identifying a tumour (e.g. biopsies) only show the **physical differences** between tumours. Now the **genetic fingerprint** of a known sarcoma (e.g. the **different mutated alleles**) can be **compared** to the genetic fingerprint of a **patient's tumour**. If there's a **match**, the sarcoma can be specifically **diagnosed** and the **treatment** can be targeted to that specific type (see page 199).

A specific mutation can be found using DNA probes and sequencing (see p. 218).

...and Animal and Plant Breeding

Genetic fingerprinting can be used on **animals** and **plants** to **prevent inbreeding**, which **decreases** the **gene pool** (the number of **different alleles** in a population, see p. 176). Inbreeding can lead to an **increased risk** of **genetic disorders**, leading to **health**, **productivity** and **reproductive problems**. Genetic fingerprinting can be used to **identify** how **closely-related** individuals are — the **more closely-related** two individuals are, the **more similar** their genetic fingerprint will be (e.g. **more bands** will **match**). The **least related** individuals will be **bred together**.

Warm-Up Questions

Q1 Why are two people unlikely to have the same genetic fingerprint?
Q2 In gel electrophoresis, which electrode do DNA fragments move towards?
Q3 Why might genetic fingerprinting be used in forensic science?

PRACTICE QUESTIONS

Exam Question

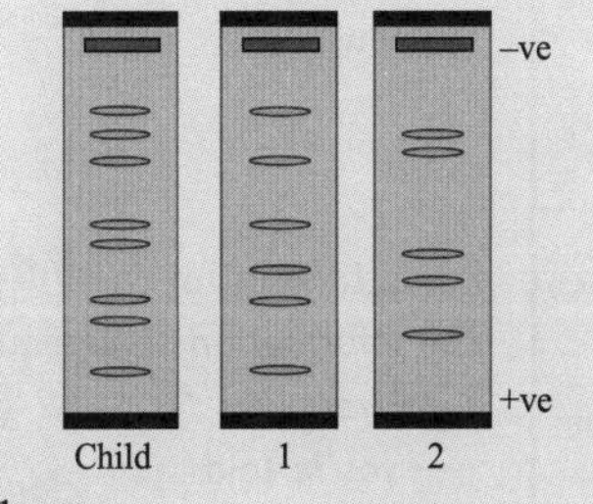

Q1 The diagram shows three genetic fingerprints — one from a child and two from possible fathers.

a) Explain how PCR enables genetic fingerprinting to be carried out. [3 marks]
b) Which genetic fingerprint is most likely to be from the child's father? Explain your answer. [1 mark]
c) Give two more uses of genetic fingerprint technology. [1 mark]

Fingerprinting — in primary school it involved lots of paint and paper...

Who would have thought that tiny pieces of DNA on a gel would be that important? Well, they are and you need to know all about them. Make sure you know the theory behind fingerprinting as well as its applications. And remember, it's very unlikely that two people will have the same genetic fingerprint (except identical twins that is).

Extra Exam Practice

Wow — Topic 8 was no picnic, that's for sure. There were lots of new facts and processes to remember. If you're like me you might understand everything perfectly while you're learning it, but find it all gets a bit jumbled up when you try to recall it. Well, there's one sure-fire way to check everything's unjumbled before exam-day — do some practice exam questions to see how much you know.

- Have a look at this example of how to answer a tricky exam question.
- Then check how much you've understood from Topic 8 by having a go at the questions on the next page.

Think you've done all the practice exam questions? Not quite — there are synoptic questions covering the whole course on p.233-241.

1 Using gene technologies, scientists have successfully identified the base sequences that code for certain enzymes.

1.1 Explain why it may be easier for scientists to identify the base sequences that code for specific enzymes in prokaryotes rather than in humans.

(3 marks)

One example of a human enzyme that scientists have identified the base sequence of is cytochrome P450 2C19. This enzyme is needed to metabolise the drug clopidogrel, which is often prescribed to patients with cardiovascular disease.
Some doctors will only prescribe clopidogrel to certain patients as part of a personalised medicine scheme, as some people have a version of cytochrome P450 2C19 with an altered sequence of amino acids.

1.2 Using this information, explain why a doctor may choose not to prescribe clopidogrel to a patient with cardiovascular disease.

(2 marks)

1.3 Explain how a doctor could determine whether or not a patient with cardiovascular disease should be prescribed clopidogrel as part of a personalised medicine scheme.

(2 marks)

You should know from earlier in the course that prokaryotes are very simple organisms — once you've worked that out, it's just a case of applying your knowledge.

1.1

Prokaryotes are **simple organisms** so **don't have much non-coding DNA**, whereas humans are more complex organisms, so their genome contains **large sections of non-coding DNA** as well as **complex regulatory genes**.

This makes it more difficult to translate the human genome into its proteome and determine the base sequences for specific enzymes, because it's **hard to find the base sequences that code for proteins** among the non-coding and regulatory DNA.

First point out the relevant differences between the genomes of the organisms. Then explain why this makes identifying the base sequences of specific enzymes more difficult in humans.

Pay attention to the information you're given — you're told that cytochrome P450 2C19 is an enzyme so think about how an altered amino acid structure could affect this specific protein.

1.2

If the patient has a version of cytochrome P450 2C19 that codes for an altered sequence of amino acids then the **specific shape** of the **enzyme's active site** may be altered, which could mean that the patient would be unable to metabolise clopidogrel efficiently.

Questions about personalised medicine are likely to involve looking at a patient's genes.

1.3

The doctor would need to have the patient **genetically screened** to determine whether they had the **allele** for **cytochrome P450 2C19** that coded for the normal sequence of amino acids, or whether they had an **allele** that coded for an **altered sequence**. If they had the allele that coded for the altered sequence, the doctor may choose not to prescribe the drug.

The altered sequence of amino acids must be caused by a different sequence of bases in the gene (a different allele).

Extra Exam Practice

2 Myelodysplastic syndrome (MDS) is a group of rare blood disorders.

The defect that causes MDS occurs in the nuclei of haematopoietic stem cells. As a result, MDS can affect the numbers of red blood cells, lymphocytes and phagocytes in the blood.

2.1 Suggest whether haematopoietic stem cells are multipotent or unipotent.
Explain your answer.

(1 mark)

2.2 Haematopoietic stem cells divide to form new cells, which can then develop into red blood cells.
Explain what happens in these new cells in order for them to become red blood cells.

(4 marks)

2.3 A possible cure for MDS is a stem cell transplant.
Suggest why this cure involves receiving cells from a donor rather than using the patient's own cells.

(2 marks)

MDS is classed as a cancer because it causes abnormal cell growth, which is caused by disruption to the normal function of tumour suppressor genes.
MDS can be treated with the drug, azacitidine. When incorporated into DNA, metabolised azacitidine binds irreversibly to DNA methyltransferases, which are the enzymes responsible for DNA methylation.

2.4 Explain how azacitidine works as a treatment for MDS.

(4 marks)

2.5 Explain why azacitidine may be described as an 'epigenetic' drug.

(1 mark)

Many studies have been done to investigate a link between smoking and development of MDS.
The results of some of these studies were analysed to produce odds ratios.
Odds ratios quantify the strength of the connection between exposure to a variable and an outcome. In this case, the greater the value of the odds ratio, the greater the odds of MDS occurring as a result of being exposed to the variable.
The odds ratios for some of the studies analysed are shown in **Table 1**.

Table 1

Study	Odds ratio	
	Current smokers	Former smokers
Lv (2011)	1.37	1.48
Ma (2009)	3.17	1.68
Dalamaga (2002)	4.03	2.60
Mele (1994)	1.20	2.70
Strom (2005), male	1.85	1.31

2.6 Some health experts advise people to stop smoking in order to reduce their risk of developing MDS.
Is this advice supported by the data shown in **Table 1**? Explain your answer.

(2 marks)

Planning an Experiment

As well as doing practical work in class, you can get asked about it in your exams too. Harsh I know.

Before You Start Planning, **Be Clear** on What You're Trying to **Find Out**

Like all scientists, you should start off by making a **prediction** or **hypothesis** — a **specific testable statement**, based on theory, about what will happen in the experiment. You then need to **plan** a good experiment that will provide **evidence to support the prediction** — or help **disprove it**.

A **Good Experiment** Gives **Results** that are...

Precise results are sometimes referred to as reliable results.

1) **Precise** — precise results **don't vary much** from the **mean**. Precision is reduced by **random error** (the unpredictable way in which all measurements vary).
2) **Repeatable and reproducible** — repeatable means that if the same person repeats the experiment using the same methods and equipment, they will get the same results. Reproducible means that if someone different does the experiment, using a slightly different method or piece of equipment, the results will still be the same.
3) **Valid** — valid results **answer** the **original question**. To get valid results you need to **control all the variables** (see below) to make sure you're only testing the thing you want to.
4) **Accurate** — accurate results are **really close** to the true answer. **Human interpretation** of a measurement (e.g. determining a colour change) can **reduce** the accuracy of results.

Here are some things you need to consider when designing a good experiment:

1) **Only one variable should be changed** — Variables are **quantities** that have the **potential to change**, e.g. pH. In an experiment you usually **change one variable** and **measure its effect** on another variable.
   - The variable that you **change** is called the **independent variable**.
   - The variable that you **measure** is called the **dependent variable**.
2) **All the other variables should be controlled** — When you're investigating a variable you need to keep everything else that could affect it **constant**. This means you can be sure that **only** your **independent** variable is **affecting** the thing you're measuring (the dependent variable).
3) **Negative controls should be used** — Negative controls are used to **check** that only the independent variable is affecting the dependent variable. Negative controls **aren't expected** to have **any effect** on the experiment.
4) **The experiment should be repeated at least three times and a mean should be calculated** — this reduces the effect of **random error** on your experiment, which makes your results **more precise**. Doing repeats and getting **similar results** each time also shows that your data is **repeatable** and makes it more likely to be **reproducible**.

EXAMPLE: Investigating the effect of **light intensity** on **rate of photosynthesis** of **Canadian pondweed**.

1) Light intensity is the **independent** variable.
2) Rate of photosynthesis is the **dependent** variable.
3) pH, temperature and the time the pondweed is left should all **stay the same** (and the quantities should be recorded to allow someone else to reproduce the experiment).
4) The experiment should be **repeated** at least **three times** for each light intensity used.
5) A **negative control**, in which the experiment is carried out in the **dark**, should also be used. No photosynthesis should happen with this control.

Examiners love getting you to **comment** on **experimental design** or **suggest improvements** to **methods** — e.g. how a method could be improved to make the results more precise. So make sure you know how to **design** a **good experiment**.

Select Appropriate **Apparatus, Equipment** and **Techniques**

1) When you're **planning** an experiment you need to decide what it is you're going to **measure** and **how often** you're going to take measurements. E.g. if you're investigating the **rate of respiration**, you could either measure the volume of **oxygen used** over time or the volume of **carbon dioxide produced** over time. You could take measurements at, e.g. 30 second intervals or 60 second intervals.
2) Then you need to choose the most **appropriate** apparatus, equipment and techniques for the experiment. E.g.
   - The **measuring apparatus** you use has to be **sensitive** enough to measure the changes you're looking for. For example, if you need to measure **small changes** in **pH**, a **pH meter** (which can measure pH to several decimal places) would be more sensitive than indicator paper.
   - The **technique** you use has to be the most **appropriate** one for your **experiment**. E.g. if you want to measure the concentration of glucose in an unknown solution, using a **colorimeter** in conjunction with **quantitative Benedict's reagent** (see page 161) will help you to get more **accurate results** than simply comparing the colour differences of the solutions by eye.

Planning an Experiment

You Need to Know How to Use **Apparatus** and **Techniques Correctly**

Examiners could ask you about a **whole range** of different apparatus and techniques. Make sure you know how to use all the instruments and equipment you've come across in class and can carry out all the techniques too. Here are some **examples** of equipment you should be able to use:

- **Measuring cylinders** and **graduated pipettes** — These have a **scale** so you can measure specific **volumes**. Whichever one you use, make sure you read the volume from the **bottom** of the **meniscus** when it's at **eye level**.

The meniscus is the curved upper surface of the liquid inside the pipette.

Read volume from here — at the bottom of the meniscus.

- **Water baths** — Make sure you **allow time** for water baths to **heat up** before starting your experiment. Don't forget that your **solutions** will need **time** to get to the **same temperature** as the water before you start the experiment too. Also, remember to **check** the **temperature** of the water bath with a **thermometer** during the investigation to make sure it **doesn't change**.
- **Data logger** — Decide **what** you are **measuring** and what **type** of **data logger** you will need, e.g. temperature, pH. Connect an **external sensor** to the data logger if you need to. Decide **how often** you want the data logger to take readings depending on the **length** of the **process** that you are measuring.

Make sure you know how to do **all** the **practical investigations** described in this book. You should be able to **apply** the techniques described to **different contexts**. For example, page 121 describes how to use a **colorimeter** and a **redox indicator dye** to investigate the rate of **dehydrogenase activity** in **chloroplasts**. You could use a similar technique (i.e. a colorimeter and a redox indicator dye) to investigate the **rate of respiration** in **yeast**.

Risk Assessments Help You to **Work Safely**

1) When you're planning an experiment, you need to carry out a **risk assessment**. To do this, you need to identify:
   - All the **dangers** in the experiment, e.g. any hazardous chemicals, microorganisms or naked flames.
   - **Who** is at **risk** from these dangers.
   - What can be done to **reduce** the **risk**, such as wearing goggles or gloves or working in a fume cupboard.
2) You also need to consider any **ethical issues** in your experiment. For example, if you're using **living animals** (e.g. insects) you must treat them with **respect**. This means **handling them carefully** and keeping them away from **harmful chemicals**, **extreme heat sources** and other things that might cause them **physical discomfort**.

Record Your **Data** in a **Table**

It's a good idea to draw a table to **record** the **results** of your experiment in.

1) When you draw a table, make sure you **include** enough **rows** and **columns** to **record all of the data** you need to. You might also need to include a column for **processing** your data (e.g. working out an average).
2) Make sure each **column** has a **heading** so you know what's going to be recorded where. The **units** should be in the **column heading**, not the table itself.

Concentration / mol dm^{-3}	*Absorbance / Absorbance Units (AU)*
0.2	*0.5*
0.4	*0.9*
0.6	*1.3*

(Labels: heading, column, units, row, data)

3) The **independent variable** should be recorded in the **left-hand** column and the **dependent variable** in the **right**.

Watch Out for **Anomalous Results**

Doing repeats makes it easier to spot anomalous results.

When you look at all the **data** in your **table**, you may notice that you have a result that **doesn't seem to fit in** with the rest at all. These results are called **anomalous results**. You should **investigate** anomalous results — if you can work out what happened (e.g. you measured something totally wrong) you can **ignore** them when **processing** your results. However, you can't just ignore a result because you don't like the look of it.

My best apparatus is the pommel horse...

It's not really, I just like the word pommel. Scientists are rightfully fussy about methods and equipment — I mean if you're going to bother doing an experiment, you should at least make sure it's going to give you results you can trust.

Processing and Presenting Data

Processing data means taking raw data and doing some calculations with it, to make it more useful.

Processing the Data Helps You to **Interpret** it

You Need to be Able to **Calculate Percentage Change** and **Ratios**

1) Calculating **percentage change** helps to **quantify** how much something has changed, e.g. the percentage change in the growth rate of pea plants when a fertiliser is added. To **calculate** it you use this equation:

$$\text{Percentage change} = \frac{\text{final value} - \text{original value}}{\text{original value}} \times 100$$

A **positive** value shows an **increase** and a **negative** value shows a **decrease.**

E.g. a person's blood glucose concentration before a meal was **4.2 mmol dm^{-3}**. Two hours after a meal it was **6.5 mmol dm^{-3}**. Calculate the percentage change.

$$\text{Percentage change} = \frac{6.5 - 4.2}{4.2} \times 100 = 55\% \text{ (2 s.f.)}$$

So the person's blood glucose concentration was 55% higher after the meal.

2) Ratios can be used to **compare** lots of different types of quantities. E.g. an organism with a **surface area to volume ratio** of **2 : 1** would theoretically have a surface area **twice as large** as its volume.

3) Ratios are usually most useful in their **simplest** (smallest) **form**. To simplify a ratio, **divide each side** by the san **number**. It's in its simplest form when there's nothing left you can divide by. To get a ratio of X : Y in the form **X : 1**, **divide both sides by Y**. E.g. to get 28 : 34 into the ratio of X : 1, divide both sides by 34. You get 0.82 :

You Need to be Able to **Use Logarithms**

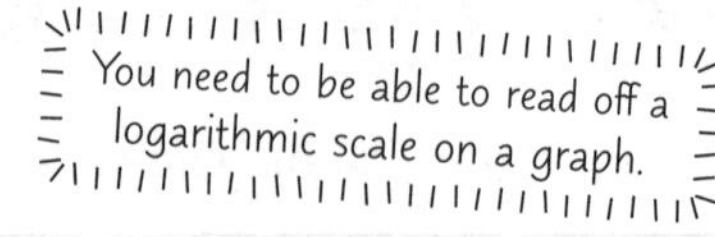

1) It's tricky to plot graphs with **very small** and **very large** numbers (e.g. both 0.1 and 1000) on the **same axis**.
2) We can make it easier by converting values to their **logarithms** and plotting them on a **logarithmic scale** (e.g. a $\log_{10}$ scale).
3) On a $\log_{10}$ **scale**, each value is **ten times larger** than the value before. This means the numbers 1, 2, 3 and 4 on a $\log_{10}$ **scale** represent 10, 100, 1000 and 10 000 on a **linear (normal) scale**.

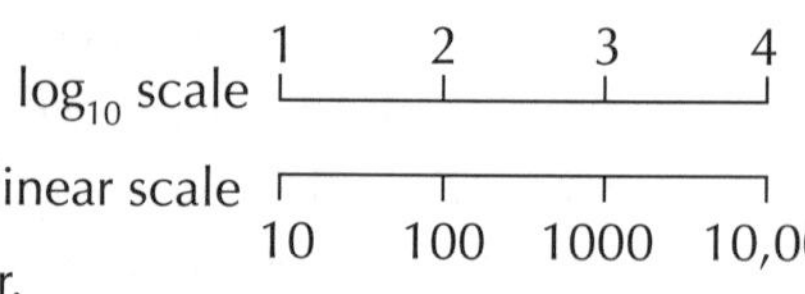

4) To calculate logarithms, you need to use the **log button** on your calculator. On most calculators 'log' will stand for $\log_{10}$, but different calculators work differently, so make sure you know how to use yours.

Averages and the **Range** Can be Used to **Summarise** Your **Data**

1) When you've done **repeats** of an experiment you should always calculate a **mean** (a type of average). To do this **add together** all the data values and **divide** by the **total** number of values in the sample.

Test tube	Mass (g)			Mean (g)	Range (g)
	Repeat 1	Repeat 2	Repeat 3		
A	28	37	32	(28 + 37 + 32) ÷ 3 = 32.3	37 – 28 = 9
B	47	51	60	(47 + 51 + 60) ÷ 3 = 52.7	60 – 47 = 13

2) You might also need to calculate the **range** (how **spread out** the data is). To do this find the **largest** data value and **subtract** the **smallest** data value from it.

3) **Standard deviation** can be more useful than the **range** because it tells you how **values** are spread about the **mean** rather than just the **total spread** of data. A **small standard deviation** means the repeated results are all **similar** and **close** to the mean, i.e. they are **precise**.

Like the mean, the **median** and **mode** are both types of average.

- To calculate the **median**, put all your data in **numerical order**. The median is the **middle value** in this list. If you have an **even number** of values, the median is **halfway** between the middle two values.
- To calculate the **mode**, count **how many times** each value comes up. The mode is the number that appears **most often**. A set of data might not have a mode — or it might have more than one.

Processing and Presenting Data

Watch Out For **Significant Figures...**

1) The **first significant figure** of a number is the **first digit** that **isn't a zero**. The second, third and fourth significant figures follow on immediately after the first (even if they're zeros).
2) When you're processing your data you may well want to round any **really long numbers** to a certain number of **significant figures**. E.g. **0.6878976** rounds to **0.69** to **2 s.f.**.
3) When you're doing **calculations** using measurements given to a certain number of significant figures, you should give your **answer** to the **lowest number** of significant figures that was used in the calculation. For example:

When rounding a number, if the next digit after the last significant figure you're using is less than five, you should round it down and if it's 5 or more you should round it up.

4) This is because the **fewer digits** a measurement has, the less **accurate** it is. Your answer can only be as accurate as the **least accurate measurement** in the calculation.

...and **Standard Form**

1) When you're processing data you might also want to change **very big** or **very small numbers** that have **lots of zeros** into something more manageable — this is called **standard form**.

   E.g. 1 000 000 can be written 1×10^6 and 0.017 can be written 1.7×10^{-2}.

2) To do this you just need to **move the decimal point** left or right. The number of places the decimal point moves is then represented by a **power of 10** — this is positive for big numbers, and negative for numbers smaller than one. For example:

A rabbit playing the piano. Definitely not standard form.

$16\,500 = 1.65 \times 10^4$ The decimal point has moved **four places** to the **left**, so the power of 10 is **+4**.

$0.000362 = 3.62 \times 10^{-4}$ The decimal point has moved **four places** to the **right**, so the power of 10 is **–4**.

You Need to Understand How and When **Statistical Tests** are Used to **Analyse Data**

Examples:

1) The **Student's t-test**. You can use the Student's t-test when you have two sets of **data** that you want to **compare**. It tests whether there is a **significant difference** in the **means** of the two data sets. The value obtained is compared to a **critical value**, which helps you decide how likely it is that the results or 'differences in the means' were **due to chance**. If the value obtained from the t-test is **greater than** the critical value at a **probability** (**P value**) of **5% or less** (≤ 0.05), then you can be 95% confident that the difference is significant and not due to chance. This is called a **95% confidence limit** — which is good enough for most biologists to **reject** the **null hypothesis**. A null hypothesis is a special type of hypothesis used with statistical tests. It states that there's no significant difference between the things you're measuring.
2) The **Chi-squared test** (see pages 174-175). You can use the Chi-squared test when you have **categorical** (grouped) **data** and you want to compare whether your **observed results** are **statistically different** from your **expected results**. You compare your result to a **critical value** — if it's **larger** than the critical value at **P = 0.05**, you can be **95% certain** the difference is significant.
3) A correlation coefficient, e.g. the **Spearman's rank correlation coefficient**. This test allows you to work out the **degree** to which **two** sets of **data** are **correlated** (see page 229 for more on correlation). It is given as a value between 1 and –1. A value of 1 indicates a **strong positive correlation**, 0 means there is **no correlation** and –1 is a **strong negative correlation**. You can then compare your result to a critical value to find out whether or not the correlation is significant.

You can be more confident in your **conclusions** (see page 229), if they're based on results that have been analysed using a statistical test.

You need to be familiar with the symbols < (less than), > (more than), << (much less than) and >> (much greater than).

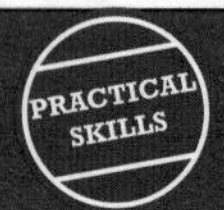

Processing and Presenting Data

Use a Suitable **Graph** or **Chart** to **Present** Your **Data**

Graphs and charts are a great way of **presenting data** — they can make results much **easier to interpret**.

1) When you have **qualitative** data (non-numerical data, e.g. blood group) or **discrete** data (numerical data that can only take certain values in a range, e.g. shoe size) you can use **bar charts** or **pie charts**.
2) When you have **continuous** data (data that can take any value in a range, e.g. height or weight) you can use **histograms** or **line graphs.**
3) When you want to show how **two variables** are **related** (or **correlated**, see next page) you can use a **scatter graph**.

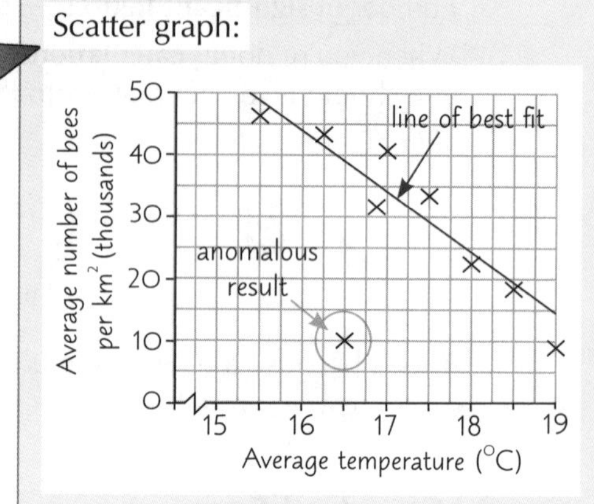

Whatever type of graph you use, you should make sure that:

- The **dependent variable** goes on the **y-axis** (the vertical axis) and the **independent** on the **x-axis** (the horizontal axis).
- You always **label** the **axes**, include the quantity and **units**, and choose a **sensible scale**.
- The graph covers **at least half** of the **graph paper**.

If you need to draw a **line** (or curve) **of best fit** on a **scatter graph**, draw the line through or as near to as many points as possible, **ignoring** any **anomalous** results.

Find the **Rate** By Finding the **Gradient**

Rate is a **measure** of how much something is **changing over time**. Calculating a rate can be useful when analysing your data, e.g. you might want to the find the **rate of a reaction**. Rates are easy to work out from a graph:

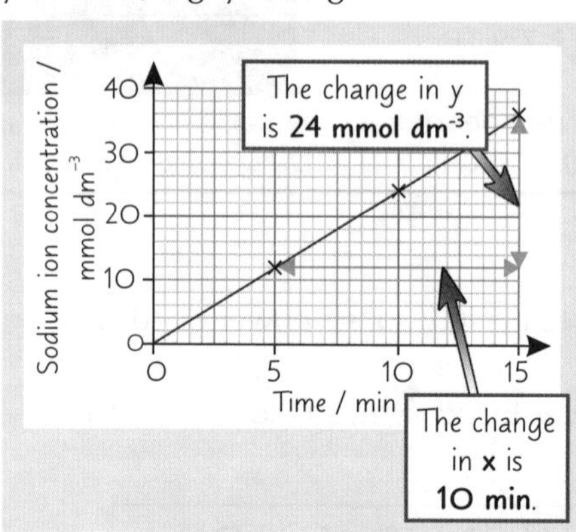

For a **linear** graph you can calculate the **rate** by finding the **gradient of the line:**

$$\text{Gradient} = \frac{\text{Change in Y}}{\text{Change in X}}$$

So in this **example**:

$$\text{rate} = \frac{24\ \text{mmol dm}^{-3}}{10\ \text{minutes}} = \mathbf{2.4\ mmol\ dm^{-3}\ min^{-1}}$$

The **equation** of a **straight line** can always be written in the form **y = mx + c**, where **m** is the **gradient** and **c** is the **y-intercept** (this is the **value of** y when the line crosses the y-axis). In this example, the equation of the line is **y = 2.4x + 0** (or just **y = 2.4x**). Knowing the equation of the line allows you to estimate results not plotted on the graph. E.g. in this case, when x (the time) is **20 min**, y (the sodium ion concentration) will be 2.4x = 2.4 × 20 = **48 mmol dm^{-3} min^{-1}**.

For a **curved** (non-linear) graph you can find the **rate** by drawing a **tangent**:

1) Position a ruler on the graph at the **point** where you want to know the **rate**.
2) **Angle** the **ruler** so there is **equal space** between the **ruler** and the **curve** on **either** side of the point.
3) **Draw** a **line** along the ruler to make the tangent.
   Extend the line right across the graph — it'll help to make your **gradient calculation easier** as you'll have **more points** to choose from.
4) **Calculate** the **gradient** of the **tangent** to find the **rate**.

Gradient = 55 m² ÷ 4.4 years = **12.5 m² year^{-1}**

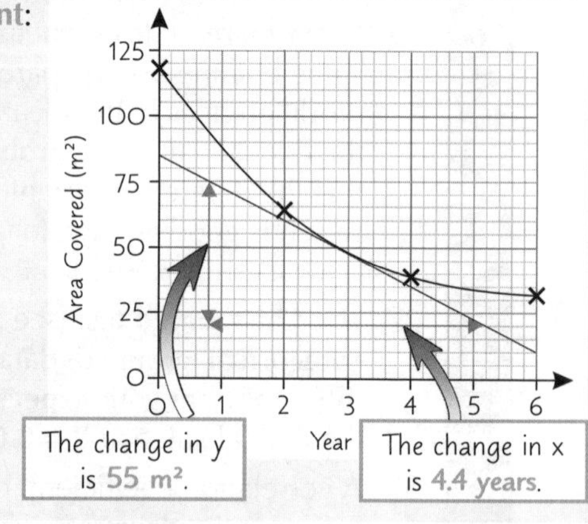

When calculating a rate (or anything else for that matter) you might have to **convert** between **units**, e.g. seconds and minutes. Make sure you can convert between common units of time, length and volume.

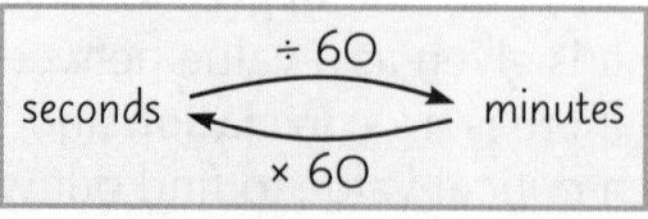

Significant figures — a result of far too many cream cakes...

Lots of maths to get your head around on these three pages, but stay calm and take your time with it all. You'll be fine.

Drawing Conclusions and Evaluating

There's no point in getting all those lovely results and just leaving it at that. You need to draw some conclusions...

You Need to be Able to **Draw Conclusions** From **Data**

1) Conclusions need to be **valid**. A conclusion can only be considered as valid if it uses valid data (see page 224).
2) You can often draw conclusions by looking at the relationship (**correlation**) between two variables:

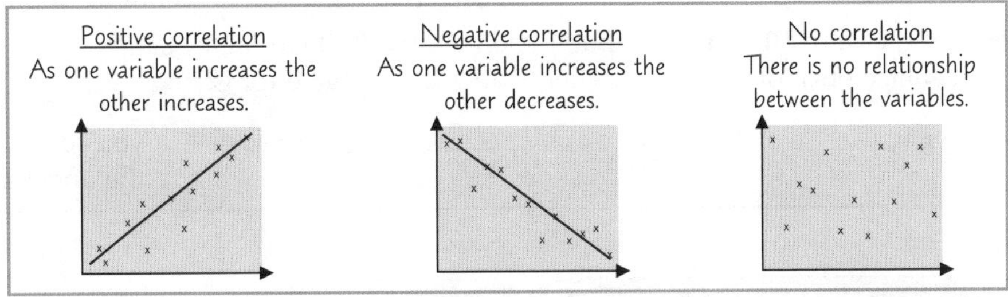

There is no correlation between the colour of your tights and the proportion of your life you spend upside down.

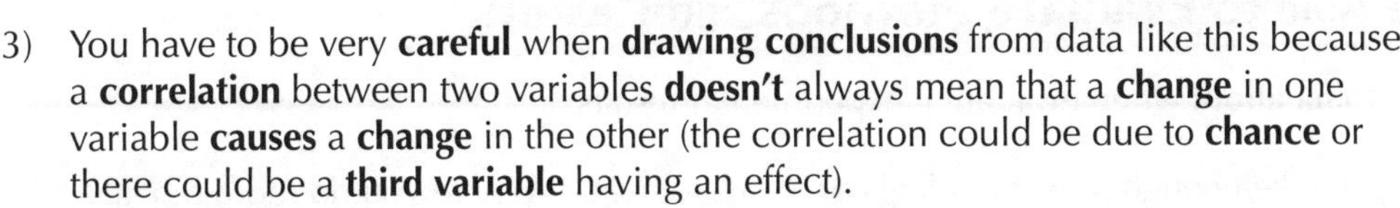

3) You have to be very **careful** when **drawing conclusions** from data like this because a **correlation** between two variables **doesn't** always mean that a **change** in one variable **causes** a **change** in the other (the correlation could be due to **chance** or there could be a **third variable** having an effect).
4) If there's a relationship between two variables and a change in one variable **does** cause a change in the other it's called a **causal relationship**.
5) It can be **concluded** that a **correlation** is a **causal relationship** if every other variable that could possibly affect the result is **controlled**.

In reality this is very hard to do — correlations are generally accepted to be causal relationships if lots of studies have found the same thing, and scientists have figured out exactly how one factor causes the other.

6) When you're making a conclusion you **can't** make broad **generalisations** from data — you have to be very **specific**. You can only **conclude** what the results show and **no more**.

Example

The graph shows the results from an investigation into the effect of concentration of plant growth factor X on the height of Plant Species A. The only **conclusion** you can draw is that as the **concentration** of **growth factor X increases**, the **height** of **Plant Species A increases**. You **can't** conclude that this is true for any other plant growth factor or any other plant species — the **results** could be **completely different**.

Height of Plant Species A

Concentration of growth factor X

Uncertainty is the Amount of **Error** Your **Measurements** Might Have

1) The results you get from an experiment won't be completely perfect — there'll always be a **degree of uncertainty** in your readings or measurements due to limits in the **sensitivity** of the apparatus you're using.
2) For example, an electronic mass balance might measure to the **nearest 0.01 g**, but the real mass could be up to **0.005 g smaller or larger**. It has an **uncertainty value** of **± 0.005 g**.
3) A ± sign tells you the **range** in which the **true value** lies (usually to within a **95% confidence level**). The range is called the **margin of error**.

A reading is when you make a judgement about one value, e.g. when you read a value off a mass balance. A measurement is when you judge two values and find the difference, e.g. when you measure length with a ruler.

You Can **Calculate** The **Percentage Error** of Your **Measurements**

If you know the **uncertainty value** of your measurements, you can calculate the **percentage error** using this formula:

$$\text{percentage error} = \frac{\text{uncertainty}}{\text{reading}} \times 100$$

Example

50 cm^3 of HCl is measured with an uncertainty value of ± 0.05 cm^3.

percentage error = $\frac{0.05}{50} \times 100 = \mathbf{0.1\%}$

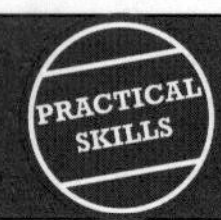

Drawing Conclusions and Evaluating

You Can **Minimise** the **Errors** in Your **Measurements**

1) One obvious way to **reduce errors** in your measurements is to buy the most **sensitive equipment** available. In real life there's not much you can do about this one — you're stuck with whatever your school or college has got. But there are other ways to **lower the uncertainty** in experiments.
2) For example, you can plan your experiment so you **measure** a **greater amount** of something:

> If you use a **500 cm³** cylinder that goes up in **5 cm³** increments, each reading has an uncertainty of **± 2.5 cm³**.
>
> So using a 500 cm³ cylinder to measure **100 cm³** of liquid will give you a percentage error of:
>
> $\frac{2.5}{100} \times 100 = \mathbf{2.5\%}$
>
> But if you measure **200 cm³** in the same cylinder, the percentage error is:
>
> $\frac{2.5}{200} \times 100 = \mathbf{1.25\%}$
>
> Hey presto — you've just **halved** the uncertainty.

You Also Need to Be Able to **Evaluate Methods** and **Results**

1) Here are some things to **think about** when evaluating experimental results:

- **Repeatability**: Did you take enough repeat readings or measurements? Would you do more repeats if you were to do the experiment again? Do you think you'd get similar data if you did the experiment again?
- **Reproducibility**: Have you compared your results with other people's results? Were your results similar? Could other scientists gain data showing the same relationships that are shown in your data?
- **Validity**: Does your data answer the question you set out to investigate? Were all the variables controlled?

2) Make sure you **evaluate** your **method** too. Is there anything you could have done to make your results more **precise** or **accurate**? Were there any **limitations** in your method, e.g. should you have taken measurements more **frequently**? Were there any **sources** of **error** in your experiment? Could you have used more sensitive **apparatus** or **equipment**? Think about how you could **refine** and **improve** your experiment if you did it again.
3) Once you've thought about these points you can decide how much **confidence** you have in your **conclusion**. For example, if your results are **repeatable**, **reproducible** and **valid** and they back up your conclusion then you can have a **high degree** of **confidence** in your conclusion.

You can apply all these questions to any results or methods you're given to evaluate in the exams too.

Solving Problems in a **Practical Context**

In the exams, you'll get plenty of questions set in a 'practical context'. As well as answering questions about the methods used or the conclusions drawn, you'll need to be able to **apply** your **scientific knowledge** to **solve problems** set in these contexts. For example:

> Q1 An experiment was carried out to investigate the role of IAA in shoot growth. The experimental set up is shown in the diagram on the right.
>
> Four shoots were then placed in the dark (experiment 1) and the other four shoots were exposed to a light source directed from the right (experiment 2). After two days, the amount of growth (in mm) and direction of growth was recorded. The results are shown in the table.
>
> a) Explain the results seen for shoot C. [3 marks]

☐ Sponge soaked in IAA and glucose
■ Sponge soaked in water and glucose

A B C D — Shoot minus the tip

	Growth / mm			
	Shoot A	Shoot B	Shoot C	Shoot D
Experiment 1 (dark)	6, right	6, left	6, straight	1, straight
Experiment 2 (light)	8, right	8, right	8, right	3, straight

You should remember from page 140 that **IAA** stimulates **cell elongation** in a shoot. In experiment 1, equal amounts of IAA diffuse down **both sides** of shoot C, making the cells elongate at the **same rate**, so the shoot grows **straight up**. In experiment 2, IAA moved to the **shaded** (left-hand side) of shoot C, so the shoot grew to the **right** — towards the **light**.

Correlation Street — my favourite programme...

Don't ever, ever assume that correlation means cause. There, I've told you again. No excuses now. A good evaluation is a sign that you really understand what makes a good experiment, so make sure your evaluation-writing-skills are top notch.

How to Do Well in Your Exams

The reason for learning all the lovely facts and diagrams in this book is so that you can ace your exams and get yourself an A-level in Biology. So, now it's a good idea to find out exactly what you'll be in for exam-wise...

Make Sure You Know the **Structure** of Your **Exams**

It seems obvious, but if you know exactly what will be **covered** in each of the exams, how much **time** you'll have to do them and how they'll be **structured**, you can be better prepared. So let's take a look at the ins and outs of all the exams you'll be facing for **A-level Biology**...

Paper	Total marks	Time	Topics assessed
1	91	2 hours	1, 2, 3, 4
2	91	2 hours	5, 6, 7, 8
3	78	2 hours	1 to 8

All three A-level papers also test you on Practical Skills — see pages 224-230 for more.

All this exam info is **only relevant** if you're taking the **A-level** in Biology. If you're taking the **AS-level**, you'll be sitting a **completely different set of papers**, which are structured in a different way. There are two AS-level papers that both test Topics 1 to 4 (and Practical Skills).

1) **Papers 1** and **2** are mainly a mixture of **short** and **long answer questions**. Some of these questions will test you on the **facts** you need to know, some will test whether you can **apply your knowledge** to unfamiliar contexts and some will test your knowledge of **practical skills**. There'll also be a few **calculation questions**.
2) **Paper 1** also contains **15 marks' worth** of **extended response questions**. These are questions that require you to write a **longer answer** with a **logical structure**. E.g. you could be asked to describe the **steps** in a particular **process**. These questions could involve an **extended calculation** too.
3) **Paper 2** also contains a **15 mark comprehension question**. You'll be given a **passage of information** to **read** and will then need to **answer** the question parts that follow using **both** the **information** you've been given and your **own scientific knowledge**.
4) **Paper 3** is split into two sections. **Section A** has lots of questions on **practical techniques** and **skills**, with 15 marks being awarded for questions that ask you for a **critical analysis** of **experimental data**. For example, you could be given some data (e.g. in a graph or table) and asked to **draw conclusions** from it or you could be given a conclusion and asked to **evaluate** how well the data **supports the conclusion**. As for Papers 1 and 2, there'll also be fact recall questions, questions that test whether you can apply your knowledge, and calculation questions.
5) **Section B** of **Paper 3** consists of a **25 mark synoptic essay question**...

Synoptic means you will need to draw together your knowledge of different areas of Biology in relation to a theme.

You Need to be Able to **Write** a **Good Essay**

You'll be given a **choice** of **two** essay titles in Section B of Paper 3 and asked to write about **one** of them. The titles are designed to get you to write about a **range of material** from **both years** of your A-level course. Writing an essay might seem like a **daunting task**, but **don't panic**. Here are some tips for getting **top marks**:

1) Before you start your essay, it's a good idea to quickly scribble down a **rough plan** — this should help you to present your ideas in a **clear, logical way**. It should also **stop you** from **repeating yourself** or **missing out** any **important bits**. You should aim to write about **at least five different topic areas**.
2) You'll need to clearly show how **all** the information you include is **relevant** to the **essay title** — don't just write down everything you know about a topic.
3) The information you include must be **detailed, scientifically correct** and of **A-level standard**. 'Plants are green and have leaves' just ain't gonna cut it I'm afraid...
4) You must use appropriate **scientific terminology**.
5) Your essay should be **well-written** and **clearly explained**.
6) To get the **very highest marks**, your answer should show evidence of **wider reading** (i.e. it should include things that aren't explicitly on the specification, but are still of a high standard and relevant to the question).

Making a quick plan of the topics you'll cover might help you decide which title you can write a better essay for.

You'll get 2 hours in total for this paper and should aim to leave yourself **about 50 minutes** to plan and write your essay. This should be enough time to write **about 3 sides of A4**.

How to Do Well in Your Exams

Command Words Tell You **What You Need to do** in a Question

Command words are just the bits of a question that tell you **what to do**. You'll find answering exam questions much easier if you understand exactly what they mean, so here's a brief summary table of the **most common** ones:

Command word:	What to do:
Give / Name / State	Give a brief one or two word answer, or a short sentence.
Describe	Write about what something's like, e.g. describe the structure of fish gills.
Explain	Give reasons for something.
Suggest	Use your scientific knowledge to work out what the answer might be.
Compare	Give the similarities and differences between two things.
Contrast	Give the differences between two things.
Calculate	Work out the solution to a mathematical problem.
Evaluate	Give the arguments both for and against an issue, or the advantages and disadvantages of something. You also need to give an overall judgement.

Even though you're taking an A-level in Biology, there will be some maths to do in these papers that's set in a biological context.

Some questions will also ask you to answer '**using the information/data provided**' (e.g. a graph, table or passage of text) or '**with reference to figure X**' — if so, you must **refer to** the information, data or figure you've been given or yo won't get the marks. Some questions may also ask you to answer '**using your calculation**' — it's the same here, you need to use your **answer** to a particular **calculation**, otherwise you won't get the marks.

Not all of the questions will have command words — instead they may just ask a which / what / how type of questio

Time Management is **Important**

1) For **Papers 1** and **2**, you get **just over a minute per mark**. So if you get stuck on a short question, it's sometimes worth moving on to another one and then coming back to it if you have time. Bear in mind that you might want to spend a **bit longer** than a minute per mark on the **extended response** and **comprehension questions**.
2) For **Paper 3**, it's a **similar story** — you'll want to **spend longer per mark** on the **essay question** than on the shorter questions, so make sure you **leave enough time** for this at the end.

Warm-Up Questions

Q1 Which two A-level papers test you on material from Topics 1 to 4?

Q2 Which A-level papers test you on Practical Skills?

Q3 What's the difference between the command words 'describe' and 'explain'?

PRACTICE QUESTIONS

Exam Question

Q1 Write an essay about the importance of proteins to living organisms. [25 marks]

You might think you need your head examined for picking A-level Biology...

...because there's a lot to learn and three big exams to do. But let me just stop you right there... Instead of worrying, just work through this book, including having a go at all of the questions and you'll be well and truly prepped for the exams. Then re-read these pages to make sure you know what's coming. After that, all there is to say is... good luck.

Synoptic Practice

Before you go any further, I'd make sure that you've gone through every single section of this book and that you know each topic like the back of your hand. Huh, I've never seen that mole before... Anyway, coming up now are pages and pages full of synoptic questions. To answer these questions you'll need to apply your knowledge of different parts of the course, which means you really need to know your stuff. Synoptic questions are tricky, but they'll make an appearance on exam day, so if I were you I'd get practising...

1 Axolotls are a species of amphibian native to the waterways around Mexico City. Unlike other amphibians, axolotls often remain living in water for their entire life.

Axolotls have external gills that branch out from their head. Blood flows into these gills through blood vessels that are very close to the skin surface. Oxygen is then able to diffuse directly through the surface of the gills into the blood. **Figure 1** shows these gills on an axolotl's head.

Figure 1

1.1 Suggest the benefit of the 'feathery' appearance of the gills.

(1 mark)

1.2 Fish gills use a counter-current system for oxygen absorption.
Suggest and explain why the gills of axolotls may provide less efficient oxygen absorption than the gills of fish.

(4 marks)

There has been a decline in wild axolotl numbers in recent years. Part of the reason for this is that toxic chemicals used by local farmers enter the waterways in which axolotls live.

1.3 A team of students want to investigate the biodiversity of one of the waterways in which axolotls live by regularly measuring species richness.
Would measuring species richness allow the students to monitor the decline in wild axolotl numbers? Explain your answer.

(2 marks)

1.4 Suggest how encouraging local farmers to use fewer toxic chemicals could help to prevent a decline in genetic diversity within the wild axolotl population.

(2 marks)

2 Scientists carried out investigations to determine the optimum conditions for the activity of two digestive enzymes, pepsin and trypsin.

2.1 Initially the scientists investigated the effect of substrate concentration on the activity of pepsin. They plotted their results on a graph with rate of reaction on the y-axis and substrate concentration on the x-axis, and produced a curve of best fit.
Explain the shape of the curve of best fit you would expect to see on the graph.

(4 marks)

Synoptic Practice

The scientists then investigated the effect of pH on the activity of both of the enzymes. Their results are shown in **Figure 2**.

Figure 2

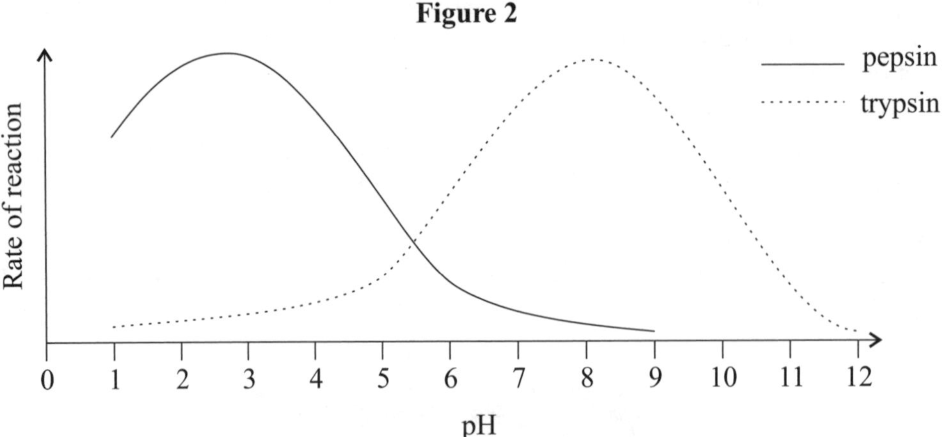

Pepsin and trypsin are both endopeptidases. Pepsin is active in the stomach and trypsin is active in the small intestine. The pH of the stomach is usually between pH 1 and pH 3 due to the presence of stomach acid.

2.2 Indigestion is a burning pain caused by stomach acid irritating the lining of the stomach, the oesophagus or the top part of the small intestine. Indigestion remedies aim to reduce this irritation by neutralising the acid in the stomach.
Using **Figure 2**, suggest how indigestion remedies might affect the rate of protein digestion in the stomach.

(2 marks)

2.3 Pancreatic juice is secreted into the small intestine, where it mixes with food from the stomach. Using **Figure 2**, suggest whether pancreatic juice is likely to be acidic or alkaline.
Explain your answer.

(3 marks)

Figure 3 shows the chemical structure of a protein.

Figure 3

$NH_2-CH_2-CO\overset{A}{-}NH-CH(CH_2OH)\overset{B}{-}CO\overset{C}{-}NH-CH(CH(CH_3)_2)\overset{D}{-}CO-NH-CH(CH_3)-COOH$

2.4 Which bond on **Figure 3** (**A-D**) would be broken by an endopeptidase?
Explain your answer.

(2 marks)

3 Mosses are simple plants that are believed to have been some of the first land plants to evolve.

3.1 Mosses absorb most of the water they need directly across cells on their outer surface. The water can then travel through the narrow spaces within the plant to where it is needed. Suggest how the cohesive property of water aids water transport in moss plants.

(1 mark)

Synoptic Practice

Moss plants can reproduce sexually or asexually. Asexual reproduction only requires one parent and produces clones of the parent plant.

3.2 Explain why sexual reproduction may be more beneficial for survival in unstable environmental conditions, compared to asexual reproduction.

(2 marks)

One stage of sexual reproduction in mosses involves sporophytes. These are structures that contain diploid cells. Some of the diploid cells divide by meiosis to produce haploid spores, which are stored in a capsule and then released from the sporophyte.

3.3 One sporophyte released around 350 000 spores from its capsule.
Calculate how many times meiosis would have occurred to produce this many spores.

(1 mark)

When the spores are released from the capsule, they go on to produce a gametophyte, which is the part of the plant most commonly seen. This stage of the life cycle produces gametes.

The chromosome number of the sporophyte of the moss species, *Bartramia pomiformis*, is 16.

3.4 Calculate the possible number of combinations of chromosomes following meiosis in *B. pomiformis*. Assume that no crossing over occurred during meiosis. Use the formula below.

number of combinations = 2^n
where n = the number of homologous chromosome pairs

(1 mark)

4 The Archaea domain contains single-celled microorganisms that often live in extreme environments. Organisms in this domain are prokaryotes, which are thought to have existed before eukaryotes.

4.1 The endosymbiotic theory proposes that eukaryotic cells arose when a large prokaryotic cell engulfed smaller prokaryotic cells. The engulfed cells carried out certain functions that benefited the larger cell and, over evolutionary time, these engulfed cells became eukaryotic organelles.
Suggest **three** features of eukaryotic organelles that give evidence for the endosymbiotic theory.

(3 marks)

Several species of the Archaea domain have been discovered near deep-sea hydrothermal vents. These vents can reach temperatures of around 400 °C. Scientists conducted phlyogenetic analysis of the archaea found near the vents by sequencing samples of their DNA.

4.2 Advances in technology, such as genome sequencing, are allowing scientists to carry out more detailed phylogenetic analyses on species.
Explain why genome sequencing helps to clarify evolutionary relationships between organisms.

(2 marks)

4.3 Using your knowledge of protein structure, suggest why it is unusual to find living organisms near deep-sea hydrothermal vents.

(2 marks)

4.4 Halophilic archaea are organisms belonging to the Archaea domain that live in environments with a high salt concentration. Some of these organisms are adapted to accumulate potassium ions in their cytoplasm.
Using your knowledge of water potential, suggest an explanation for how this adaptation may help cells to survive in environments with a high salt concentration.

(3 marks)

Synoptic Practice

5 **Figure 4** shows the structure of human immunodeficiency virus (HIV).

Figure 4

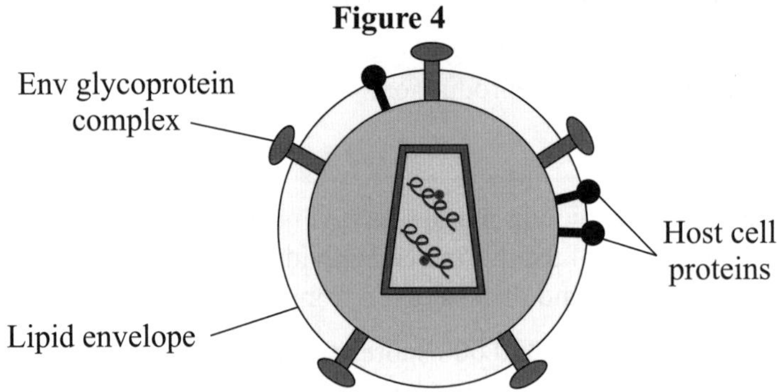

5.1 Suggest the role of the Env glycoprotein complex on the surface of HIV.

(1 mark)

5.2 Using your knowledge of the way in which HIV replicates, suggest why host proteins are present in the lipid envelope of HIV.

(2 marks)

5.3 The way that genetic information is stored in HIV differs from the way it is stored in the host cell. Compare and contrast the structures of the nucleotides used to store genetic information in HIV and in the host cell.

(4 marks)

5.4 HIV releases its genetic material into the host cell's cytoplasm.
Describe how viral proteins are made in the host cell from HIV's genetic material.

(3 marks)

5.5 Suggest why it is more difficult to develop drugs to prevent the replication of HIV and other viruses, than it is to develop drugs to prevent the reproduction of bacteria.

(2 marks)

HIV can be tested for using an ELISA test. In this test, HIV antigens are bound to the bottom of a well and a sample of a patient's blood is added. The well is washed out and a secondary antibody, that has a specific enzyme attached to it, is added to the well. The well is washed out again and a solution containing a substrate that reacts with the enzyme is added. The reaction between the enzyme and substrate produces a coloured product, which indicates that the person has HIV.

5.6 Suggest what the secondary antibodies bind to when they are added to the well.

(1 mark)

5.7 The intensity of the colour change in the well can be measured with an instrument called a spectrophotometer.
Suggest why a more intense colour change may indicate that the person has had HIV for longer than if a less intense colour change was detected.

(2 marks)

5.8 Scientists are hoping to develop a vaccine against HIV in the near future.
Suggest how the results of an ELISA test could be affected if the patient had been vaccinated against HIV. Explain your answer.

(3 marks)

Synoptic Practice

6 A pyrogen is any substance that causes a fever, which is an increase in the body temperature above 38 °C. Substances produced by some bacteria can act as pyrogens in an infected host.

6.1 During a fever, the body shivers, which is the involuntary, rapid contraction and relaxation of muscles. Explain how shivering may affect the rate of glycolysis in the body.

(2 marks)

6.2 Shivering increases the oxygen requirement of the muscle cells.
Explain how this increase in oxygen demand could cause heart rate to increase.

(5 marks)

Scientists investigated the effects of several substances in rodents in order to determine if they could act as pyrogens in humans. They tested each substance (**A-E**) on 20 different rodents and recorded the average increase in body temperature they caused. Their results are shown in **Table 1**.

Table 1

Substance	Average increase in body temperature / °C	Range / °C	Standard deviation / °C
A	0.3	0.5	± 0.2
B	1.1	0.6	± 0.3
C	1.0	0.9	± 0.7
D	1.2	1.1	± 0.6
E	0.9	0.7	± 0.3

6.3 The range and standard deviation are both values that indicate how the results are dispersed. Explain which measure is better for judging the precision of the results shown in **Table 1**.

(2 marks)

6.4 The greater the increase in body temperature recorded, the more dangerous the scientists considered the substance to be. Using data from **Table 1**, calculate the percentage difference in average increase in body temperature between the most and the least dangerous substance.

(1 mark)

7 Neuromuscular blockers are a group of muscle relaxant drugs often used to ensure patients do not move during surgery.

7.1 One such drug prevents acetylcholine from binding to nicotinic cholinergic receptors.
Explain how this neuromuscular blocker would prevent muscle contraction.

(5 marks)

7.2 Another type of neuromuscular blocker works by binding to nicotinic cholinergic receptors and generating an action potential. The drug remains bound to the receptors for a long time, resulting in extended depolarisation of the postsynaptic membrane.
Using your knowledge of how an action potential works, explain how this drug prevents a person from moving.

(2 marks)

7.3 Other than binding to nicotinic cholinergic receptors, suggest **one** way in which a drug could act at a neuromuscular junction to prevent a person from moving.

(1 mark)

Synoptic Practice

Scientists investigated how the dose of a neuromuscular blocker given to patients affected the force of muscle contraction. In their investigation, the scientists measured the force of contraction of one muscle in the body before the patient had been given any of the drug and again once they had received a dose of the drug. 80 patients were included in the investigation. The results are shown in **Figure 5**.

Figure 5

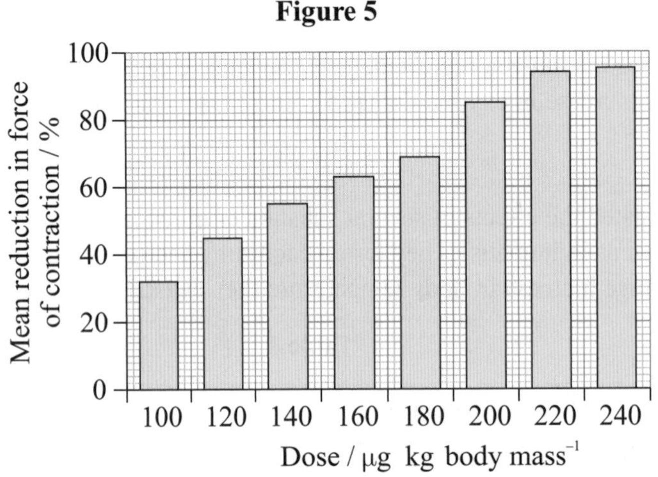

7.4 For the drug to be classed as suitable for use in surgery, the scientists recommended that the patient should have no more than 15% of their usual force of contraction remaining. Using this information and **Figure 5**, calculate the minimum amount of the drug in mg that should be given to a patient with a body mass of 75 kg.

(2 marks)

7.5 A student wanted to determine if the difference in the results for 200 μg kg body mass^{-1} and 220 μg kg body mass^{-1} were statistically significant.
Suggest which statistical test they could use to determine this. Explain your choice.

(2 marks)

7.6 The kidneys play a role in the excretion of drugs from the body. Some neuromuscular blocker molecules entering the kidneys from the renal arteries can pass into the Bowman's capsules to be excreted. However, the number of functioning glomeruli a person has tends to decline with age, meaning doctors need to consider a patient's age before administering neuromuscular blockers.
Using the information provided, suggest why a neuromuscular blocker may have a prolonged effect in an elderly patient.

(2 marks)

7.7 After surgery, patients can be given anticholinesterases to help overcome the effects of the neuromuscular blockers. Anticholinesterases have an inhibitory effect on the enzyme, acetylcholinesterase, which normally breaks down acetylcholine in the synaptic cleft.
Suggest how anticholinesterases work to overcome the effects of the neuromuscular blockers.

(2 marks)

8 Familial hypercholesterolaemia (FH) is an inherited medical condition that is characterised by a high blood cholesterol level. Normally, low density lipoproteins (LDLs) carry cholesterol molecules through the blood and into body cells, by binding to LDL receptors present on cells. FH can be caused by many mutations that affect the LDL receptor gene and result in defective LDL receptors, causing cholesterol to remain in the blood.

8.1 FH can be described as an autosomal dominant disease.
Describe what is meant by the term 'autosomal dominant'.

(2 marks)

Synoptic Practice

8.2 There are several mutations that can lead to the LDL receptor being defective. An individual was screened for mutations in their LDL receptor gene that would put them at risk of FH. They were found to have a mutation that resulted in a change from the triplet code CCG to CTG. No other mutations were detected. Explain whether this mutation would cause a frameshift.

(2 marks)

Figure 6 shows the incidence of FH in some members of a family. Horizontal lines between two individuals link mating parents and vertical lines link parents and offspring.

Figure 6

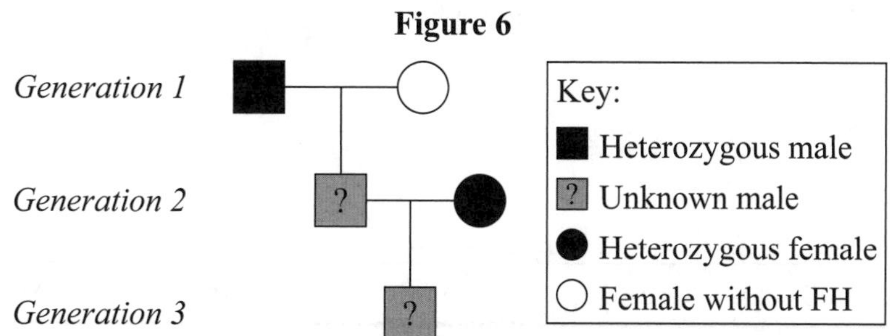

8.3 One student concluded that the Generation 3 male has a 50% chance of inheriting FH. Evaluate this conclusion.

(3 marks)

It is estimated that around 1 in every 1 000 000 people in the UK is homozygous for the FH allele.

8.4 A city in the UK has an estimated population size of 9.62×10^6 people. Calculate an estimate of how many people in this UK city are homozygous for the FH allele.

(1 mark)

8.5 Using the Hardy-Weinberg equations, calculate the frequency of the heterozygous genotype in the UK. Show your working.

(2 marks)

People with FH can be given drugs, called statins, to reduce their blood cholesterol level. However, there are some side-effects associated with statins, such as an increased risk of developing Type II diabetes.

8.6 Statins can increase the expression a protein complex called NLRP3 inflammasome. This protein complex can lead to the inflammation of β cells in the pancreatic islets of Langerhans. Suggest an explanation for how this could lead to the development of Type II diabetes.

(4 marks)

Scientists hope that gene therapy could eventually be used as an alternative treatment for FH.

8.7 Explain how gene therapy could be used to increase the uptake of cholesterol from the blood of a person with FH.

(4 marks)

8.8 Suggest **one** possible complication of the use of gene therapy as a treatment for FH.

(1 mark)

9 Light is an important abiotic factor for optimum crop plant growth.

9.1 Over several years, large areas of a farmer's crop fields have become more shaded. Rabbits are a pest species that feed on the farmer's crop plants. Explain how the increase in shade may have changed the carrying capacity of the rabbits living in the fields.

(3 marks)

Synoptic Practice

Plant leaves contain photosynthetic pigments, which can absorb different wavelengths of light. **Figure 7** shows the absorption spectra for two photosynthetic pigments found in plant leaves, chlorophyll a and chlorophyll b.

Figure 7

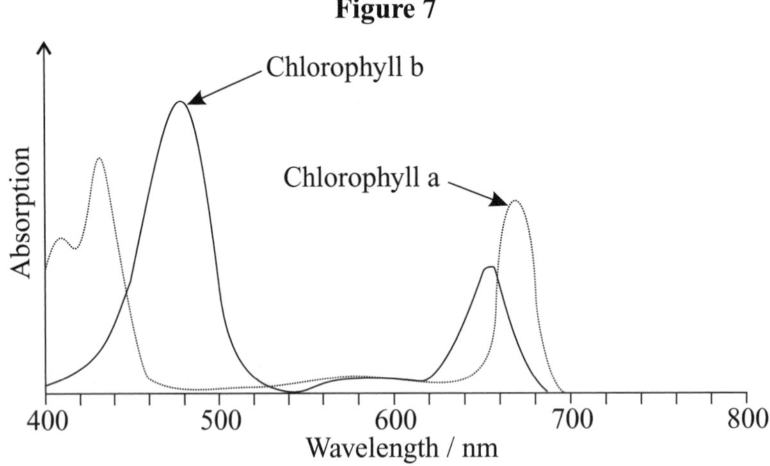

Far-red light is light with a wavelength of around 700-800 nm.

9.2 Using **Figure 7**, explain why exposure to only far-red light could limit plant growth.

(3 marks)

9.3 Far-red light is reflected off the leaves of plants. Plants can detect an increase in the amount of far-red light, which indicates that their environment may have become more shaded. Far-red light is detected by a photoreceptor called phytochrome. When phytochrome is exposed to far-red light, it is converted from an active form (Pfr) into an inactive form (Pr). Pfr inhibits the activity of a protein called TTA1, which is involved in the synthesis of indoleacetic acid.
Suggest and explain how the detection of far-red light could help a plant to maximise growth as its environment becomes more shaded.

(3 marks)

9.4 When different individuals of the same plant species have been grown in either shaded or brightly lit environments, they have been observed to develop different phenotypes.
Explain how transcription factors can lead to differences in phenotypes between two organisms that have the same genome.

(3 marks)

10 Read the following passage.

Huntington's disease is an inherited disorder for which there is currently no cure. The gene responsible for the disease, *HTT*, contains a sequence of CAG triplet repeats. An autosomal dominant mutation in the *HTT* gene can lead to many additional CAG triplet repeats being added to this DNA sequence. The huntingtin protein produced as a result of this mutation can lead to the degradation of neurones in the brain, which can lead to a reduction in the secretion of inhibitory neurotransmitters by these neurones.

Symptoms of Huntington's disease often do not develop until adulthood, and many affected individuals have children prior to a diagnosis. Those who are at risk from Huntington's disease, such as those with parents who have the disease, may decide to have a genetic test to determine if they have the mutation.

Synoptic Practice

During the genetic test, a blood sample is taken from a person and the DNA isolated. PCR can then be used to create many copies of the *HTT* gene present in the DNA. Electrophoresis can then be carried out on the products of PCR, and the results used to help determine whether the person is likely to develop Huntington's disease.

A new drug has been developed that shows promise of slowing down the progression of Huntington's disease. The drug functions by relying on the basic principles of RNA interference.

Using information from the passage and your own knowledge, answer the following questions.

10.1 Huntington's disease is caused by an autosomal dominant mutation (line 3).
A man is heterozygous for the *HTT* gene and his wife is homozygous recessive.
Give the probability that their child will inherit Huntington's disease.

(1 mark)

10.2 The mutation in the *HTT* gene can lead to many additional CAG triplet repeats being added to the DNA sequence (lines 3-4).
Suggest why this mutation leads to the formation of a defective huntingtin protein.

(2 marks)

10.3 Huntington's disease can lead to a reduction in the secretion of inhibitory neurotransmitters in the brain (lines 5-6).
Explain how inhibitory neurotransmitters can prevent the transmission of a nerve impulse.

(2 marks)

10.4 People may decide to have a genetic test to determine if they have the mutation responsible for Huntington's disease (lines 9-10).
Give **one** reason for and **one** reason against having the genetic test for Huntington's disease.

(2 marks)

10.5 PCR can be used to create many copies of the *HTT* gene (line 12).
Explain why DNA needs to be temporarily heated to 95 °C during the PCR procedure.

(2 marks)

10.6 Electrophoresis can be carried out on the products of PCR in the genetic test for Huntington's disease (line 13).
Suggest how electrophoresis could be used to determine if a person is at risk of developing Huntington's disease.

(2 marks)

10.7 The new drug developed for Huntington's disease functions by relying on the basic principles of RNA interference (lines 16-17).
Explain how RNA interference could prevent the development of Huntington's disease.

(4 marks)

11 Write an essay on the topic below.

The importance of enzymes in living organisms.

(25 marks)

Answers

Topic 1A — Biological Molecules

Page 5 — Carbohydrates

1 Two molecules of alpha-glucose ***[1 mark]*** are joined by a glycosidic bond ***[1 mark]***. A molecule of water is released/a condensation reaction takes place ***[1 mark]***.

2 Take a new sample of the test solution, add dilute HCl and heat it in a water bath that's been brought to the boil ***[1 mark]***. Neutralise it with sodium hydrogencarbonate ***[1 mark]***. Then add blue Benedict's solution and heat it in a water bath that's been brought to the boil ***[1 mark]***. If the test is positive for a non-reducing sugar, a brick red precipitate will form ***[1 mark]***. If the test is negative, the solution will stay blue ***[1 mark]***.
The question asks to describe a test for a non-reducing sugar. Remember, the test for a reducing sugar has to be performed first. If this gives a negative result, a non-reducing sugar may still be present. That's when the test for non-reducing sugars should be performed.

3 a) Because it is made up of chains of a monosaccharide/ N-acetylglucosamine ***[1 mark]***.
b) Cellulose and chitin are both polysaccharides ***[1 mark]***, made up of long and unbranched chains ***[1 mark]***. The chains are linked together by weak hydrogen bonds ***[1 mark]***.
c) A molecule of water ***[1 mark]*** is used to break the glycosidic bond between the monosaccharides in the chain ***[1 mark]***.
d) Secretion of chitinases would protect plants against attack by insects ***[1 mark]*** and fungal infection ***[1 mark]***, by breaking down the chitin in the exoskeleton of insects and the cell walls of fungi ***[1 mark]***, which would kill the invading organisms ***[1 mark]***.

Page 7 — Lipids

1 The hydrophobic tails force them to clump together in the cytoplasm as insoluble droplets ***[1 mark]***. This means they can be stored in cells, as a source of energy, without affecting the cell's water potential ***[1 mark]***.

2 a) Two fatty acid molecules ***[1 mark]*** and a phosphate group ***[1 mark]*** attached to one glycerol molecule ***[1 mark]***.
Don't get phospholipids mixed up with triglycerides — a triglyceride has three fatty acids attached to one glycerol molecule.
b) Saturated fatty acids don't have any double bonds between their carbon atoms ***[1 mark]***. Unsaturated fatty acids have one or more double bonds between their carbon atoms ***[1 mark]***.

Page 9 — Proteins

1 A peptide bond ***[1 mark]*** forms between the carboxyl group of one amino acid and the amino group of the other amino acid ***[1 mark]***. A molecule of water is released / a condensation reaction takes place ***[1 mark]***.
If you find it difficult to explain a process, such as a dipeptide forming, learn the diagrams too because they may help you to explain the process.

2 The secondary structure is coiled and folded further to form the protein's final 3D structure ***[1 mark]***. More bonds, including hydrogen bonds, ionic bonds and disulphide bridges, form between different parts of the polypeptide chain ***[1 mark]***.

Page 11 — Enzyme Action

1 The complementary substrate binds to the active site of the enzyme ***[1 mark]*** to form an enzyme-substrate complex ***[1 mark]***. As the substrate binds, the active site changes shape slightly, which provides a better fit ***[1 mark]***. The substrate is broken down / joined together to form the product(s) ***[1 mark]***.

2 A change in the amino acid sequence of an enzyme may alter its tertiary structure ***[1 mark]***. This changes the shape of the active site so that the substrate can't bind to it ***[1 mark]***.

Page 13 — Factors Affecting Enzyme Activity

1 a) Competitive inhibitor molecules have a similar shape to the substrate molecules ***[1 mark]***. They compete with the substrate molecules to bind to the active site of an enzyme ***[1 mark]***. When an inhibitor molecule is bound to the active site it stops th substrate molecule from binding ***[1 mark]***.
b) Non-competitive inhibitor molecules bind to enzymes away from their active site ***[1 mark]***. This causes the active site to change shape so the substrate molecule can no longer fit ***[1 mark]***.

Page 15 — Enzyme-Controlled Reactions

1 65 °C gradient = $40\ cm^3 \div 4\ s =$ **$10\ cm^3s^{-1}$**
(accept between $8\ cm^3s^{-1}$ and $13\ cm^3s^{-1}$) ***[1 mark]***.

Topic 1B — More Biological Molecules

Page 17 — DNA and RNA

1

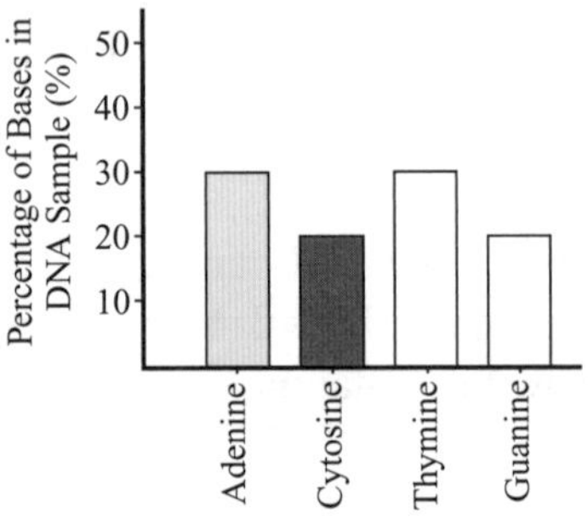

[1 mark for a bar drawn for thymine at 30%.
1 mark for a bar drawn for guanine at 20%]
Remember, thanks to complementary base pairing, there are always equal amounts of adenine and thymine in a DNA sample and equal amounts of cytosine and guanine. Double-check your answer by making sure the percentages of all four bases add up to 100%.

2 a) Nucleotides are joined between the phosphate group of one nucleotide and the (deoxyribose) sugar of the next ***[1 mark]*** by phosphodiester bonds ***[1 mark]*** in a condensation reaction ***[1 mark]***.
b) Two polynucleotide strands join through hydrogen bonding between the base pairs ***[1 mark]***. Base pairing is complementa (e.g. A always pairs with T and C always pairs with G) ***[1 mark]***. The two antiparallel polynucleotide strands twist to form a DNA double helix ***[1 mark]***.

Page 19 — DNA Replication

1 Any five from: e.g. DNA helicase breaks the hydrogen bonds between the two DNA strands and the DNA helix unwinds ***[1 mark]***. / Each strand acts as a template for a new strand ***[1 mark]***. / Individual free DNA nucleotides join up along the template strand by complementary base pairing ***[1 mark]***. / DNA polymerase joins the individual nucleotides together, so that the sugar-phosphate backbone forms ***[1 mark]***. / Hydrogen bonds then form between the bases on each strand and the strands twis to form a double-helix ***[1 mark]***. / Two identical DNA molecules are produced ***[1 mark]***. / Each of the new molecules contains a single strand from the original DNA molecule and a single new strand ***[1 mark]***. ***[Maximum of 5 marks available.]***

Page 21 — Water

1 a) As the water evaporates from the surface of the elephant's body ***[1 mark]***, some of the elephant's heat energy is used to break the hydrogen bonds which hold the water molecules together ***[1 mark]***. This cools the surface of the elephant's body ***[1 mark]***
b) There is strong cohesion between water molecules ***[1 mark]***. Th results in water having a high surface tension when in contact with air, causing it to form droplets ***[1 mark]***.

Answers

ge 23 — Inorganic Ions

A condensation reaction *[1 mark]* occurs between this molecule (ADP) and inorganic phosphate/P_i *[1 mark]*. The reaction is catalysed by ATP synthase *[1 mark]*.

a) Iron ions are a key component of haemoglobin *[1 mark]*. The iron ions in haemoglobin bind to oxygen *[1 mark]*. The haemoglobin is transported around the body in red blood cells *[1 mark]*.

b) Phosphate ions form the phosphate groups of ATP *[1 mark]*. Breaking the bonds between the phosphate groups in ATP releases energy *[1 mark]*.

xtra Exam Practice for Topic 1

ges 24-25

1 glycosidic bond *[1 mark]*

2 E.g.

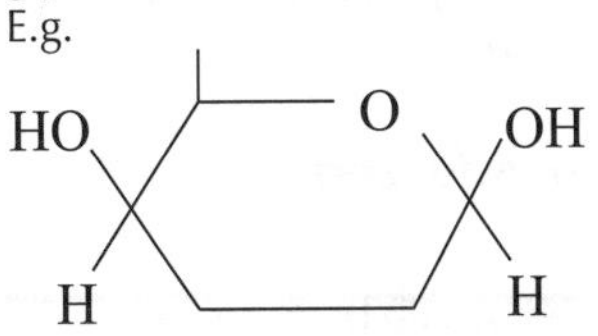

[1 mark]

You should know that lactose is made up of the monomers glucose and galactose joined by a glycosidic bond. You should also know the structure of glucose, so you should be able to work out the structure of galactose from there.

3 The rate of flow of milk needs to be controlled because if it is too slow, then the concentration of galactose in the column would increase *[1 mark]*. The galactose molecules would compete with the lactose molecules for the active sites of the lactase enzyme *[1 mark]*, which would mean that fewer enzyme-substrate complexes would form, and less lactose would be broken down *[1 mark]*.

Lactose starts being broken down as soon as the milk enters the column. If the rate of flow is too slow then it takes more time for the galactose to leave the column at the other end, so its concentration in the column increases.

4 E.g. pH / enzyme concentration / substrate concentration / the length of time the reaction was left for *[1 mark]*.

5 If immobilisation strengthens the tertiary structure, then the immobilised lactase would have been less likely than free lactase to lose its shape/denature as temperature increased *[1 mark]*. The relative yield of glucose produced by immobilised lactase declined less rapidly than for free lactase at increasing temperatures *[1 mark]*, which suggests that the tertiary structure/shape of immobilised lactase was less affected by increasing temperature, supporting the student's suggestion *[1 mark]*.

opic 2A — Cell Structure and Division

age 29 — Eukaryotic Cells and Organelles

a) E.g. helps maintain pressure inside the cell/keeps the cell rigid *[1 mark]* / isolates unwanted chemicals inside the cell *[1 mark]*.

b) E.g. cell wall *[1 mark]*, chloroplasts *[1 mark]*

Ciliated epithelial cells have lots of mitochondria *[1 mark]* because they need lots of energy *[1 mark]*.

Any four (in order) from: e.g. ribosomes *[1 mark]* / rough endoplasmic reticulum *[1 mark]* / Golgi apparatus *[1 mark]* / Golgi vesicle *[1 mark]* / cell-surface membrane *[1 mark]*.

This question really tests how well you know what each organelle does. The rough endoplasmic reticulum transports proteins that have been made in the ribosomes to the Golgi apparatus. At the Golgi apparatus the proteins are packaged and sent in Golgi vesicles to be secreted at the cell-surface membrane.

Page 31 — Prokaryotic Cells and Viruses

1 a) murein *[1 mark]*

b) Any three from: e.g. *Vibrio cholerae* replicates its circular DNA and its plasmids *[1 mark]*. / The cell gets bigger and the DNA moves to opposite poles *[1 mark]*. / New cell walls begin to form *[1 mark]*. / The cytoplasm divides to make two daughter cells *[1 mark]*. / This process is called binary fission *[1 mark]*.

Vibrio cholerae is a prokaryotic organism, so its cell wall must be made from murein and it must replicate by binary fission.

c) Having a capsule may help to protect *Vibrio cholerae* from attack by the immune system cells of the people it infects *[1 mark]*.

Page 33 — Analysis of Cell Components

1 mitochondrion *[1 mark]* and nucleus *[1 mark]*

The resolution of light microscopes is not good enough to show objects smaller than 0.2 μm *[1 mark]*.

2 It should be kept ice-cold to reduce the activity of enzymes that break down organelles *[1 mark]*. It should be kept isotonic to prevent damage to the organelles through osmosis *[1 mark]*.

Page 35 — Cell Division — Mitosis

1 a) A — Metaphase *[1 mark]*, B — Telophase *[1 mark]*, C — Anaphase *[1 mark]*.

b) X — Chromosome/Chromatid *[1 mark]*, Y — Centromere *[1 mark]*, Z — Spindle fibre *[1 mark]*.

Page 37 — Cell Division — Investigating Mitosis

1 32 ÷ 42 = **0.76** ***[2 marks for the correct answer or 1 mark for the correct calculation.]***

Topic 2B — Cell Membranes

Page 39 — Cell Membrane Structure

1 The membrane is described as fluid because the phospholipids are constantly moving *[1 mark]*. It is described as a mosaic because the proteins are scattered throughout the membrane like tiles in a mosaic *[1 mark]*.

2 a) Cut five equal sized pieces of beetroot and rinse them to remove any pigment released during cutting *[1 mark]*. Make up five test tubes with alcohol concentrations at 0, 25, 50, 75 and 100% *[1 mark]*. Place a piece of beetroot in each test tube for the same length of time *[1 mark]*. Remove the pieces of beetroot from each tube and use a colorimeter to measure how much light is absorbed by each of the remaining solutions *[1 mark]*.

b) As the concentration of alcohol increased, the absorbance also increased *[1 mark]*. This means that more pigment was released by the beetroot as the alcohol concentration increased, so the membrane became more permeable at higher concentrations of alcohol *[1 mark]*.

Page 41 — Exchange Across Cell Membranes — Diffusion

1 a) channel protein(s) *[1 mark]*

Channel proteins transport charged particles, such as ions.

b) E.g. ions are water soluble *[1 mark]* and the centre of the phospholipid bilayer is hydrophobic *[1 mark]*.

c) The rate of facilitated diffusion will slow down/level off *[1 mark]*. As diffusion progresses, the concentration gradient of the chloride ions will decrease/the concentration of chloride ions inside and out side of the cell will become the same (an equilibrium will be reached) *[1 mark]*.

Answers

Page 43 — Exchange Across Cell Membranes — Osmosis

1 a) The water potential of the sucrose solution was higher than the water potential of the potato ***[1 mark]***. So water moved into the potato pieces by osmosis, increasing their mass ***[1 mark]***.
b) The water potential of the potato and the water potential of the solution was the same ***[1 mark]***.
c) –0.4 g ***[1 mark]***. The difference in water potential between the solution and the potato is the same as with the 1% solution, so the mass difference should be about the same, but negative ***[1 mark]***.
The potato has a higher water potential than the solution, so it will lose water and mass.

Page 45 — Exchange Across Cell Membranes — Active Transport

1 a) Solute X. E.g. because the concentration of solute X inside the cell continues to increase over time, showing uptake against a concentration gradient ***[1 mark]***. / Because the concentration of solute Y levels off, which does not happen in active transport ***[1 mark]***.
Solute Y is being transported by some form of diffusion. Once the concentration of solute Y inside the cell reaches equilibrium with the concentration outside the cell, the rate levels off. This doesn't happen with active transport.
b) Energy is needed because the solute is being transported against its concentration gradient ***[1 mark]***.
c) Energy is released by the hydrolysis of ATP ***[1 mark]*** into ADP and P_i/inorganic phosphate ***[1 mark]***.

Topic 2C — Cells and the Immune System

Page 47 — The Immune System

1 Antibodies bind pathogens together / agglutinate pathogens ***[1 mark]***. This allows phagocytes to engulf many pathogens at once ***[1 mark]***.
2 A secondary immune response is a faster and stronger response than the primary response ***[1 mark]***. This is because memory cells are produced during the primary response, which are able to recognise the foreign antigen when it is encountered again ***[1 mark]***. During the second infection, memory cell B-cells can quickly divide to form plasma cells, which secrete the correct antibody to the antigen ***[1 mark]***. Memory T-cells quickly divide into the right type of T-cells to kill the cell carrying the antigen ***[1 mark]***.
You'll only get the full marks for this question if you explain (as well as describe) why the secondary response differs.

Page 49 — Immunity and Vaccines

1 When some individuals in a population receive the vaccine, the occurrence of the disease in the population is reduced ***[1 mark]***. This means that those in the population who haven't been vaccinated are less likely to become infected ***[1 mark]***. This is called herd immunity ***[1 mark]***.
2 The flu virus is able to change its surface antigens/shows antigenic variation ***[1 mark]***. This means that when you're infected for a second time with a different strain, the memory cells produced from the first infection will not recognise the new/different antigens ***[1 mark]***. The immune system has to carry out a primary response against these new antigens ***[1 mark]***. This takes time and means you become ill ***[1 mark]***.
3 a) Active immunity involves the production of memory cells speci[fic] to a particular antigen. This means the immune system is able to mount a secondary immune response if the same antigen is detected again ***[1 mark]***. Passive immunity only offers short-ter[m] protection because the antibodies given are broken down in the body. / Memory cells are not produced, so the body can't moun[t a] secondary immune response ***[1 mark]***.
b) It takes time for the body to produce antibodies/memory cells against the antigens in the vaccine ***[1 mark]***.

Page 51 — Antibodies in Medicine

1 Monoclonal antibodies are made against antigens specific to cancer cells/tumour markers ***[1 mark]***. An anti-cancer drug is attached to the antibodies ***[1 mark]***. The antibodies bind to the antigens/tumour markers on cancer cells because their binding sites have a complementary shape ***[1 mark]***. This delivers the anti-cancer drug to the cells ***[1 mark]***.

Page 53 — Interpreting Vaccine and Antibody Data

1 a) Fewer people were being infected by Hib because they had been vaccinated against it ***[1 mark]*** or were benefiting from her[d] immunity ***[1 mark]***.
b) E.g. fewer people received the vaccine. / A new strain of Hib appeared, which the vaccine was less effective against ***[1 mark]***.

Page 55 — HIV and Viruses

1 HIV has a core that contains the genetic material (RNA) and some proteins ***[1 mark]***. It has an outer layer called the capsid, which is made of protein ***[1 mark]***, surrounded by an envelope that is made from the membrane of the host cell ***[1 mark]***. There are attachment proteins sticking out from the envelope ***[1 mark]***.

Extra Exam Practice for Topic 2

Pages 56-57

2.1 Image size = size of real object × magnification
Size of real object = 75 μm ÷ 1000
= 0.075 mm
Image size = 0.075 × 200
= **15 mm**
[2 marks for the correct answer, otherwise 1 mark for using the correct rearrangement of the magnification formula]
2.2 Glycoproteins ***[1 mark]*** because these are receptor proteins in th[e] cell-surface membrane that have a carbohydrate chain attached ***[1 mark]***.
2.3 The mitotic index of the root tip cells of the onion plants grown in lectin solution would be greater than those grown in distilled water ***[1 mark]***. This is because the exposure to the lectin solutio[n] would have resulted in the cells having a reduced duration of interphase, so more of these cells would be undergoing mitosis ***[1 mark]***.
3.1 Scale factor = 40% ÷ 30%
= 1.333...
$20\ cm^3 \div 1.333... = 15\ cm^3$
So he needs to use **15 cm³** of the original solution ***[1 mark]*** and 20 – 15 = **5 cm³** of distilled water ***[1 mark]***.

Answers

If betalain accumulates around the beetroot discs, the betalain concentration gradient between the discs and the solution surrounding them reduces ***[1 mark]***. By regularly mixing the solution, the concentration gradient would be maintained, allowing more betalain to leave the beetroot discs, increasing the accuracy of the results ***[1 mark]***.

3 The absorbance values may be lower ***[1 mark]***. This is because the sitosterol molecules would fit between the phospholipids/ increase the stability of the cell-surface membrane ***[1 mark]***, potentially reducing the damage caused by ethanol to the cell-surface membrane, so reducing the amount of betalain that could leak out ***[1 mark]***.

opic 3A — Exchange nd Transport Systems

age 59 — Size and Surface Area

A small mammal has a bigger surface to volume ratio than a large mammal ***[1 mark]***. This means that heat is lost more easily from a small mammal ***[1 mark]***. So a smaller mammal needs a relatively high metabolic rate, in order to generate enough heat to maintain a constant body temperature ***[1 mark]***.

age 61 — Gas Exchange

Any one from: gaseous exchange surfaces have a large surface area ***[1 mark]***, e.g. mesophyll cells in a plant (or any other suitable example) ***[1 mark]***. / Gaseous exchange surfaces are thin, which provides a short diffusion pathway ***[1 mark]***, e.g. the walls of tracheoles in insects (or any other suitable example) ***[1 mark]***. / A steep diffusion gradient is constantly maintained across gaseous exchange surfaces ***[1 mark]***, e.g. the counter-current system in fish gills (or any other suitable example) ***[1 mark]***. ***[Maximum of 2 marks available]***

Sunken stomata and hairs help to trap any moist air near to the stomata ***[1 mark]***, reducing the concentration gradient from leaf to air, which reduces water loss ***[1 mark]***.

age 63 — Gas Exchange in Humans

Any two from: e.g. the lungs contain millions of tiny air sacs called alveoli, creating a large surface area for gas exchange ***[1 mark]***. / The alveolar epithelium is only one cell thick, which means there is a short diffusion pathway ***[1 mark]***. / The alveoli are surrounded by a dense network of capillaries, which maintains a steep concentration gradient of oxygen and carbon dioxide between the alveoli and the blood ***[1 mark]***.

The external intercostal muscles and diaphragm contract ***[1 mark]***. This causes the ribcage to move up and out and the diaphragm to flatten ***[1 mark]***, increasing the volume of the thoracic cavity ***[1 mark]***. The air pressure in the lungs decreases and air flows down the pressure gradient into the lungs ***[1 mark]***.

age 65 — The Effects of Lung Disease

a) Emphysema involves the loss/break down of elastin in the walls of the alveoli ***[1 mark]***. This means the alveoli can't recoil to expel air as well ***[1 mark]***.

b) 1.7 / 3.2 × 100 = **53%** *(to 2 s.f.)* ***[1 mark]***

c) Both FEV_1 and FVC are reduced, so the ratio between them stays the same as in a healthy person ***[1 mark]***.

Page 67 — Interpreting Lung Disease Data

1 a) The daily death rate increased rapidly after 4th December ***[1 mark]*** peaking around the 7th, then decreasing afterwards ***[1 mark]***. Both pollutants followed the same pattern ***[1 mark]***.

You could also get the marks by saying it the other way round — the pollutants rose and peaked around the 7th then decreased, with the death rates following the same pattern.

b) There is a link/correlation between the increase in sulfur dioxide and smoke concentration and the increase in death rate ***[1 mark]***.

Don't go saying that the increase in sulfur dioxide and smoke caused the increase in death rate — there could have been another reason for the trend, e.g. there could have been other pollutants responsible for the deaths.

Page 69 — Dissecting Gas Exchange Systems

1 E.g. the liquid preservative has entered the grasshopper's tracheae, so they are no longer filled with air (and they would appear silver in colour if filled with air) ***[1 mark]***.

2 a) The lung tissue will float as it/the alveoli still contain(s) some air ***[1 mark]***.

b) E.g. make sure the dissecting instruments are clean, sharp and free from rust ***[1 mark]***. / Carry out the dissection on a cutting board ***[1 mark]***. / Cut downwards and away from the body when using a scalpel ***[1 mark]***. / Wash hands/disinfect work surfaces after carrying out the dissection ***[1 mark]***.

Topic 3B — More Exchange and Transport Systems

Page 71 — Digestion and Absorption

1 a) lactase ***[1 mark]***

b) The digestion products of lactose/glucose and galactose are absorbed across the epithelial cells of the ileum by active transport with sodium ions ***[1 mark]*** via a co-transporter protein ***[1 mark]***.

Page 73 — Haemoglobin

1 a) It is composed of more than one polypeptide chain ***[1 mark]***.

The reason that haemoglobin has a quaternary structure is because it has more than one polypeptide chain. The fact that it's made up of four polypeptides isn't important.

b) i), ii)

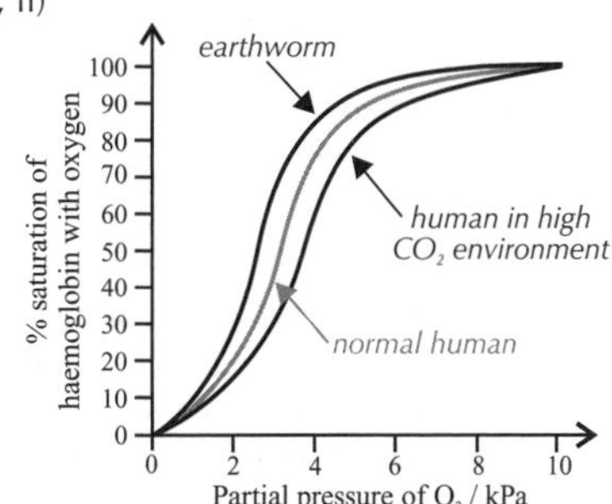

i) The curve for a human in a high carbon dioxide environment should look like a normal human dissociation curve that has shifted right (see graph above) ***[1 mark]***. This is the Bohr effect ***[1 mark]***. High concentrations of carbon dioxide increase the rate of oxygen unloading and the saturation of blood with oxygen is lower for a given pO_2 ***[1 mark]***.

ii) The curve for the earthworm should be drawn to the left of the human one (see graph above) ***[1 mark]***.

The earthworm lives in an environment with a low partial pressure of oxygen, so it needs haemoglobin with a higher affinity for oxygen than human haemoglobin.

Answers

Page 75 — The Circulatory System

1 E.g. they have elastic tissue in the walls *[1 mark]* so they can stretch and recoil as the heart beats, which helps maintain the high pressure *[1 mark]*. The inner lining (endothelium) is folded *[1 mark]* so that the artery can expand when the heartbeat causes a surge of blood *[1 mark]*.

2 The hydrostatic pressure in the capillary is greater than the hydrostatic pressure in the spaces around the cells *[1 mark]*, so fluid moves out of the capillary and into spaces around the cells *[1 mark]*.

Page 78 — The Heart

1 a) 0.2 - 0.4 seconds *[1 mark]*.
The AV valves are shut when the pressure is higher in the ventricles than in the atria.

b) 0.3 - 0.4 seconds *[1 mark]*.
When the ventricles relax the volume of the chamber increases and the pressure falls. The pressure in the left ventricle was 16.5 kPa at 0.3 seconds and it decreased to 7.0 kPa at 0.4 seconds, so it must have started to relax somewhere between these two times.

c) $16.5 - 0.5 = 16$
$(16 \div 0.5) \times 100 =$ **3200%** ***[1 mark for the correct answer.]***
In this question you need to calculate the percentage increase from 0.5 kPa (blood pressure at 0.0 s) to 16.5 kPa (blood pressure at 0.3 s). To do this you find the difference between the two blood pressures (16 kPa), divide this by the starting blood pressure (0.5 kPa), and multiply the whole thing by 100.

Page 81 — Cardiovascular Disease

1 a) A large sample size was used *[1 mark]*.
The sample included many countries *[1 mark]*.

b) E.g. a large waist measurement could indicate that someone is overweight *[1 mark]*. Being overweight can be linked to high blood pressure *[1 mark]*. High blood pressure is a risk factor for cardiovascular disease because it increases the risk of damage to artery walls *[1 mark]*.

Page 83 — Transport in Plants — Xylem

1 a) The evaporation of water from plant surfaces *[1 mark]*.

b) Transpiration from the leaves at the 'top' of the xylem creates tension, which pulls more water into the leaf *[1 mark]*. Water molecules are cohesive, so when some are pulled into the leaf others follow *[1 mark]*. This means the whole column of water in the xylem, from the leaves down to the roots, moves upwards, pulling water into the stem through the roots *[1 mark]*.

Page 85 — Transport in Plants — Phloem

1 a) Leaves can act as a source because they are a part of a plant where solutes/products of photosynthesis are made *[1 mark]*.

b) Radioactive solutes/products of photosynthesis have been translocated to the fruits because the fruits are acting as a sink *[1 mark]*.

Extra Exam Practice for Topic 3

Pages 86-87

2.1 When the heart segment is contracting the pressure inside it will increase *[1 mark]*. When the pressure becomes greater than the pressure in the body cavity, the valve will shut, preventing blood from re-entering the contracting segment of the heart *[1 mark]*.

2.2 In mammals, haemoglobin binds to oxygen molecules to transport them in the blood to respiring cells *[1 mark]*. However, in insects oxygen molecules diffuse directly into respiring cells from the tracheoles, so they don't require haemoglobin *[1 mark]*.

2.3 Solutes/sugars from the leaves are loaded into the sieve tubes near site 2 *[1 mark]*, meaning the water potential at this site is relatively low *[1 mark]*. The water potential increases as the phloem sap moves towards site 3 *[1 mark]* because the solutes/sugars are being removed from the sieve tubes to be used by cells *[1 mark]*.

2.4 The stem at site 2 is connected to more leaves than the stem at site 1 *[1 mark]*. Therefore, the phloem sap at site 2 is likely to contain more solutes/sugars than the phloem sap at site 1 *[1 mark]*. This causes more water to move into the sieve tubes from the xylem and companion cells by osmosis *[1 mark]*. This results in the pressure of the phloem sap at site 2 being greater than at site 1, which forces a greater volume out through the style *[1 mark]*.

2.5 Amylase enzymes are needed to hydrolyse starch/large carbohydrates into smaller molecules *[1 mark]* so they can be digested *[1 mark]*. If insects ingest a compound that inhibits amylase enzymes, then the plant will become a less valuable food source, meaning that they may stop feeding on it *[1 mark]*.

Topic 4A — DNA, RNA and Protein Synthesis

Page 89 — DNA, Genes and Chromosomes

1 Any five points from: e.g. in the nucleus of eukaryotic cells, DNA is stored as chromosomes *[1 mark]*. It is linear *[1 mark]*. It is wound around proteins called histones *[1 mark]*. Mitochondria and chloroplasts in eukaryotic cells also contain DNA *[1 mark]*. In mitochondria and chloroplasts, the DNA is short and circular *[1 mark]*. The DNA in mitochondria / chloroplasts is not associated with histones *[1 mark]*.

2 $672 \div 3 =$ **224** amino acids
[2 marks for the correct answer, 1 mark for the correct calculation.]
Remember, only the exons actually code for amino acids. Three nucleotides code for each amino acid, so you need to divide the number of nucleotide pairs in the exons by three.

Page 91 — RNA and Protein Synthesis

1 The drug binds to DNA, preventing RNA polymerase from binding, so transcription can't take place and no mRNA can be made *[1 mark]*. This means there's no mRNA for translation and so protein synthesis is inhibited *[1 mark]*.

Page 93 — The Genetic Code and Nucleic Acids

1 a) GUG = valine
UGU = cysteine
CGC = arginine
GCA = alanine
Correct sequence = **valine, cysteine, arginine, alanine.**
[2 marks if all four amino acids are correct and in the correct order. 1 mark if three amino acids are correct and in the correct order.]

b) valine = GUG
arginine = CGC
alanine = GCA
mRNA sequence = GUG CGC GCA
DNA sequence = **CAC** *[1 mark]* **GCG** *[1 mark]* **CGT** *[1 mark]*.

Answers

) The mRNA sequence is 18 nucleotides long and the protein produced is 6 amino acids long ***[1 mark]***. 18 ÷ 6 = 3, suggesting three nucleotides code for a single amino acid ***[1 mark]***.

) E.g. The sequence produced began leucine-cysteine-glycine. This would only be produced if the code is non-overlapping, e.g. UUGUGUGGG = UUG-UGU-GGG = leucine-cysteine-glycine ***[1 mark]***.
If the code was overlapping, the triplets would be, e.g. UUG-UGU-GUG-UGU, which would give a sequence starting leucine-cysteine-valine-cysteine.
Also, this part of the DNA sequence produces 6 amino acids. This is only correct if the code is non-overlapping — the sequence of amino acids would be longer if the code overlapped ***[1 mark]***.

opic 4B — Diversity, Classification nd Variation

ge 96 — Meiosis and Genetic Variation

)

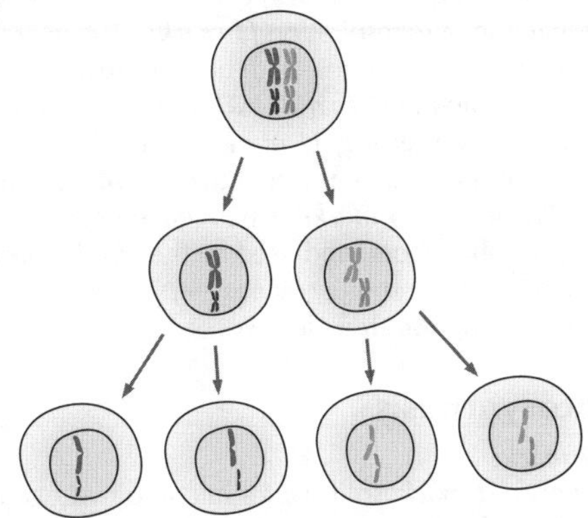

[1 mark for 2 single-stranded chromosomes (not sister chromatids) in each daughter cell.]

) During meiosis homologous pairs of chromosomes come together ***[1 mark]***. The chromatids twist around each other and bits swap over ***[1 mark]***. The chromatids now contain different combinations of alleles ***[1 mark]***. This means each of the four daughter cells will contain chromatids with different combinations of alleles ***[1 mark]***.

) Independent segregation means the homologous chromosome pairs can split up in any way ***[1 mark]***. So, the daughter cells produced can contain any combination of maternal and paternal chromosomes with different alleles ***[1 mark]***.
Chromosome non-disjunction may mean that the sex chromosomes fail to separate during meiosis ***[1 mark]***. This could mean that one of the daughter cells/gametes ends up without a copy of the X chromosome, whilst another daughter cell/gamete gets two X chromosomes ***[1 mark]***. If the gamete without an X chromosome is fertilised, the resulting zygote will be missing one X chromosome, resulting in Turner syndrome ***[1 mark]***.

ge 97 — Mutations

a) substitution ***[1 mark]***

b) The second amino acid will be arginine for the mutated gene, rather than serine (as for the original gene) ***[1 mark]***. The rest of the sequence of amino acids produced will not be affected ***[1 mark]***.

Page 99 — Genetic Diversity and Natural Selection

1 a) E.g. the brown owls may be better camouflaged/blend in with the landscape better than the grey owls when there's no snow cover ***[1 mark]***. This makes them less likely to be eaten by predators ***[1 mark]***.
Snow makes everything white, so lighter coloured owls blend in better when there's snow around. They stick out more when there's no snow though.

b) The brown owls are more likely to survive and reproduce when there's less snow cover ***[1 mark]*** and pass on the allele for darker/brown colouring to their offspring ***[1 mark]***. Over time, the allele for darker/brown colouring will become more common in the population ***[1 mark]***.

Page 101 — Investigating Selection

1 a) This is an example of stabilising selection ***[1 mark]***. The initial sample shows a fairly wide range of shell colours from light to dark ***[1 mark]***. Over time, the average colour of oyster shell has shifted towards the middle of the range, so more oysters have a mid-range coloured shell in the final sample than in the initial sample ***[1 mark]***.

b) Oysters at the extremes of light and dark are less likely to survive because they can be more easily seen by predators against the sand ***[1 mark]***. This means that the mid-range coloured oysters have an advantage and are more likely to survive and reproduce ***[1 mark]***. The advantageous alleles for mid-range coloured oysters are more likely to be passed on to the next generation ***[1 mark]*** leading to an increase in mid-range coloured oysters in the population ***[1 mark]***.

Page 103 — Classification of Organisms

1 a)

Domain	**Kingdom**	Phylum	**Class**	**Order**	**Family**	Genus	Species
Eukarya	Animalia	Chordata	Actinopterygii	Salmoniformes	**Salmonidae**	**Salmo**	**trutta**

[1 mark for 4 or more answers correct. 2 marks for all 7 answers correct.]

b) They are unable to reproduce to give fertile offspring ***[1 mark]***.
Although brook trout and brown trout do sometimes mate to produce offspring, those offspring are infertile.

Page 105 — DNA Technology, Classification and Diversity

1 a) Mouse and rat ***[1 mark]***.

b) Chicken ***[1 mark]*** because all of the amino acids in the protein sequence for the chicken are different to the amino acids for the other species ***[1 mark]***.

Page 107 — Investigating Variation

1 a) species A mean $= \frac{8 + 11 + 9 + 10 + 7 + 9}{6} = \frac{54}{6} =$ **9 days** ***[1 mark]***

species B mean $= \frac{12 + 10 + 6 + 12 + 15 + 11}{6} = \frac{66}{6} =$ **11 days** ***[1 mark]***

b) The standard deviation for species B is higher than that of species A suggesting that the values are more spread out from the mean ***[1 mark]***. This indicates that there is more variety in development time for species B ***[1 mark]***.

Answers

Page 109 — Biodiversity

1 a) The number of different species *[1 mark]* and the number of individuals/population size of each species in a community *[1 mark]*.

b) Site 1 —
N (N – 1) = 51 (51 – 1) = 2550
Σn (n – 1) = 15 (15 – 1) + 12 (12 – 1) + 24 (24 – 1) = 894
Use of N (N – 1) ÷ Σn (n – 1) to calculate diversity index of 2550 ÷ 894 = **2.85**
[2 marks for correct answer, 1 mark for incorrect answer but correct working.]
Site 2 —
N (N – 1) = 132 (132 – 1) = 17292
Σn (n – 1) = 35 (35 – 1) + 25 (25 – 1) + 34 (34 – 1) + 12 (12 – 1) + 26 (26 – 1) = 3694
Use of N (N – 1) ÷ Σn (n – 1) to calculate diversity index of 17292 ÷ 3694 = **4.68**
[2 marks for correct answer, 1 mark for incorrect answer but correct working.]
It's always best if you put your working — even if the answer isn't quite right you could get a mark for correct working.

c) The diversity of bumblebee species is greater at site 2 *[1 mark]*. This suggests there's a link between enhanced field margins and an increased diversity of bumblebee species *[1 mark]*.

Extra Exam Practice for Topic 4

Pages 110-111

2.1 There are more possible combinations of base triplets than there are amino acids, so some amino acids are coded for by more than one base triplet *[1 mark]*. Therefore, some base substitutions will not change the amino acid that's incorporated into the polypeptide chain during translation *[1 mark]* and so a polypeptide with the same primary structure, able to fold into the correct shape/develop the same tertiary structure, will still be formed even if there's a mutation *[1 mark]*.

2.2 In eukaryotic cells, the DNA must be dissociated from histone proteins before transcription of a gene can take place/mRNA can be made *[1 mark]*. Also in eukaryotic cells, in order to produce mRNA the pre-mRNA that is produced during transcription will need to undergo splicing to remove introns *[1 mark]*. In prokaryotic cells the DNA is not associated with histones and doesn't contain introns, meaning the production of mRNA is likely to take less time *[1 mark]*.

2.3 The tetracycline molecule prevents translation from taking place *[1 mark]*. This is because it prevents the anticodon of a tRNA molecule attaching to a codon on the mRNA *[1 mark]*. This means that the amino acid carried by the tRNA can't be added to the polypeptide chain *[1 mark]*.

2.4 It would reduce genetic diversity because only individuals with the alleles for tetracycline-resistance would survive and reproduce *[1 mark]*, meaning these alleles would become more common and the alleles that don't give tetracycline-resistance would be lost *[1 mark]*.

3.1 Rhizostomae *[1 mark]*
Phylogenetic trees show a hierarchy, with the taxa getting progressively smaller at each level. You should have realised that *Cassiopea andromeda* was a species due to its binomial name. By working backwards from its position on the tree you can identify the order it belongs to.

3.2 The scientists could test how well antibodies from each species bind to the proteins in each of the other species *[1 mark]*, as antibodies will only bind to proteins that have a similar structu to those in the species that the antibodies came from *[1 mark]*. The two species that show the most binding of antibodies and proteins are likely to be more closely related, and so should be placed closer together on the tree *[1 mark]*.

3.3 Chromosome non-disjunction could mean that after meiosis, some gametes could end up with an extra copy of a chromosor and some could end up with none *[1 mark]*. If one of these cel was involved in fertilisation, the resulting zygote would have ar abnormal number of chromosomes, meaning its genome woulc be different from other offspring *[1 mark]*.

Topic 5A — Photosynthesis and Respiration

Page 113 — Photosynthesis, Respiration and ATP

1 Any six points from: e.g. in the cell, ATP is synthesised from ADP and inorganic phosphate/P_i *[1 mark]* using energy from an energy-releasing reaction, e.g. respiration *[1 mark]*. The energy stored as chemical energy in the phosphate bond *[1 mark]*. ATP synthase catalyses this reaction *[1 mark]*. ATP then diffuses to the part of the cell that needs energy *[1 mark]*. Here, it's broke down back into ADP and inorganic phosphate/P_i *[1 mark]*, whic is catalysed by ATP hydrolase *[1 mark]*. Chemical energy is released from the phosphate bond and used by the cell *[1 mark*
Make sure you don't get the two enzymes confused — ATP **syn**thase **syn**thesises ATP, and ATP hydrolase breaks it down.

Page 117 — Photosynthesis

1 a) Photosystem II *[1 mark]*.

b) Photolysis/light energy *[1 mark]* splits water into two hydrogen ions and oxygen *[1 mark]*. The electrons from the water replace the electrons lost from chlorophyll *[1 mark]*.
The question asks you to explain the purpose of photolysis, so make sure you include why the water is split up — to replace the electrons lost from chlorophyll.

c) Excited electrons are transferred to reactant D/NADP from photosystem I/object C *[1 mark]* along with a proton/H^+ ion fro the stroma *[1 mark]*.

2 a) Any five points from: e.g. ribulose bisphosphate/RuBP and carbon dioxide/CO_2 join together to form an unstable 6-carbon compound *[1 mark]*. This reaction is catalysed by the enzyme rubisco *[1 mark]*. The compound breaks down into two molecules of a 3-carbon compound called glycerate 3-phospha GP *[1 mark]*. Two molecules of glycerate 3-phosphate are then converted into two molecules of triose phosphate/TP *[1 mark]*. The energy for this reaction comes from ATP *[1 mark]* and the H ions come from reduced NADP *[1 mark]*.

b) Ribulose bisphosphate is regenerated from triose phosphate/TP molecules *[1 mark]*. ATP provides the energy to do this *[1 mark*
This question is only worth two marks so only the main facts are needed, without the detail of the number of molecules.

c) No glycerate 3-phosphate/GP would be produced *[1 mark]*, so no triose phosphate/TP would be produced *[1 mark]*. This means there would be no glucose produced *[1 mark]*.

Answers

age 119 — Limiting Factors in Photosynthesis

a) Any two points from: e.g. by burning propane to increase air CO_2 concentration ***[1 mark]***. / By adding heaters to increase temperature ***[1 mark]***. / By adding coolers to decrease temperature ***[1 mark]***. / By adding lamps to provide light at night ***[1 mark]***.

b) Potatoes ***[1 mark]*** because the yield showed the smallest percentage increase of 25% (850 – 680 = 170, 170 ÷ 680 × 100 = 25%) ***[1 mark]***.

age 121 — Photosynthesis Experiments

a) Dehydrogenase enzymes catalyse the reaction that produces reduced NADP ***[1 mark]***.

b) Redox indicator dyes take the place of NADP as an electron acceptor ***[1 mark]***. This means that dehydrogenase activity reduces the dye instead of NADP ***[1 mark]***. The reduction is coupled with a colour change, which can be easily observed ***[1 mark]***.

age 125 — Aerobic Respiration

a) Reduced NAD ***[1 mark]***

b) The regenerated NAD is needed for glycolysis to continue ***[1 mark]*** and ATP to be produced under anaerobic conditions, providing the energy to keep running ***[1 mark]***.

a) The transfer of electrons down the electron transport chain stops ***[1 mark]***. So there's no energy released to phosphorylate ADP/produce ATP ***[1 mark]***.

b) The Krebs cycle stops ***[1 mark]*** because there's no oxidised NAD/FAD coming from the electron transport chain ***[1 mark]***.

Remember that when the electron transport chain is inhibited, the reactions that depend on the products of the chain are also affected.

Any six points from: e.g. glucose is phosphorylated using a molecule of ATP ***[1 mark]***. This creates one molecule of glucose phosphate ***[1 mark]*** and one molecule of ADP ***[1 mark]***. ATP is used to add another phosphate to glucose phosphate ***[1 mark]***, forming hexose bisphosphate ***[1 mark]***, which is then split into two molecules of triose phosphate ***[1 mark]***. Triose phosphate is oxidised/loses hydrogen to form two molecules of pyruvate ***[1 mark]***. NAD collects the hydrogen ions, forming two molecules of reduced NAD ***[1 mark]***.

age 127 — Respiration Experiments

a) To stop oxygen getting into the solution, which forces the yeast to respire anaerobically ***[1 mark]***.

b) Both ethanol and CO_2 are products of anaerobic respiration ***[1 mark]***. Measuring how fast CO_2 is produced would indicate how fast ethanol is being produced ***[1 mark]***.

c) Any two from: e.g. the temperature the investigation is being carried out at — could be controlled by putting the test tubes in a water bath at a set temperature ***[1 mark]***. / The mass of yeast used — could be controlled by weighing out a set amount of yeast to use in each test tube ***[1 mark]***. / The volume/concentration of the glucose solution used — could be controlled by measuring out a known volume of glucose solution for use in each test tube/using a fixed concentration of glucose solution in each test tube ***[1 mark]***.

d) A control tube should be set up for each pH being investigated, which contains glucose solution but no yeast ***[1 mark]***. No CO_2 should be produced. This will allow the student to check that any CO_2 being released in the other tubes is actually being produced by the yeast ***[1 mark]***.

Topic 5B — Energy Transfer and Nutrient Cycles

Page 129 — Energy Transfer in Ecosystems

1 a) Not all of the energy available from the grass is taken in by the Arctic hare ***[1 mark]***. This is because some parts of the grass aren't eaten, so the energy they contain isn't taken in ***[1 mark]***, and some parts of the grass are indigestible, so they'll pass through the hare and come out as waste ***[1 mark]***. Also, some energy is lost to the environment when the Arctic hare respires ***[1 mark]***.

b) $N = I - (F + R)$
$2345 = 18\,905 - (F + R)$
$F + R = 18\,905 - 2345 =$ **16 560 kJ m^{-2} y^{-1}**
[2 marks for correct answer, otherwise 1 mark for the correct calculation]

2 a) He could dry out one of his cabbages, e.g. in an oven ***[1 mark]***. He could then burn a known mass of dry tissue in a calorimeter ***[1 mark]*** and use the change in water temperature to calculate the chemical energy stored in the dry biomass of the cabbage ***[1 mark]***.

b) Net primary production, because some of the chemical energy converted by the plant through photosynthesis is immediately used for respiration ***[1 mark]*** so it does not get stored as the biomass of the cabbages ***[1 mark]***.

Page 131 — Farming Practices and Production

1 a) They eat the crop, reducing the amount of energy available for crop growth ***[1 mark]***.

b) That for the crop shown, the pesticide was most effective at reducing the percentage crop loss to pest 2 ***[1 mark]*** but that it had no effect on reducing the crop loss to the other two pests ***[1 mark]***.

c) Any two from: e.g. use an insecticide that kills multiple pests ***[1 mark]***. / Use another pesticide in conjunction with the first one ***[1 mark]***. / Use biological controls as well as chemical insecticides ***[1 mark]***.

d) Keep them in pens, so respiratory losses through movement are reduced ***[1 mark]***. Keep them warm, so less energy is wasted in generating body heat ***[1 mark]***.

Page 133 — Nutrient Cycles

1 a) A — ammonification ***[1 mark]***, B — nitrogen fixation ***[1 mark]***, C — denitrification ***[1 mark]***

b) i) Saprobionts convert nitrogen compounds in dead organisms, faeces and urine ***[1 mark]*** into ammonia ***[1 mark]***.

ii) They secrete enzymes and digest their food externally ***[1 mark]***, then absorb the nutrients they need ***[1 mark]***.

Page 135 — Fertilisers and Eutrophication

1 a) The control river helps to determine whether it is the fertiliser added to the adjacent field that is causing the observed changes in algal and oxygen content in the river or another variable ***[1 mark]***.

b)
$$\text{percentage change} = \frac{\text{final value} - \text{original value}}{\text{original value}} \times 100$$
$$= \frac{95\,000 - 10\,000}{10\,000} \times 100$$
= **850%** ***[1 mark]***

Answers

c) There's a negative correlation between the algal content and the oxygen content of the water / as the algal content increases, the oxygen content decreases, and vice versa *[1 mark]*.

d) The increasing algal content could have prevented light from reaching plants below *[1 mark]*, causing them to die and be decomposed by bacteria *[1 mark]*. The increased numbers of bacteria use up oxygen in the river when carrying out aerobic respiration, resulting in a reduction in dissolved oxygen content *[1 mark]*. Where algal content is lower, there's less dead plant matter/decomposition and oxygen content is higher *[1 mark]*.

Extra Exam Practice for Topic 5

Pages 136-137

2.1 Light energy is absorbed by chlorophyll which leads to photoionisation/electrons being released *[1 mark]*. Some of the energy released from this process is used for the photophosphorylation of ADP/to add a phosphate group to ADP, which generates ATP *[1 mark]*.

2.2 The constant light would allow the light-dependent reaction of photosynthesis to produce ATP and reduced NADP continuously *[1 mark]*. These products could then be used continuously by the light-independent reaction/Calvin cycle *[1 mark]*, to reduce glycerate 3-phosphate/GP to triose phosphate/TP *[1 mark]*. This would lead to an increase in the production of organic substances, since these are made from triose phosphate/TP *[1 mark]*.

2.3 Any five from: saprobionts break down the organic compounds in the fish waste *[1 mark]*. This releases phosphate ions, which can be used by the plants to make molecules containing phosphorus *[1 mark]*. Saprobionts also turn nitrogen compounds from the fish waste into ammonia *[1 mark]*, which goes on to form ammonium ions *[1 mark]*. Nitrifying bacteria then convert the ammonium ions into nitrites *[1 mark]* and then nitrates, which are used by the plants to make molecules containing nitrogen *[1 mark]*.

3.1 Rubisco is an enzyme involved in the Calvin cycle/ light-independent reaction of photosynthesis *[1 mark]*. An increase in rubisco's activity would mean that organic substances could be produced more rapidly *[1 mark]*. This would result in the producer generating more chemical energy/biomass in a given time, which would increase its *NPP* *[1 mark]*.
Remember, *NPP* (net primary productivity) is the chemical energy stored in the biomass of a producer in a given area, in a given time, once respiratory losses have been taken into account.

3.2 The efficiency of energy transfer from the producer to primary consumer = 7%
N of primary consumer = $(64\ 375 \div 100) \times 7$
$= 4506.25$ kJ m^{-2} yr^{-1}
Energy ingested by secondary consumer (I) is 90% of 4506.25
$I = (4506.25 \div 100) \times 90 = 4055.625$ kJ m^{-2} yr^{-1}
N of secondary consumer = $I - (F + R)$
$= 4055.625 - (2343 + 1261)$
$=$ **451.6 kJ m^{-2} yr^{-1}** (to 4 s.f.)
[3 marks for the correct answer, otherwise 1 mark for correctly calculating 'N' of the primary consumer and 1 mark for correctly calculating 'I' of the secondary consumer]
You should know that F and R both represent ways in which energy is lost — F is chemical energy lost in faeces and urine, and R is energy lost through respiration.

3.3 By breathing faster, more oxygen will be available in the cells of the secondary consumer *[1 mark]*. Oxygen is needed as a final electron acceptor in the oxidative phosphorylation stage of aerobic respiration *[1 mark]*. By increasing the amount of oxygen that is available, oxidative phosphorylation will be able to occur at a faster rate, increasing the rate of ATP synthesis / oxidative phosphorylation will be able to occur, reducing the need for anaerobic respiration, which produces ATP at a slower rate than aerobic respiration *[1 mark]*.

Topic 6A — Stimuli and Responses

Page 139 — Nervous Communication

1 Receptors detect stimuli *[1 mark]*. Effectors bring about a response to a stimulus to produce an effect *[1 mark]*.

2 a) Touch receptors on the surface of the eye (A) are stimulated *[1 mark]*. An electrical impulse is sent along the sensory neuro (B) to a relay neurone (C) *[1 mark]*. The impulse is then passed a motor neurone (D) *[1 mark]*, which stimulates effector muscle (E) causing them to contract and the person's eyelids to close *[1 mark]*.

b) Damage to the CNS could interrupt the transmission of the refle preventing the reflex response from occurring *[1 mark]*.

3 Motor neurones carry electrical impulses from the CNS to effectors which then respond *[1 mark]*. Damage to the motor neurones means the CNS can't communicate with effectors suc as muscles *[1 mark]*, so muscles don't respond and move/are paralysed *[1 mark]*.

Page 141 — Responses in Plants and Animals

1 a) The data shows that the plants provided with auxins grew more than those not given auxins *[1 mark]*. This is because auxins stimulate plant growth (by cell elongation) *[1 mark]*.

b) Providing tomato plants with auxins could, potentially, be used to increase the height of tomato plants, which might increase th yield of tomatoes/number of tomatoes grown *[1 mark]*.

c) Auxin is redistributed to the shaded side of the shoot *[1 mark]*. Auxin stimulates cell elongation on the shaded side *[1 mark]* so the shoot bends to grow towards the light *[1 mark]*.

Page 143 — Receptors

1 When a Pacinian corpuscle is stimulated, the lamellae are deformed and press on the sensory nerve ending *[1 mark]*. This causes the sensory neurone's cell membrane to stretch and the deformation of stretch-mediated sodium ion channels *[1 mark]*. The sodium ion channels open and sodium ions diffuse into the cell creating the generator potential *[1 mark]*.

2 In the retina/fovea, cones are close together and each cone join one bipolar neurone *[1 mark]*. When light from two points hits two cones, action potentials from each cone go to the brain *[1 mark]*. This means you can distinguish two points that are close together as two separate points *[1 mark]*.

Page 145 — Control of Heart Rate

1 a) The sinoatrial node acts as a pacemaker/sets the rhythm of the heartbeat *[1 mark]*.

b) The Purkyne tissue conducts electrical impulses through the ventricle walls *[1 mark]*.

2 a) E.g. chemoreceptors in the aorta/carotid artery/medulla detect the high CO_2 concentration *[1 mark]*. Impulses are sent from the receptors to the medulla *[1 mark]*, which sends impulses along sympathetic neurones to the sinoatrial node (SAN) *[1 mar* These neurones secrete noradrenaline, which binds to receptors on the SAN *[1 mark]*. This increases the SAN activity, which increases heart rate *[1 mark]*. ***[Up to 3 marks for explaining how impulses get to the SAN, 1 mark for linking increased SAN activity to increased heart rate. Maximum of 4 marks available.]***

b) Low blood O_2 level *[1 mark]*, low blood pH level *[1 mark]*.
The low blood pH level is caused by the increased CO_2 level.

Answers

The AVN passes waves of electrical activity on to the bundle of His and the Purkyne tissue to make the ventricles contract ***[1 mark]***. By stopping the AVN from functioning, the rapid irregular impulses from the atria aren't transmitted via the bundle of His and the Purkyne tissue to the ventricles, so they can't affect the heart rate (i.e. make it high and/or irregular) ***[1 mark]***.

Without a functioning AVN the heart can't beat normally/ ventricles can't contract normally ***[1 mark]***. A pacemaker is needed to generate electrical impulses that cause the heart to beat normally/ventricles to contract normally ***[1 mark]***.

pic 6B — Nervous Coordination

ge 148 — Neurones

A stimulus causes sodium ion channels in the neurone cell membrane to open ***[1 mark]***. Sodium ions diffuse into the cell, so the membrane becomes depolarised ***[1 mark]***.

The first action potential fired at 0.5 ms. If the second one fired at 4.5 ms, this means an action potential is fired every (4.5 – 0.5 =) 4 ms.
Number of ms in one hour = 60 × 60 × 1000 = 3 600 000.
There is one action potential every 4 ms, so in one hour there will be 3 600 000 ÷ 4 = 900 000 = **9 × 10^5** action potentials.
[2 marks for the correct answer, allow 1 mark for the correct calculation of 3 600 000 ÷ 4.]
There's a lot to do to get the marks here, but that's A-level Biology for you. Just take your time and make sure you write down your calculations — that way you might pick up a mark even if you don't get the final answer right.

) 30 mV ***[1 mark]***
This is the same as the maximum potential difference shown on the graph. Remember, action potentials always fire with the same change in voltage no matter how big the stimulus is.

ge 151 — Synaptic Transmission

) It is the threshold that needs to be reached for an action potential to fire ***[1 mark]***.

) Any four from: before the action potential fired, the potential difference across the membrane increased three times in quick succession ***[1 mark]***. The increases in potential difference were caused by nerve impulses arriving at the synapse and releasing neurotransmitter ***[1 mark]***, which caused sodium ion channels to open on the postsynaptic membrane ***[1 mark]***. This allowed an influx of sodium ions into the postsynaptic membrane, which increased the potential difference across the membrane ***[1 mark]***. It was not until the arrival of the third impulse that enough neurotransmitter was acting on the membrane to allow the threshold level to be reached and the action potential to be fired ***[1 mark]***. ***[Maximum of 4 marks available.]***

There will be fewer receptors for acetylcholine/ACh to bind to ***[1 mark]***, so fewer sodium ion channels will open at neuromuscular junctions ***[1 mark]***, making it less likely that action potentials will be generated in the muscle cells ***[1 mark]***.
Galantamine would stop acetylcholinesterase/AChE breaking down acetylcholine/ACh, so there would be more acetylcholine/ ACh in the synaptic cleft ***[1 mark]*** and it would be there for longer ***[1 mark]***. This means more nicotinic cholinergic receptors would be stimulated ***[1 mark]***.

ge 153 — Muscle Contraction

) A = sarcomere ***[1 mark]***.
B = Z-line ***[1 mark]***.
C = H-zone ***[1 mark]***.

) The A-bands stay the same length during contraction ***[1 mark]***. The I-bands get shorter ***[1 mark]***.

c) Drawing number 3 ***[1 mark]*** because the M-line connects the middle of the myosin filaments ***[1 mark]***. The cross-section would only show myosin filaments, which are the thick filaments ***[1 mark]***.
The answer isn't drawing number 1 because all the dots in the cross-section are smaller, so the filaments shown are thin actin filaments — which aren't found at the M-line.

Page 155 — Muscle Contraction

1 Muscles need ATP to relax because ATP provides the energy to break the actin-myosin cross bridges ***[1 mark]***. If the cross bridges can't be broken, the myosin heads will remain attached to the actin filaments ***[1 mark]***, so the actin filaments can't slide back to their relaxed position so the muscle stays contracted ***[1 mark]***.

2 The muscles won't contract ***[1 mark]*** because calcium ions won't be released into the sarcoplasm, so tropomyosin will continue to block the actin-myosin binding sites ***[1 mark]***. This means no actin-myosin cross bridges can be formed ***[1 mark]***.

Topic 6C — Homeostasis

Page 157 — Homeostasis Basics

1 a) Statement A because body temperature continues to increase from the normal level and isn't returned ***[1 mark]***.

b) It makes metabolic reactions less efficient ***[1 mark]*** because the enzymes that control metabolic reactions may denature ***[1 mark]***.

2 Multiple negative feedback mechanisms give more control over changes in the internal environment than just having one feedback mechanism ***[1 mark]***. This is because you can actively increase or decrease a level so it returns to normal ***[1 mark]***.

Page 159 — Control of Blood Glucose Concentration

1 a) Negative feedback because the pancreas secretes hormones that return blood glucose concentration to normal if it is detected as being too high or too low ***[1 mark]***.

b) Insulin binds to specific receptors on muscle cells causing them to become more permeable to glucose, so more is absorbed from the blood ***[1 mark]***. / Insulin activates glycogenesis, so that glucose can be stored as glycogen ***[1 mark]***. / Insulin causes the rate of respiration of glucose to increase, so that more glucose is used up ***[1 mark]***.

2 When adrenaline and glucagon bind to the receptors on the cell membrane they activate an enzyme called adenylate cyclase ***[1 mark]***. Activated adenylate cyclase converts ATP into cAMP, a second messenger ***[1 mark]***. cAMP activates protein kinase A, which activates a cascade that breaks down glycogen into glucose ***[1 mark]***.

Page 161 — Control of Blood Glucose Concentration

1 a) Any two from: Person A's blood glucose concentration is initially at a higher level than person B's blood glucose concentration ***[1 mark]***. / Person A's blood glucose concentration reaches a much higher level than person B's blood glucose concentration ***[1 mark]***. / It takes longer for person A's blood glucose concentration to start to decrease than it does for person B's blood glucose concentration to start to decrease ***[1 mark]***. / Person A's blood glucose concentration decreases at a much slower rate than person B's blood glucose concentration ***[1 mark]***. ***[Maximum of 2 marks available.]***

Answers

b) The insulin receptors on person A's cell membranes don't work properly, so the cells don't take up enough glucose ***[1 mark]***. This means their blood glucose concentration remains higher than normal ***[1 mark]***.

Page 163 — The Kidneys

1 a) The efferent arteriole has a smaller diameter than the afferent arteriole, so the blood in the glomerulus is under high pressure ***[1 mark]***. The high pressure forces liquid and small molecules into the Bowman's capsule (point A), forming the glomerular filtrate ***[1 mark]***.

b) Point C, because glucose is reabsorbed in the proximal convoluted tubule/PCT, so by the time the filtrate reaches point C there will be less glucose remaining ***[1 mark]***.

c) If there is 0 mg of glucose in the urine, all the glucose filtered out of the blood must be reabsorbed. So:
$6300 \times 0.9 = 5670$ mg hour^{-1}
$5670 \div 60 =$ **94.5 mg min^{-1}** ***[1 mark]***

Page 165 — Controlling Blood Water Potential

1 a) Strenuous exercise causes more sweating, so more water is lost ***[1 mark]***. This decreases the water potential of the blood ***[1 mark]***. This is detected by osmoreceptors in the hypothalamus ***[1 mark]***, which stimulates the posterior pituitary gland to release more ADH ***[1 mark]***.

b) The ADH increases the permeability of the walls of the distal convoluted tubule and collecting duct ***[1 mark]***. This means more water is reabsorbed into the medulla and into the blood by osmosis ***[1 mark]***.

2 A longer descending limb, means more water can be reabsorbed into the blood from the nephron in the descending limb ***[1 mark]***. A longer ascending limb means more ions are actively pumped out into the medulla ***[1 mark]***, which creates a really low water potential in the medulla ***[1 mark]***. This means more water moves out of the collecting duct into the capillaries, giving a low volume of urine ***[1 mark]***.

Extra Exam Practice for Topic 6

Pages 166-167

2.1 Each cone cell in the eye connects to the optic nerve with one neurone, whereas one neurone connects several rod cells to the optic nerve ***[1 mark]***. This means that cone cells give higher visual acuity/clearer vision than rod cells ***[1 mark]*** because information about light hitting each cone cell can be interpreted separately by the brain, whereas information about light hitting each rod cell is combined before it reaches the brain ***[1 mark]***.

2.2 If glucose transporter proteins are blocked it means that glucose can't be reabsorbed into the blood from the kidneys ***[1 mark]***. The unabsorbed glucose will remain in the glomerular filtrate, meaning the glucose concentration of the urine will increase ***[1 mark]***.

2.3 Absorbance $= 0.5 \pm 0.05$
$= 0.45 - 0.55$
Range of glucose concentration = **2.2-3.2 mM**
[2 marks for the correct answer, otherwise 1 mark for correctly calculating the range of absorbance values]
The uncertainty is in the absorbance of the colorimeter, so firstly you need to work out the range of absorbance values that would give a reading of 0.5. Then read off the glucose concentrations at these two absorbance values to get the range in glucose concentration.

3.1 A phosphate group taken from phosphocreatine can be added to ADP to make ATP ***[1 mark]***, to provide the energy needed for muscle contraction ***[1 mark]***.

3.2 The calcium ions leave their binding site on the protein attache to tropomyosin ***[1 mark]***. This causes the tropomyosin molecu to move back so they block the myosin head binding sites on actin again ***[1 mark]***. Because there are no actin-myosin cross-bridges, the actin filaments slide back to their relaxed position, lengthening the sarcomere ***[1 mark]***.

3.3 A person with diabetes insipidus would be more at risk of dehydration, because not as much ADH would be released wh the fall in blood water potential was detected ***[1 mark]***. This would mean that the permeability of the DCT and collecting d wouldn't be increased as much as normal ***[1 mark]***, so not as much water would be reabsorbed into the blood by osmosis fr the glomerular filtrate ***[1 mark]***. With less water being reabsor into the blood, it would be more difficult to increase the blood water potential, meaning the person would be more at risk of dehydration ***[1 mark]***.

Topic 7A — Genetics

Page 170 — Inheritance

1 Parents' genotypes identified as RR and rr ***[1 mark]***. Correct genetic diagram drawn with gametes' alleles identified as R, R and r, r ***[1 mark]*** and gametes crossed to show Rr as the only possible genotype in the offspring ***[1 mark]***.
The question specifically asks you to draw a genetic diagram so make su that you include one in your answer, e.g.

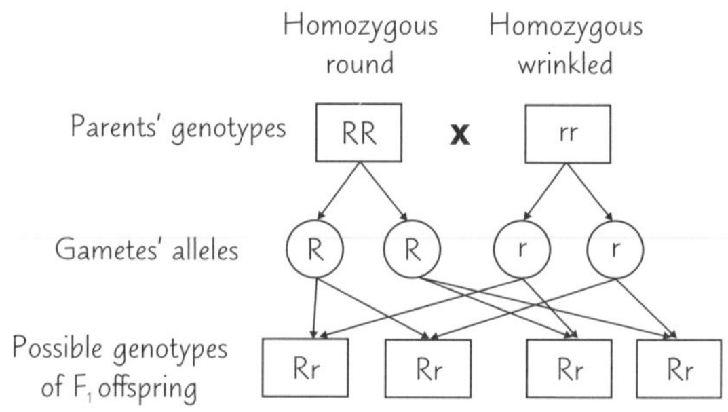

2 a) Because the alleles for red and white coats are codominant, so they are both expressed in the phenotype ***[1 mark]***.

b) Parents' genotypes identified as C^WC^W and C^RC^W ***[1 mark]***. Correct genetic diagram drawn with gametes' alleles identified as C^W, C^W and C^R, C^W ***[1 mark]*** and gametes crossed to show two offspring with genotype C^WC^W and two with genotype C^RC ***[1 mark]***. The phenotypes of the offspring are stated as two wh and two roan ***[1 mark]***.
The question specifically asks you to draw a genetic diagram so make su that you include one in your answer, e.g.

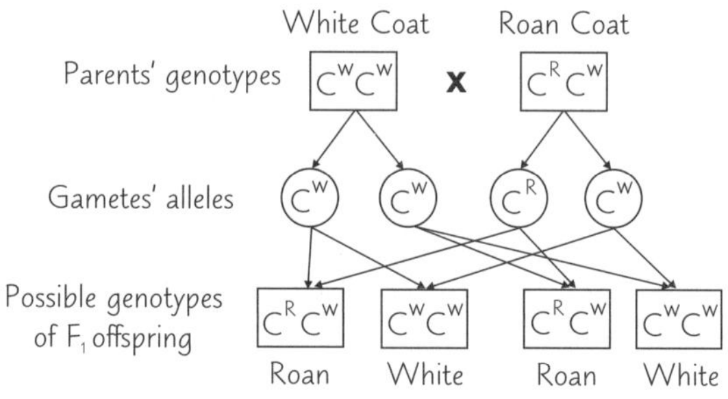

Answers

ge 173 — Linkage and Epistasis

) Parents' genotypes identified as X^HX^h and X^hY ***[1 mark]***. Correct genetic diagram drawn with gametes' alleles identified as X^H, X^h and X^h,Y ***[1 mark]*** and gametes crossed to show X^HX^h, X^HY, X^hX^h and X^hY as the possible genotypes of the offspring ***[1 mark]***.
The question specifically asks you to draw a genetic diagram, so make sure that you include one in your answer, e.g.

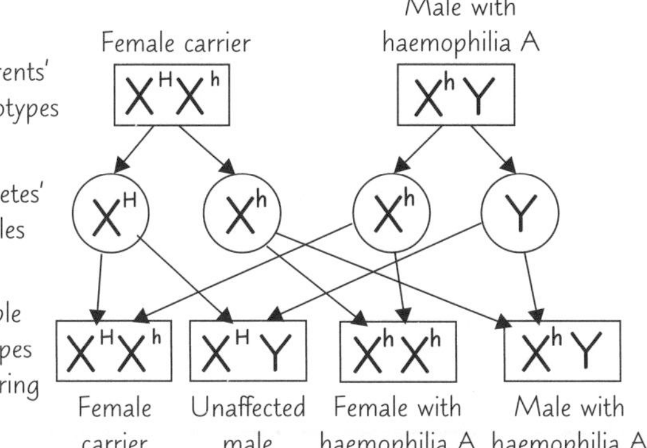

) Men only have one copy of the X chromosome (XY) but women have two (XX) ***[1 mark]***. Haemophilia A is caused by a recessive allele, so females would need two copies of the allele for them to have haemophilia A ***[1 mark]***. As males only have one X chromosome they only need one recessive allele to have haemophilia A, which makes them more likely to have haemophilia A than females ***[1 mark]***.

The table shows that a cross between hhss and HHSS produces a 36 : 9 : 3 or 12 : 3 : 1 phenotypic ratio in the F_2 generation of bald : straight hair : curly hair ***[1 mark]***. This is because the hair gene has a dominant epistatic allele (H) ***[1 mark]***, which means having at least one copy of the dominant epistatic gene (Hh or HH) will result in a bald phenotype that masks the expression of the type of hair gene ***[1 mark]***.

ge 175 — The Chi-Squared Test

) There's no difference between the observed and expected results ***[1 mark]***.

) Yes, the data supports the theory that petal colour in the flower is controlled by dihybrid inheritance. The χ^2 value is smaller than the critical value ***[1 mark]*** so the scientist is unable to reject the null hypothesis ***[1 mark]***.

pic 7B — Populations and Evolution

ge 177 — The Hardy-Weinberg Principle

) Frequency of genotype TT = $p^2 = 0.14$
So the frequency of the dominant allele = $p = \sqrt{0.14} = 0.37$
The frequency of the recessive allele = q
$q = 1 - p$
$q = 1 - 0.37 =$ **0.63** ***[2 marks for the correct answer or 1 mark for 1 – √0.14]***

) Frequency of homozygous recessive genotype tt = $q^2 = 0.63^2$
= **0.40** ***[1 mark. Allow 1 mark for evidence of correct calculation using incorrect answer to part a]***.

) Those that don't have a cleft chin are homozygous recessive tt = 40%, so the percentage that do have a cleft chin, Tt or TT, is 100% – 40% = **60%** ***[1 mark]***.
There are other ways of calculating this answer, e.g. working out the value of 2pq and adding it to p^2. It doesn't matter which way you do it as long as you get the right answer.

2 Frequency of allele F^B = 43% = 0.43
So the frequency of allele $F^W = 1 - 0.43 = 0.57$
The frequency of the heterozygous genotype = 2pq
$2pq = 2(0.43 \times 0.57) =$ **0.49** ***[2 marks for the correct answer or 1 mark for 2(0.43 × 0.57)]***
In the Hardy-Weinberg equations, 'p' is usually the dominant allele and 'q' is usually the recessive allele, but this doesn't have to be the case. In this scenario, there's no recessive allele, so you can just make 'p' represent one of the alleles and 'q' represent the other — it doesn't matter which way round you do it either. If you're not told which is the dominant and which is the recessive allele in an exam question, you can do the same thing.

Page 179 — Variation and Selection

1 a) As temperature decreases from 22 °C to 16 °C the frequency of h, the long hair allele, increases from 0.11 to 0.23 ***[1 mark]***. This could be because the allele for long hair is more beneficial at colder temperatures ***[1 mark]***. Hamsters with the h allele will have a greater chance of surviving, reproducing and passing on their genes, including the beneficial h allele ***[1 mark]***. So a greater proportion of the next generation will inherit the beneficial allele and the frequency of the h allele will increase ***[1 mark]***.

b) Directional selection ***[1 mark]***.

Page 181 — Speciation and Genetic Drift

1 a) E.g. The new species could not breed with each other ***[1 mark]***.

b) Different populations of flies were physically/geographically isolated and experienced different selection pressures (different food) ***[1 mark]***. This led to changes in allele frequencies between the populations ***[1 mark]***, which made them reproductively isolated/unable to interbreed and produce fertile offspring, and eventually resulted in speciation ***[1 mark]***.

Topic 7C — Populations in Ecosystems

Page 183 — Ecosystems

1 Their wings are light and flexible, which allows them to catch fast and manoeuvrable insects. This increases their chances of catching enough food to survive ***[1 mark]***. They use echolocation so they can catch insects that come out at night. This also increases their chances of catching enough food to survive ***[1 mark]***. They make unique mating calls so they only attract a mate of the same species. This increases their chance of reproduction by making successful mating more likely ***[1 mark]***.
This question is only asking about the biotic conditions (the living features of the ecosystem), so you won't get any marks for talking about abiotic conditions (the non-living features of the ecosystem).

Page 185 — Variation in Population Size

1 a) $\text{Rate} = \dfrac{\text{Change in } y}{\text{Change in } x} = \dfrac{25}{4}$
= **6.25 thousand year^{-1} / 6250 year^{-1}**
[2 marks for the correct answer or 1 mark for the correct calculation.]
'Year^{-1}' means 'per year'.

b) The predator population increased as the prey population increased because there was more food available for the predators ***[1 mark]***. The population of prey then fell because many prey were eaten by the large population of predators ***[1 mark]***. The predator population then fell because there was less prey for the predators to eat ***[1 mark]***.

c) Between 25 and 30 years ***[1 mark]***. The prey population starts to decline but predator numbers have been very low for several years, suggesting that the prey are competing with one another for space and food / space and food have become the limiting factors for the prey population size ***[1 mark]***.

Answers

Page 187 — Investigating Populations

1 a) E.g. the field could be divided into a grid a random number generator could be used to select random coordinates on the grid ***[1 mark]***. frame quadrats could be placed on the ground at these random coordinates ***[1 mark]***. The percentage of each frame quadrat that's covered by clover plants could be recorded ***[1 mark]***. The percentage cover for the whole field could then be estimated by taking a mean of the data collected in all of the frame quadrats ***[1 mark]***.

Clover plants are small and grow very close together, so it's much easier to estimate their population size using percentage cover, rather than trying to count individual plants.

b) E.g. including plant species that aren't clover plants could increase the estimate of percentage cover / ignoring clover plants could reduce the estimate of percentage cover ***[1 mark]***.

Page 189 — Succession

1 a) Primary succession ***[1 mark]*** because there is no soil or organic matter ***[1 mark]***.

b) When the grass dies, microorganisms decompose the dead organic material, forming a soil ***[1 mark]***. The formation of soil helps to retain water and makes the conditions less hostile, which allows larger plants, like shrubs, to move in ***[1 mark]***.

Page 191 — Conservation

1 a) There's a link between fishing mortality rate and the cod stock size ***[1 mark]***. As the fishing mortality rate increases, the cod stock size decreases/there's a negative correlation between fishing mortality rate and cod stock size ***[1 mark]***.

b) E.g. it could be used by governments to make decisions about cod fishing quotas (the amount of cod allowed to be removed from the sea by fishermen each year) ***[1 mark]***.

c) The conservationists will want to limit the amount of fishing to a sustainable level to maintain fish stocks for future generations ***[1 mark]***. However, limiting the amount of fishing may reduce the incomes of people employed in the fishing industry ***[1 mark]***.

2 a) It provides wood for people to use whilst preserving some trees which can continue to grow and provide wood in the future ***[1 mark]***.

b) Any two from: e.g. it maintains the woodland habitat for other organisms ***[1 mark]***. / It allows new trees to grow from seeds produced by the mature standards ***[1 mark]***. / The mature standard can be used to produce larger logs at a later date ***[1 mark]***.

c) The canopy of mature standard trees will block out the light that the coppiced trees need to grow ***[1 mark]***.

Extra Exam Practice for Topic 7

Pages 192-193

2.1 Primary succession occurs when pioneer species colonise new land surfaces that don't contain soil ***[1 mark]***. As the grasses in Figure 2 require soil and begin growing from time = 0, soil must already be in the environment and so Figure 2 cannot be showing primary succession ***[1 mark]***.

2.2 Removing trees at point A would prevent succession ***[1 mark]*** and therefore maintain more shrubs and grasses in the woodland, which are likely to be important habitats ***[1 mark]***.

2.3 E.g. at point C the biomass of trees is greater than it is at point B, which suggests that the shrub species face greater competition from trees at point C ***[1 mark]***. Greater competition is likely to reduce the carrying capacity of shrub species, as there are not enough resources to maintain the population size of each species ***[1 mark]***.

Remember, the carrying capacity is the maximum stable population size of a species that an ecosystem can support.

2.4 Directional selection is likely to have acted on the trees ***[1 mar*** Trees with the allele(s) for thicker bark would have been more likely to survive wildfires and to reproduce, passing on the allele(s) for thicker bark ***[1 mark]***. Over time this increased the frequency of the allele(s) for thicker bark in the population of tr frequently exposed to wildfires ***[1 mark]***.

2.5

$$\text{Total population size} = \frac{\text{Number caught in first sample} \times \text{Number caught in second sample}}{\text{Number marked in second sample}}$$

Number caught in first sample after wildfire = 28
Number caught in second sample after wildfire = 24
Number marked in second sample after wildfire = 12
Total population size after wildfire = (28 × 24) ÷ 12
= 56 individuals

$$\%\ \text{change} = \frac{\text{population size after wildfire} - \text{population size before wildfire}}{\text{Population size before wildfire}} \times 100$$

$$= \frac{56 - 336}{336} \times 100 = \frac{-280}{336} \times 100$$

= **-83%** (to 2 s.f.)

[3 marks for the correct answer, otherwise 1 mark for evidenc of the correct formula for calculating population size and 1 m for 56 individuals after wildfire]

The population size has decreased, so make sure you include a minus sig your answer or else clearly say that your answer is a percentage decrease.

Topic 8A — Mutations and Gene Expression

Page 195 — Mutations

1 a) AG<u>G</u>TAT<u>G</u>AGGC<u>C</u> ***[1 mark]***.

b) The original gene codes for the amino acid sequence serine-tyrosine-glutamine-alanine and the mutated gene codes for the amino acid sequence arginine-tyrosine-glutamic acid-alanine ***[1 mark]***. Even though there are three mutations, there are only two changes to the amino acid sequence ***[1 mark]***. This is because of the degenerate nature of the DNA code, which mea more than one codon can code for the same amino acid ***[1 mar*** So the substitution mutation on the last triplet doesn't alter the amino acid (GCT and GCC both code for alanine) ***[1 mark]***.

Page 197 — Cancer

1 a) Any two points from: e.g. malignant tumours are cancers. Beni tumours are not cancerous ***[1 mark]***. Malignant tumours usuall grow rapidly. Benign tumours usually grow slower than malign tumours ***[1 mark]***. Malignant tumours can invade and destroy surrounding tissues/spread to other parts of the body. Benign tumours can't ***[1 mark]***.

b) If a mutation occurs in a tumour suppressor gene ***[1 mark]***, proteins that stop cells dividing and cause cell death might not be produced ***[1 mark]***. If a mutation occurs in a proto-oncogene ***[1 mark]***, it can turn it into an oncogene (an overactive version of the proto-oncogene) causing the productic of too many proteins that cause cells to divide ***[1 mark]***. In bot cases, the mutation allows cells to grow and divide uncontrolla ***[1 mark]***.

c) Oestrogen can stimulate some breast cells to divide and replica ***[1 mark]***. Because more replication is taking place, the chance of new cancer-causing mutations being introduced increases ***[1 mark]***. This stimulation could also help already cancerous c replicate ***[1 mark]***. Some research also suggests that oestrogen can directly cause mutations in certain breast cells, which again increases the chance of cancer-causing mutations being introduced ***[1 mark]***.

Answers

age 199 — Interpreting Data on Cancer

E.g. if the screen revealed that a woman had the BRCA1 mutation, she could be screened for signs of breast cancer more regularly than the rest of the population, so the cancer could be diagnosed early if it does develop *[1 mark]*. She would also be aware that she had a higher risk of developing breast cancer, so would know to be more vigilant when checking for signs of the disease *[1 mark]*. She could also choose to take steps to reduce the risk developing breast cancer, such as having a mastectomy *[1 mark]*. If the disease did develop, knowing the mutation that has caused it could also help to determine the specific treatment used to give the best chance of survival *[1 mark]*.

age 203 — Stem Cells

a) E.g. stem cell therapies are currently being used for some diseases affecting the blood and immune system *[1 mark]*. Bone marrow contains stem cells that can become specialised to form any type of blood cell *[1 mark]*. Bone marrow transplants can be used to replace faulty bone marrow in patients with leukaemia (a cancer of the blood or bone marrow) *[1 mark]*. The stem cells in the transplanted bone marrow divide and specialise to produce healthy blood cells *[1 mark]*.

b) Obtaining embryonic stem cells involves the destruction of an embryo *[1 mark]*. Some people believe that embryos have a right to life and that it's wrong to destroy them *[1 mark]*.

c) E.g. induced pluripotent stem cells are produced by 'reprogramming' specialised adult body cells to become pluripotent *[1 mark]*. To do this, the adult body cells are made to express a series of transcription factors that are normally associated with pluripotent stem cells *[1 mark]*. The genes that code for the transcription factors are introduced to the adult cell's DNA *[1 mark]* using a modified virus that has the genes within its own DNA *[1 mark]*.

age 206 — Regulation of Transcription
nd Translation

a) The results of tubes 1 and 2 suggest that oestrogen affects the expression of the gene for the Chi protein *[1 mark]* because mRNA and active protein production only occur in the presence of oestrogen *[1 mark]*.

b) The mutant could have a faulty oestrogen receptor *[1 mark]*. Oestrogen might not bind to the receptor / the oestrogen-oestrogen receptor complex might not work as an activator *[1 mark]*. This would mean even in the presence of oestrogen transcription wouldn't be activated, so no mRNA or protein would be produced *[1 mark]*.

This is a pretty tricky question — drawing a diagram of how oestrogen controls transcription would help you figure out the answer.

c) E.g. the siRNA and associated proteins would attach to the mRNA of the Chi protein and cut it up into smaller portions *[1 mark]*, resulting in no full length mRNA *[1 mark]*. No mRNA would be available for translation, so no protein would be produced *[1 mark]*.

ge 208 — Epigenetic Control
Gene Expression

a) Histones are proteins that DNA wraps around to form chromatin, which makes up chromosomes *[1 mark]*.

b) When acetyl groups are removed from the histones in chromatin, the chromatin becomes highly condensed *[1 mark]*. This means that the enzymes/proteins needed for transcription cannot access the DNA *[1 mark]* and the DNA cannot be transcribed *[1 mark]*.

c) E.g. acetyl groups are removed from histones by histone deacetylase (HDAC) enzymes *[1 mark]*. Drugs can be used to inhibit these enzymes *[1 mark]*. This means that the histones remain acetylated and the DNA associated with them can be transcribed as normal *[1 mark]*.

Page 209 — Evaluating Data on Phenotypes

1 That genetic factors have a bigger influence than environmental factors on stuttering *[1 mark]*.

Topic 8B — Genome Projects and Gene Technologies

Page 212 — Genome Projects and Making DNA Fragments

1 a) There's a BamHI recognition sequence at either side of the DNA fragment, so you could use this restriction endonuclease to isolate the fragment *[1 mark]*. BamHI would be incubated with the bacterial DNA, so that it cuts the DNA at each of these recognition sequences *[1 mark]*.

b) The genetic code is universal/all organisms use the same genetic code *[1 mark]*. Transcription and translation mechanisms are similar in different species *[1 mark]*.

c) Simple organisms, like bacteria, have fewer non-coding regions than more complex organisms such as plants *[1 mark]*. Plants also have regulatory genes and bacteria don't *[1 mark]*. This makes it harder to find the parts that code for proteins in the plant's DNA than in the bacteria's DNA *[1 mark]*.

You're effectively being asked to compare the difficulty of translating the genome into the proteome for two different organisms here — when answering any comparison question, make sure you talk about both of the things you're comparing in your answer.

d) mRNA that's complementary to the DNA fragment is isolated from the cells *[1 mark]* and mixed with free DNA nucleotides and reverse transcriptase *[1 mark]*. The reverse transcriptase uses the mRNA as a template to synthesise a new strand of cDNA *[1 mark]*.

Page 214 — Amplifying DNA Fragments

1 a) Colony A is visible/fluoresces under UV light, but Colony B isn't visible/doesn't fluoresce *[1 mark]*. So only Colony A contains the fluorescent marker gene, which means it contains transformed cells *[1 mark]*.

b) The plasmid vector DNA would have been cut open with the same restriction endonuclease that was used to isolate the DNA fragment containing the target gene *[1 mark]*. The plasmid DNA and gene (DNA fragment) would have been mixed together with DNA ligase *[1 mark]*. DNA ligase joins the sticky ends of the DNA fragment to the sticky ends of the plasmid DNA *[1 mark]*.

Page 217 — Using Recombinant DNA Technology

1 a) The drought-resistance gene could be inserted into a plasmid *[1 mark]*. The plasmid is then inserted into a bacterium *[1 mark]*, which is used as a vector to get the gene into the plant cells *[1 mark]*.

b) The transformed wheat plants could be grown in drought-prone regions *[1 mark]*, where they would reduce the risk of famine and malnutrition *[1 mark]*.

c) They could be concerned that the large agricultural company will have control over the recombinant DNA technology used to make the drought-resistant plants, which could force smaller companies out of business *[1 mark]*.

Answers

Page 219 — Gene Probes and Medical Diagnosis

1 a) The allele that you want to screen for is sequenced ***[1 mark]***. Multiple complementary copies of parts of the allele are made by PCR to be used as DNA probes ***[1 mark]***.

b) Microscopic spots of different DNA probes are attached in series to a glass slide, producing a microarray ***[1 mark]***. A sample of the person's labelled DNA is washed over the array and if any of the DNA matches any of the probes, it will stick to the array ***[1 mark]***. The array is washed and visualised, under UV light/X-ray film ***[1 mark]***. Any spot that shows up means that the person's DNA contains that specific allele ***[1 mark]***.

2 a) Because the patient tested negative for the mutated allele (KRAS oncogene) that the drug specifically targets ***[1 mark]***.

b) So the results of the patient's screening can be explained to them ***[1 mark]*** and so the treatment options can also be explained ***[1 mark]***.

Page 221 — Genetic Fingerprinting

1 a) Genetic fingerprinting is based on comparing the length of variable number tandem repeats/VNTRs at particular points on the genome ***[1 mark]***. PCR is used to make copies of/amplify the areas of DNA that contain the VNTRs ***[1 mark]***. This produces many DNA fragments for analysis with gel electrophoresis, which produces a genetic fingerprint ***[1 mark]***.

b) Genetic fingerprint 1 is most likely to be from the child's father because five out of six of the bands on his genetic fingerprint match that of the child's, compared to only one on fingerprint 2 ***[1 mark]***.

c) Any two from: e.g. it can be used to link a person to a crime scene (forensic science). / To prevent inbreeding between animals or plants. / To diagnose cancer or genetic disorders. / To investigate the genetic variability of a population. ***[1 mark for two correct answers.]***

Extra Exam Practice for Topic 8

Pages 222-223

2.1 Multipotent because they can produce different types of cell, whereas unipotent cells can only become one type of cell ***[1 mark]***.

2.2 Transcription factors bind to the DNA in the cell nucleus ***[1 mark]***. These control which genes are expressed ***[1 mark]***, so that only the genes that code for proteins specific for red blood cells are transcribed and translated ***[1 mark]***. These proteins modify the cell, resulting in red blood cells ***[1 mark]***.

2.3 The stem cells that are transplanted would need to be haematopoietic stem cells, since these are the cells which develop into the cells that are affected by MDS ***[1 mark]***. As the defect that causes MDS is found in haematopoietic stem cells, a donor would be needed to obtain healthy versions of these cells ***[1 mark]***.

2.4 Metabolised azacitidine binds irreversibly to DNA methyltransferases, so there will be a reduced amount of methylation (at CpG sites) on tumour suppressor genes ***[1 mark]***. As a result, tumour suppressor genes are expressed ***[1 mark]*** and so proteins that stop cells from dividing / cause cells to self-destruct/apoptosis are produced ***[1 mark]***. This means that the rate of cell division/mitosis is controlled ***[1 mark]***.

2.5 Azacitidine counteracts the epigenetic change (abnormal methylation of tumour suppressor genes) that causes the abnormal cell growth seen in MDS ***[1 mark]***.

2.6 E.g. yes, as for three out of the five studies shown in Table 1 ***[1 mark]*** the odds ratio is lower for former smokers than it is for current smokers ***[1 mark]***.

Do Well In Your Exams

Page 232 — How to Do Well in Your Exams

Q1 ***21-25 marks:***

The answer includes material from a variety of different topic areas and clearly shows its link to the question title.
No irrelevant material is included. The answer includes a range of detailed and accurate biological facts that are all of A-level standard. No incorrect material is included. Appropriate scientific terminology is used. Explanations are clear and the overall essay is very well written.
(To get top marks, evidence of wider reading beyond the specification must be shown.)

16-20 marks:

The answer includes material from several relevant topic areas and links these to the question title. An irrelevant topic may be included. The answer includes a range of biological facts that ar accurate and of A-level standard but may sometimes be lacking i detail. There may be one significant error in the scientific conter Appropriate scientific terminology is used. Explanations are clea

11-15 marks:

The answer includes material from several relevant topic areas b doesn't link them to the question title. More than one irrelevant topic may be included. The biological facts included in the answer are mostly correct and of A-level standard but material is lacking in detail. There may be a few significant errors in the scientific content. Appropriate scientific terminology is usually used. Explanations are usually clear.

6-10 marks:

The answer includes material from one or two relevant topic areas but doesn't link them to the question title.
Several irrelevant topic areas may be included. Some A-level content may be included but it will be lacking in detail and may contain several significant scientific errors. There may be limitec use of scientific terminology. Explanations lack clarity.

1-5 marks:

The answer includes material that is only vaguely linked to the question title. Material is presented as a series of facts.
Most of the material is irrelevant. The content is below A-level standard and contains a large number of scientific errors.
Scientific terminology is not used or is below A-level standard.
Explanations are poor or absent.

0 marks:

Nothing relevant is included in the answer or nothing has been written.

Here are some topic areas you might write about:

- enzymes catalysing important cellular reactions (e.g. in photosynthesis and respiration);
- carrier and co-transport proteins aiding facilitated diffusion and active transport of materials across cell membranes;
- antigens and antibodies in the immune response against pathogens;
- proteins producing beneficial phenotypes in natural selection
- protein ion channels in cell membranes allowing action potentials to be generated and nervous responses to stimuli to take place;
- the role of actin and myosin proteins in muscle contraction and movement;
- receptor proteins on the surface of cells allowing hormonal responses to stimuli to take place (e.g. insulin and glucagon receptors in the control of blood glucose concentration);
- proteins produced by tumour suppressor genes and proto-oncogenes controlling cell division.

This is not a full list of all the topic areas you could write about — it's jus to give you an idea. Remember, you should aim to write about at least fi of these topic areas. Whatever topic areas you include, you must relate them to the essay title — so in this case, don't just write about proteins, make it really clear how proteins are important to living organisms.

Answers

ʼnoptic Practice

ges 233-241

The gills are 'feathered' to give them a large surface area, which increases the rate of gas exchange at their surface ***[1 mark]***.

In the gills of the axolotl, as oxygen is absorbed into the gills the concentration of oxygen in the water immediately around the gills is likely to fall ***[1 mark]***. This may reduce the oxygen concentration gradient between the water and the blood, which may lead to oxygen absorption not being very efficient ***[1 mark]***. However, the counter-current system in fish gills means that the flow of blood through the lamellae occurs in one direction and water flows over in the opposite direction ***[1 mark]***. This maintains a large concentration gradient of oxygen between the water and the blood, resulting in more efficient oxygen absorption than in the gills of axolotls ***[1 mark]***.

No, because when measuring species richness the students would only be counting the number of different species in the waterway ***[1 mark]***, so it would only measure whether or not axolotls were present rather than the size of the population ***[1 mark]***.

Using fewer toxic chemicals could prevent large numbers of axolotls dying before they reproduce ***[1 mark]***, which could help to prevent a decline in the number of different alleles present in the population ***[1 mark]***.

The curve of best fit would start from the origin and increase linearly with increasing substrate concentration ***[1 mark]***, as the presence of more substrate molecules means it's more likely that enzyme-substrate complexes will form ***[1 mark]***. Eventually the line would flatten/plateau ***[1 mark]***, as all the available active sites of the pepsin molecules are occupied by the substrate ***[1 mark]***.

Indigestion remedies will increase the pH of the stomach towards pH 7, meaning that the activity of pepsin will be lower than it is within the normal pH range of the stomach ***[1 mark]***. A reduction in the activity of pepsin will mean that the rate of protein digestion is likely to decrease ***[1 mark]***.

Pancreatic juice is likely to be alkaline, as this will raise the pH of the food that enters the small intestine from the stomach ***[1 mark]***, which will increase the activity of trypsin ***[1 mark]*** and therefore increase the rate of protein digestion ***[1 mark]***.

C, because an endopeptidase would hydrolyse a peptide bond ***[1 mark]*** within the protein ***[1 mark]***. ***[Only award marks for explanation if correct bond identified.]***

The cohesive property of water means that water molecules stick together, so many water molecules travel together/water is able to flow through the plant ***[1 mark]***.

Sexual reproduction increases genetic variation within a species, whereas asexual reproduction doesn't ***[1 mark]***. The more genetic variation there is in a species, the more likely it is that some individuals will have alleles that allow them to survive and reproduce in unstable environmental conditions ***[1 mark]***.

350 000 ÷ 4 = **87 500** ***[1 mark]***

Each occurrence of meiosis produces four daughter cells — in this case, the daughter cells are called spores.

Total number of chromosomes = 16.
So, number of homologous pairs = 8, which is *n*.
The possible number of different combinations of chromosomes following meiosis = $2^n = 2^8 =$ **256** ***[1 mark]***

E.g. organelles are smaller than their host cells, like prokaryotic cells are smaller than eukaryotic cells ***[1 mark]***. Some organelles contain short, circular DNA, like prokaryotic cells ***[1 mark]***. Some organelles have their own membranes, like prokaryotic cells ***[1 mark]***.

Genome sequencing allows the entire base sequence of an organism's DNA to be determined ***[1 mark]***. Comparing the entire DNA base sequence of two different organisms allows all the genetic differences between them to be identified, and helps to clarify how closely related they are ***[1 mark]***.

4.3 Usually, high-temperatures cause proteins to denature by breaking the hydrogen / disulfide / ionic bonds holding the tertiary structure together ***[1 mark]***. In turn, this would cause essential proteins to lose their 3D shape and cease to function, which could be fatal ***[1 mark]***.

4.4 By accumulating potassium ions in their cytoplasm, the halophilic archaea reduce the water potential of their cytoplasm ***[1 mark]***. This reduces the difference between the water potential of their cytoplasm and the water potential of their high salt environment ***[1 mark]***, which means they are less likely to lose water by osmosis and so are more likely to survive ***[1 mark]***.

5.1 The Env glycoprotein complex may be important for allowing HIV to bind to a receptor molecule on the membrane of the host helper T-cell ***[1 mark]***.

5.2 After the HIV proteins have been replicated in the host cell, the new virus buds from the host cell ***[1 mark]*** and uses the cell membrane of the host cell as its own, resulting in some host proteins being present in the new virus's lipid envelope ***[1 mark]***.

5.3 E.g. the nucleotides in HIV's genetic material/RNA contain a sugar called ribose, whereas the nucleotides in the host cell's genetic material/DNA contain a sugar called deoxyribose ***[1 mark]***. The nucleotides in HIV's genetic material/RNA can contain the base uracil, whereas the nucleotides in the host cell's genetic material/DNA can contain the base thymine ***[1 mark]***. However, the nucleotides in the genetic material of both HIV and the host cell contain a phosphate group ***[1 mark]*** and can contain the bases adenine/A, cytosine/C, or guanine/G ***[1 mark]***.

5.4 Reverse transcriptase is used to make a complementary strand of DNA, using the viral RNA/genetic material as a template ***[1 mark]***. Using this complementary strand, double-stranded DNA is made and inserted into the host cell's DNA ***[1 mark]***. Host cell enzymes are then used to make viral proteins from the viral DNA that has been inserted into the host cell's DNA ***[1 mark]***.

5.5 Drugs that prevent the reproduction of bacteria can target molecules/processes that are only found in bacterial cells ***[1 mark]***. However, HIV and other viruses use host cells and components to replicate, so any drugs designed to prevent viral replication could also damage the host cells ***[1 mark]***.

5.6 They bind to any antibodies from the patient's blood sample that bonded to the HIV antigens ***[1 mark]***.

5.7 A more intense colour change would suggest that there were more HIV antibodies present in the patient's blood for the secondary antibodies to bind to ***[1 mark]***, meaning that the person's immune system may have been producing antibodies against HIV for a longer period of time ***[1 mark]***.

5.8 A HIV vaccine would contain HIV antigens ***[1 mark]***. This would cause the immune system of the patient to produce antibodies against HIV ***[1 mark]***. This may produce a positive result in the ELISA test even if the patient did not have a HIV infection ***[1 mark]***.

6.1 Shivering is likely to increase the rate of glycolysis as the muscles need more ATP for contraction ***[1 mark]***, which is produced during glycolysis ***[1 mark]***.

6.2 As the muscles use more oxygen, the oxygen concentration in the blood may fall ***[1 mark]***, which would be detected by chemoreceptors in the aorta/carotid artery/medulla ***[1 mark]***. Impulses would be sent from the chemoreceptors to the medulla ***[1 mark]***, which would send impulses along the sympathetic neurones ***[1 mark]***. These would secrete noradrenaline onto receptors of the SAN in the heart, causing heart rate to increase ***[1 mark]***.

6.3 Standard deviation is better for judging the precision of the results as this shows the spread of data around the mean ***[1 mark]***, whereas the range just shows the total spread of the data ***[1 mark]***.

6.4 $$\text{Percentage difference} = \frac{\text{final value} - \text{original value}}{\text{original value}} \times 100$$

1.2 – 0.3 = 0.9
(0.9 ÷ 0.3) × 100 = **300%** ***[1 mark]***

Answers

7.1 The drug will prevent acetylcholine from binding to receptors on the postsynaptic membrane/motor end plate, meaning that sodium ion channels will not open ***[1 mark]***. As there is no influx of sodium ions, the postsynaptic membrane/motor end plate will not become depolarised ***[1 mark]*** and an action potential will not be generated in the target/muscle cells ***[1 mark]***. This will mean that the sarcoplasmic reticulum will not be triggered to release calcium ions ***[1 mark]*** and so the actin-myosin binding sites will remain blocked by tropomyosin, preventing muscle contraction ***[1 mark]***.

7.2 The drug prevents the membrane from repolarising / returning to its resting potential ***[1 mark]***, which prevents the generation of a new action potential and any further muscle contraction ***[1 mark]***.

7.3 E.g. it could prevent the influx of calcium ions into the presynaptic neurones so the vesicles containing acetylcholine are not triggered to release it. / It could increase the rate of acetylcholine breakdown in the synaptic cleft, so that it can't bind to receptors on the postsynaptic membrane/motor end plate and generate an action potential. / It could bind to acetylcholine in the synaptic cleft to prevent it from binding to receptors on the postsynaptic membrane/motor end plate and generating an action potential. ***[1 mark]***

7.4 Minimum dose at which no more than 15% of usual force of contraction remaining (the mean reduction in force of contraction is no less than 85%) = 200 µg kg body mass^{-1}.
200 × 75 = 15 000 µg
15 000 µg ÷ 1000 = **15 mg**
[2 marks for the correct answer, otherwise 1 mark for 15 000 µg or for a correct conversion from µg to mg.]

7.5 A Student's t-test ***[1 mark]*** because mean values for two data sets are being compared ***[1 mark]***.

7.6 As elderly patients are likely to have fewer functioning glomeruli than younger patients, fewer drug molecules will enter the Bowman's capsules every time blood passes through the kidneys ***[1 mark]***. This means that drug molecules will remain in the blood for longer, where they can continue to act at neuromuscular junctions and affect muscle contraction ***[1 mark]***.

7.7 If acetylcholinesterase was inhibited by anticholinesterases then more acetylcholine would remain in the synaptic cleft ***[1 mark]***, and be available to bind to receptors on the postsynaptic membrane/motor end plate to overcome the action of the neuromuscular blocker ***[1 mark]***.

8.1 The disease is caused by a gene carried on a chromosome that is not a sex chromosome ***[1 mark]*** and the person will have the disease phenotype even if they inherit just one copy of the disease allele ***[1 mark]***.

8.2 As this mutation is a substitution mutation, it will not cause the number of bases in the gene to change ***[1 mark]***. This means that the base triplets that follow the mutation will still be read in the same way, so it won't cause a frameshift ***[1 mark]***.
Remember, a frameshift can be caused by a deletion, addition or duplication mutation. These mutations cause the sequence of amino acids downstream to change.

8.3 The Generation 2 male is either heterozygous or homozygous for the healthy allele ***[1 mark]***. If the Generation 2 male is homozygous for the healthy allele then there is a 50% chance that the Generation 3 male will inherit the disease, meaning the student's conclusion is correct ***[1 mark]***. However, if the Generation 2 male is heterozygous then there is a 75% chance that the Generation 3 male will inherit the disease, meaning the student's conclusion is incorrect ***[1 mark]***.
It might be helpful to draw out a genetic diagram (e.g. a Punnett square) in your exam to help you get your head around what's going on.

8.4 Number of people in the city = 9.62×10^6 = 9 620 000.
Number of people in the city homozygous for the FH allele =
9 620 000 ÷ 1 000 000 = 9.62 = **10 people** ***[1 mark]***
Here you're working out the number of people that would be homozygous for the FH allele, so you need to round your answer to the nearest whole number (as you can't have 0.62 of a person).

8.5 $p^2 + 2pq + q^2 = 1$
Frequency of the homozygous dominant genotype = p^2
so p^2 = 1 in 1 000 000 = 1 ÷ 1 000 000 = 1×10^{-6}
so $p = \sqrt{(1 \times 10^{-6})} = 0.001$
$p + q = 1$, so $q = 1 - p = 1 - 0.001 = 0.999$
Frequency of the heterozygous genotype = 2pq
= 2 × 0.001 × 0.999 = **0.001998**
[2 marks for the correct answer, otherwise 1 mark for recognising that the frequency of the heterozygous genotype = 2pq]

8.6 Inflammation of the β cells in the pancreatic islets of Langerha could reduce the secretion of insulin into the blood ***[1 mark]***. Insulin normally lowers the blood glucose level by increasing uptake of glucose by cells ***[1 mark]*** and increasing the conversi of glucose into glycogen ***[1 mark]***. If insulin secretion is reduc then the blood glucose level can't be controlled properly/remai high, which could lead to Type II diabetes ***[1 mark]***.

8.7 E.g. a DNA fragment containing a copy of the normal LDL receptor allele / DNA fragments that would disrupt the mutated LDL receptor allele could be inserted into a vector ***[1 mark]***. The vector could then be inserted into the patient's cells ***[1 mar*** where the DNA fragment(s) would become incorporated into tl patient's DNA ***[1 mark]***. This would mean the patient produce the normal LDL receptor protein, which would increase the uptake of cholesterol from the blood ***[1 mark]***.

8.8 E.g. it could result in the overproduction of the LDL receptor protein. / The inserted DNA fragment could be incorporated int another gene / a regulatory region of a gene and disrupt gene expression elsewhere in the body. ***[1 mark]***

9.1 The increase in shade means that the crop plants may not have access to as much light as before, so their growth rate may be reduced ***[1 mark]***. This could mean that there is less food available for the rabbits ***[1 mark]***, so the environment can no longer support the same number of rabbits as before, meaning that the carrying capacity will have been reduced ***[1 mark]***.

9.2 Chlorophyll a or b can't absorb light beyond a wavelength of 700 nm, which could mean that ATP and reduced NADP aren't produced in the light-dependent reaction of photosynthesis ***[1 mark]***. This could mean that there is less ATP and reduced NADP available for the light-independent reaction/Calvin cycle ***[1 mark]***, so fewer products needed for plant growth (e.g. gluco are produced ***[1 mark]***.

9.3 When far-red light is detected, TTA1 activity is no longer inhibi so the amount of indoleacetic acid/IAA produced will increase ***[1 mark]***. Indoleacetic acid/IAA stimulates the growth of plants shoots ***[1 mark]***, so an increase in the indoleacetic acid/IAA concentration could cause the plant to grow taller/bend further that it is exposed to more light ***[1 mark]***.

9.4 Transcription factors control the rate of transcription of their tar genes ***[1 mark]***. If two organisms with the same genome have different transcription factors acting on their genes then there will be different amounts of various proteins produced ***[1 mark*** which will lead to the organisms having different phenotypes ***[1 mark]***.

10.1 50% / 0.5 / 1 in 2 ***[1 mark]***

10.2 As there are excess base triplets in the gene, there will be more than the usual number of amino acids in the protein that's produced ***[1 mark]***. This will mean that the protein can't fold i its unique tertiary shape, which will affect its function ***[1 mark]***

10.3 Inhibitory neurotransmitters bind to receptors on the postsynap membrane at a synapse, which can cause hyperpolarisation ***[1 mark]***. This would prevent an action potential from being fi and therefore stop the transmission of a nerve impulse. / This m prevent excitatory neurotransmitters from being able to depolar the membrane and trigger an action potential, which would sto the transmission of a nerve impulse ***[1 mark]***.

Answers

0.4 For: e.g. the person may want to know whether they carry the mutated allele before they decide whether or not to have children. / Knowing that they are likely to develop the disease may give the person time to prepare for the difficulties it may cause. ***[1 mark]*** Against: e.g. because there's currently no cure, knowing that they are likely to develop the disease may be difficult to cope with emotionally ***[1 mark]***.

0.5 The DNA is heated to 95 °C to break the hydrogen bonds between the two strands of DNA and therefore separate them ***[1 mark]***. This means that single DNA strands are created which act as templates to create the complementary strands of DNA ***[1 mark]***.

0.6 Electrophoresis separates out DNA fragments based on size, so copies of a *HTT* gene that have a larger number of CAG repeats would move a shorter distance than those that have a smaller number of CAG repeats ***[1 mark]***. Therefore, if the PCR products of the genetic test only move a short distance (e.g. compared to a control/sample taken from a person known not to have a mutated *HTT* gene) it suggests that the person is at risk of developing Huntington's disease ***[1 mark]***.

0.7 A small, double-stranded RNA molecule in the cytoplasm of a cell could unwind ***[1 mark]***. One strand of this RNA could then bind to the mRNA produced from the mutated *HTT* gene ***[1 mark]***. This could cause the mRNA to be cut into fragments / blocked ***[1 mark]***, which would stop the gene from being translated, meaning that the defective protein that causes Huntington's disease wouldn't be produced ***[1 mark]***.

1 ***21-25 marks:***
The answer includes material from a variety of different topic areas and clearly shows its link to the question title.
No irrelevant material is included. The answer includes a range of detailed and accurate biological facts that are all of A-level standard. No incorrect material is included. Appropriate scientific terminology is used. Explanations are clear and the overall essay is very well written.
(To get top marks, evidence of wider reading beyond the specification must be shown.)

16-20 marks:
The answer includes material from several relevant topic areas and links these to the question title. An irrelevant topic may be included. The answer includes a range of biological facts that are accurate and of A-level standard but may sometimes be lacking in detail. There may be one significant error in the scientific content. Appropriate scientific terminology is used. Explanations are clear.

11-15 marks:
The answer includes material from several relevant topic areas but doesn't link them to the question title. More than one irrelevant topic may be included. The biological facts included in the answer are mostly correct and of A-level standard but material is lacking in detail. There may be a few significant errors in the scientific content. Appropriate scientific terminology is usually used. Explanations are usually clear.

6-10 marks:
The answer includes material from one or two relevant topic areas but doesn't link them to the question title.
Several irrelevant topic areas may be included. Some A-level content may be included but it will be lacking in detail and may contain several significant scientific errors. There may be limited use of scientific terminology. Explanations lack clarity.

1-5 marks:
The answer includes material that is only vaguely linked to the question title. Material is presented as a series of facts. Most of the material is irrelevant. The content is below A-level standard and contains a large number of scientific errors. Scientific terminology is not used or is below A-level standard. Explanations are poor or absent.

0 marks:
Nothing relevant is included in the answer or nothing has been written.

Here are some topic areas you might write about:

- the importance of ATP synthase in catalysing the formation of ATP from ADP and inorganic phosphate to allow energy to be stored and transported to cells (e.g. during photosynthesis/respiration);
- the importance of ATP hydrolase in breaking down ATP to release energy for energy-requiring reactions in cells (e.g. active transport across cell membranes/muscle contraction), and releasing inorganic phosphate for the phosphorylation of other compounds;
- the importance of DNA helicase in unwinding the DNA double-helix and of DNA polymerase in joining nucleotides in the new strand, during DNA replication;
- the importance of digestive enzymes such as amylases, lipases and peptidases in the hydrolysis of large food molecules into smaller molecules for absorption;
- the importance of the roles of rubisco and dehydrogenase enzymes in photosynthesis;
- the importance of enzymes involved in the control of blood glucose concentration (i.e. in glycogenesis, glycogenolysis and gluconeogenesis) including the importance of adenylate cyclase and protein kinase A in the second messenger model of adrenaline and glucagon action.

Remember, you should try to write about at least five different topic areas. You don't have to have written about only those listed here though — these are just to give you an idea. Make sure you link all the topic areas you have written about to the essay title. It's no good just writing about enzymes — you need to make it clear how they're important to living organisms.

Acknowledgements

Cover image © duncan1890/iStockphoto.com

MMR graph on page 52 adapted from H. Honda, Y. Shimizu, M. Rutter. No effect of MMR withdrawal on the incidence of autism: a total population study. Journal of Child Psychology and Psychiatry 2005; 46(6):572-579.

Data used to construct Herceptin® graph on page 52 from M.J. Piccart-Gebhart, et al. Trastuzumab after Adjuvant Chemotherapy in HER2-positive Breast Cancer. NEJM 2005; 353: 1659-72.

Data used to construct the Hib graph on page 53 from the Health Protection Agency. Reproduced under the terms of the Open Government Licence https://www.nationalarchives.gov.uk/doc/open-government-licence/version/3/

Data for graph showing ethanol concentration vs absorbance on page 57 was obtained using a Mystrica colorimeter © Mystrica Ltd. www.mystrica.com

Data used to construct the smoking graph on page 66 from Cancer Research UK, http://www.cancerresearchuk.org/cancer-info/cancerstats/causes/tobacco-statistics/#Smoking, January 2015

Data used to construct the lung cancer graph on page 66 from Cancer Research UK, http://www.cancerresearchuk.org/cancer-info/cancerstats/types/lung/mortality/uk-lung-cancer-mortality-statistics, January 2015

Data used to construct asthma and sulfur dioxide graphs on page 67. Source: National Statistics. Crown copyright material is reproduced under the terms of the open government licence http://www.nationalarchives.gov.uk/doc/open-government-licence/version/3/

Exam Question graph on page 67, The Relationship Between Smoke And Sulphur Dioxide Pollution And Deaths During The Great London Smog, December 1952, Source: Wilkins, 1954

Data used to construct the graph on page 81 from P.M. Ridker, et al. Comparison of C-reactive protein and low densitylipoprotein cholesterol levels in the prediction of first cardiovascular events. NEJM 2002; 347: 1557-65.

With thanks to Science Photo Library for permission to reproduce the images on pages 36 and 68.

Data for graphs showing glucose concentration vs absorbance on pages 161 and 167 was obtained using a Mystrica colorimeter © Mystrica Ltd. www.mystrica.com

Data used to construct graph on page 191 showing the stock of spawning cod in the North Sea and the rate of mortality caused by fishing since 1960 from the International Council for the Exploration of the Sea.

Graph to show incidence of breast cancer defined by age and family history on page 198 reprinted from The Lancet, Volume 358, Issue No 9291. Familial Breast Cancer: collaborative reanalysis of individual data from 52 epidemiological studies including 58 209 women with breast cancer and 101 986 women without the disease, pgs 1389-1399 © Oct 2001 with permission from Elsevier.

Second graph on page 198 — Alcohol, tobacco & breast cancer — collaborative reanalysis of individual data from 53 epidemiological studies, including 58 515 women with breast cancer and 95 067 women without the disease. Reprinted by permission from Macmillan Publishers Ltd on behalf of Cancer Research UK: British Journal of Cancer © Nov 2002

Data used to construct the table on the link between smoking and MDS on page 223 © 2013 Tong et al. A Meta-Analysis of the Relationship between Cigarette Smoking and Incidence of Myelodysplastic Syndromes. PLoS ONE 8(6): e67537. https://doi.org/10.1371/journal.pone.0067537. Reproduced under the terms of the Creative Commons Attribution License.

Every effort has been made to locate copyright holders and obtain permission to reproduce sources. For those sources where it has been difficult to trace the originator of the work, we would be grateful for information. If any copyright holder would like us to make an amendment to the acknowledgements, please notify us and we will gladly update the book at the next reprint. Thank you.

Index

Index

D

E

F

G

Index

Index

M

N

O

P

Index

S

T

Index